CALORIC VALUES OF PURE FOODS

FOOD	CALORIES, kcal/100 g*	FOOD	CALORIES, kcal/100 g*
Milk products:			
Milk (whole, about ½ cup)	75	Peas	60
Cheese	400	Beets	40
Cottage cheese	100	Spinach	60
Meats		Lettuce (raw)	20
Fish	100–200	Tomatoes (raw)	20
Liver	125	**Fruit:**	
Steak	200	Apple (raw, ½ of a fruit)	75
Pork (loin)	250–300	Banana (raw, ½ banana)	100
Bacon	600–700	Pear and peach (raw, ½–¾ of a fruit)	50–60
Chicken	100	Orange (½ orange)	50
Turkey	150	Grapefruit (⅓–½ of a fruit)	50
Hot dogs (about 2–3)	250	Strawberries (raw)	40
Breads:†		Raisins	300
White (5–7 slices)	250	**Desserts:**	
Pancakes (about 5)	200	Chocolate fudge	400
Biscuits (about 5)	400	Peanut brittle	550
Crackers (soda, about 35)	425	Lemon meringue pie	250
Cereals:‡		Apple pie	300
Cornflakes	400	Tapioca pudding	100
Wheat flakes	350	Sugar cookies	425
Cream of wheat	350	Angel food cake	300
Oatmeal	400	Devil's food cake	400
Starches:		Gelatin dessert	375
Potatoes (boiled, 1 small)	100	Miscellaneous	
Rice (boiled)	100	Nuts (most kinds)	700
Spaghetti, macaroni, noodles	350	Peanuts, chestnuts	400–500
Popcorn	400	Jelly	300
Vegetables:		Butter, margarine (8 tblsp)	800
Corn	100	Oils	900
Carrots	50	Mayonnaise	750
Green beans	40	Granulated sugar (sucrose)	400
Lima beans	130	Honey	325
Navy beans (dried)	350		

* All values are for 100-g (about ¼-lb) serving and are a composite value taken from many sources including *The Heinz Handbook of Nutrition* (see footnote to Table 15-6) and the *CRC Handbook of Chemistry and Physics*, CRC Press, Inc., Cleveland, Ohio. More familiar units are given in some cases.

† No spread.

‡ Dry—no milk or sugar.

McGRAW-HILL BOOK COMPANY
New York | St. Louis | San Francisco | Auckland
Bogotá | Hamburg | London | Madrid
Mexico | Montreal | New Delhi | Panama
Paris | São Paulo | Singapore | Sydney | Tokyo | Toronto

FRANK L. WISEMAN
Georgetown College

CHEMISTRY IN THE MODERN WORLD
Concepts and Applications

Chemistry in the Modern World: Concepts and Applications

3 4 5 6 7 8 9 0 VNHVNH 8 9 8 7

ISBN 0-07-071150-X

This book was set in Times Roman by York Graphic Services, Inc.
The editors were Stephen Zlotnick and Susan Gamer;
the designer was Nicholas Krenitsky;
the production supervisor was Charles Hess.
The drawings were done by Wellington Studios Ltd.
Von Hoffman Press, Inc., was printer and binder.

Library of Congress Cataloging in Publication Data

Wiseman, Frank L.
Chemistry in the modern world.

Includes bibliographical references and index.
1. Chemistry. I. Title.
QD33.W836 1985 540 84-10039
ISBN 0-07-071150-X

This book is dedicated to three people who have been most influential in helping me develop in the profession of teaching and thus, directly or indirectly, in bringing this book to reality.

To my wife Ellen, without whose encouragement I would not have had the courage to begin or the stamina to complete the writing of this text.

To W. Richard Stephens (then academic dean of Greenville College), who at a critical time in my career presented the mastery concept of education—the idea that the aim of teaching is to help all students learn the material at an acceptable level of accomplishment.

To Diane Russo, whose advice and friendship have been so very valuable, both professionally and personally. She exemplifies the many good students I have had, yet stands out among them as special. Her positive response to my teaching methods encouraged me to develop even better ways by which students could learn chemistry. The programmed method used in this text was a result.

Contents in Brief

Contents

To the Student

You are about to begin your study of chemistry. It is not only central to the sciences in general but important in our culture, since it provides many of the necessities and conveniences available to each one of us.

Many students taking introductory chemistry are natural science majors; but an increasing number of students with majors outside the natural sciences are choosing chemistry as part of their general education and as a means of becoming more effective in their chosen professions. For some nonmajors, the idea of taking a chemistry course is alarming; these students frequently feel intimidated by the mathematical tools, the scientific principles, or simply the unfamiliarity of the subject. This book is written with this naturally felt apprehension in mind. I have tried hard to produce *learning devices* that are appropriate for the student who is just beginning; I will assume no prior knowledge of the subject at hand.

The learning devices include:

1 Lists of terms to learn, in a study guide at the end of each chapter. The meanings can be found in the text; and, in most cases, a concise definition can be found in the glossary.

2 Practice problems and questions, as well as questions requiring you to think about a controversial topic or state an opinion. These are also in the end-of-chapter study guides.

3 Programmed learning units. These are sections within chapters, designed to develop skill in various technical areas. Examples are worked out in a stepwise fashion, and each step is explained. Then the student is given the chance to work a similar problem and the work can be checked, since most problems are worked for you immediately in the program.

The following are the goals for this book:

1 To give an understanding of the scientific process and the place of science in our culture.

2 To give an understanding of the nature of matter and the basic principles of chemistry *in a way that can be mastered by the beginner*. Chemistry need not be the awful course it is reputed to be. With this goal in mind, games, animations, and careful stepwise procedures for skill development have been provided.

3 To give an understanding of the current technical aspects of our culture so that the student as a citizen and consumer will be better able to make decisions when voting and purchasing goods and services.

4 To provide a reasonable amount of practical information that can be used by the student to solve common household problems.

I wish you the best as you begin this new adventure in knowledge and hope that you will learn to appreciate this field of chemistry and find here some of what will make you the person you are to become as well as information that will be of value to you at many points in your future.

Frank L. Wiseman

To the Instructor

My desire in writing this book was to produce a text that would be as versatile as possible, since there seems to be a great diversity in how chemistry courses for nonmajors are taught and at what level they are taught.

The order of the chapters is always debatable in a text of this type, and each instructor has different ideas about what is appropriate. I have started the text with a chapter on the nature of chemistry which contains some subject material that should be mastered. It is followed by a short chapter on how to study chemistry, with the idea that the student will be more motivated to pay attention to Chapter 2 *after* seeing what kinds of material are to be covered.

Chapter 3, ''Chemistry and Society,'' was placed near the beginning of the book because it was felt that an overview at the start would be more beneficial than a review at the end of the text. It should be pointed out, however, that both Chapter 2 and Chapter 3 can be skipped, or used later or earlier, without doing damage to the flow of the more standard chemical coverage.

The more complex material (which some instructors prefer to have and others to skip) has been separated as advanced study (as in Chapter 3) or as appendixes (Appendixes B–F, for Chapters 4, 6, and 8). It can be referred to at the appropriate time, if desired, or not used at all. Future chapters do not build on the material covered in Appendixes B–F.

Chapter 3 contains material on educational institutions, because the author has observed that most college students don't even know what *Ph.D.* stands for, much less how one is earned.

The section on religious thought in Chapter 3 was included because many students have religious convictions of some strength. Religion is an important force in culture and does indeed touch science at a few points. I have tried to treat this topic briefly and factually, so that no violation of students' personal beliefs or problems of ''separation of church and state'' will occur. However, that section can be ignored in class discussion if it is not desired by the instructor (religion is not mentioned again in the text).

I have included sections on inorganic and organic nomenclature as a part of the regular text (in Chapters 10 and 11, respectively), even though some instructors do not use it. I have used programmed learning units to help the students with mastery in these areas. It was felt that these are areas that can be mastered relatively easily by the students taking this course and thus give the student a sense of accomplishment. Again, little if any building is done on these sections, and they could be omitted.

It is hoped that these "blocks" of material will make the text a flexible one.

An Instructor's Manual is available; it includes such material as alternative chapter sequences, lecture games, demonstrations, and student animations. There are also test questions and references to articles for outside reading assignments. Laboratory exercises are also given; some of these can be used in a formal laboratory period and others can be given as special assignments.

Acknowledgments

I would like to acknowledge the valuable help of the following people in preparing this text: John Blackburn, Georgetown College (general advice); J. D. Fox, University of Kentucky (Chapter 15); Frank Harris (suggestions on Chapters 1–5); Beth Kleppinger, University of Kentucky (Chapter 13); Nicholas Martin, M.D., Lexington, Kentucky (Chapter 16); and Diane Russo (suggestions on the original outline). The following people reviewed the manuscript and gave many helpful suggestions: James Coke, University of North Carolina at Chapel Hill; Danette Dobyns, California State—Dominguez Hills; Paul F. Endres, Bowling Green State University; Pat Garvey, Des Moines Area Community College; Ray Johnson, Hilldale College; and Charles W. Owens, University of New Hampshire.

Frank L. Wiseman

Chemistry in the Modern World

Concepts and Applications

Goals

To develop ideas about the role of chemistry in our society

To understand the nature, strengths, and limitations of science

To develop skill in classifying matter

To identify various types of changes in matter

To understand how the properties of each type of matter determine its use

1

INTRODUCTION TO CHEMISTRY: WHAT IS CHEMISTRY?

During the American Civil War, many soldiers were, of course, killed in battle. But an even larger number died of infected wounds or of dysentery resulting from unsanitary conditions. This situation occurred again during World War I, although to a less serious extent. In 1928, however, when World War II was not yet even on the horizon, a British scientist named Alexander Fleming isolated a material called *penicillin* from mold growing on a culture medium. His discovery was to have a worldwide impact—touching the lives and well-being of millions of people, including virtually everyone reading this book. In 1939, Flory and Chain recognized the value of penicillin in stopping the growth of certain kinds of bacterial infections in human beings. The first clinical use of this material was in 1941, and in 1942 the same scientists treated bacterially caused meningitis with penicillin and demonstrated the first cure of this often fatal disease. The large-scale isolation of penicillin from mold cultures was then begun, and this very valuable material was used during World War II to stop wound-related infections and many illnesses caused by bacteria. This saved lives that in previous wars would have been lost.

Penicillin was being produced by living mold cells at that time, and its exact nature was not known. A large cooperative effort by British and American chemists was begun for the purpose of determining the chemical nature of this material. It involved about 300 chemists from both countries and was certainly one of the largest and most intensive scientific endeavors ever attempted. As a result of this effort, the structure of the penicillin molecule was discovered in the early 1950s by Kenneth Henry-Logan. This discovery led to the large-scale production of partially synthetic material and greatly reduced the cost and increased the availability of this drug. Knowledge of the structure of penicillin allowed investigation of the mechanism by which it stopped the growth of bacteria—which in turn helped in the development of more useful antibiotics.

In the case of penicillin, nature produced a substance; a human being found it; others recognized its potential usefulness; and still others worked to understand why it acted as it did. All this came about through the application of *chemistry*. That is the subject of this book—how human beings have *discovered* knowledge of chemistry; what the underlying *chemical principles* are; and how they have been, and will continue to be, *applied* to benefit human beings.

One way to begin a study of chemistry is to define this field of study and understand something of its scope. There are two ways of approaching this task. The first is to give a descriptive definition and discuss what chemists do and how chemistry affects our lives, to gain a "sense" or "feel" for this field of study. The second is to give a formal, concise, easily remembered definition. This chapter will do both, beginning with the descriptive approach.

WHAT DO CHEMISTS DO?

The effect of chemistry on our lives can be determined by thinking about the products you yourself have used today that were made possible by the work of chemists. It is perhaps even more dramatic to try to think of products you have

TABLE 1-1

Role of chemists in the modern world

MATERIALS	ROLE OF CHEMISTS
Aspirin, vitamin pills, prescription medications	Discovery, medical testing, manufacture, checking for purity prior to marketing
Deodorant, paint, perfume, wood filler, glue, varnish remover, soap, shampoo	Original preparation of active ingredients, determining what proportions of each component to use for the best product, manufacture, testing final product for correct amounts of each substance
Polyester clothing, plastic ink pen, Formica kitchen counter top, vinyl floor tile, rubber motor vehicle tires	Original discovery of materials and a recognition of their useful properties, improvement of products to provide better uses, addition of color to make more decorative products, development of inexpensive methods of production
Foods such as white bread, instant coffee, synthetic "cream" (nondairy cream), bacon	Discovery of additives such as flavor enhancers and preservatives: development of processes to produce these materials
Natural foods such as eggs, steak, milk, whole-grain breads	Treatment of animals and plants with food supplements, medications, and fertilizers to keep them healthy and productive, testing for pollutants such as lead
Gasoline, fuel oil, natural gas, iron	Refining from natural source, understanding the physical properties of each material, determining the energy released in various kinds of changes
Your own body	Discovery and production of medicines without which perhaps you would not be healthy, mentally able, or even alive
Air you breathe and water you drink	Development of chemical processes to purify air or water polluted by, for example, a manufacturing facility or motor vehicle; or prevention of pollution by an understanding and application of chemistry

used in which chemistry was *not* involved. Table 1-1 is a list of some materials and of the roles chemists played in our having them.

Have you thought of anything yet that chemistry did not *help* to bring to us? Whether we like it or not, we are accustomed to a high standard of living which is possible only because of synthetic and processed natural materials. Our comfortable lifestyle is largely the result of the activity of chemists, both past and present; and the phrase used by Dupont in its promotional activities—"Better things for better living, through chemistry"—is more than an advertising slogan; it is a statement of philosophy and practice for our culture. Table 1-2 lists ways in which chemistry affects our daily lives.

Now let's take a look at chemists' activities in a more systematic way, beginning with synthesis.

TABLE 1-2

Ways chemistry affects your daily life

MATERIAL OR ACTIVITY	USEFULNESS	PRODUCT OR RESULT
Aluminum chlorhydrate	Stops perspiration	Deodorant
Sodium benzoate	Prevents microorganism growth and preserves freshness of foods	Fresh dry cereal
Chlorine	Kills bacteria	Drinkable water
Cholesterol determination	Tells amount of cholesterol in one's blood	Diet adjustment to maintain good health
Polyethylene terphthalate	Does not wrinkle easily	"Wash and Wear" clothes
Aspirin	Relieves aching pain	Relief of headache pain
Cadmium sulfide	Yellow pigment	Part of the color in a beautiful painting
Silicon	Computer chips	Video game, pocket calculator
Tungsten	Light bulb filament	Light with which to read, eat, etc.
Silicon dioxide	Component of glass	High-quality eyeglasses
Tetraethyl lead	Antiknock compound in gasoline	Smooth-running powerful automobile engines
2,4-D*	Weed killer	Lower-cost food from more easily maintained farm fields
Sodium hypochlorite	Bleach	Household "chlorine" bleach
Sodium bicarbonate	Source of carbon dioxide bubbles	Light, fluffy pancakes
Taking insulin by injection	Replaces a lack of natural insulin	The continuation of a happy, productive life
Maleic hydrazide	Growth inhibitor	Potatoes that do not sprout and thus do not become ruined quickly
Stannous fluoride	Enamel hardener	Healthy teeth
Ferric ferrocyanide	Blue pigment	Blue ink for pens
Milk	Dietary source of protein and calcium	Healthy bodies and minds
Sodium lauryl sulfate	Cleaning agent	Clean, bouncy, shiny hair

*(2,4-Dichlorophenoxy) acetic acid.

Synthesis

Chemists are involved in the synthesis of pure materials. Research chemists in college and university laboratories as well as in industrial laboratories are constantly attempting to prepare new materials (ones that have not been made before) from materials that are already available. They are also attempting to prepare *synthetically* materials which are found in nature (Figure 1-1) but which for some reason cannot be isolated in large enough quantities or are too

FIGURE 1-1
A chemist at work in a laboratory producing new synthetic materials to solve human problems or meet human needs.

expensive. Such pure materials include, for example, aspirin, aluminum chlorhydrate, sodium benzoate, and polyethylene terphthalate. Do not be concerned with learning these names (they will be referred to again later); just remember that they are pure substances which are synthesized in a laboratory by chemists.

Determination of properties

Testing for (or discovering) the *properties* of these new materials is done either to find a new product for a specific purpose (as in Figure 1-2) or as a

FIGURE 1-2
Animals are used to test drugs that are potentially useful for improving humans' health.

matter of academic interest (curiosity). Properties may also be discovered by accident as a chemist is working with the material. Examples of properties are the ability of aspirin to stop or reduce pain in the human body, of aluminum chlorhydrate to stop perspiration, and of sodium benzoate to prevent growth of microorganisms such as bacteria and mold; and the nature of polyethylene terphthalate as a flexible solid with a high melting point. We can think of *physical properties* (how substances look, smell, and taste), *chemical properties* (how they interact with other substances), and *physiological properties* (how they affect living systems).

Formulation

Once a substance is prepared and its properties are known, perhaps someone will decide that it could be used to make life more convenient or solve some problem. At this point (which usually occurs in an industrial laboratory) chemists begin to work on the form of the substance that can best be used. Perhaps it would be best used in solid form or perhaps in dissolved form. Some materials are best used when mixed with other substances. The procedure used to determine the most useful form for a substance is called *formulation*. Following are some examples.

Aspirin (see Figure 1-3) is best used in solid tablet form, since the powder does not taste good. It decomposes when dissolved in water and would not last long on the medicine shelf if sold as a solution.

The deodorant aluminum chlorhydrate could be dissolved in water and rubbed on the underarms with one's fingers (as in Figure 1-4*a*). That would stop perspiration, but it would be messy. Someone involved in formulation thought of dissolving it in water and placing it in a pressurized can (Figure 1-4*b*). Alternatively, it could be mixed with a soft waxy solid so that it could be rubbed on without a mess; or a thickener could be added to a water solution of the compound for use in a ''roll-top'' container (Figure 1-4*b*).

It was found that sodium benzoate is not toxic to humans, even though it will prevent bacteria and mold from growing, and that about one-tenth of

FIGURE 1-3

Different forms of aspirin. *Left,* powder; *center,* solution; *right,* tablets.

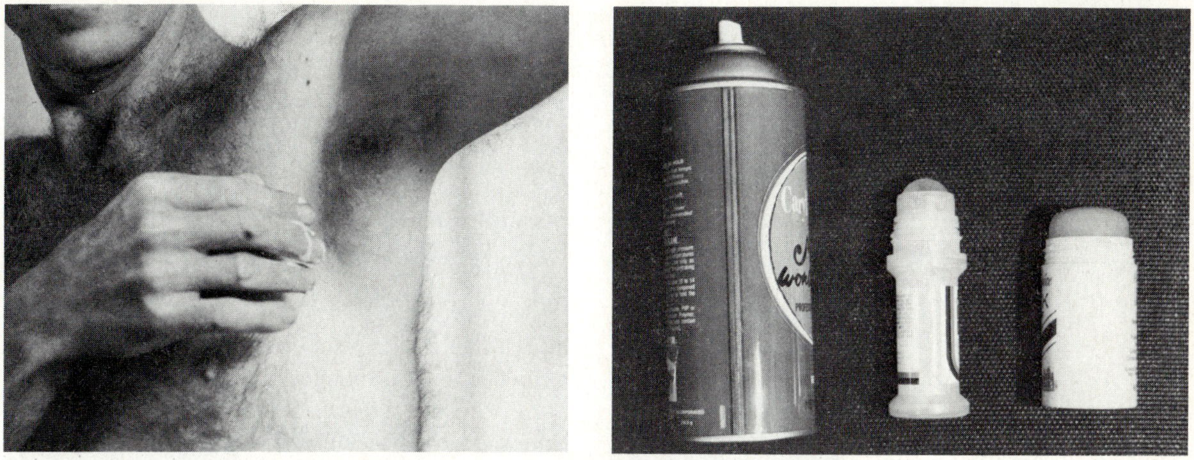

(a) (b)

FIGURE 1-4
Different formulations of deodorant. (a) Solution, to be applied with the fingers. (b) *Left,* aerosol can; *center,* roll-on; *right,* stick.

FIGURE 1-5
Old pieces of bread. The piece at *left,* untreated, is spoiled and moldy; the piece at *right,* treated beforehand with sodium benzoate, shows little spoilage.

1 percent can safely be added to breads and beverages to prevent them from spoiling (see Figure 1-5).

It was likewise found that polyethylene terphthalate, because it is a strong and flexible solid, can be made into threads and woven into cloth. This cloth has the especially desirable characteristics of comfort, long wear, and easy care and is produced by Dupont as Dacron. It can be made into thin ribbons and impregnated with iron compounds and used for magnetic sound and video tape recording. (See Figure 1-6.)

Manufacture

Chemists are involved directly in the *manufacturing* of a product (as in Figure 1-7). They make sure that all the correct procedures are followed to ensure the quality of the product, the safety of the workers, and the most profitable use of all raw materials.

Analysis

Chemists are also involved in analysis of materials (as in Figure 1-8). Chemical analysis involves determining *what* substances are present (this is called *qualitative analysis*) and *how much* of each material is present (*quantitative*

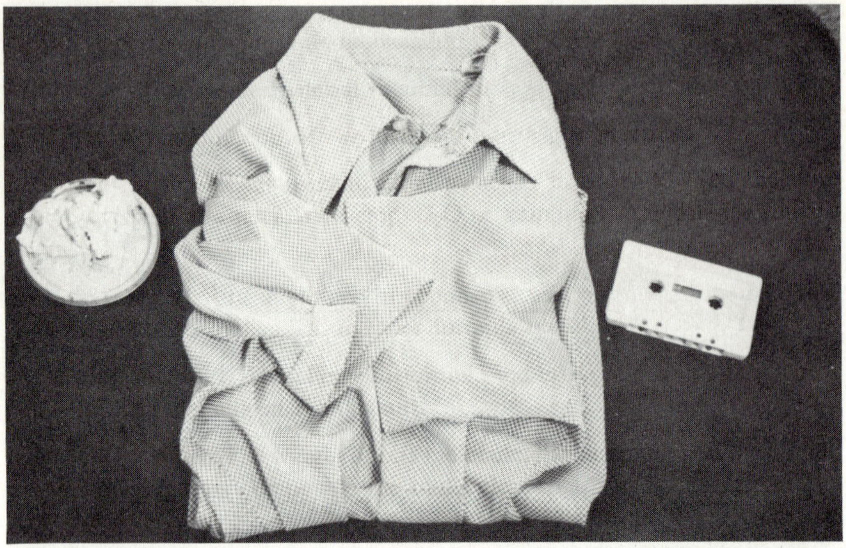

FIGURE 1-6
Different forms of polyethylene terphthalate. *Left,* unformed material; *center,* polyester shirt; *right,* sound recording tapes.

FIGURE 1-7
A manufacturing chemist at work, seeking better ways to prepare a substance.

analysis). As a consumer product is manufactured, an analytical chemist must make sure that it has the correct amount of each component or, if it is a pure substance, determine that no undesirable impurities are present. For example, lead tends to collect in the milk of cows that eat grass in fields along highways (see Chapter 12). When it is manufactured into condensed milk, the concentration of lead tends to increase; therefore, condensed milk must be analyzed for lead content before distribution. Figure 1-9 is a schematic representation of how products are developed.

Analysis is also required in areas other than quality control. Air and water should be constantly analyzed for potentially harmful materials (pollutants), and some natural foods tested for toxic materials. For example, tuna has to be

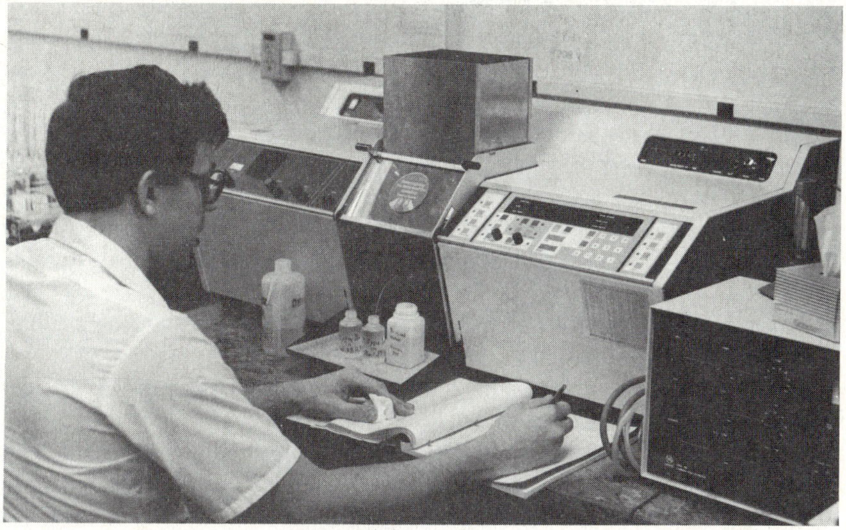

FIGURE 1-8
An analytical chemist doing a lead analysis of milk to ensure its safety.

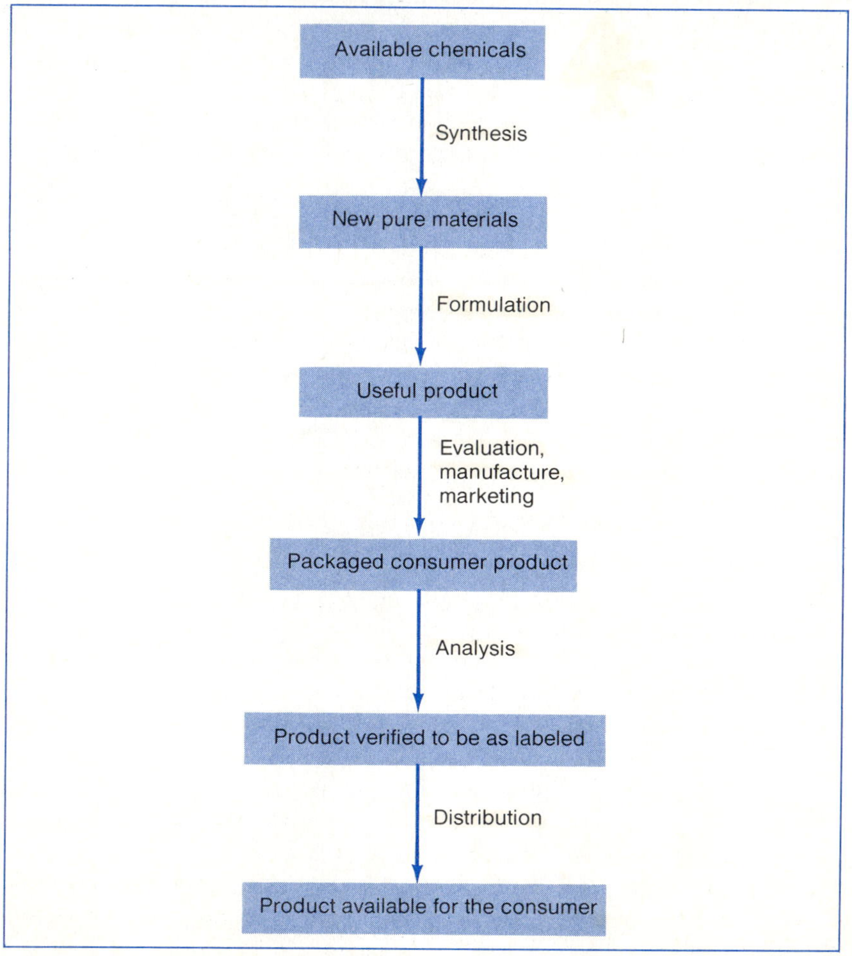

FIGURE 1-9
How a chemical product is developed.

checked for mercury (a toxic material). Much of this work is done in government laboratories. Also, body fluids are analyzed for various components, and the results of these tests are used by physicians to identify medical problems. Blood, urine, and spinal fluid have normal amounts of various components, and deviations from these norms may be indicative of particular maladies. An increasingly common practice among physicians is to analyze a patient's blood to be sure that a prescribed dose of a medication is actually resulting in the desired level of the medication in the blood. (This is expensive for the patient. It is necessary, however, because in the treatment of some diseases the level of medication in the bloodstream is critical but varies among patients receiving equal dosages. In the treatment of epilepsy, for example, such monitoring is important in the control of seizures.)

In addition to those who actually carry out analytical procedures (practicing chemists and laboratory technicians), there are *research* chemists who must find an accurate and convenient method of determining the presence and amount of various materials for which there is a need. For example, glucose in the blood is determined by mixing a treated blood sample with a chemical that will produce a color when it mixes with glucose; the intensity of the color is proportional to the amount of glucose present and is measured by an instrument called a *spectrometer*. Mercury in tuna provides another example. The amount of mercury in tuna is found by chemically digesting the fish and trapping the mercury vapor that results. Light is then passed through the gas sample; if mercury is present, the light is stopped in proportion to the concentration of mercury.

Theory

Last, but by no means least, many chemists are concerned with gaining a better understanding of chemical substances. Only out of theories about matter—only out of our understanding of how chemistry works—can we develop synthetic procedures, methods of analysis, and manufacturing procedures. This fundamental knowledge is gained primarily in university laboratories by instructors and graduate students in the basic sciences. Figure 1-10 is a schematic diagram of how chemical knowledge is developed.

CHEMISTRY—A SCIENCE

So far, we have been taking a descriptive view of chemistry. Now, let's take a more technical or formal view.

Science and scientific method

To begin with, chemistry is a science. A science is a field of knowledge based on observations and on the organization of those observations. The process by which truth is obtained in a science is diagramed in Figure 1-11.

Before the several aspects of this process, which is a powerful one, are discussed, several possible misconceptions should be mentioned and corrected. The scientific process is not something that only professional scientists do. It is actually the same *rational thinking* that people do in solving any of

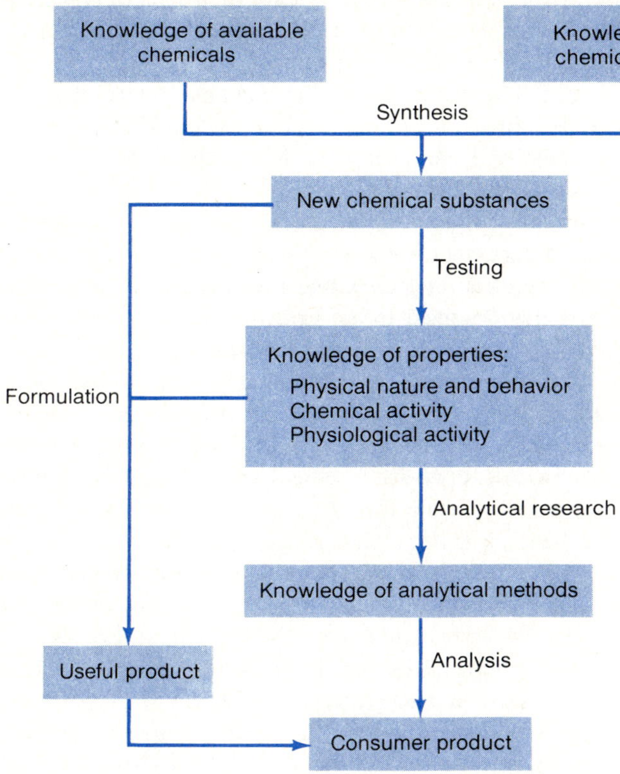

FIGURE 1-10
How chemical knowledge is developed.

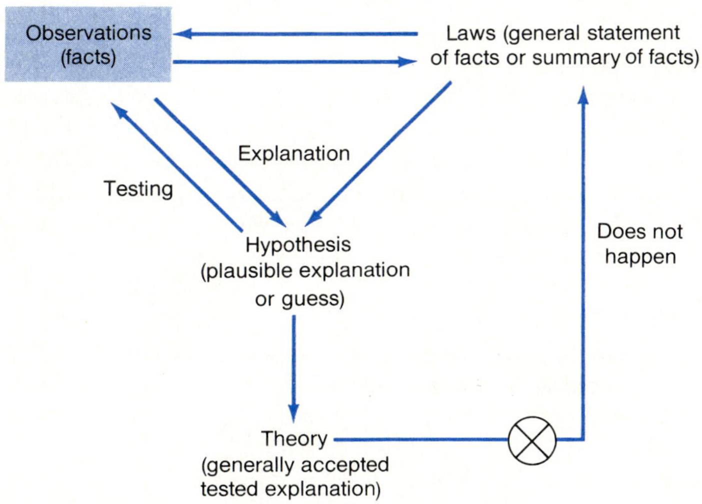

FIGURE 1-11
Scientific method.

life's problems. Also, science is not a purely mechanical activity in which one puts in facts and automatically obtains theories. It involves *creativity* as well as technical skills and knowledge. There is no science without art (there is no scientific advancement without creativity) and no art without science (no one can create without a technical knowledge of the medium being used).

Major advances in all fields of science have their serendipitous aspects as well—the right circumstances encountered by the right person at the right time. But although chance can account for part of some major discoveries, it should also be noted that chance favors the prepared mind. Others—not so well prepared, or less brilliant, or less creative—faced with the same set of circumstances would not have seen the correct relationships and therefore would not have formed a valuable hypothesis or theory.

Having noted these points, we are prepared to examine Figure 1-11: the general flow of reasoning that is the basis for science.

Observations are made by perceptive people, either by accident or—more commonly—in a designed *experiment*. These may be a collection of facts or a single fact. Sometimes, when many people under varying circumstances observe similar kinds of events, and there is reason to think a certain fact is universally true, it is said to be a *law*. An example would be the law of gravity: objects heavier than air always fall toward the earth when support is removed and never "fall" upward. Another example would be the universal gas law, a mathematical formula which describes how gases behave; it summarizes or generalizes many individual facts about gases.

Creative, curious minds want to understand or *explain* what is perceived, and so explanations are frequently sought for both simple observations and laws. At first, a proposed explanation may be only an educated guess, thought to be a possible explanation. This initial explanation is called a *hypothesis*. One can test a hypothesis by using it to predict the outcome of a particular experiment. If the prediction is correct, the hypothesis is given added credibility. If the prediction does not work, the hypothesis must be discarded and another one proposed. This process is not necessarily done by one scientist over a short period of time, but it may be.

When a hypothesis stands the test of many experiments by a number of workers over a period of time, it is accepted by all or most of those who are professionals in the field and is called a *theory*. A theory, then, is a generally accepted, tested explanation. A theory cannot be proved beyond a shadow of a doubt, since we can never be sure that another explanation—which no one has thought of yet—will not be superior. Some theories become so widely accepted that they are thought of as fact. If new evidence contradicts an established theory, however, it must be discarded in favor of a better one. Science is, therefore, a *process* by which knowledge is gained—*not* a final statement of truth.

The general flow from specific facts to generalization is called *inductive reasoning*. A flow in the opposite direction, from generalization (or explanation) to specific facts, is *deductive reasoning;* it is used when testing a hypothesis. Since the first approach used in science is inductive (from observation to theory or laws), scientific method is frequently called the *inductive method*. However, both induction and deduction are used in science. An example of each may help to clarify the meaning of these terms.

Imagine that I am a creature from another planet. I am trying to determine how to tell men from women, having heard there are these two kinds of people. I ask several people which kind they are and notice that Jill, Kelly, and Barbara (who claim to be women) all have long hair, while John, Steve, and Mike (who say they are men) all have rather short hair. I conclude, from these specific facts, that all men have short hair (a generalization). This is *inductive* reasoning. I then meet Diane, who has short hair, and conclude that Diane is a man. I have used my generalization about gender and hair length to predict a specific fact. This is *deductive* reasoning. Now Diane informs me that she is a woman, and I must discard my generalization and search for another one. I have used both kinds of reasoning in my search and have taken a truly scientific approach to the problem.

Another illustration or two will suffice to show how this works. (Later we shall encounter a number of examples from the field of chemistry.) First, an example from everyday life. Mrs. Jones comes home from work and finds the family car missing from the driveway (observation). Her first explanation is that the car has been stolen (hypothesis). It is a possible explanation, but there is not any real evidence to support it. She proceeds to check out (test) this hypothesis by calling her husband to be sure that he does not know where the car is. He tells her that he sent one of his assistants to get it, since he needed it for business reasons. Her first hypothesis had to be discarded and another idea, for which there was evidence, put in its place. Remember that this process is really just rational human thinking.

Another example—a real one from the field of health science—has to do with smoking and cancer. Some years ago, it was noticed that tobacco smokers frequently developed cancer (observation). The natural hypothesis was that something in tobacco smoke caused cancer. This was not as obvious as it may seem, since a correlation does not necessarily imply a cause-and-effect relationship. Perhaps those who smoke also have a propensity for eating fried eggs, and it is really the eggs that cause the cancer. It turned out, however, after this hypothesis was tested by isolating the various compounds of smoke and seeing how each one affected animals, that the hypothesis was sound: there now appears to be a firm case that one of the components, benzopyrene, causes cancer. Benzopyrene reacts with chemicals in the body to produce new materials which in turn cause cell growth to proceed in a grotesque and uncontrolled manner—cancer. This is now generally accepted by the scientific community and by society in general (except possibly the tobacco industry, which sometimes refers to the "health allegation" against smoking). It is, therefore, a theory at this point. Further experiments may cause it to be modified or replaced.

A theory is useful if it helps us to explain or if it helps us to organize a large body of otherwise incoherent facts. Atomic theory (the heart of modern chemistry) not only explains *all* known chemical facts but also helps a chemist to remember many isolated observations because they now make sense as a coherent whole. The theory of evolution serves similar functions in biology. Facts about plants and animals can be organized in one's mind according to the evolutionary scheme.

These two theories illustrate another point that should be made about scientific method. While all theories are useful, not all are equally well supported by evidence and therefore not all are equally likely to be correct.

Atomic theory is very strong because no known evidence is contrary to it, and it is most likely actually the case that atoms do exist as real physical entities (we will examine this theory in greater detail in Chapter 5, along with the evidence for it).

Evolution is quite a useful model, since many facts support it and it helps in the study of biology; but there are some observations that are not consistent with this view. There is, therefore, some question as to whether it is historically correct, at least in its classical form. It is, then, a weaker theory than atomic theory. Perhaps it is totally wrong, or perhaps only parts of it need to be modified. Further experimentation and thinking are needed to clarify the picture. Let us examine this somewhat more closely.

The classic theory of evolution states that all living forms come about by gradual changes over a very long period of time (uniformitarianism). This implies that species intermediate between reptiles and birds (for example) should have existed at one time and one should be able to find their fossils; scientists used to believe that they would find these fossils. Extensive searching has not revealed them, however. It it also true that an organ such as the eye would at some point have to be only partially developed and unable to function; but then it would be of no advantage to the organism possessing it, and that species would not be in line for further selective adaptation. The classic theory is now being modified. Perhaps, instead of a uniform change, there were short periods of very rapid change in living systems followed by long periods during which relatively small changes occurred. This is an excellent example of a current use of the scientific method.

Even with this modification, the evolution of an organ like the eye still causes problems for the rational mind; and there are also reported anomalies, such as human and dinosaur footprints in the same rock strata, as well as single petrified trees traversing rock layers that were supposed to have been laid down over a span of millions of years. Further studies will perhaps incorporate these findings into a modified evolutionary scheme similar to the present one or may cause the present theory to be discarded in favor of a totally different one—or perhaps these anomalous ''facts'' will not be supported by other researchers. (Remember that science is based on repeated observations—not on single incidents.)

No scientific theory should be accepted by scientists or laypeople as representing final, proven truth. One should always keep an open, critical mind when dealing with science. As a matter of fact, since science is a process rather than dogma, anyone who fails to keep an open mind is not a scientist in the true sense of the word.

Another comment about the scientific method is in order here. *An explanation cannot possibly become a fact.* Therefore, *a theory never becomes a law.* Explaining something and reporting it as fact are two entirely different concepts, not different aspects of the same idea. The facts that matter behaves as if it is composed of atoms, that there is a correlation between smoking and cancer, and that the animals on one island are different from those on another will never become explanations. Likewise, atomic theory, the theory that smoking causes cancer, and the theory of evolution will never become statements of absolute fact.

Natural science

The large field of science is divided into at least two areas: the social sciences and natural sciences. The social sciences include the study of human behavior; the natural sciences are the study of the ''natural'' world. Chemistry falls into the area of natural science. Figure 1-12 outlines the more important natural sciences and their subdisciplines, on the basis of how fundamental knowledge in that area is. Mathematics is generally considered to be the most

FIGURE 1-12

How chemistry is related to the other academic disciplines.

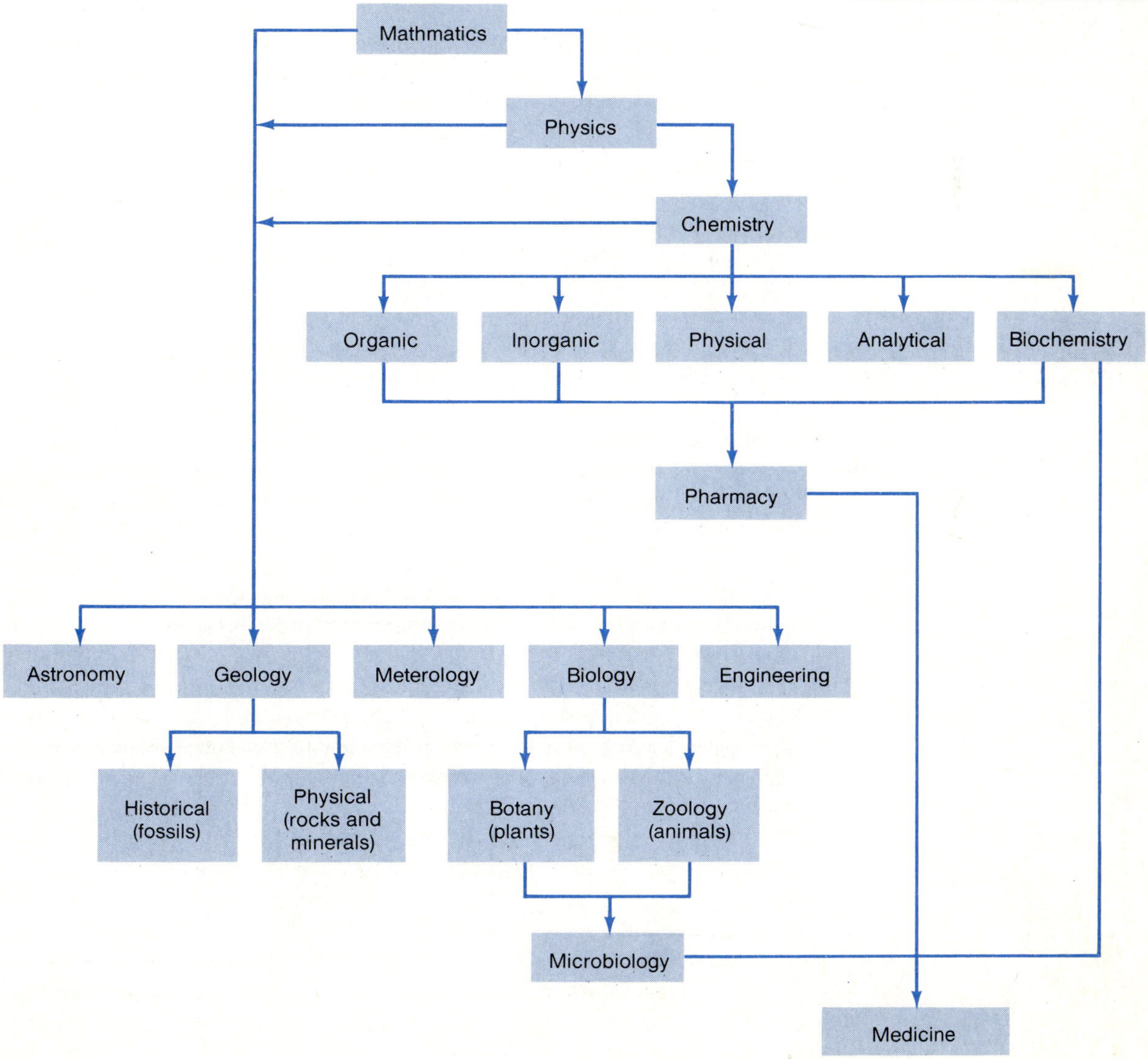

basic of the fields. Chemistry and physics are the most basic of the physical sciences. Most other sciences are some combination of, or are derived from, the principles and facts of chemistry and physics.

Formal definition of chemistry

Now we have come to the point where a formal definition of chemistry can be given. Chemistry is *a natural science which involves the study of the matter of the universe*. There is interest in three areas:

1 Understanding all individual *substances* and mixtures

2 Understanding the *changes* those substances can undergo and how they interact with each other

3 Studying the *energy* changes associated with the changes in matter

Matter, in turn, can be formally defined as *that which has mass and occupies space*. However, we intuitively know what matter is: it is material, or substance, or "stuff."

Areas of chemistry

The field of chemistry has traditionally been divided into several subdisciplines (see Figure 1-12). Most chemistry departments at colleges and universities are thus divided into organic, inorganic, physical, and analytical chemistry and biochemistry on the basis of the exact nature of the chemistry being studied and the kinds of skills being developed in the students.

Organic chemistry Organic chemistry is the study of carbon-containing substances and the synthesis of such materials. Since so many consumer products are organic compounds, there is a large demand for organic chemists; and organic chemistry is one of the larger areas in terms of number of students.

Inorganic chemistry Inorganic chemistry is the study of the properties and synthesis of materials containing all elements other than carbon. This is an area that is growing rapidly now; and many exciting discoveries will probably be made in it in the near future.

Physical chemistry Physical chemistry is the measurement of various quantitative physical properties of matter, along with an effort to correlate large amounts of data. These activities are aimed at understanding the underlying principles of chemistry rather than developing methods of synthesis. Physical chemistry is especially concerned with understanding energy changes in chemical processes.

Analytical chemistry Analytical chemistry is the study of methods used to determine the presence of materials and to establish how much of a material is present. Although this is a well-established discipline, methods for new materials must be found, as well as easier ways of doing previously known analyses. With increasing concern over the environment, we also must find ways of detecting smaller and smaller amounts of harmful materials.

Biochemistry Biochemistry is the study of all substances and processes in living systems. Many of the skills used in the other areas of chemistry are used here in specialized ways. It is generally thought that many serious medical problems can be solved only by a more complete understanding of the chemical processes in the normal organism and by an understanding of the chemical basis for malfunctions. This is an area which is in its early stages; present knowledge is slight compared with the vast areas that still remain unknown.

CLASSIFICATION OF CHEMICAL SUBSTANCES

We have seen that chemistry is defined, basically, as the study of the substances in the universe. This study is aided by dividing all materials into categories—classes or types. Classification helps both the beginning student and the professional scientist to learn and understand differences in behavior among the many different materials to be studied. There are at least three ways of dividing matter into different types, which are discussed below.

Gases, liquids, and solids: States of matter

Matter exists in three *states* or *phases:* gas, liquid, and solid. Most substances can exist in all three states at appropriate temperatures. Water, for example, can be a solid (ice), liquid (ordinary water), and gas (steam); see Figure 1-13. These are *different forms of the same substance* rather than different substances.

(a) (b) (c)

FIGURE 1-13
The three phases of matter. (a) Solid: ice. (b) Liquid: water. (c) Gas: steam.

TABLE 1-3

Taxonomy of types of matter and methods of identifying each

TYPE	PRACTICAL DISTINGUISHING CHARACTERISTICS	FORMAL DEFINITION	EXAMPLES
Pure matter	Cannot be separated into more than one substance	Material in which only one substance is present	
Element	Found in a complete list of the elements	One of the basic substances from which all other material is composed	Lead, oxygen
Compound	Composed of a pure substance which is not an element	Substance composed of two or more elements	Water, salt, limestone*
Mixture	Can be separated physically, is known to be prepared by mixing, is known to contain at least two substances, can have different composition with different samples	Material composed of two or more pure substances	
Homogeneous	Is clear, is composed of one phase, has no visible "pieces," looks like a pure substance	Mixture that has the same composition throughout (uniform)	Sugar water, Kool-Aid (after preparation)
Heterogeneous	Is cloudy, has visible "pieces," shows differentiation in color, sometimes has two or more layers, appears "cruddy looking"	Mixture that has a different composition at different locations within it (nonuniform)	Homogenized milk, asphalt

*Limestone is nearly pure calcium carbonate, but natural samples do vary somewhat in color and texture (e.g., fossil imprints can sometimes be seen).

Pure substances: Elements and compounds

It is even more useful to classify matter according to its purity and fundamental nature: is it a combination of other materials, or is it a basic material? Table 1-3 outlines this system. It gives the formal definition (the concept involved) and a practical way a beginning student can tell classes of substances apart. Figure 1-14 shows an example of each class.

Pure materials fall into two broad groups: (1) elements, which are those few substances basic to all other materials; and (2) compounds, which contain two or more elements.

FIGURE 1-14
Examples of different classes of matter. (a) Element: lead balls. (b) Compound: table salt. (c) Homogeneous mixture: sugar water. (d) Heterogeneous mixture: asphalt.

Elements Elements are not formed from other materials, nor can they be decomposed into any more basic substances. They are inherently themselves; and they are the matter from which all other kinds of materials are made. There are 92 naturally occurring elements and (currently) 13 produced in artificial nuclear processes.

Each element is given a name and a letter symbol composed of the first letter or first letter and one other letter in the name of the element in English, German, or Latin. For example, cobalt, one of the elements, has the symbol Co (from the English). Likewise, hydrogen is symbolized by H (English), sodium is Na (from the German *natrium*), and iron is Fe (from the Latin *ferris*). Table 1-4 is a list of all the elements and includes the name and symbol of each along with other data which will be referred to later.

TABLE 1-4

Atomic numbers and atomic weights of the elements

ELEMENT	SYMBOL	ATOMIC NUMBER	ATOMIC WEIGHT
Actinium	Ac	89	(227)*
Aluminum	Al	13	26.98
Americium	Am	95	(243)
Antimony	Sb	51	121.75
Argon	Ar	18	39.95
Arsenic	As	33	74.92
Astatine	At	85	(210)*
Barium	Ba	56	137.34
Berkelium	Bk	97	(247)†
Beryllium	Be	4	9.01
Bismuth	Bi	83	208.98
Boron	B	5	10.81
Bromine	Br	35	79.90
Cadmium	Cd	48	112.40
Calcium	Ca	20	40.08
Californium	Cf	98	(251)†
Carbon	C	6	12.01
Cerium	Ce	58	140.12
Cesium	Cs	55	132.91
Chlorine	Cl	17	35.45
Chromium	Cr	24	52.00
Cobalt	Co	27	58.93
Copper	Cu	29	63.55
Curium	Cm	96	(247)†
Dysprosium	Dy	66	162.50
Einsteinium	Es	99	(254)†
Erbium	Er	68	167.26
Europium	Eu	63	151.96
Fermium	Fm	100	(253)†
Fluorine	F	9	19.00
Francium	Fr	87	(223)*
Gadolinium	Gd	64	157.25
Gallium	Ga	31	69.72
Germanium	Ge	32	72.59
Gold	Au	79	196.97
Hafnium	Hf	72	178.49
Hahnium	Ha	105	(262)†
Helium	He	2	4.00
Holmium	Ho	67	164.93
Hydrogen	H	1	1.01
Indium	In	49	114.82
Iodine	I	53	126.90
Iridium	Ir	77	192.22
Iron	Fe	26	55.85
Krypton	Kr	36	83.80

Continued

ELEMENT	SYMBOL	ATOMIC NUMBER	ATOMIC WEIGHT
Lanthanum	La	57	138.91
Lawrencium	Lr	103	(257)†
Lead	Pb	82	207.20
Lithium	Li	3	6.94
Lutetium	Lu	71	174.97
Magnesium	Mg	12	24.31
Manganese	Mn	25	54.94
Mendelevium	Md	101	(256)†
Mercury	Hg	80	200.59
Molybdenum	Mo	42	95.94
Neodymium	Nd	60	144.24
Neon	Ne	10	20.18
Neptunium	Np	93	(237)†
Nickel	Ni	28	58.71
Niobium	Nb	41	92.91
Nitrogen	N	7	14.01
Nobelium	No	102	(253)†
Osmium	Os	76	190.20
Oxygen	O	8	16.00
Palladium	Pd	46	106.40
Phosphorus	P	15	30.97
Platinum	Pt	78	195.09
Plutonium	Pu	94	(244)†
Polonium	Po	84	(209)*
Potassium	K	19	39.10
Praseodymium	Pr	59	140.91
Promethium	Pm	61	(147)*
Protactinium	Pa	91	231.04
Radium	Ra	88	226.03
Radon	Rn	86	(222)*
Rhenium	Re	75	186.20
Rhodium	Rh	45	102.91
Rubidium	Rb	37	85.47
Ruthenium	Ru	44	101.07
Rutherfordium	Rf	104	(261)†
Samarium	Sm	62	150.40
Scandium	Sc	21	44.96
Selenium	Se	34	78.96
Silicon	Si	14	28.09
Silver	Ag	47	107.87
Sodium	Na	11	22.99
Strontium	Sr	38	87.62
Sulfur	S	16	32.06
Tantalum	Ta	73	180.95
Technetium	Tc	43	98.91

TABLE 1-4

Continued

ELEMENT	SYMBOL	ATOMIC NUMBER	ATOMIC WEIGHT
Tellurium	Te	52	127.60
Terbium	Tb	65	158.93
Thallium	Tl	81	204.37
Thorium	Th	90	232.04
Thulium	Tm	69	168.93
Tin	Sn	50	118.69
Titanium	Ti	22	47.90
Tungsten	W	74	183.85
Uranium	U	92	238.03
Vanadium	V	23	50.94
Xenon	Xe	54	131.30
Ytterbium	Yb	70	173.04
Yttrium	Y	39	88.91
Zinc	Zn	30	65.38
Zirconium	Zr	40	91.22

*These elements are rare and radioactive (see Chapter 13), and thus very little is known about them. The numbers in parentheses are the mass numbers of the most stable type of atom found in nature or produced in experiments.

†These elements are artificial (not found in nature—produced in nuclear processes). The numbers in parentheses are the mass numbers of the most stable type of atoms produced in experiments.

Figure 1-15 is a more standard presentation of the elements in what is called the *periodic chart*—or the *periodic table—of the elements*. (It appears on the inside back cover of this book, for ease of reference.) Not only does the periodic chart give the name and symbol of each element, along with pertinent data, but the arrangement of the elements communicates a great deal of chemical information. This will be discussed in more detail in Chapter 6.

One can identify a material as being an element by finding its name on a complete list of elements; one should read over the list of elements and become familiar with them, so that when one encounters the name or symbol of an element it will be recognized as such. It will be helpful also if the name-symbol relationships of the more common elements are memorized. Table 1-5 lists 28 of the most commonly encountered elements along with a brief description.

FIGURE 1-15
Opposite page:
The periodic chart.

I A	II A	III B	IV B	V B	VI B	VII B	VIII B	VIII B	VIII B	I B	II B	III A	IV A	V A	VI A	VII A	VIII
1 **H** Hydrogen 1.01																	2 **He** Helium 4.00
3 **Li** Lithium 6.94	4 **Be** Beryllium 9.01											5 **B** Boron 10.8	6 **C** Carbon 12.0	7 **N** Nitrogen 14.0	8 **O** Oxygen 16.0	9 **F** Fluorine 19.0	10 **Ne** Neon 20.2
11 **Na** Sodium 23.0	12 **Mg** Magnesium 24.3											13 **Al** Aluminum 27.0	14 **Si** Silicon 28.1	15 **P** Phosphorus 31.0	16 **S** Sulfur 32.1	17 **Cl** Chlorine 35.5	18 **Ar** Argon 39.9
19 **K** Potassium 39.1	20 **Ca** Calcium 40.1	21 **Sc** Scandium 45.0	22 **Ti** Titanium 47.9	23 **V** Vanadium 51.0	24 **Cr** Chromium 52.0	25 **Mn** Manganese 54.9	26 **Fe** Iron 55.9	27 **Co** Cobalt 58.9	28 **Ni** Nickel 58.7	29 **Cu** Copper 63.5	30 **Zn** Zinc 65.4	31 **Ga** Gallium 69.7	32 **Ge** Germanium 72.6	33 **As** Arsenic 74.9	34 **Se** Selenium 79.0	35 **Br** Bromine 79.9	36 **Kr** Krypton 83.8
37 **Rb** Rubidium 85.5	38 **Sr** Strontium 87.6	39 **Y** Yttrium 88.9	40 **Zr** Zirconium 91.2	41 **Nb** Niobium 92.9	42 **Mo** Molybdenum 95.9	43 **Tc** Technetium 98.9	44 **Ru** Ruthenium 101.1	45 **Rh** Rhodium 102.9	46 **Pd** Palladium 106.4	47 **Ag** Silver 107.9	48 **Cd** Cadmium 112	49 **In** Indium 115	50 **Sn** Tin 119	51 **Sb** Antimony 122	52 **Te** Tellurium 128	53 **I** Iodine 127	54 **Xe** Xenon 131
55 **Cs** Cesium 133	56 **Ba** Barium 137	57 **La** Lanthanum 139	72 **Hf** Hafnium 178	73 **Ta** Tantalum 181	74 **W** Tungsten 184	75 **Re** Rhenium 186	76 **Os** Osmium 190	77 **Ir** Iridium 192	78 **Pt** Platinum 195	79 **Au** Gold 197	80 **Hg** Mercury 201	81 **Tl** Thallium 204	82 **Pb** Lead 207	83 **Bi** Bismuth 209	84 **Po** Polonium (209)	85 **At** Astatine (210)	86 **Rn** Radon (222)
87 **Fr** Francium (223)	88 **Ra** Radium 226	89 **Ac** Actinium (227)	104 **Rf** Rutherfordium (261)	105 **Ha** Hahnium (262)													

58 **Ce** Cerium 140.1	59 **Pr** Praseodymium 140.9	60 **Nd** Neodymium 144.2	61 **Pm** Promethium (147)	62 **Sm** Samarium 150.3	63 **Eu** Europium 152.0	64 **Gd** Gadolinium 157.2	65 **Tb** Terbium 158.9	66 **Dy** Dysprosium 162.5	67 **Ho** Holium 164.9	68 **Er** Erbium 167.3	69 **Tm** Thulium 168.9	70 **Yb** Ytterbium 173.0	71 **Lu** Lutecium 175.0
90 **Th** Thorium 232.0	91 **Pa** Protactinium 231.0	92 **U** Uranium 238.0	93 **Np** Neptunium 237.0	94 **Pu** Plutonium (242)	95 **Am** Americium (243)	96 **Cm** Curium (247)	97 **Bk** Berkelium (247)	98 **Cf** Californium (251)	99 **Es** Einsteinium (254)	100 **Fm** Fermium (253)	101 **Md** Mendelevium (256)	102 **No** Nobelium (254)	103 **Lr** Lawrencium (257)

TABLE 1-5

Commonly encountered elements

NAME	SYMBOL	DESCRIPTION
Aluminum	Al	Light-gray metal
Barium	Ba	Silvery metal
Bromine	Br	Red liquid*
Calcium	Ca	Reactive, silver metal
Carbon	C	Black solid
Chlorine	Cl	Green gas
Copper	Cu	Brown metal
Fluorine	F	Toxic, colorless gas
Gold	Au	Yellow, precious metal
Hydrogen	H	Colorless, flammable gas
Iodine	I	Violet solid
Iron	Fe	Heavy, strong metal
Lead	Pb	Very heavy, soft-gray metal
Lithium	Li	Silvery metal
Magnesium	Mg	Light, gray metal used in airplane construction
Mercury	Hg	Silvery liquid metal*
Nitrogen	N	Colorless gas which is 80% of air
Oxygen	O	Colorless gas needed for combustion; 20 percent of air
Phosphorus	P	Soft (claylike) material, several colored forms
Platinum	Pt	Precious metal used as a catalyst
Potassium	K	Reactive soft metal
Silicon	Si	Gray, brittle element
Silver	Ag	Shiny precious metal
Sodium	Na	Metal which is a component of salt
Sulfur	S	Yellow, brittle solid
Tin	Sn	Shiny metal used to coat steel in "tin cans"
Uranium	U	Radioactive heavy metal
Zinc	Zn	Gray metal used to galvanize steel

*Bromine and mercury are the only two elements that are liquid at ordinary temperatures. Gallium and cesium melt at 30°C, which is only slightly above room temperature.

Compounds The second class of pure substances are called *compounds*. A compound is composed of more than one element and can be prepared from the individual elements or, at least in theory, broken down into them. Salt, for example, can be decomposed by chemical methods into sodium and chlorine (two elements), and limestone can be broken down into carbon, calcium, and oxygen (each of which is an element). Salt and limestone are classified as compounds because each is composed of more than one element. Compounds are not just a mixture of elements; they are new materials that behave completely differently from the elements of which they were made.

Sodium is a soft, silvery metal that reacts violently with water. Chlorine is a poisonous, green gas. Yet the salt formed from these two elements is a white solid that does not react with water and, so far from being poisonous, is actually essential to life. Likewise, hydrogen is flammable, and oxygen supports (is necessary for) a flame; but the compound made from them, water, is used to put out fire. A *mixture* of hydrogen and oxygen is highly explosive; but the *compound* formed from them is not flammable and will not explode. Salt and water are new materials completely different in all respects from the elements of which they are composed.

In classifying a material as a compound, one should determine that it is not an element and that it is a pure material. The latter criterion, purity, is often difficult for a beginning student to grasp. Usually, a material is pure if one cannot imagine it being "separated" by ordinary physical methods, such as evaporation or filtering or picking out pieces with tweezers. If one asks, "Can this material be separated—not decomposed chemically, but separated physically?" and the answer is *no,* then it is probably a pure material. Salt is salt; one cannot imagine "separating" it. Likewise, limestone is limestone and cannot be separated. The questions hardly even make sense in these cases. In contrast, peanut butter can be separated into a solid brown substance and an oil, and so it is not a pure material. If different samples of a substance are somewhat different (like different brands of peanut butter), it is probably a mixture.

Mixtures: Homogeneous and heterogeneous

Pure materials, both elements and compounds, can be blended to form two categories of *mixtures*. Unlike compounds, mixtures have the properties of each component rather than totally new properties. A *homogeneous* mixture is the same throughout, is in one phase or state, and is usually clear. A *heterogeneous* mixture is not the same throughout and is usually characterized by cloudiness, or the visible presence of "chunks" or pieces that are different in color, texture, or phase from the rest of the mass.

To compare the two kinds of mixtures, consider sugar added to water. At first, the solid sugar is visible on the bottom. You can see the pieces of solid sugar, and there are two phases (liquid and solid); this is, therefore, a heterogeneous mixture (one place is visibly different from another, not the same throughout). If you stir this mixture, the sugar dissolves throughout the entire volume of the water and the mixture has the same appearance everywhere. It is now homogeneous. Not only does it look the same everywhere; it also has the same composition everywhere. There are, at this point, no visible pieces, and there is only one phase (liquid).

This mixture can be separated again. For example, the water can be evaporated, and the solid sugar would remain (see Figure 1-16).

Table 1-3 (refer back to page 18) summarizes the types of matter and the methods one can use to classify a given material. No one set of rules will always work; but by using these rules you will develop an intuition for this classification system, and that will help you to recognize the exceptions. The short program starting on page 27 will help you develop skill in this procedure.

(a)

(b)

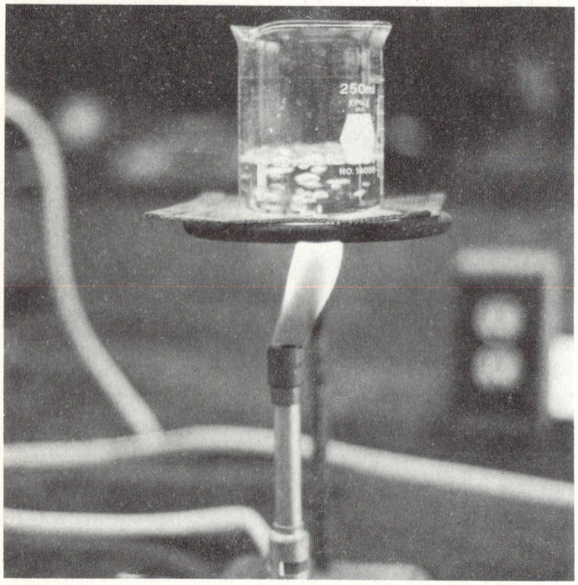
(c)

(d)

FIGURE 1-16
An example of physical separation. (a) Granulated sugar and water. (b) Sugar solution. (c) Water is removed by boiling. (d) Sugar is separated again as a pure material.

CLASSIFYING MATTER

Worked examples

MATERIAL	CLASSIFICATION	REASONS
Water	Compound	Is a pure material (cannot be separated and is not found on list of elements).
Aluminum	Element	Found on element list.
Sealed carbonated beverage	Homogeneous mixture	From general knowledge, one knows a beverage contains water and sugar (once two substances can be named, it is a mixture). It is homogeneous, not cloudy.
Homogenized milk	Heterogeneous mixture	Contains water and some solids (therefore a mixture). Since it is cloudy, it is heterogeneous.

Problem

Fill in the missing columns. See Table 1-3 for methods of classification.

MATERIAL	CLASSIFICATION	REASONS
Piece of concrete		
Seawater		
Orange juice		
Table sugar		
Whiskey		
Iron		

Answers

MATERIAL	CLASSIFICATION	REASONS
Concrete	Heterogeneous mixture	Contains visible pieces of rock, sand, and ''gray material''

MATERIAL	CLASSIFICATION	REASONS
Seawater	Homogeneous mixture	Clear, contains water and salt (from general knowledge)
Orange juice	Heterogeneous mixture	Cloudy, contains water, sugar, and vitamin C
Table sugar	Compound	Is not separable, is not on list of elements
Whiskey	Homogeneous mixture	Clear (not cloudy), contains alcohol and water, brown color
Iron	Element	Found on list of elements

CHANGES IN MATTER: PHYSICAL AND CHEMICAL CHANGES

It has already been stated that chemistry includes the study not only of matter but also of the changes that occur in matter and of the way various kinds of matter interact with each other. These changes can be divided into two types—physical and chemical.

In a *physical change,* a material (an element or a compound) does not change; only its form changes. This category includes changes in texture, shape, and state, as well as any dissolving or other mixing. The freezing of liquid water to ice, the molding of clay, and the dissolving of salt in water are examples of physical changes.

A *chemical change* involves the formation of new substances; but chemical changes are often accompanied by physical changes such as formation of new colors, smells, or tastes or a change of state. A chemical change can usually be identified by one or more of the following criteria:

1 It is *not* one of the common physical changes, such as change of state or texture, or dissolving.

2 There is a difference in color, taste, or odor. While these observable changes are themselves physical, they show that a new material is present.

3 It cannot be reversed by ordinary physical methods.

4 A new, easily identifiable material is formed, or a material is destroyed.

While no list of rules is infallible, these will suffice in most cases to allow the identification of chemical changes.

CLASSIFYING CHANGES IN MATTER

Worked examples

CHANGE	CLASSIFICATION	REASON
Changing water to steam	Physical	The substance is the same; the change can be reversed (steam back to water). One of the standard physical changes.
Cooking an egg	Chemical	New color (white), odor, taste. It cannot be reversed.
Grinding rock to powder	Physical	Material is not changed. (This change cannot easily be reversed, however.)
Dissolving sugar in coffee	Physical	Change can be reversed. One of the standard physical changes.
Formation of carbon dioxide from baking soda and vinegar	Chemical	New material, change cannot be reversed.

Problem

Classify each of the following changes as chemical or physical and give your reason.

CHANGE	CLASSIFICATION	REASON
Souring of milk		
Turning of leaf colors in the fall		
Honey turning to solid sugar		
Corrosion of a copper wire		
Burning of wood		

Answers

CHANGE	CLASSIFICATION	REASON
Souring of milk	Chemical	Change in taste; therefore a new material must be present.
Turning of leaf colors in the fall	Chemical	New color, therefore new material, not reversible.
Honey turning to solid sugar	Physical	Change can be reversed by heating or by adding water.
Corrosion of a copper wire	Chemical	Change not reversible, change in color (brown to green).
Burning of wood	Chemical	Wood changes to gases and ashes, not reversible.

PROPERTIES OF MATTER: PHYSICAL AND CHEMICAL PROPERTIES

It has already been noted that the properties of a substance are a means of identifying it. These are natural characteristics of a particular element, compound, or mixture.

Physical properties are characteristics describing the physical nature of the material or indicating something about the physical changes it can undergo. These include physical descriptions such as color, hardness, shine, smell, and taste, as well as melting and boiling temperatures, the conduction of electricity, and the ability to dissolve other materials.

Chemical properties are those which determine the types of chemical changes a substance can undergo. They include the tendency of an element to combine with other elements to form compounds and the tendency of a compound to decompose back to elements or react with another compound to form a third material. Table 1-6 shows some of the chemical and physical properties of three materials.

The physical properties of a substance determine its usefulness in fabricating products or formulating solvents. For example, mercury (Hg) is the only metal that exists as a liquid at room temperature. It also conducts electricity. This combination of physical properties makes it useful in fabricating silent electric switches. Figure 1-17 shows how these work. Another substance, called *Wood's metal*—which is a mixture of the elements bismuth, tin, lead, and cadmium—has a melting point of 71°C.* This is well above

*C stands for *Celsius*. See Chapter 4.

TABLE 1-6

Some chemical and physical properties of sodium, neon, and limestone

	SODIUM	NEON	LIMESTONE
Physical properties			
Melting point	97°C	−249°C	Decomposes but never melts
Boiling point	880°C	−246°C	Does not boil
State at room temperature	Solid	Gas	Solid
Color	Silvery (metallic)	Colorless	Dull gray
Chemical properties			
Reaction with oxygen	Combines with oxygen easily	Does not combine with oxygen	Does not react with oxygen
Changes when heated	Simply melts	Becomes a warmer gas	Decomposes to lime and carbon dioxide

room temperature (25°C) but below the temperature produced by a flame. Wood's metal is used in automatic sprinklers. An automatic sprinkler is an iron pipe with holes drilled in it; the holes are filled and sealed with Wood's metal, and the pipes are run in the ceiling of a building. If a fire should start, the Wood's metal will melt and the water will put the fire out automatically. A third example is ethylene dichloride, which will dissolve grease. It is included in "spot removers," since most hard-to-remove stains contain grease.

The chemical properties of a substance determine its usefulness in preparing other materials, in various products such as cleaners and medicines, and in analytical procedures. For example, penicillin, a natural substance which can be made synthetically, has the ability to combine chemically with certain materials in bacteria. This action stops bacterial growth and makes penicillin a useful antibiotic. Hydrochloric acid (the main component of liquid toilet bowl cleaners) reacts chemically with hard-water scale (like that which forms in teapots) and changes it to carbon dioxide and other materials that can be easily dissolved and washed away. This makes hydrochloric acid a good scale-removing agent.

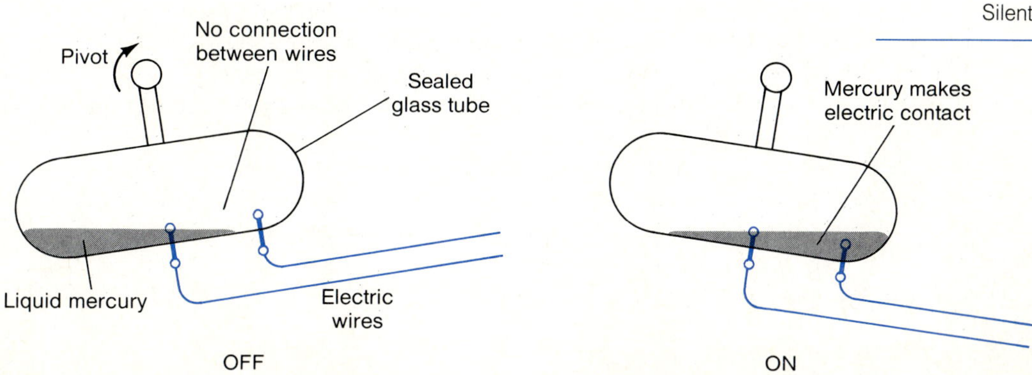

FIGURE 1-17
Silent mercury switch.

STUDY GUIDE

Terms to learn

Analysis
Analytical chemistry
Biochemistry
Chemical change
Chemical properties
Chemistry
Compound
Deduction
Element
Formulation
Heterogeneous
Homogeneous
Hypothesis
Induction
Inorganic chemistry
Law
Mixture
Observation
Organic chemistry
Physical change
Physical chemistry
Physical properties
Properties
Pure material
Qualitative analysis
Quantitative analysis
Science
Synthesis (chemical)
Testing (of a substance)
Theory

Skill practice problems

1 Give a concise formal definition of chemistry.

2 List the five major subcategories within the field of chemistry.

3 List ways chemists were involved in producing the following materials.
(a) Typewriter
(b) Shampoo
(c) Hair spray

4 Which of the following statements are true? Which are false?
(a) A theory is useful if it allows unification and prediction, even if it is not a correct representation of physical reality.
(b) Natural science is a statement of absolute truth.
(c) All theories are equally strong.
(d) Theories, if well established, eventually become laws.
(e) For a theory to be useful it must be agreed upon by *all* professionals in the field.
(f) The average nonscientist uses the scientific method to solve problems.
(g) Once an idea is established in a scientific field, it is almost impossible to change.

5 Which of the following are elements, which are compounds, and which are mixtures?

(a) Marijuana
(b) Aluminum
(c) Epsom salts
(d) Rose plant
(e) Perfume
(f) Butter
(g) Iron
(h) Cola drink
(i) Pure alcohol

6 Classify each of the following changes as chemical or physical.

(a) Cooking a steak
(b) Evaporating water
(c) Digesting a cookie
(d) Washing sweat off of one's dirty body
(e) Burning a cigarette
(f) Preparing instant tea
(g) Drying fingernail polish
(h) Melting ice cream

Questions for discussion

What was your view or definition of chemistry before starting this course? What are the essential differences between that and what you now understand?

Goals

To gain a realistic perspective on some problems encountered in the study of chemistry and on how they can be overcome

To understand the value of studying chemistry

To begin the development of general study skills

2

THE STUDY OF CHEMISTRY: HOW TO SUCCEED BY TRYING

You have begun the study of chemistry. For many of you this is a new field, and whatever your reason for becoming involved in this study, you may be wondering about the benefits that will be yours when you have mastered it. This chapter will answer that question and will also take a look at some of the reasons why beginning students are sometimes apprehensive about studying chemistry. Some practical suggestions about how these fears can be overcome will also be given.

REASONS FOR STUDYING CHEMISTRY

First, let it be said that learning the basic facts of chemistry and mastering its principles and the methods of thinking involved are not reserved for science majors who intend to become professionals in the basic or applied sciences. Some of you are taking this course as a preliminary to a more advanced general chemistry course. Some of you may become interested in a career in chemistry through reading this book and change your major as a result. Most of you, however, will want to use the material as part of your quest for knowledge in general, or to give yourself a broader base on which to build expertise in your chosen field.

Since science in general—and chemistry in particular—has a prominent place in our culture, every citizen should understand the basic ideas of science. This is probably best accomplished by a thorough study of one of the sciences rather than by a course in general science. A study of one science can help the student appreciate both the capabilities of science and its limitations.

We all work with substances every day: we use water to clean; we work with soil in our gardens; we use inked typewriter ribbons; we eat food; we sit at a wooden table (or is it plastic?); we wear clothing. The study of chemistry can help us understand what these materials are like and why they behave as they do. They will be a little less mysterious in one sense, and yet the universe will still hold many fascinating mysteries to be explored. Chemistry explains some mysteries and reveals others. The study of chemistry can give us a feeling for the intricate organization of matter and a perspective on the place of humans in the universe.

In addition to basic chemistry, we will be considering some of the chemical processes which are important to our standard of living and the products of the chemical industries to which we have become so accustomed. You may learn to be even more thankful for these products; at the same time, you will probably understand that all good things have potential for misuse as well. You will be a better consumer, a better citizen, and a more rational person, less likely to be swayed by the emotional aspects of an issue. This is not to deny the important place emotions have in our lives, for they greatly enrich our human experiences. They should, however, be carefully controlled when critical decisions are being made. Decisions ranging from what brand of a product to buy to how to control a local source of pollution to how to establish national policies on energy—all these should be based on reason. These are, of course, lofty goals, which no one course of study can accomplish, but it is hoped that this text will move each reader toward them.

FIGURE 2-1
Dissolving unwanted materials is one of the everyday problems chemistry can deal with.

We will also examine some information that will aid in solving practical problems encountered in everyday life (Figure 2-1 is an example). This information will appear mostly in the later chapters and will be based on the fundamental concepts covered in earlier chapters. Real-life problems are sometimes complex; but there are certain standard procedures which can be used even by nonscientists to save money and time and which sometimes can do a better job than a commercially available product can do—or do a job for which no commercial product is available. References will also be given to sources containing more detailed information of this type.

WHY THE STUDY OF CHEMISTRY IS CHALLENGING

For several reasons, beginning students in chemistry face a number of challenges.

First, the student encounters many new ideas and words for the first time, and any new material, even if it is relatively simple, takes a certain amount of time to be assimilated by the mind.

Second, in chemistry one is required to think about abstract concepts and about real physical entities, such as atoms and molecules, which one cannot see or otherwise experience. Related mathematical ideas and manipulations are also abstract. In this book, most of the material is presented in a nonmathematical way; only Chapters 4 and 8 use mathematics, and even there the procedures are carefully explained. No mathematical knowledge except elementary arithmetic is assumed. Most people develop the ability to think about abstractions sometime between the ages of 15 and 20 years. Thus some entering college students (who are usually 18 years old) have this ability and others—perhaps 40 or 50 percent—do not, or at least have not developed it to a high level. If you are in the latter group, you will have to work a little harder and pay more attention to the illustrations and analogies; but do not think that just because you have a more difficult time, you will not be able to master the material.

A third problem students have with chemistry is the widespread but erroneous notion that science is hard to learn—unlike history, English, sociology, and literature, which can be learned by anyone. This is partly a result of previous conditioning in primary and secondary schools, and partly a result of a widespread idea that scientists are ''eggheads'' with unusually high IQs. Whatever the reason, this idea is not correct. The sciences are no more difficult than medieval history, Shakespeare, foreign languages, or English grammar. If a student perceives the sciences as hard, however, the anxiety that results can in itself become a barrier.

The activities in the next two sections of this chapter are designed to show that the instructional techniques used in this book will aid the beginning student, and to show that there are specific things to do which will contribute to direct instruction and help reduce ''science anxiety.''

AIDS TO THE STUDY OF CHEMISTRY

In this text, every attempt will be made to give a careful, stepwise development of each reasoning procedure or skill. Also, short programmed learning segments will help you develop the necessary skills. In these programmed units, you will find examples of each type of skill worked out, with each step explained. You should follow these steps carefully in your own mind. After several worked examples are presented, a problem will be presented for you to work. You should attempt to work it using the procedure of the worked examples. Only after you have tried it should you turn to the solution. After repeating this procedure several times—and perhaps making several errors, from which a fuller understanding should result—you will be asked to work problems for which only the answers are given. Any skill needs practice; therefore, you should work all the examples given.

In some cases, the flow of human events leading to a theory is presented—not only to teach the history of science but also to allow you to go through the same stages of understanding that humanity in general has gone through to reach our current state of knowledge. This is an effective learning technique.

The mastery of abstract ideas can be facilitated by the use of analogies and diagrams, for these help us to develop useful mental images. It is also true that the more experience one has with models and with thinking about abstract ideas, the easier it becomes to think about new ones. Therefore, many analogies and diagrams are included, since everyone must base a new abstraction on some tangible example, imperfect though the connection may be. Later, as the mental image becomes more highly developed, the imperfect analogy can be at least partly discarded. Table 2-1 lists examples of abstract ideas and Table 2-2 lists examples of concrete experiences. One should attempt to recognize abstract ideas and try to think about ways of developing concrete props to serve as models of abstractions or to take advantage of the illustrations given in this text.

TABLE 2-1

Examples of abstract concepts

CONCEPT	EXAMPLES
Beauty	Beautiful sunset
Confidence	Feeling of certainty that your automobile brakes will stop your car in an emergency
Love	Feelings and actions toward a special friend
Faith	Confidence in the existence of God or in the ability of the democratic system to stand stress and survive even though neither can be proved
Energy	The vigorous exertion of a person running
Atom	Atom of iron
Symbol	Fe as the symbol for iron

TABLE 2-2

Examples of concrete experiences

EXPERIENCE	EXAMPLE
Color	Red.
Fact	The car did stop when I stepped on the brake.
Expression of affection	A kiss.
Action	Voting.
Material	Iron.
Hardness	The touch of a piece of iron.

ACTIVITIES TO FACILITATE LEARNING

This textbook has been organized to facilitate learning, but of course it is also necessary for students to apply themselves diligently. For most chemistry courses, one should study 2 to 3 hours (h)* outside class for every hour in class. Study could include reading the text, reading lecture notes, working problems, and memorizing facts. No teachers, text, or method will work without this most important of all factors. It is also assumed that one will attend class and do all assigned work.

At times, even a student who is spending enough time and applying enough effort will encounter difficulties, or a student may find that the hours spent studying do not seem to be productive. In this situation, students should feel free to approach the instructor for help, either as individuals or as small groups, or seek assistance from the resources within the school. *Peer teaching,* which takes place in study groups of three to five students from the same course on a regular bases, can be of great help (see Figure 2-2). Study groups are not intimidating, and one member of the group may understand a given concept and be able to explain it to the others. Each member can prepare practice test questions for the others; and preparing one's own questions and answering those of the other students can be of value in mastering the material.

*In this textbook, standard scientific abbreviations will be used for units of measurement. The first time any unit appears, the spelled-out form will be followed by the scientific abbreviation in parentheses. Thus the abbreviation for *hours* is h.

FIGURE 2-2
Students studying together to reinforce learning.

A lack of confidence during lectures, study, or examinations can be a tremendous barrier to learning or to demonstrating knowledge, or both. Such anxiety may have several sources, including lack of problem-solving skills or lack of self-confidence. (This text is organized to develop both skills and self-confidence.) Anxiety may also be caused by the misconceptions about science which are so prevalent in our culture. These include the following:

Only very bright people can learn chemistry.

It is all right to fail chemistry, because it is so difficult and some of my best friends failed it.

The opposite view, the one you should adopt from the beginning, is that by carefully applying yourself and having confidence in your ability to learn, you can master chemistry.

General physical and emotional health also has a bearing on academic achievement. A student who is involved in a troubling emotional situation may find it hard to concentrate and study effectively. Unfortunately, one does not always have control over such circumstances. Often general physical well-being is more easily attainable than emotional equilibrium. A good exercise program improves muscle tone and blood circulation, both of which help to reduce anxiety. Physical activity as well as proper diet helps one feel better, and a student who is more comfortable usually learns more easily. Neither the mind nor the body can function well if not properly nourished, and excesses of various foods (including carbohydrates) can cause mental sluggishness. Some attention to these details is, therefore, of value. (We will examine nutrition in more detail in Chapter 15.) In addition, most academic institutions offer help in such areas as general study skills, note taking, reading, and test anxiety. One should take full advantage of these if other resources fail to produce results.

Questions 5, 6, and 7 in the Study Guide at the end of this chapter are designed to give practice in some of the techniques of learning you will find useful in a chemistry course (or any other course); these techniques include developing a good system for memorizing facts, developing mental images of things you have never seen, and thinking about a problem for several different short periods of time and letting your subconscious mind help solve the problem while you are not working on it.

It should be pointed out that careful attention to new vocabulary is important here, as in any discipline. It is also true that reading in science usually takes longer than reading in other areas. One cannot expect to ''speed-read'' most of this text. To understand it well, you should read it slowly and carefully. Some people prefer two readings—one for a rapid overview and a second for more attention to detail.

Figure 2-3 shows the organization of the processes involved in mastering a new subject; Table 2-3 lists problems commonly encountered in a chemistry course and some suggested remedies.

FIGURE 2-3
Processes involved in learning new subject matter.

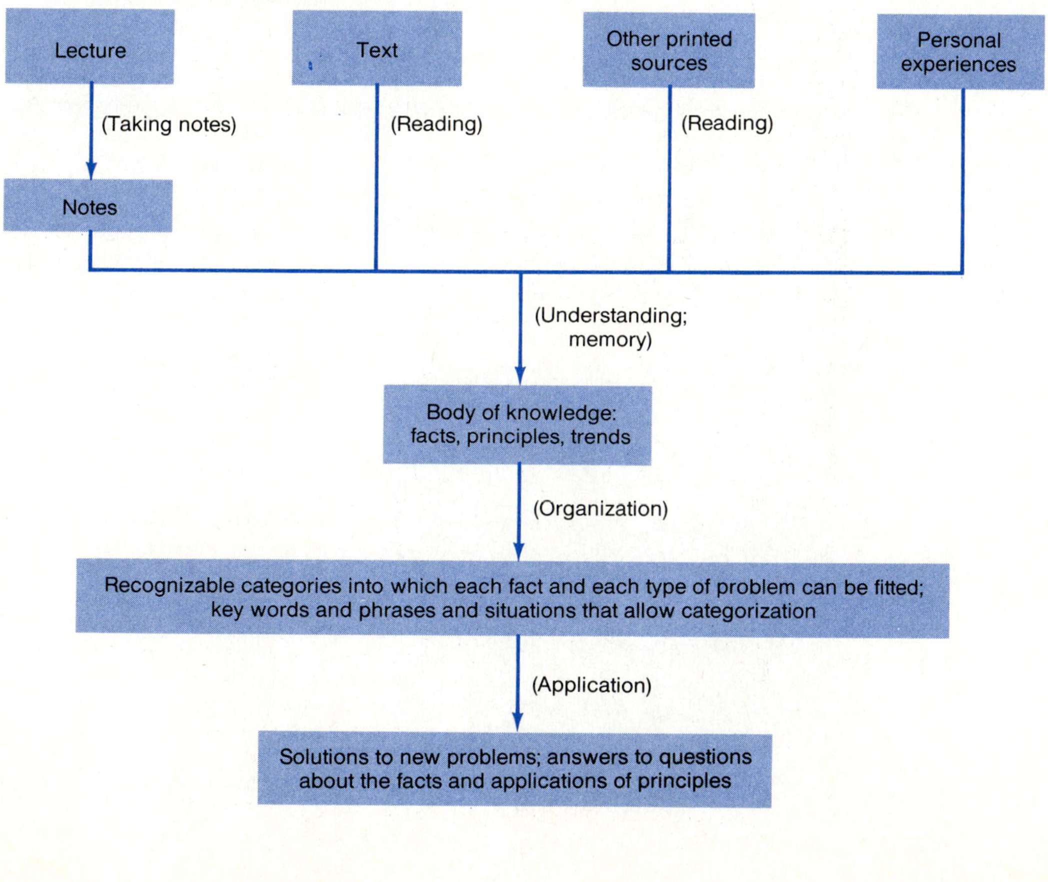

TABLE 2-3

Commonly encountered difficulties and suggested remedies

PROBLEM	AID
Many new facts to learn	Study 2 to 3 h outside of class for each hour in class.
	Prepare lists of facts or flash cards, or both, for short, frequent drills.
Abstract concepts	Pay close attention to diagrams, models, and analogies.
	Consciously try to develop a mental picture of each abstraction.
Mathematical concepts	Learn the procedures as outlined and memorize each mathematical formula after understanding it.
Lethargy	Exercise regularly and follow a nutritious diet.
Attitude	Develop confidence in your own ability. Try simple tasks, and the success gained will help in the more complex ones.
Lack of understanding	Reread the text, and when necessary get your instructor to reexplain the concept in question.
Study skills such as reading comprehension, note taking effective studying, test taking	Seek advice from your instructor, a fellow student who has been successful, or the study skills office if your college or university has one. Outline the material. Prepare practice exams and take them.
Lack of precision in learning	Arrange peer study groups to answer questions, explain concepts, and prepare practice tests.

The whole purpose in this course is for you to learn chemistry. Do not be afraid to ask questions. There is no value in hiding a lack of understanding behind a facade of indifference or false confidence. If you don't know something, admit it. The words in Figure 2-4 should be the motto of all beginning chemistry students. May you enjoy this study and may it be to your benefit for many years to come.

Go ahead—
ask dumb questions.
They're easier
to cope with than
stupid mistakes.

FIGURE 2-4
The chemistry student's motto.

STUDY GUIDE

Questions and activities

1 What are your personal goals for this course (other than fulfilling academic requirements)?

2 What are some fears you have about this course? Express these to your instructor or a fellow student.

3 Identify another class member (or several classmates) whom you know (or would like to know) with the same expectations and anxieties as yours. Set up a regular time each week to study together.

4 Write a weekly schedule with definite times to study each course you are taking. Include 5 h or more for chemistry. Also be sure to include time for relaxing—life is more than academics!

5 Using flash cards or lists, memorize the following phone numbers over the next 2 or 3 days.

Diane	701-0966
Dawn	710-7642
Ellen	606-1365
Edward	502-4005
Jonathan	258-1615
Joel	173-8332

The same techniques you use to memorize familiar facts such as telephone numbers can also be used to learn facts and principles in chemistry, and other new information. Take note of any difficulties you had and how you overcame them. Apply these same techniques to the subject matter of this course.

6 There is an object on a planet far, far away. It is spherical and made of a transparent material. It is hollow with a small cluster of even smaller opaque balls at the center. You have never seen this object. Can you "see" it in your mind? Make a drawing of it. Close your eyes and envision this object without looking at the drawing given later. This same technique will be useful in dealing with some abstract concepts you will encounter in this course.

7 Using the numbers in squares A and B on page 44, determine the way the unboxed numbers can be combined (+, −, ÷, ×) to give the number in the small boxes. Apply this same pattern to calculate the numbers that should be in the small boxes in squares C and D. If you do not see the solution right away, use trial and error. If you have not gotten the answer after 5 to 10 minutes (min), leave the problem and return to it in a few hours. Repeat this procedure the next day. This is *not* a test of intelligence. It is an exercise in

how to solve problems. These same techniques (trial and error and returning to the problem later) can serve you well both on chemistry homework and on tests.

Box	Top left	Top right	Bottom left	Bottom right
A	2	6	4	32
B	5	1	3	18
C	3	1	1	?
D	6	2	2	?

Answers to Questions 6 and 7

6

7 C—4; D—16. Method: (Top left + top right) × bottom left = bottom right.

Goals

- To understand how chemistry affects, and is affected by, various segments of our society
- To see why chemistry at times raises moral issues
- To see how chemistry helps solve some of society's problems
- To encourage thought about some personal issues related to the study of science

3

CHEMISTRY AND SOCIETY

We have already considered some aspects of how chemistry affects various aspects of our lives and our society. Figure 3-1 shows some of the interactions within our culture. In this chapter, we will see how a few of these influence and are influenced by science in general and chemistry in particular. In Chapter 1, the effects of chemistry on our individual well-being and its contributions to our standard of living were noted. In this chapter we will consider the relationship of chemical knowledge to various institutions and organizations in our culture, including the federal government, educational institutions, the chemical industry, and religion. Figure 3-1 is not meant to imply that chemistry is the center of society; chemistry is placed in the middle of the diagram only because it is the focus of this text. It should be noted that most of the arrows are two-way: chemistry both influences and is influenced by other aspects of our culture. Mathematics and physics, being more basic than chemistry, contribute to chemistry, but chemistry has little influence on them.

LAW ENFORCEMENT: FORENSIC CHEMISTRY

As we will see later in this text, there are many ways in which society uses and benefits from the results of chemical activities. One way is solving crimes. If samples of material are gathered from a victim, a suspected criminal, or the scene of a crime, these materials can then be analyzed to determine their chemical and physical properties. Such properties can be evidence for the defense or prosecution of a suspect. This is called *forensic chemistry* (chemistry related to all legal matters); it primarily uses the techniques of analytical chemistry.

The simplest type of forensic analysis involves identifying a substance such as a drug. A white substance suspected of being heroin can be treated with certain chemicals; if it is indeed heroin, a definite color will be seen. For each material commonly encountered in criminal cases, there is a test that can be run to determine its identity.

More complex materials such as blood or ink can be used as evidence. Each person's blood has a slightly different chemical content. The various materials present in differing amounts can be used for ''fingerprinting'' blood. For example, a murder victim may have scratched the killer during the attack and have blood under the fingernails. This blood can be analyzed thor-

FIGURE 3-1

The relationship of chemistry to various aspects of society.

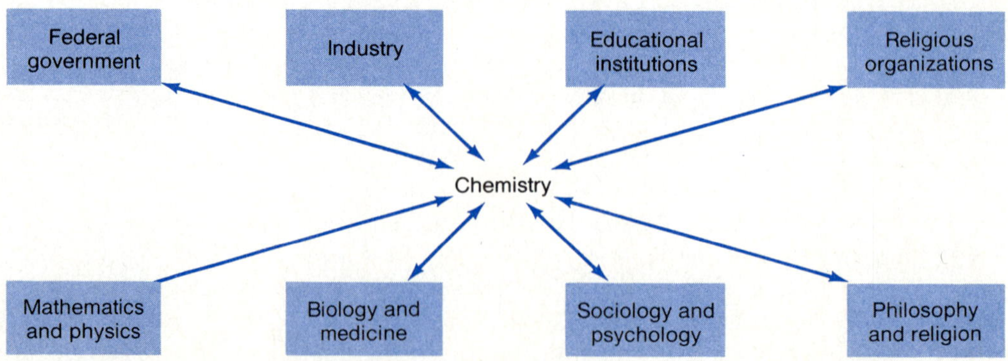

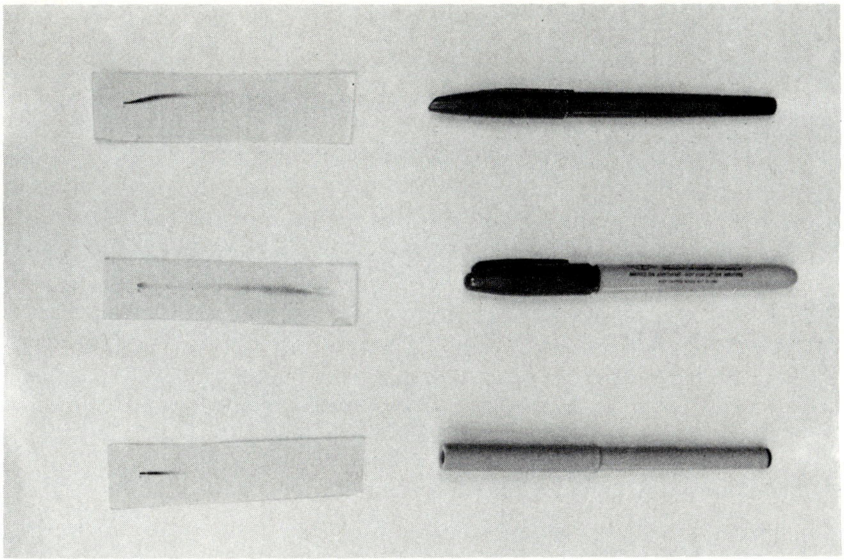

FIGURE 3-2
A typical thin-layer chromatogram of inks from several pens. Note that they are all different.

oughly; if it is found to match the blood of a suspect in all respects, it virtually proves that the suspect was present at the scene of the crime (although it does not necessarily show that the suspect is the murderer).

Ink used to write a ransom note can be analyzed by a technique called *thin-layer chromatography* (*TLC*). Figure 3-2 shows what happens when an ink spot is placed on a plate coated with silica and a solvent is allowed to work its way up the plate. The various dyes in the ink travel at different rates and are separated. Each batch of ink has a slightly different dye composition, and if the pattern of ink in the note matches that of ink in a pen found on a suspect, it is evidence (though not proof) that the suspect wrote the note. (It is possible, of course, that the suspect, although innocent, had a pen made by the same company as the pen used by the actual kidnapper.) Separation by the TLC method is based on the greater solubility in the solvent of some of the colored components of the ink. These components travel faster up the plate, which effects a separation and produces the characteristic pattern for each ink mixture.

In cases of suspected arson, the presence of trace amounts of the material used to start the fire can be detected by careful analysis. Even if a material such as gasoline was used and the building was virtually consumed in the blaze, a small amount of nonflammable material will remain in, say, charred wood or a rug, or even in cement. The exact nature of this material can be determined. The chemical substances composing it and the relative amount of each can be found by means of a technique called *gas chromatography*. Figure 3-3 shows the instrument used in gas chromatography, and Figure 3-4 shows the graph produced. Each ''peak'' in Figure 3-4 represents a different material, and the sizes of the peaks indicate the amount of each material. This graph can be matched with flammable material available from various local suppliers to gain clues about the arsonist.

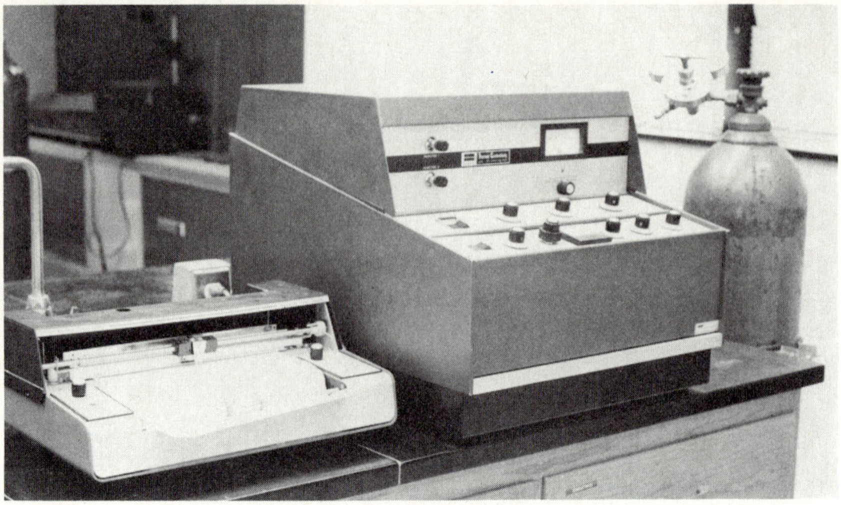

FIGURE 3-3
A modern gas chromatograph. This is the instrument used to obtain the graph in Figure 3-4.

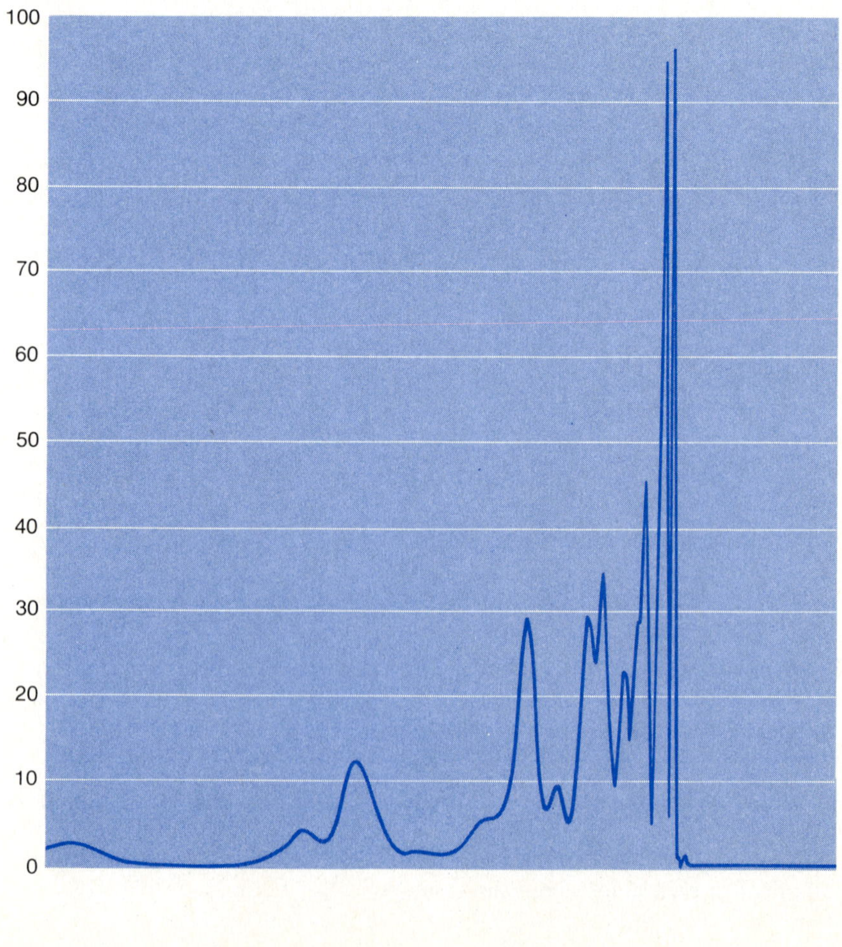

FIGURE 3-4
A gas chromatograph of regular gasoline. Each peak represents a different component.

In cases of suspected rape, vaginal material can be examined under a microscope for the presence of sperm cells. Even if none can be seen, sensitive tests involving color formation can be run to detect small amounts of enzymes (biochemical materials, see Chapter 15) that are always present in seminal fluid. Their presence indicates that rape could have occurred.

Hair is a repository of toxic metals—such as lead, mercury, and arsenic—and because it does not decay, it provides a source of evidence long after a crime has occurred. An interesting example is the death of a man on an early Arctic expedition. Fifty years later, a historian suspected foul play and analyzed hair from the long-buried remains; large amounts of toxic metals were found, indicating that a poisoning had occurred. Another example—a famous one—has to do with the death of Napoleon in 1821 while he was in exile on the Island of Saint Helena in the South Atlantic. The exact cause of his death, at the relatively early age of 51 years, was in dispute for many years. Analysis of hair taken from his head at his death and analyzed 140 years later showed that it contained large amounts of arsenic—evidence (though not proof) that he had been murdered by arsenic (As) poisoning. Several methods are now available that allow determination of extremely small amounts of toxic metals from trace amounts of body materials. In the case of hair, a single piece of hair ¼ inch (in) long (or even less) is all that is needed to do a reliable analysis for mercury, arsenic, or chromium. One can do analyses of various segments of a hair and determine if several poisonings occurred and, if so, approximately when (by knowing the rate of hair growth).

Thus one of society's major problems, that of controlling criminal activities, is helped toward a solution with the aid of chemistry. The study of the physical and chemical properties of materials, if properly used by careful scientists using modern instruments, can be invaluable in providing evidence to convict the guilty and acquit the innocent.

FEDERAL GOVERNMENT: REGULATORY AND TASK-ORIENTED AGENCIES

Table 3-1 lists the various federal agencies related to the field of chemistry, and Table 3-2 summarizes the involvement of federal agencies with various aspects of chemistry. The section "Topics for Advanced Study" later in this chapter gives a fuller description of the activities of these agencies. The following paragraphs illustrate how the federal government is involved in some aspects of chemistry in our society.

FDA: Artificial sweeteners

In about 1968, artificial sweeteners called *cyclamates* were shown to cause cancer in test animals when large doses were given. The Delaney clause of the Food and Drug Act states that any compound which causes cancer in any test animal at any dose level must be banned for human consumption. Once the tests on cyclamates had established the facts and they were published, the Food and Drug Administration (FDA) had no choice but to forbid the sale of

TABLE 3-1

Government agencies related to chemistry

ABBREVIATION	NAME	TASK
FDA	Food and Drug Administration	Regulate sale of food, drugs, and cosmetics as related to safety and effectiveness of products
FTC	Federal Trade Commission	Regulate transport of chemicals
EPA	Environmental Protection Agency	Enforce environmental laws
NSF	National Science Foundation	Direct and fund various science projects
NIH	National Institutes of Health	Identify and solve health problems
NBS	National Bureau of Standards	Keep physical standards and report standard methods of analysis
DOD	Department of Defense	Fund projects related to weapons development or equipment maintenance
OSHA	Occupational Safety and Health Agency	Ensure a safe work environment for workers
NASA	National Aeronautics and Space Administration	Production of new materials for use in space

TABLE 3-2

Summary of federal government involvement with chemistry

1 Control over chemical activity

Industry:
- Regulation of production (e.g., production of oil) (FTC)
- Safety procedures in manufacturing (OSHA)
- Environmental control (EPA)
- Transportation of chemicals (FTC)

Academia:
- Safety in student laboratories (OSHA)
- Use of hazardous materials (OSHA)
- Funding of chemical research (NSF, NIH, NBS, NASA, EPA, DOD)
- Funding of education for individual science students (NSF)

2 Control over use

Human use:
- Regulation of all foods, drugs, and cosmetics (FDA)

Industrial use:
- Waste disposal (EPA)
- Safety (OSHA)
- Toxicity standards (EPA, NBS)

any product containing cyclamates—which it did in 1970. This cost some companies millions of dollars in profits. Actually, cyclamates used in the small doses needed to sweeten various foods probably have very little carcinogenicity, but the law is very explicit.

Since that time, saccharin—another artificial sweetener—has also been shown to be slightly carcinogenic in test animals in large doses; and it too would have been removed from the market, except for special action. Because it was the only artificial sweetener available commercially at that time, and because the health problems of some people would have been even greater without saccharin, Congress voted to exempt saccharin from the law. Some people have medical problems that make ingestion of natural sugars unsafe, and they may need saccharin to make certain foods palatable. The risk from using sugar is higher for these people than the very low risk of getting cancer from saccharin; and Congress wisely acted to protect them. Saccharin is a weak carcinogen: it has been estimated that probably fewer than 100 deaths are due to its use each year. This means that the risk of cancer associated with saccharin is less than that associated with high-altitude air flights, where cosmic ray bombardment is increased. Most people consider this an acceptably small risk.

This example serves to indicate the regulatory power of FDA, the law under which it operates, and the power of Congress to change that law in general or for a specific compound.

EPA: The Times Beach story

In 1974, it was discovered that the soil in Times Beach, Missouri, had been treated with oil containing small amounts of a substance called *dioxin*. This caused concern among the residents, since dioxin is known to be highly toxic to humans.

An investigation produced the following information. A chemical company had inadvertently produced this toxin in trace amounts while manufacturing herbicides. The waste material from the industrial process was known to contain dioxin, and an attempt was made to dispose of this waste material properly. A waste oil hauling company obtained possession of some of it but did not know it contained dioxin. Not knowing of the danger, this company sprayed several horse arenas with this dioxin-containing waste oil to help reduce dust in the arenas. After one child and a number of animals died, the oil was suspected as a likely cause and the soil containing it was removed. Again, rather than being disposed of in an uninhabited or otherwise safe place, some of this soil was used as fill dirt for road construction in residential areas of Times Beach.

This took place in 1971, but not until 1974 did officials identify the culprit as dioxin. As little as 10 parts dioxin per billion parts soil (ppb) is considered dangerous; the samples from Times Beach had from 100 ppb to 2000 ppb. In 1982 the Environmental Protection Agency (EPA) carried out an intense investigation and found the amount and distribution of dioxin to be far above the safe level. Relatively swiftly, EPA took the unprecedented action (in February 1984) of declaring the area contaminated and offering to purchase all the property in the town. This would allow the owners to move and buy homes elsewhere in a safer environment. The money came from a "superfund" set up by Congress for this purpose, and amounted to about $33 million.

NIH: Search for new drugs

The standard treatment for a person with a chronic seizure disorder is constant chemotherapy. A drug or combination of drugs is taken regularly, and thus a blood level sufficient to prevent seizures is maintained. The levels of drugs in a person's blood depend not only on the amount of medication taken but also on the presence of other medicine and on the rate at which the body destroys (metabolizes) drugs. Since each person has a unique biochemistry and each person with a seizure disorder has a unique physiological malfunction, chemotherapy is intensively individualized and presents a real challenge for the physician.

Unwanted side effects also vary from one person to another taking the same drug. These side effects range from low-level discomfort in the case of gingivitis (a side effect of phenytoin) to inability to function normally, as in the case of sedation from phenobarbital, or inability to remember and make decisions in the case of toxic levels of phenytoin (Dilantin). Not all persons can have their seizures controlled by the 15 drugs currently available in the United States. About 20 to 30 percent of the 2 million persons who have seizures in this country are not completely controlled or have serious side effects, and so there is a recognized need for the development of new anticonvulsant drugs.

The National Institutes of Health (NIH), Institute of Neurological and Communicative Disorders and Stroke (NINCDS), began the anticonvulsant screening program (ASP) in 1974 and has been testing several thousand chemical compounds each year since then. Materials submitted to this project by medicinal chemists may be previously known but as yet untested. They may also be new compounds which are synthesized for this program because their molecular structures are similar to known anticonvulsant drugs or which are for some other reason suspected of possessing anticonvulsant properties.

This approach is one in which testing of materials is done without knowledge of the biochemical mechanism of seizures or of the exact mechanism of action of known anticonvulsants. While it is true that similar chemical structures sometimes have similar physiological activity, it is also true that almost all major drugs in use today have been found by chance or by systematic screening rather than by inductive reasoning based on a complete understanding of the disease. Some of the materials submitted to the ASP are chemically similar to drugs such as phenobarbital, Dilantin, primidone, Tegretol, or other drugs currently on the market, but many are not. Approximately 50 percent of the compounds submitted for screening show some anticonvulsant activity; but most have a low activity, are toxic, or both.

The testing is done for NIH at the University of Utah, where potential drugs are given to test animals that are subsequently subjected to a seizure-producing stimulus such as corneal electroshock or metrazol (a chemical that causes convulsions). If the animal does *not* have a seizure when given the seizure-producing stimulus, the compound has been shown to be an anticonvulsant. It is also noted if the test animals die, get sick, or show lack of coordination (indicating central nervous system poisoning). To be considered a possible drug, a material should stop convulsions in a large proportion of the animals tested at a fairly low dosage:—100 milligrams (mg) per kilogram (kg) of body weight or less—and show no toxic side effects. Even after succeeding to this level, a compound must be subjected to several years of long-term testing on animals and then clinical testing on humans before it can be considered for medical use. Several compounds, out of the several thousand screened each year, have reached the long-range testing stage, but to date none has been approved as a new drug. The only new antiseizure drug to be marketed in this country in this decade is valproic acid. It had been developed and used abroad before the NIH project began but only recently received FDA approval for use in the United States.

Since seizure disorder (epilepsy) is not a "killer disease," funds from government and private sources are scarce—especially compared with funds available for research on cardiovascular disorders and cancer. As with any other research, more progress can be made with more funds; but in spite of this, a number of medicinal chemists are working in colleges, in universities, and in industry throughout the country and are helping to solve this important medical problem.

INDUSTRY

Our free enterprise system strives, ideally, for minimal involvement of government with industry except in the areas of safety regulation and antitrust laws. But as a new industry develops or an old one expands or takes on new roles, it is sometimes necessary to establish a new government agency to ensure safety. When scientists developed nuclear energy, a new nuclear power industry developed; because of its new, and sometimes dangerous, potential the Atomic Energy Commission (AEC) was originally established to encourage and regulate it. This agency has since been changed and reorganized into the Nuclear Regulatory Commission (NRC).

The Department of Defense and other governmental groups frequently need supplies of a chemical nature (such as explosives and medications) and offer contracts to private industry to prepare these materials. Some industries, such as the aircraft industry, derive a large portion of their business from government contracts.

Industry is primarily interested in producing salable products, and the chemical industry is no exception. Research chemists are hired by chemical companies to find better or less-expensive ways of producing products and to synthesize or isolate new materials potentially of value. The desire for profit in the free enterprise system is a powerful motivator toward production of better pure materials and formulations for widespread use. Society benefits a great deal from the products of the chemical industry. In fact, the majority of practicing chemists in the United States work for industry in some capacity. The chemical industry itself is (not surprisingly) the largest employer of chemists; it also provides many jobs for nonchemists at almost all levels of training.

EDUCATIONAL INSTITUTIONS

Colleges and universities are the main source of new chemical knowledge and the primary means for passing that knowledge to successive generations of students. While some pure chemical research* is done in industrial or government laboratories, most is done by instructors and graduate students at universities. In such schools, graduate education is the main focus, and most instructors are interested not only in scholarly research but also in publishing new knowledge in professional journals. Both master's degrees (M.S. and M.A.) and doctoral degrees (Ph.D. and D.A.) require conducting independent research, writing it up, and defending its scientific merit before a committee of faculty members.

Typically, an entering graduate student will take 20 or 25 credit hours of advanced course work for a master's degree and 35 to 40 credit hours for a doctoral degree in the field of specialization (such as organic chemistry). Meanwhile, the student chooses an instructor with whom to do research; usually the student is interested in a project that the instructor is working on. The student then works on this research for 1 to 2 years for a master's degree or 2 to 4 years for a doctorate. After the research is complete, it is carefully written up in a compilation called a *thesis* or *dissertation*. The student meets with a committee of faculty members who ask questions about the work reported in the thesis or about the student's knowledge. If this committee agrees that the student has done work at a sufficiently high level, it recommends that the appropriate degree be awarded. Clearly, to be awarded a degree, the student will have to learn proper techniques of research and writing.

Much new chemical knowledge is first discovered by a team consisting of an instructor and a graduate student, is first reported in a dissertation, and is later published in a professional journal, such as the *Journal of the American Chemical Society* or any one of hundreds of similar publications throughout the world. Thus, graduate schools contribute immeasurably to the field of chemistry.

RELIGIOUS THOUGHT

Perhaps the best way to begin this section is to define religion and compare it with science. Religion is a system of beliefs and practices involving forces or beings beyond oneself. It is based on revelation, or supposed revelation, from a teacher or a deity, and requires a certain amount of faith (confidence based on evidence but not on proof). Religion may be simply one's perception of what is true and one's personal system of what is fair and right; it need not be a result of belief in a god. Nonreligious faith may have to do with one's own ability to know the truth.

The personal convictions of scientists—like those of other people—vary a great deal. Some scientists are devout Christians or Jews and thus believe in a personal god. Others follow the pantheistic Asian religions, in which there

**Pure* research is aimed at knowledge rather than at product development.

is no personal god but rather a "ground of all being." Some are simply not interested in religion. Some are atheists, believing that there is no god. Some are agnostics, believing that the existence or nonexistence of a god is unknowable. The author has known many scientists in each of these categories.

A science, by definition, is based on observation and includes organization and explanation of observed facts. This is a human activity, dependent on human observation and creativity. However, it would be erroneous to say that faith (confidence based on evidence) is not part of each scientist's personal world view. One must have confidence in the ability of the scientific process to reveal truth and in the ability of the human mind to think correctly. Thus, contrary to popular opinion, science, like religion, is based on faith—at least to some extent. It is sometimes said that religion is based entirely on faith (unprovable statements) and that science is based entirely on fact (provable statements). This implies that science and religion are at opposite poles of human experience. But we can test the validity of some religious beliefs, such as the historical accuracy of scriptures. And scientists cannot always find strong evidence for their hypotheses, even those that remain useful. Human judgment and intuition are involved in creating both scientific theories and religious doctrine.

It is true that science and religion deal with separate areas of life and take different approaches to discovering truth. A person who is both a scientist and religious must come to grips with this fact and seek to integrate, not dichotomize, the two areas. Moreover, scientists and religious people should respect each other. People who believe in a god have not necessarily committed intellectual suicide; and scientists are not forces against morality and religion (extreme as these ideas may sound, some people do hold them). While science and religion are not always interactive—or, at this stage of our understanding, even compatible—they are not mutually exclusive; and it is only proper that each area give due respect to the other.

Observations in the realm of nature (life on earth, for example) can help us to understand only nature; they say nothing about other realms and can tell us nothing about the supernatural. It would be arrogant of scientists to claim a total understanding of nature, let alone of anything beyond nature. If a spiritual or supernatural realm—a god, for example—exists, it is not subject to scientific detection, observation, or study. Scientists, therefore, should not make statements about it; they can only plead ignorance. Because science cannot answer questions about the supernatural, scientists should at all times maintain humility.

Morality—ethics—is an area not unlike religion and is in some ways related to it. Here too there are issues for science and scientists. Most scientists, regardless of their religious convictions, agree that there are moral decisions to be made about the uses of new discoveries. (The use and development of nuclear bombs is a most important example.) But decisions about such matters are usually not in the hands of the scientists who discover and develop new technologies, but rather in the hands of the general population or —more likely—of governments. The basis on which people or governments make judgments of right and wrong is sometimes purely pragmatic, but it may also be a matter of personal values or of an externally imposed system of religion. Such judgments are often difficult and must be made with care.

Issues such as the use of drugs, lethal gases, pesticides, contraceptives, and pollutants face many of us today. Issues such as genetic manipulation of populations and chemical manipulation of individuals will probably face us in the future. Religious organizations do not always agree among themselves on a given issue, nor can they be counted upon to make the correct decisions in all cases. However, religious or ethical principles should be part of the process by which important issues are decided. There must be an interaction between science and religion. Science can extend our knowledge; ethics or religion (defined in the broadest sense) can help us use our knowledge wisely. Scientists, then (like others), must abide by a reasonable moral code.

Let us now take a brief look at some conflicts that have arisen between the scientific community and religious groups. Science produces new technology and gives rise to theories which affect moral decisions and statements of truth. At times, some people find that scientific statements conflict with their religious beliefs. Religious objections to a scientific discovery may arise simply because of its novelty and because of the natural human tendency to resist change—including changes in ideas. Many discoveries are no longer considered a threat after people have had time to get used to them. Chemical contraception may fall into this category as far as the general population is concerned, but there are still many people who do not accept it because of their religious convictions. In the United States today we believe that people have a right not to practice contraception, but that belief is not universally accepted (in the Republic of China, contraception is enforced).

Another subject of considerable debate now—even after a century of discussion—is evolution and "creationism." Here, the disagreement is not simply a matter of opposition to a new idea; it involves a more basic difference in fundamental presuppositions. Many Christians and Jews believe that the Bible is a source of absolute truth; pure science believes that only observation and human thought are the source of knowledge.

Despite occasional disagreements, science and religion sometimes support each other. A striking example is the verification of the age of the Dead Sea scrolls (see below, and Chapter 13). Another is the investigation of the shroud of Turin—a piece of cloth believed by some to be the burial cloth of Christ. Scientists cannot prove the authenticity of the shroud; but chemical analysis is consistent with the claim: the shroud is, at least, not a recent fraud. Archeology provides a third example; it has often verified the historical correctness of biblical stories once assumed by scholars to be mythical.

Let us now look in some more detail at our first example: the Dead Sea scrolls.

Example: The Dead Sea scrolls

One of the most interesting and constructive interactions between science and religion took place in the early 1950s. A shepherd boy in Israel discovered several parchment scrolls in a cave at Qumran near the Dead Sea. These had been stored in pottery jars, and in the dry climate had been preserved for what seemed to archeologists to be a very long time.

Scholars had been interested in, and arguing about, the authenticity of ancient biblical manuscripts: were the oldest extant writings accurate copies of the originals? The oldest writings available at that time had been written about A.D. 300. The Qumran scrolls contained parts of all but one book of the Old Testament, and the entire text of some of the books. If they were older than A.D. 300, their content would be valuable evidence in the textual debates going on at the time.

A method developed by Willard Libby—a nuclear chemical process called *carbon-14 dating*—indicated that the parchments dated from around 100 B.C. This method (explained in Chapter 13) had already been used to determine the ages of many archeological samples. It now showed that the Qumran scrolls were 400 years older than the oldest ones known beforehand. The text of the Qumran scrolls was so similar to the ones from which previous translations had been made that scholars now had a high degree of confidence that they had texts very close to the original. Thus the thought patterns and historical facts in the Bible could be analyzed as those of the ancients themselves rather than modifications by more recent copiers or translators.

In conclusion, the various segments of society are all affected by science, and in turn affect it. No matter what profession you choose or what your personal lifestyle or beliefs are, you will be affected by chemical knowledge and products. You will also—through community, governmental, or industrial policy-making processes—be able to influence society. It is hoped that this text will help to prepare you to act wisely in that capacity.

Let us now proceed to some topics for more advanced study: several agencies of the federal government.

TOPICS FOR ADVANCED STUDY: FEDERAL AGENCIES

Food and Drug Administration (FDA)

Most state and local governments are not involved in science to any great extent, but many federal agencies are very much involved. The Food and Drug Administration (FDA) is responsible for enforcing the Food, Drug, and Cosmetic Act of 1938, along with subsequent amendments; it must give or deny approval of all materials sold for human consumption (foods and drugs) or skin application (cosmetics). Before a material can be marketed, both its safety and its utility for the purpose for which it will be sold must be demonstrated. The present standards are higher than former ones; some materials which were approved earlier would most likely not be allowed under current laws and regulations. However, such materials (aspirin, for example) are usually not taken off the market unless it can be shown that they do specific harm, especially if their use is deeply rooted in culture and common practice. The FDA does have the power, under the law, to forbid the sale of any compound even after it has become established, or to require a warning on the label if there is some danger involved.

FIGURE 3-5
Large-scale shipment of chemicals is one of the areas regulated by the Federal Trade Commission.

Federal Trade Commission (FTC)

The Federal Trade Commission (FTC) affects chemistry insofar as it regulates and specifies methods of transporting chemicals (Figure 3-5). Because the chemical industry is such a large one, a great deal of transportation, sometimes over great distances, is required. The primary concerns of the FTC are to ensure free competition and to protect consumers from misleading advertising. However, the safety of the general public is also one of its concerns. Tankloads of chemicals are shipped by truck and rail in close proximity to homes, workplaces, and automobile traffic. Some of these materials are toxic, corrosive, or flammable, and special safety precautions are required by the FTC (although it is virtually impossible to eliminate all danger). Thus, the FTC partially controls the chemical industry, and chemistry affects the FTC by requiring new regulations for new products; the two interact at many points.

Environmental Protection Agency (EPA)

The Environmental Protection Agency (EPA) was set up by Congress in 1970 to enforce environmental laws. The fact that science can develop methods of detecting pollutants and identifying their harmful effects led to the formation of this agency. Thus, science contributed to the expansion of government, and now the EPA affects the way chemical processes are carried out.

The EPA not only enforces laws related to pollution and prosecutes offenders but also is engaged in research aimed at determining what materials are harmful and at what levels they can be safely tolerated. The agency provides funds to scientists and local governments to conduct research and monitor pollution levels. Methods of analysis and tests with animals and humans are typical of the ongoing research in this area. Industries usually bear the cost of any pollution they have caused, but local governments frequently receive aid from the EPA to determine and maintain air, water, and land quality.

Some of the data in Chapter 12 are taken from research performed by the EPA or were obtained through funding provided by EPA. It is essential that such funding continue if the quality of our environment—including the detection of yet unknown harmful substances—is to be improved. And it is certainly true that the general public benefits greatly from expenditures in this area of government involvement.

Occupational Safety and Health Agency (OSHA)

The Occupational Safety and Health Agency (OSHA) is responsible for environmental standards in workplaces and all areas of safety for employees of chemical (and other) industries.

National Science Foundation (NSF)

The National Science Foundation (NSF) is responsible for helping to direct, primarily by funding, some of the science activities done in the United States. This includes the biological sciences, the physical sciences (chemistry, physics, mathematics), engineering, and the social sciences. Not only does NSF supply money for technical scientific research, it also encourages science education at all levels (elementary through doctoral). Some individuals are given scholarships (gifts toward educational expenses) or fellowships (funds used to pay students to teach or assist during their years of training). For many years, NSF also had a program of undergraduate summer research (URP) for promising college students majoring in the sciences (see Figure 3-6). An instructor who was doing research could request a grant from NSF for a salary, chemicals and equipment, and salaries and housing for a small group of student assistants during the summer months. This gave firsthand experience to the students, which might encourage them to pursue careers in science or might suggest to them that this was not what they wanted to do after all. Many of today's scientists (including the author of this text) were greatly

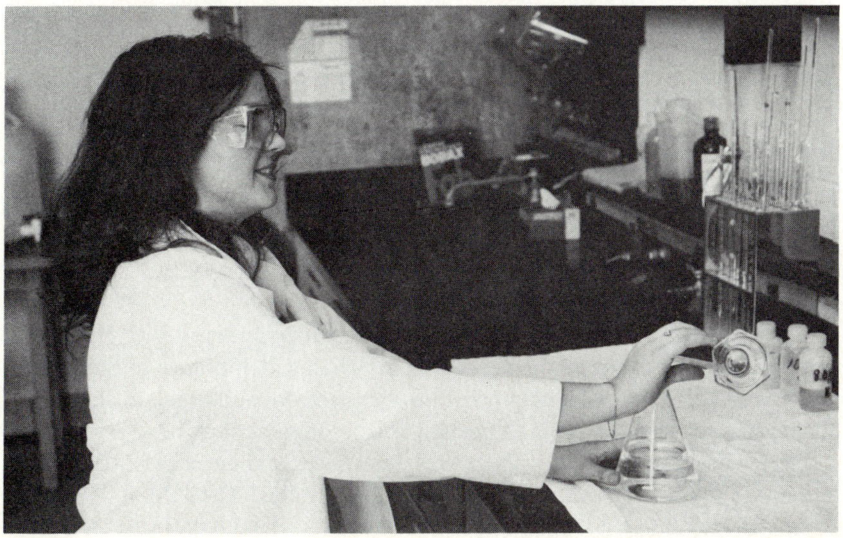

FIGURE 3-6
Undergraduate summer research is an important part of scientific training.

helped by this program and other programs similar to it. These programs are initiated or stopped by each administration.

Research on science needs in the United States is also a part of NSF's activities. It is important to predict whether there will be an excess or a deficiency of scientists and science teachers in a given area of study 10 or 20 years from a given time, and to reduce excesses or improve areas of deficiency in order to keep a strong national science capability. (This is very difficult to do in actual practice, however.) NSF also is responsible for several important science research centers, such as the National Radio Astronomy Observatory in West Virginia and the National Center for Atmospheric Research in Colorado.

National Institutes of Health (NIH)

The National Institutes of Health (NIH) are a large federal agency under the Department of Health and Human Resources (formally the Department of Health, Education, and Welfare) which since 1948 have been working to identify and solve major national health problems. Many kinds of medical research are carried out at the main facility in Bethesda, Maryland, and NIH funds one-third to one-half of the health-related research in the United States. It is also interested in monitoring the demand for and supply of physicians and in encouraging their education.

This agency is composed of 10 institutes, each of which is responsible for a given area. They are:

1 Institute of Allergy and Infectious Diseases
2 Institute of Arthritis and Metabolic Diseases
3 Cancer Institute
4 Institute of Child Health and Human Development
5 Dental Institute
6 Eye Institute
7 Institute of Environment Health Sciences
8 Institute of General Medical Sciences
9 Heart Institute
10 Institute of Neurological and Communicative Disorders and Stroke

The NIH program for finding better treatment of epilepsy was discussed earlier in this chapter (see Chapter 16 for the chemical approach to this problem). Eventually, this screening program is expected to aid in the treatment of epilepsy, which affects about 2 percent of the population.

Major programs are also under way for cancer and heart disease, which are the two major causes of death in the United States. The NIH is involved not only in trying to find cures or better treatment of health problems but also in distributing information to scientists working in health services and in providing to the general public aid with, and information on, disease prevention.

National Bureau of Standards (NBS)

There are several other bodies within the federal government that control or aid science. These include the National Bureau of Standards (NBS), which holds physical standards of measurement against which others can be compared. If one wishes to check the accuracy of a laboratory balance, one can get a set of weights from the NBS and compare the weight of each one with the value given by the instrument in question, so that any differences can be taken into account during experimentation. Once in a while NBS publishes a standard method of analyzing for a material (such as the analysis of lead leached by acidic foods from cookware). If chemists want to check their own analytical accuracy for a particular determination, they can obtain materials with known amounts of that substance from NBS. (For example, if one is doing analysis of iron ore commercially, one can get an ore known to have, say, 10 percent iron, Fe. This ore is then analyzed for Fe, and the chemist's skill or the validity of the method being used can thus be checked: if the value is not 10 percent, something is wrong and must be corrected.) This helps to ensure reliability in the field of analysis.

Other agencies

There are also other agencies, such as the Department of Defense (DOD) and the National Aeronautics and Space Administration (NASA), which are mission- or project-oriented (in many cases the projects are scientific in nature or in implementation). These agencies have scientists on their permanent staffs and provide funds for both industrial and academic scientists to work on specific problems that are theoretical or practical in nature. These are important sources of money through which academic institutions can purchase much-needed equipment and provide training to graduate students. These students will become the next generation of professional scientists. Their training, which requires time and money, is essential if we are going to continue to be a scientifically progressive society.

STUDY GUIDE

Terms to learn

Applied research	Faith
Delaney clause	Fellowship (monetary)
Dissertation	Forensic chemistry
DOD	FTC
EPA	NASA

NBS	Pure research
NIH	Regulation
NRC	Research
NSF	Scholarship
OSHA	Thesis

Questions for discussion

1 What problem areas in your hometown or college community seem to call for government regulation?

2 What products have you used safely today that involved both scientific discoveries and government regulation?

3 What is your opinion about government regulation versus freedom of industries in the areas of pollution and safety?

4 Do you have any personal (or religious) beliefs that conflict with a statement of science? What are they? Discuss these problems openly with your instructor, a religious leader, or a friend until you can state the conflict clearly and give your own opinion on it.

5 Considering the dangers of *not* having saccharin available, do you agree that 100 deaths per year due to its use are tolerable? Why or why not?

6 Do you think EPA acted correctly in buying the Times Beach property? Why?

7 Ask your instructor to tell you about his or her graduate school experience. What were some of the problems encountered? What were some of the joys? What was the research topic on which his or her degree was earned?

Goals

To develop skill in use of the metric system

To develop skill in converting one unit to another in chemical and household situations

To understand the concept of energy

To develop skill in temperature conversions and percent calculations in everyday applications

To develop the idea of precision and recognize the misleading use of numbers in advertising

4

MEASUREMENTS, CONVERSIONS, AND ENERGY

The heart of modern chemistry (and other natural sciences) is observation and careful measurement. In our daily lives, we also rely on measurements of time, temperature, length, weight, and volume, and we sometimes must change a measurement into terms other than those in which it is given. In this chapter we will study various aspects of measurement systems and conversions used in chemistry, as well as those that can be used by the nonscientist in everyday life.

SCIENTIFIC NOTATION

Scientists frequently use numbers that are very large or very small, and it is convenient to have a system of expressing these numbers more briefly than in standard notation, with its many zeros.

First, consider large numbers. The number 100 can be expressed as 10×10 or 10^2. The superscript 2 is called an *exponent;* it indicates the number of times one should multiply 10 by itself to get the number one is expressing.

$$100 = 10 \times 10 = 10^2 \longleftarrow \text{Exponent}$$

One reads this "10 to the second power" or "10 squared." This is called *exponential notation* or *scientific notation*. Other numbers expressed in this way are:

$$1000 = 10 \times 10 \times 10 = 10^3$$
$$10{,}000 = 10 \times 10 \times 10 \times 10 = 10^4$$
$$1{,}000{,}000 = 10 \times 10 \times 10 \times 10 \times 10 \times 10 = 10^6$$

Numbers that are not whole powers of 10 can also be expressed in similar terms. For example:

$$227 = 2.27 \times 100 = 2.27 \times 10^2$$
$$5{,}270{,}000 = 5.27 \times 1{,}000{,}000 = 5.27 \times 10^6$$
$$0.01 = 1/100 = 1/10^2 = 10^{-2}$$

Now consider small numbers. The decimal number 0.01 could be expressed as the fraction 1/100. Since $100 = 10^2$, 1/100 could be expressed as $1/10^2$. This is awkward to write, and so the expression 10^{-2} is used instead. The negative sign on the exponent means that the number is actually in the denominator (or that it is a small number). Thus, 10^2 means 100, but 10^{-2} means one one-hundredth. Likewise,

$$0.001 = \frac{1}{1000} = \frac{1}{10^3} = 10^{-3}$$

$$0.00001 = \frac{1}{100{,}000} = \frac{1}{10^5} = 10^{-5}$$

Small numbers that are not whole powers of 10 can be treated with a combination of standard and exponential notation. For example:

$$0.027 = 2.7 \times 0.01 = 2.7 \times \frac{1}{100}$$

$$= 2.7 \times \frac{1}{10^2} = 2.7 \times 10^{-2}$$

$$0.00051 = 5.1 \times 0.0001 = 5.1 \times \frac{1}{10{,}000}$$

$$= 5.1 \times \frac{1}{10^4} = 5.1 \times 10^{-4}$$

Electronic calculators express numbers in scientific notation as follows:

5.1×10^{-4}	is	5.1 E − 4	or	5.1 − 4
3.15×10^{7}	is	3.15 E 7	or	3.15 7

where the E or the empty space stands for "×10" or exponential notation.

EXPRESSING NUMBERS IN SCIENTIFIC NOTATION

Since we do not want to go through the whole procedure used in the preceding section each time we change a number from standard to scientific notation, an easier and faster method will be presented. (See Appendix B for operations using exponents.)

The usual way of expressing a number in this system is to place the decimal to the right of the leftmost digit (ignoring zeros), but it is not essential to place the decimal there.

Preferred	Correct	2.89×10^5
Less preferred	Correct	28.9×10^4
	Correct	0.289×10^6

The following steps are used in changing a number from standard to scientific notation.

1 Write the number in standard fashion (as you normally would, not using scientific notation), placing the decimal in its usual place.

2 Write the number in standard fashion with the decimal to the right of the leftmost digit.

3 Count the number of places you had to move the decimal in going from step 1 to step 2.

4 Write the number (without any accompanying zeros) with the decimal to the right of the leftmost digit.

5 Follow this with a "× 10."

6 Place the exponent on the 10 according to these results:
(a) If you moved the decimal (in step 3) to the left, the exponent is the number of places it was moved (a positive number).
(b) If you moved the decimal to the right, the exponent is the negative of the number of places moved.

Worked example

Express 283,000 in scientific notation.

1 Write: 283,000.

Note The decimal is understood to be to the right of the rightmost digit when no decimal is actually written.

2 Write 2.83000 with the decimal to the right of the left digit.

3 Count the places the decimal moved.

2.83,000.
↑ ↑
Finish Begin

In this case it moved *five* places to the *left*.

4 Write: 2.83, ignoring the zeros

5 Write: 2.83 × 10

6 Since the decimal was moved five places, the exponent is 5, and the complete expression is

$$2.83 \times 10^5$$

Worked example

Express 0.00000501 in scientific notation.

1 Write: 0.00000501

2 Write: 0.000005.01

3 Count the places the decimal moved. Six places to the *right* were required.

0.000005.01
↑ ↑
Begin Finish

4 Write: 5.01

5 Write: 5.01 × 10

6 Since the decimal moved six places to the *right,* the exponent is −6, and the expression is correctly written:

$$5.01 \times 10^{-6}$$

Problem

Express 2110 in scientific notation.

Solution

1 Write: 2110.

2 Write: 2.110.

3 Count the places the decimal moved: 2.110. Three places to the left were required.

4 Write: 2.11

5 Write: 2.11 × 10

6 Write: 2.11×10^3

Problem

Express 0.000153 in scientific notation.

1 Write: 0.000153

2 Write: 0.0001.53

3 Count the places the decimal moved: 0.0001.53. Four places to the right were required.

4 Write: 1.53

5 Write: 1.53 × 10

6 Write: 1.53×10^{-4}

Problem

Express each of the following in scientific notation:

a 37,000

b 98

c 7,347,000,000

d 0.0000893

e 0.000000000172

Answers

a 3.7×10^4

b 9.8×10^1 or 9.8×10

c 7.347×10^9

d 8.93×10^{-5}

e 1.72×10^{-10}

Now let's try going the other way—converting scientific numbers into standard notation. The following steps are used:

1 Write the standard notation portion.

2 Move the decimal the number of places indicated by the exponent: to the right if the exponent is positive or to the left if negative.

Worked example

Write 3.74×10^{-8} in standard notation.

1 Write: 3.74

2 Move the decimal eight places to the left, putting in any needed zeros as place holders.

0.00000003.74
↑ (Finish) ↑ (Start)

3 Answer: 0.0000000374

Worked example

Express 1.073×10^5 in standard notation.

1 Write: 1.073

2 Move the decimal five places to the right (adding zeros as place holders).

1.07300.
↑ (Start) ↑ (Finish)

3 Answer: 107,300

Problems

Express each of the following in standard notation:

a 3.85×10^{-3}

b 3.76×10^{4}

c 1.002×10^{-1}

d 8.6×10^{1}

Answers

a 0.00385

b 37,600

c 0.1002

d 86

THE METRIC SYSTEM OF MEASUREMENT

All standards of physical measurement are arbitrary. It is important that everyone use the same standards and that these be very carefully defined, so that there is clarity of communication. However, the origin or nature of a standard is not critical. The system of measurement developed in England is the one with which we are most familiar in the United States; but the rest of the world (now including England) uses the metric system (see Figure 4-1), which is more convenient. Because of the greater convenience of the metric system, and for purposes of world trade, the United States too is slowly but surely adopting it. Table 4-1 shows the kind of measurements commonly used in both systems and the standard in each case. Appendix A gives both the metric and the United States system in more detail.

The metric system has been used in Europe—and in most of the rest of the world—for a long time, and it is used by scientists in all countries. More recently, the International System of Measurement (Système International d'Unités—SI), which is based on the older metric system, has become the official universal system for all scientific applications.

For simplicity in changing the magnitude of measurements (from large to small, for example) a system of prefixes has been developed; it is shown in Table 4-2. For measuring objects of ordinary (human) size, the meter (m) is most conveniently used. For greater distances (such as distances between cities), the kilometer (km) is used. (It is more convenient to express the distance from New York City to Washington, D.C., as 375 km than as 375,000 m.) The distance between two atoms of carbon is easier to express as 0.154 nanometers (0.154 nm) than as 1.54×10^{-10} m or 0.000000000154 m.

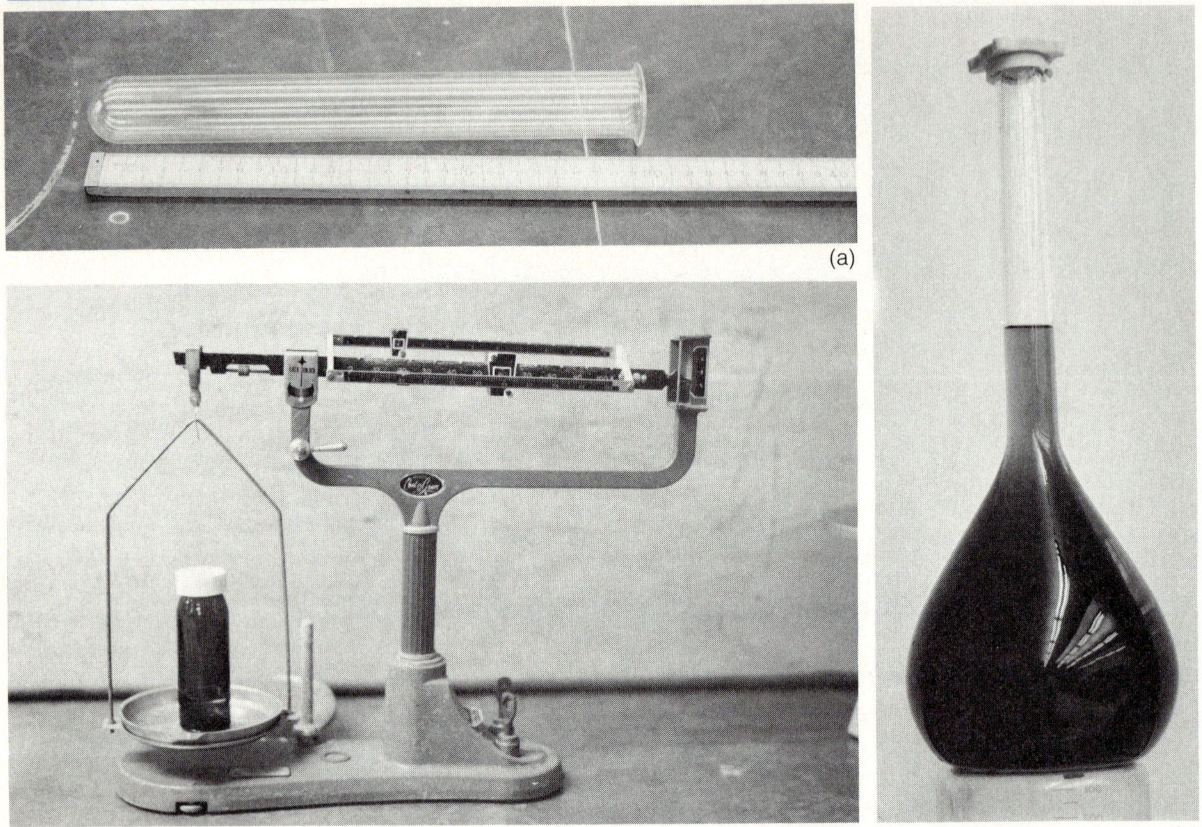

(a) (b) (c)

FIGURE 4-1

Metric measuring devices.
(a) Meter stick: length.
(b) Laboratory balance: mass.
(c) 1-L flask: volume.

Clearly, conversions are easier in the metric system than in the United States system. It is, for instance, a simple matter to change from meters to kilometers: the latter is exactly 1000 times the former. All conversions within the SI system are whole powers of ten (see Table 4-2)—in contrast to the United States system, where 1 foot (ft) is 12 times the size of 1 inch (in); 1 mile (mi) is 5280 ft; 1 pound (lb) is 16 ounces (oz); etc. Thus, most of these conversions are too cumbersome to be done in one's head. Conversions in SI units, on the other hand, are performed by simply moving the decimal point.

It should be pointed out, however, that while the metric system is more convenient, it does *not* allow one to measure with greater accuracy or precision. A given length expressed in centimeters (cm), millimeters (mm), inches, or tenths of an inch is equally meaningful.

It is frequently desirable to change a measurement of length, volume, mass, etc., from one kind of units into another. For instance, one may know one's own height in feet or inches and want to express it in terms of centimeters or meters. Converting a measurement from one system of measurement (United States, say) to another system (metric or SI) is called *intersystem conversion*. Or one may want to change a measurement from one unit in a system to another in the same system—for example, to change meters to kilometers (both are in the SI system). This is called *intrasystem conversion*. See Table 4.3 for conversion factors.

TABLE 4-1

Kinds of standards of measurements

MEASUREMENT	UNITED STATES STANDARD	METRIC STANDARD
Length	foot (ft)	meter (m)
Mass	pound (lb)	kilogram (kg)
Volume	gallon (gal)	liter (L)
Time	second (s)	second (s)
Energy	British thermal unit (Btu)	calorie (cal) or joule (j)

TABLE 4-2

Commonly used prefixes in the SI system

PREFIX	ABBREVIATION	MEANING	EXAMPLE
mega	M	$1{,}000{,}000 = 10^6$	$\text{ML} = 10^6\ \text{L}$
kilo	k	1000	kg = 1000 g
deka	da	10	dam = 10 m
deci	d	$\frac{1}{10}$	dg = 0.10 g
centi	c	$\frac{1}{100}$	$\text{cm} = \frac{1}{100}\ \text{m}$
milli	m	$\frac{1}{1000}$	$\text{mg} = \frac{1}{1000}\ \text{g}$
micro	μ*	$\frac{1}{1{,}000{,}000} = \frac{1}{10^6}$	$\text{g} = 10^6\ \mu\text{g}$
nano	n	$\frac{1}{10^9} = 10^{-9}$	$\text{nm} = 10^{-9}\ \text{m}$

*The Greek letter mu.

TABLE 4-3

Relationships between the International System (SI) and the United States System*

	SI	UNITED STATES SYSTEM RELATIONSHIP TO SI
Length	meter (m) millimeter (mm) = 1/1000 m centimeter = 1/100 m kilometer = 1000 m	1 yd = 0.914 m 1 in = 2.54 cm 1 km = 0.621 mi
Mass (weight)	gram (g) milligram (mg) = 1/1000 g microgram (μg) = 1/1,000,000 g kilogram (kg) = 1000 g 1 metric tonne = 1000 kg megagram (Mg) = 1,000,000 g	1 oz = 28.35 g 1 lb = 454 g 1 lb = 0.454 kg English ton = 907 kg 1 tonne = 2200 lb
Volume	liter (L) milliliter (mL) = 1/1000 L	1 qt = 0.946 L = 946 mL 1 fluid oz = 29.57 mL

*A good working knowledge of the United States system is assumed.

Sometimes such conversions are complicated. For example, one might want to convert the density of mercury from 13 grams per milliliter (g/mL) to pounds per cubic foot (lb/ft^3). In other words, "If 1 mL weighs 13 g, how many pounds would 1 ft^3 weigh?" Another typical conversion involves changing 500 years into seconds (s): "How many seconds are in 500 years?"

Table 4-3 shows the relationship between United States and SI measurements. It is recommended that at least those measurements enclosed in boxes (one in each dimension) be committed to memory, along with all the prefixes in Table 4-2. Once this is done, the following learning program will enable you to develop skill in unit conversions. (More-complex unit conversions and those using exponents are covered in Appendix C.)

UNIT CONVERSIONS

Simple (one-step) conversion

Worked example

Convert 15 in to centimeters.

1 Write 15.0 in (the original measurement given in the problem).

$$15.0\ \text{in}$$

2 Place parentheses after the 15.0 in.

$$15.0\ \text{in}\left(\qquad\right)$$

3 Draw a horizontal solid line inside the parentheses.

$$15.0\ \text{in}\left(\frac{\qquad}{\qquad}\right)$$

4 Place a dotted vertical line in the parentheses.

$$15.0\ \text{in}\left(\frac{\quad\vdots\quad}{\quad\vdots\quad}\right)$$

5 In the lower right, place the unit symbol associated with the original measurement.

$$15.0\ \text{in}\left(\frac{\quad\vdots\quad}{\quad\vdots\ \text{in}}\right)$$

Note Since the 15 in is *not* written as part of a fraction, it can be thought of as:

$$\frac{15.0\ \text{in}}{1}$$

Note Unit symbols (such as in, ft, and cm) can be multiplied and canceled just as though they were numbers. This is called *dimensional analysis* or *factor analysis* and is a valuable tool in determining if the correct numerical operations have been performed.

Note One writes inches in the *lower* portion of the fraction so that it will be in a canceling relationship with inches in the 15-in term.

$$\frac{15.0\ \cancel{\text{in}}}{1}\left(\frac{\quad\vdots\quad}{\quad\vdots\ \cancel{\text{in}}}\right)$$

Note The fraction we are developing in the parentheses will have the following numerical meanings:

> Any numbers and units placed on *top* (numerator) will be multiplied by the 15.0 in.
>
> Any numbers and units placed on the *bottom* (denominator) will be divided into the 15.0 in.

6 Cancel the inches units by drawing a diagonal line through each.

$$15.0\ \cancel{\text{in}}\left(\frac{\quad\vdots\quad}{\quad\vdots\ \cancel{\text{in}}}\right)$$

7 In the upper right, place the units one wants to change to (centimeters in this case).

$$15.0\ \cancel{\text{in}}\left(\frac{\quad\vdots\ \text{cm}}{\quad\vdots\ \cancel{\text{in}}}\right)$$

Note In the parentheses one always places the unit to be changed (inches in this case) in a canceling relationship to the unit in the original data and the unit desired (centimeters in this case) in the desired position.

8 Refer to a table of conversion factors (or eventually to one's memory) and place the correct conversion number in front of the centimeter and inch terms.

$$15.0\ \cancel{\text{in}}\left(\frac{2.54\ \vdots\ \text{cm}}{1\ \vdots\ \cancel{\text{in}}}\right)$$

Note The term in parentheses means, "There are 2.54 cm per inch," and is called the *conversion factor*.

9 Multiply 15.0 × 2.54.

Answer: 38.1

10 Place the uncanceled unit after this number.

38.1 cm

11 To summarize the entire operation:

$$15.0\ \not{in}\left(\frac{2.54\ \text{cm}}{1\ \not{in}}\right) = 38.1\ \text{cm}$$

Note This means that if a 15.0-in object were measured with a metric ruler, the ruler would read 38.1 on the centimeter scale.

Worked example

Convert 120 cm to inches.

1 Write 120 cm.

$$120\ \text{cm}$$

2 Construct parentheses with lines next to the 120 cm.

$$120\ \text{cm}\left(\frac{\quad\vdots\quad}{\quad\vdots\quad}\right)$$

3 Since centimeters are to be changed and inches are the desired unit, place centimeters in lower right (so it will cancel with centimeters in the 120-cm term), and place inches in the upper right.

$$120\ \not{cm}\left(\frac{\quad\vdots\ \text{in}}{\quad\vdots\ \not{cm}}\right)$$

4 Put the proper conversion numbers in front of each unit in the conversion factor.

$$120\ \not{cm}\left(\frac{1\ \vdots\ \text{in}}{2.54\ \vdots\ \not{cm}}\right)$$

5 Divide 120 by 2.54 and place inches after the answer.

$$47.2\ \text{in}$$

6 Summary:

$$120\ \not{cm}\left(\frac{1\quad \text{in}}{2.54\quad \not{cm}}\right) = 47.2\ \text{in}$$

Problem

Convert 16.0 lb to grams (g). The method and answer are shown below—but do not look at the solution until you have attempted it yourself.

Solution

1 Write 16.0 lb.

$$16.0\ \text{lb}$$

2 Construct parentheses.

$$16.0\ \text{lb}\left(\overline{\qquad\qquad}\right)$$

3 Place pounds in the lower right and grams in the upper right and cancel pounds.

$$16.0\ \cancel{\text{lb}}\left(\frac{\qquad\text{g}}{\qquad\cancel{\text{lb}}}\right)$$

4 Place the proper conversion numbers with each unit.

$$16.0\ \cancel{\text{lb}}\left(\frac{454\quad\text{g}}{1\quad\cancel{\text{lb}}}\right)$$

5 Perform the arithmetic (multiply 16 × 454).

$$16.0\ \cancel{\text{lb}}\left(\frac{454\quad\text{g}}{1\quad\cancel{\text{lb}}}\right) = 7264\ \text{g}$$

Problem

Convert 3.00 quarts (qt) to milliliters.

Solution

1 Write 3.00 qt.

$$3.00\ \text{qt}$$

2 Construct parentheses.

$$3.00\ \text{qt}\left(\overline{\qquad\qquad}\right)$$

3 Place quarts in lower right and milliliters in upper right.

$$3.00\ \text{qt}\left(\frac{\qquad\text{mL}}{\qquad\text{qt}}\right)$$

4 Place the proper conversion numbers with each unit.

$$3.00\ \cancel{\text{qt}}\left(\frac{946\quad\text{mL}}{1\quad\cancel{\text{qt}}}\right)$$

5 Obtain the numerical answer (3 × 946).

$$3.00\ \text{qt}\left(\frac{946\quad\text{mL}}{1\quad\cancel{\text{qt}}}\right) = 2838\ \text{mL}$$

Problem

Convert 2168 g to pounds.

Solution

1 Write 2168 g.

$$2168 \text{ g}$$

2 Construct parentheses.

$$2168 \text{ g} \left(\frac{\quad}{\quad}\right)$$

3 Place grams in the lower right and pounds in the upper right.

$$2168 \not{g} \left(\frac{\text{lb}}{\not{g}}\right)$$

4 Place the correct conversion numbers with each unit.

$$2168 \not{g} \left(\frac{1 \quad \text{lb}}{454 \quad \not{g}}\right)$$

5 Do the arithmetic $\left(2168 \div 454 \text{ or } \frac{2168}{454}\right)$.

$$2168 \not{g} \left(\frac{1 \quad \text{lb}}{454 \quad \not{g}}\right) = 4.77 \text{ lb}$$

Problems

a Convert 200 cm to inches.

b Convert 526 in to centimeters.

c Convert 865 mL to quarts.

d Convert 469 mL to liters.

e Convert 46 lb to ounces.

f Convert 42,700 g to kilograms.

g Convert 263 qt to liters.

Solutions

a
$$200\ \cancel{\text{cm}}\left(\frac{1\quad \text{in}}{2.54\quad \cancel{\text{cm}}}\right) = 78.7\ \text{in}$$

b
$$526\ \cancel{\text{in}}\left(\frac{2.54\quad \text{cm}}{1\quad \cancel{\text{in}}}\right) = 1336\ \text{cm}$$

c
$$865\ \cancel{\text{mL}}\left(\frac{1\quad \text{qt}}{946\quad \cancel{\text{mL}}}\right) = 0.914\ \text{qt}$$

d
$$469\ \cancel{\text{mL}}\left(\frac{1\quad \text{L}}{1000\quad \cancel{\text{mL}}}\right) = 0.469\ \text{L}$$

e
$$46\ \cancel{\text{lb}}\left(\frac{16\quad \text{oz}}{1\quad \cancel{\text{lb}}}\right) = 736\ \text{oz}$$

f
$$42{,}700\ \cancel{\text{g}}\left(\frac{1\quad \text{kg}}{1000\quad \cancel{\text{g}}}\right) = 42.7\ \text{kg}$$

g
$$263\ \cancel{\text{qt}}\left(\frac{0.946\quad \text{L}}{1\quad \cancel{\text{qt}}}\right) = 249\ \text{L}$$

Multistep single-unit conversions

In this type of problem one unit is changed to another; but because a single conversion factor is not readily available in these cases, several steps are needed.

Worked example

Convert 27 ft to centimeters.

1 Write 27 ft.

$$27\ \text{ft}$$

2 Construct parentheses.

$$27\ \text{ft}\left(\frac{\qquad}{}\right)$$

Note Since we do not know the direct conversion from feet to centimeters, a two-step procedure using what we do know is required.

3 Place feet in the lower right (so that feet in the original measurement will cancel).

$$27\ \cancel{\text{ft}}\left(\frac{\qquad}{\cancel{\text{ft}}}\right)$$

4 In the upper right, place a unit that is related to feet. Inches would be convenient here.

$$27\ \cancel{\text{ft}}\left(\frac{\text{in}}{\cancel{\text{ft}}}\right)$$

5 Place the proper conversion numbers with each unit.

$$27\ \cancel{\text{ft}}\left(\frac{12\ \text{in}}{1\ \cancel{\text{ft}}}\right)$$

Note At this point we have the number of inches in 27 ft (i.e., 27 × 12 = 324 in), but we want the measurement in centimeters. Another conversion factor is therefore needed.

6 Construct another parentheses.

$$27\ \cancel{\text{ft}}\left(\frac{12\ \text{in}}{1\ \cancel{\text{ft}}}\right)\left(\frac{\qquad}{}\right)$$

7 In the lower right within the new parentheses, place the unit remaining after step 5 (in this case inches) to produce a canceling relationship.

$$27\ \cancel{\text{ft}}\left(\frac{12\ \cancel{\text{in}}}{1\ \cancel{\text{ft}}}\right)\left(\frac{\qquad}{\text{in}}\right)$$

8 In the upper right, place the desired unit (centimeters in this case).

$$27\ \cancel{\text{ft}}\left(\frac{12\ \cancel{\text{in}}}{1\ \cancel{\text{ft}}}\right)\left(\frac{\text{cm}}{\cancel{\text{in}}}\right)$$

9 Place the correct conversion numbers with each unit.

$$27\ \cancel{\text{ft}}\left(\frac{12\ \cancel{\text{in}}}{1\ \cancel{\text{ft}}}\right)\left(\frac{2.54\ \text{cm}}{1\ \cancel{\text{in}}}\right)$$

10 Do the arithmetic by multiplying and labeling with the remaining units (in this case centimeters).

$$27 \times 12 \times 2.54 = 823 \text{ cm}$$

11 Summary:

$$27 \cancel{\text{ft}} \left(\frac{12 \quad \cancel{\text{in}}}{1 \quad \cancel{\text{ft}}}\right)\left(\frac{2.54 \quad \text{cm}}{1 \quad \cancel{\text{in}}}\right) = 823 \text{ cm}$$

Note Being able to use this multistep method is better than memorizing all possible one-step conversion factors.

Note Working problems by placing a series of conversion factors in tandem is sometimes called the *chain method*.

Worked example

How many seconds are in 50 years?

1 Write 50 years.

$$50 \text{ years}$$

2 Construct parentheses.

$$50 \text{ years} \left(\frac{\quad}{\quad}\right)$$

Note Since we do not know the direct conversion from years to seconds, we will have to use the multistep method.

3 Place years in the lower right in order to cancel years in the original measurement.

$$50 \cancel{\text{years}} \left(\frac{\quad}{\cancel{\text{year}}}\right)$$

4 Choose a unit related to years—days would be convenient here—and place it in the upper right.

$$50 \cancel{\text{years}} \left(\frac{\text{days}}{\cancel{\text{year}}}\right)$$

5 Place the proper conversion numbers with each unit.

$$50 \cancel{\text{years}} \left(\frac{365 \quad \text{days}}{1 \quad \cancel{\text{year}}}\right)$$

6 Construct another parentheses.

$$50\ \cancel{\text{years}}\left(\frac{365\quad \text{days}}{1\quad \cancel{\text{year}}}\right)\left(\frac{\qquad}{\qquad}\right)$$

7 In the lower right, place the unit left after the previous cancellations.

$$50\ \cancel{\text{years}}\left(\frac{365\quad \cancel{\text{days}}}{1\quad \cancel{\text{year}}}\right)\left(\frac{\qquad}{\qquad \cancel{\text{day}}}\right)$$

8 In the upper right, place a unit related to days—for example, hours.

$$50\ \cancel{\text{years}}\left(\frac{365\quad \cancel{\text{days}}}{1\quad \cancel{\text{year}}}\right)\left(\frac{\qquad \text{h}}{\qquad \cancel{\text{day}}}\right)$$

9 Place the proper conversion numbers with each unit.

$$50\ \cancel{\text{years}}\left(\frac{365\quad \cancel{\text{days}}}{1\quad \cancel{\text{year}}}\right)\left(\frac{24\quad \text{h}}{1\quad \cancel{\text{day}}}\right)$$

10 Repeat steps 6 to 9 to get rid of hours and obtain minutes.

$$50\ \cancel{\text{years}}\left(\frac{365\quad \cancel{\text{days}}}{1\quad \cancel{\text{year}}}\right)\left(\frac{24\quad \cancel{\text{h}}}{1\quad \cancel{\text{day}}}\right)\left(\frac{60\quad \text{min}}{1\quad \cancel{\text{h}}}\right)$$

11 Repeat steps 6 to 9 to get rid of minutes and obtain seconds—which is the unit we want here.

$$50\ \cancel{\text{years}}\left(\frac{365\quad \cancel{\text{days}}}{1\quad \cancel{\text{year}}}\right)\left(\frac{24\quad \cancel{\text{h}}}{1\quad \cancel{\text{day}}}\right)\left(\frac{60\quad \cancel{\text{min}}}{1\quad \cancel{\text{h}}}\right)\left(\frac{60\quad \text{s}}{1\quad \cancel{\text{min}}}\right)$$

12 Obtain the numerical answer by multiplying and labeling with the remaining units.

$$50 \times 365 \times 24 \times 60 \times 60 = 1{,}580{,}000{,}000$$

$$50\ \cancel{\text{years}}\left(\frac{365\quad \cancel{\text{days}}}{1\quad \cancel{\text{year}}}\right)\left(\frac{24\quad \cancel{\text{h}}}{1\quad \cancel{\text{day}}}\right)\left(\frac{60\quad \cancel{\text{min}}}{1\quad \cancel{\text{h}}}\right)\left(\frac{60\quad \text{s}}{1\quad \cancel{\text{min}}}\right)$$

$$= 1{,}580{,}000{,}000\ \text{s} \quad \text{or} \quad 1.58 \times 10^{9}\ \text{s}$$

Worked example

Convert 90 kg to pounds.

1 Write 90 kg.

$$90\ \text{kg}$$

2 Construct parentheses.

$$90\ \text{kg}\left(\frac{\qquad}{\qquad}\right)$$

3 Place kilograms in a canceling relationship.

$$90\ \not{kg}\left(\frac{\qquad}{\not{kg}}\right)$$

4 In the upper right, place the units related to kilograms, such as grams (g).

$$90\ \not{kg}\left(\frac{\qquad g}{\not{kg}}\right)$$

5 Place the proper conversion numbers with each unit.

$$90\ \not{kg}\left(\frac{1000\quad g}{1\quad \not{kg}}\right)$$

6 Construct another parentheses.

$$90\ \not{kg}\left(\frac{1000\quad g}{1\quad \not{kg}}\right)\left(\frac{\qquad}{}\right)$$

7 Place grams in the new parentheses to set up a canceling relationship.

$$90\ \not{kg}\left(\frac{1000\quad \not{g}}{1\quad \not{kg}}\right)\left(\frac{\qquad}{\not{g}}\right)$$

8 Place the unit desired in the upper right.

$$90\ \not{kg}\left(\frac{1000\quad \not{g}}{1\quad \not{kg}}\right)\left(\frac{\qquad lb}{\not{g}}\right)$$

Note If we had not known the relationship of pounds to grams directly, we would have used whatever unit we did know in relation to grams and would have continued with the chain method until we got the desired units.

9 Place the proper conversion numbers with each unit.

$$90\ \not{kg}\left(\frac{1000\quad \not{g}}{1\quad \not{kg}}\right)\left(\frac{1\quad lb}{454\quad \not{g}}\right)$$

10 Obtain the numerical answer with units.

$$\frac{90 \times 1000}{454} = 198\ lb$$

11 Summary:

$$90\ \not{kg}\left(\frac{1000\quad \not{g}}{1\quad \not{kg}}\right)\left(\frac{1\quad lb}{454\quad \not{g}}\right) = 198\ lb$$

Problem

How many weeks would be equal to 1,000,000 s?

Solution

1 Write 1,000,000 s.

$$1{,}000{,}000\ \text{s}$$

2 Construct parentheses and place seconds in the parentheses to set up a canceling relationship. Place a unit related to seconds, such as minutes, in the upper right.

$$1{,}000{,}000\ \cancel{\text{s}}\left(\frac{\quad\ \text{min}}{\quad\ \cancel{\text{s}}}\right)$$

3 Place the correct conversion numbers with each unit.

$$1{,}000{,}000\ \cancel{\text{s}}\left(\frac{1\ \ \text{min}}{60\ \ \cancel{\text{s}}}\right)$$

4 Construct another parentheses and place minutes in a canceling relationship and hours in the top part of the fraction.

$$1{,}000{,}000\ \cancel{\text{s}}\left(\frac{1\ \ \cancel{\text{min}}}{60\ \ \cancel{\text{s}}}\right)\left(\frac{\quad\ \text{h}}{\quad\ \cancel{\text{min}}}\right)$$

5 Place the correct conversion numbers with each unit.

$$1{,}000{,}000\ \cancel{\text{s}}\left(\frac{1\ \ \cancel{\text{min}}}{60\ \ \cancel{\text{s}}}\right)\left(\frac{1\ \ \text{h}}{60\ \ \cancel{\text{min}}}\right)$$

6 Repeat steps 4 and 5 to convert hours to days.

$$1{,}000{,}000\ \cancel{\text{s}}\left(\frac{1\ \ \cancel{\text{min}}}{60\ \ \cancel{\text{s}}}\right)\left(\frac{1\ \ \cancel{\text{h}}}{60\ \ \cancel{\text{min}}}\right)\left(\frac{1\ \ \text{day}}{24\ \ \cancel{\text{h}}}\right)$$

7 Repeat steps 4 and 5 to convert days to weeks.

$$1{,}000{,}000\ \cancel{\text{s}}\left(\frac{1\ \ \cancel{\text{min}}}{60\ \ \cancel{\text{s}}}\right)\left(\frac{1\ \ \cancel{\text{h}}}{60\ \ \cancel{\text{min}}}\right)\left(\frac{1\ \ \cancel{\text{day}}}{24\ \ \cancel{\text{h}}}\right)\left(\frac{1\ \ \text{week}}{7\ \ \cancel{\text{days}}}\right)$$

8 Obtain the answer by:

$$\frac{1{,}000{,}000}{60 \times 60 \times 24 \times 7} = 1.65\ \text{weeks}$$

9 Summary:

$$1{,}000{,}000\ \cancel{\text{s}}\left(\frac{1\ \ \cancel{\text{min}}}{60\ \ \cancel{\text{s}}}\right)\left(\frac{1\ \ \cancel{\text{h}}}{60\ \ \cancel{\text{min}}}\right)\left(\frac{1\ \ \cancel{\text{day}}}{24\ \ \cancel{\text{h}}}\right)\left(\frac{1\ \ \text{week}}{7\ \ \cancel{\text{days}}}\right) = 1.65\ \text{weeks}$$

Problems

a Convert 18 in to kilometers.

b How many kilograms of coal are in 5 tons of coal?

c Convert 25.3 gal to liters.

Solutions

a $$18\ \cancel{\text{in}}\left(\frac{2.54\ \cancel{\text{cm}}}{1\ \cancel{\text{in}}}\right)\left(\frac{1\ \cancel{\text{m}}}{100\ \cancel{\text{cm}}}\right)\left(\frac{1\ \text{km}}{1000\ \cancel{\text{m}}}\right) = 0.00046\ \text{km} \quad \text{or} \quad 4.6 \times 10^{-4}\ \text{km}$$

b $$5\ \cancel{\text{tons}}\left(\frac{2000\ \cancel{\text{lb}}}{1\ \cancel{\text{ton}}}\right)\left(\frac{454\ \cancel{\text{g}}}{1\ \cancel{\text{lb}}}\right)\left(\frac{1\ \text{kg}}{1000\ \cancel{\text{g}}}\right) = 4540\ \text{kg}$$

c $$25.3\ \cancel{\text{gal}}\left(\frac{4\ \cancel{\text{qt}}}{1\ \cancel{\text{gal}}}\right)\left(\frac{0.946\ \text{L}}{1\ \cancel{\text{qt}}}\right) = 95.7\ \text{L}$$

SIGNIFICANT FIGURES: PRECISION OF MEASUREMENT

While we need not go into great detail about it, a word ought to be said about precision (or carefulness) of measurement. A crude measurement is one that is made to the nearest unit, such as to the nearest foot (say, 5 ft) or nearest millimeter (say, 3 mm). Each of these measurements has one significant figure. More careful measurements have more significant figures; 5.361 ft or 3.013 mm, for example, has four significant figures. The greater the number of *significant figures,* the more precisely the measurement has been made.

If two or more measurements are used to calculate a number, one must be aware of the precision of each measurement used. If a crude number and a precise one are used to determine a quantity, the answer must be expressed crudely also.

For example, one can calculate the weight of a chemistry class as follows:

Exactly 1612 students (approximately 150 lb per student) = 241,800 lb

↑	↑	↑
This is known exactly to the nearest student (four significant figures) and is a precise number.	This is an imprecise number (two significant figures). We can only estimate the average weight plus or minus 5 lb.	This is the answer one gets mechanically on a calculator.

Since we know "student weight" only to the nearest 5 lb, we have no justification for giving "class weight" to the nearest 100 lb. We should round our answer off to two significant figures, since the *least* precise number used to get the answer had only two significant figures (this rule applies only to multiplication and division). To put it another way, we should express the weight of the class only to the precision determined by the possible error in the weight of each student times the number of students. Because we do not know the average weight any closer than 5 lb, the possible error in the weight of 1612 students would be about 8000 lb. The following is a general rule to use when multiplying or dividing numbers: Express the answer to the same number of significant figures as the least precise figure (lowest significant figure) used in the operation. Thus 241,800 should be *rounded off* and written as 240,000 lb. This is to the nearest 10,000 lb, which is justified. You should keep this rule in mind when doing calculations or interpreting numbers others present to you.

Another example is in order here. Suppose one is calculating the average gas mileage on two consecutive automobile trips. On the first trip one uses 24 gal to go 691 mi. The calculated mileage is therefore 691/24 = 28.79 mi/gal, which rounded off to two significant figures is 29. On the second trip, it took 23 gal to travel 665 mi. The mileage is 665/23 = 28.91 mi/gal, which rounded off is 29. Did one actually get *slightly better* gas mileage on the second trip? The precision with which the gallons of gas are reported is ±0.5 gal (a rather crude measurement, since there are only two significant figures). Since the ±0.5-gal error would have made about a ±0.6 mi/gal difference in our calculation, the actual mileage is actually known to be between 28.20 and 29.40 in the first case and 28.30 and 29.56 on the second trip. The small 0.13 mi/gal difference in gas mileage between the two trips is insignificant compared with the large error in measurement of the gallons of gasoline. Even though the number is slightly larger for the second trip, we really did *not* get better mileage. The rounded-off results (29 in both cases) indicate this.

The following rules govern significant figures in a number:

1 All digits 1 to 9 are significant.

2 Any 0 between the digits 1 to 9 is significant.

3 Any 0 not needed to place the decimal is significant.

The following examples illustrate these rules:

123.7	Four significant figures
200.13	Five significant figures (three digits of 1 to 9, plus two 0s between these digits)
2300	Two significant figures (the two 0s are used to place the decimal—to put the 2 in the thousands place and the 3 in the hundreds place)

25.1200	Six significant figures (four digits of 1 to 9, plus two 0s *not* needed to place the decimal)
0.00013	Two significant figures (the three 0s are used to place the decimal; they are place holders and are not significant in indicating precision)

The following examples show the use of the "lowest-significant-figure rule" given above:

$$\frac{27.201}{\boxed{2.2}} = 12.36409 = 12$$

Two significant figures

$$\boxed{25{,}100}\ (4.812) = 120{,}781.2 = 121{,}000$$

Three significant figures

$$\frac{(8.231)(2.400)}{\boxed{4.1}} = 4.8181463 = 4.8$$

Two significant figures

During any calculation, one should express the answer to approximately the correct number of significant figures so that the proper degree of precision is expressed. An example of incorrect use of significant figures is an advertisement comparing the sugar in a product with that in a peanut butter and jelly sandwich. It was stated that half a sandwich contained 12.3 g sugar. Now, the jelly is the main source of sugar here (although the peanut butter also has some), and different people will put different amounts of jelly and peanut butter on a sandwich and will use brands containing different amounts of sugar. While the average may be about 12.3 if calculated mechanically, expressing it in this fashion indicates that one knows the amount of sugar to the nearest 0.1 g—which is a very small amount of sugar. This is not true: we actually know it only to the nearest several grams, considering the factors just mentioned. A better way of expressing this would have been to say that it contained about 12 g of sugar, or better still, 9 to 15 g. Consumers should not draw unjustified conclusions from the improper use of significant figures.

ENERGY

Energy is formally defined as the ability to do work. It is found in many different *forms,* including electrical, heat, chemical, and mechanical energy. Energy can generally be thought of as existing in two *states*.

The first state, called *potential energy* (Figure 4-2a), is possessed by systems which have the *ability* to do work but which are not actually doing any at the moment (stored energy). Examples would be electricity in a wall

(a) (b)

FIGURE 4-2
Kinetic versus potential energy. (a) A car parked on a hill has potential energy. (b) A moving car has kinetic energy.

socket with nothing plugged into the socket, a tank of gasoline, and an automobile parked on a hill (the car by virtue of its *position* on the hill has the ability to do work).

The second state, called *kinetic energy* (Figure 4-2*b*), is exhibited by something actually doing work. Electricity running a vacuum cleaner, gasoline burning in a running automobile engine, and a car moving down a hill all have kinetic energy. The car has kinetic energy by virtue of its *motion*. The kinetic energy of a moving object is calculated as follows:

$$\text{Kinetic energy} = 1/2\ mv^2$$

where m is the mass of an object and v is its velocity. Since velocity is squared in the equation, it is a more important contributor to energy than mass is.

Doubling the mass of a car going 25 mi/h will simply double its kinetic energy, but doubling its velocity to 50 mi/h will increase its energy by *four* times ($50^2 = 2500$; $25^2 = 625$). We will not be doing any calculations with this equation, but the concept it embodies should be kept in mind when we examine the energy possessed by atoms and molecules. Atoms and molecules are real physical entities with mass; when they move, they possess kinetic energy, just as a moving car does.

TEMPERATURE

Temperature is a measure of the average kinetic energy of the atoms or molecules of a substance (i.e., energy possessed by virtue of *motion*). This is sometimes referred to as the *thermal energy* of the substance; it is used to measure how hot something is. This means that molecules of water (a common substance) are moving faster (or have a greater thermal energy) at higher temperature.

The three temperature scales

The scales used to designate temperatures (i.e., give numerical values to different thermal energy states) are purely arbitrary. Historically it made sense to take as reference points the thermal energy of water at its boiling and freezing points (at sea level).* The two scales chosen on this basis are the following:

Fahrenheit (°F)
Water freezes at 32°F
Water boils at 212°F

Centigrade or Celsius (°C)
Water freezes at 0°C
Water boils at 100°C

It was later thought desirable to develop a temperature scale on which the numbers were actually proportional to various physical quantities such as the volume of a gas at various temperatures. It was also desired to have a scale where zero corresponded to no thermal energy (i.e., no motion of atoms or molecules from one place to another—no translational motion). The Kelvin scale was developed:

Kelvin or absolute (K)†

$$0\ K = -273°C$$

Comparison of the three scales

	FAHRENHEIT, °F	CELSIUS, °C	KELVIN, K
Boiling H_2O	212 } 180	100 } 100	373 } 100
Freezing H_2O	32	0 } 273	273 } 273
Absolute zero (no thermal energy)	−459	−273	0

Since there are 180°F and 100°C between boiling and freezing water, the ratio of the number of degrees is 180/100 = 9/5. This means that for a given change in thermal energy there are 9°F for every 5°C. Water freezes at 32°F, but at 0°C; therefore the two scales are offset by 32°. Putting these two facts together, the relationships needed to convert between the Fahrenheit and Celcius scales are as follows:

°C = 5/9(°F − 32)	In changing Fahrenheit to Celsius, *first* subtract 32 from the Fahrenheit temperature, *then* multiply by 5 and divide by 9.
°F = [9/5(°C)] + 32	In changing Celsius to Fahrenheit, *first* multiply the Celsius temperature by 9 and divide by 5, *then* add 32 to that number.

*At lower pressures (high elevations) water boils at less than 100°C (212°F).
†By convention, the degree sign (°) is not used with measurements on the Kelvin scale.

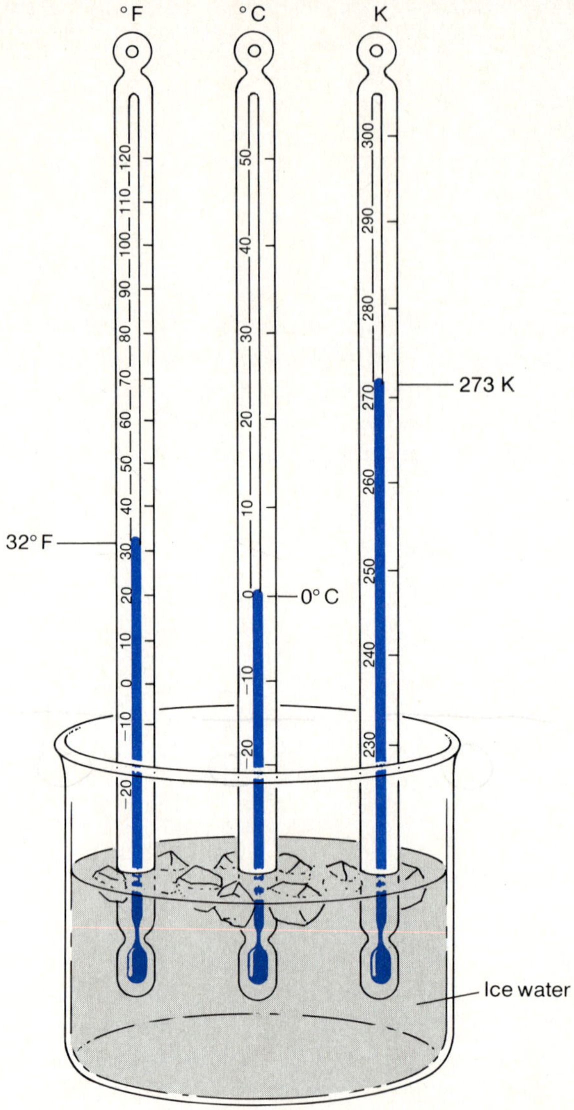

FIGURE 4-3

Temperature of ice water on three different scales.

The *size* of the degree on the Kelvin scale is the *same* as that on the Celsius scale, but zero is offset by 273°. Therefore, the conversion relationships are:

$$K = {}^{\circ}C + 273$$
$${}^{\circ}C = K - 273$$

Using these four relationships, the temperature on any one of the three scales can be converted to that on any of the others. Figure 4-3 compares the temperature of ice water on the three scales.

TEMPERATURE CONVERSIONS

Worked example

Convert 50°C to Fahrenheit.

1 Write the correct mathematical relationship from memory. Since we want Fahrenheit, we place it on the left.

$$°F = [9/5(°C)] + 32$$

2 Put 50 in place of the symbol °C.

$$°F = [9/5(50)] + 32$$

3 Work the arithmetic.
(a) Divide 50 by 5.

$$50/5 = 10$$

(b) Multiply 10 by 9.

$$9(10) = 90$$

Note The order of multiplying and dividing is *not* important.

(c) Add 90 to 32.

$$90 + 32 = 122$$

Note One must perform the operations within the parentheses *before* adding the 32.

4 This means that 50°C is equal to 122°F.

Worked example

Convert 200°F to Celsius.

1 Write the correct formula from memory. Since we want Celsius, it should be on the left.

$$°C = 5/9(°F - 32)$$

2 Put 200 in place of the symbol °F.

$$°C = 5/9(200 - 32)$$

3 Work the arithmetic.
(a) Subtract 32 from 200.

$$200 - 32 = 168$$

Note The subtraction must be done first.

(b) Multiply 168 by 5.

$$168(5) = 840$$

Note The order of multiplying and dividing is *not* important.

(c) Divide 840 by 9.

$$840/9 = 93$$

4 This means that 200°F equals 93°C.

Worked example

Convert −80°C to Fahrenheit.

1 Write the correct formula. We want degrees Fahrenheit.

$$°F = [\tfrac{9}{5}(°C)] + 32$$

2 Put −80 in place of °C.

$$°F = [\tfrac{9}{5}(-80)] + 32$$

3 Do the arithmetic.
(a) Divide −80 by 5.

$$-80/5 = -16$$

Note Do not forget the minus sign!

(b) Multiply by 9.

$$9(-16) = -144$$

(c) Add 32 to −144.

$$-144 + 32 = -112$$

4 This means that −80°C is the same as −112°F.

Worked example

Convert 40°C to Kelvin.

1 Write the correct formula.

$$K = °C + 273$$

2 Put 40 in place of °C.

$$K = 40 + 273$$

3 Do the arithmetic by adding 40 to 273.

$$40 + 273 = 313$$

4 Therefore 40° on the Celsius scale is 313 on the Kelvin scale.

Problem

Convert 26°C to Fahrenheit. (Try to work this problem on your own. Then compare the method and answer with your own.)

Solution

1 Write the appropriate formula.

$$°F = [\tfrac{9}{5}(°C)] + 32$$

2 Substitute 26 for the symbol °C.

$$°F = [\tfrac{9}{5}(26] + 32$$

3 Do the arithmetic.

$$\begin{aligned} °F &= [\tfrac{9}{5}(26)] + 32 \\ &= 47 + 32 = 79°F \end{aligned}$$

Problem

Convert 150°F to Celsius. (Do not look at the solution until you have tried to work this problem. Then check your work against the solution.)

Solution

1 Write the formula.

$$°C = \tfrac{5}{9}(°F - 32)$$

2 Substitute 150 for the symbol °F.

$$°C = \tfrac{5}{9}(150 - 32)$$

3 Do the arithmetic.

$$\tfrac{5}{9}(150 - 32) = \tfrac{5}{9}(118) = 66°C$$

Problem

Convert −100°F to Celsius.

Solution

1 Write the formula.

$$°C = \frac{5}{9}(°F - 32)$$

2 Substitute −100 for the symbol °F.

$$°C = \frac{5}{9}(-100 - 32)$$

3 Do the arithmetic.

$$\frac{5}{9}(-132) = -73°C$$

Problem

Convert 600°C to Kelvin.

Solution

1 Write the formula.

$$K = °C + 273$$

2 Substitute 600 for the symbol °C.

$$K = 600 + 273$$

3 Do the arithmetic.

$$K = 873$$

Problems

Work these problems. Only the answers are given.

a Convert 500°F to Celsius.

b Change 5°C to Fahrenheit.

c What is −95°F on the Celsius scale?

d If a Celsius thermometer reads −60°, what would a Fahrenheit thermometer read?

e Change 250 K to Celsius.

f What is 98°C on the Kelvin scale?

g Convert 20°F to Kelvin. (Use two steps.)

Answers

a 260°C

b 41°F

c −71°C

d −76°F

e −23°C

f 371 K

g 20°F is −7°C, which is 266 K.

PERCENT CALCULATIONS

In everyday life as well as in scientific endeavors, the use of percent (%) is universal. This term means "parts per hundred" and is calculated using the following mathematical relationship:

$$\text{Percent A} = \frac{\text{amount A}}{\text{total}} \times 100$$

The percent of a component A in a mixture is found by dividing the amount of A (in terms of number, weight, volume, area, etc.) by the total amount of the mixture (total number, weight, volume, area, etc.) and then multiplying this by 100. The following program shows several examples of how this is done.

CALCULATING PERCENT

Worked example

What is the percent of white sheep in a flock with 90 sheep if 15 are white?

1 There are 15 white sheep.

2 There are 90 sheep total.

3 Place these numbers into the formula:

$$\% = 15/90 \times 100$$

(a) First divide 15 by 90, which gives 0.167.
(b) Multiply by 100, which gives 16.7.

4 Summary:

$$\% = 15/90 \times 100 = 16.7\% \text{ white sheep}$$

5 This means that about 17 sheep out of every 100 are white. This is an example of percent by number.

Note The original measurements (15 and 90) determined the significant figures in the answer. The 100 is defined and thus has an infinite number of significant figures.

Worked example

What is the percent of water in a sample of meat if 15 g of the meat upon being heated gives up 12 g of water?

1 The weight of the water is 12 g.

2 The total weight is 15 g.

3

$$\% \text{ water} = 12 \text{ g}/15 \text{ g} \times 100 = 80\%$$

4 This means that there is 80 g of water for every 100 g of meat. This illustrates percent by weight (or weight percent).

Worked example

What is the percent of aspirin in a tablet that has 325 mg aspirin, 100 mg caffeine, and 150 mg phenacetin?

1 The aspirin weight is 325 mg.

2 The total weight has to be calculated by adding the weight of all the compounds:

$$325 + 100 + 150 = 575 \text{ mg}$$

3

$$\% \text{ aspirin} = 325/575 \times 100 = 56.5\%$$

Problem

What is the percent of sodium hypochlorite in a chlorine bleach if analysis shows that 250 g of bleach contains 15 g of sodium hypochlorite?

Solution

1 The amount of sodium hypochlorite is 15 g.

2 The total weight is 250 g.

3 Then:

$$\% = 15/250 \times 100 = 6\%$$

Problem

If one mixes 20 mL of alcohol, 40 mL of water, and 5 mL of glycerin and the total volume after mixing is 60 mL, what is the percent of glycerin?

Solution

1 The amount of glycerin is 5 mL.

2 Total is 60 mL.

3 Then:

$$\% = 5/60 \times 100 = 8.3\%$$

Problem

If a bottle of cold tablets is labeled as containing 200 tablets and actually has 190, what was the percent error in packaging?

Solution

1 The error was 200 − 190 = 10 tablets.

2 The total was supposed to be 200.

3 Then:

$$\% = 10/200 = 100 = 5\% \text{ error}$$

APPLICATIONS

Unit conversions and temperature conversions are frequently needed in everyday life; the ability to perform these conversions is a skill every educated person should possess. On mastering this skill, one becomes more familiar with both systems of measurement and therefore more comfortable with them. The concept of percent and the calculation of percentages are important in almost every area of life. And scientific notation is also used frequently, sometimes even in the popular press. The concept of precision and the incorrect use of significant figures are important to consumers, and to voters, as they consider data presented to them (in advertisements, for instance).

Moreover, these ideas, and the terms related to them, are part of the vocabulary of the rest of this text and should therefore be mastered. At times you will be asked in later chapters of this text to do simple calculations using the methods presented in this chapter.

STUDY GUIDE

Terms to learn

Absolute temperature
Celsius
centi-
Centigrade
Energy
Exponent
Exponential notation
Fahrenheit
Intersystem
Intrasystem
Kelvin
kilo-
Kinetic energy
Length
Mass
Metric system
milli-
Percent
Percent by number
Percent by volume
Percent by weight
Potential energy
Precision
SI
Scientific notation
Significant figures
Temperature
Unit
Volume
Weight

Skill practice problems

1 Express each of the following in scientific notation.
(a) 5,170,000,000
(b) 0.0000000129
(c) 3,000,000
(d) 0.0000253
(e) 218.1

2 Express each of the following in standard notation.
(a) 8.16×10^{-3}
(b) 3.06×10^{-6}
(c) 6.71×10^{5}
(d) 1.3716×10^{10}
(e) 2.5×10^{2}
(f) 6.02×10^{23}

3 Make the following conversions.
(a) 18 in to centimeters
(b) 15 ft to centimeters
(c) −15°C to kelvin
(d) 28°F to kelvin
(e) −40°C to Fahrenheit
(f) 10 m to feet
(g) 500 lb to kilograms
(h) 3 oz to grams

4 If 40 lb of apples, upon being squeezed, produces 30 lb of apple juice, what is the percent of juice in the apples?

5 Vinegar is 5% acetic acid in water. How much acetic acid is in 2 L of vinegar?

6 An administrator says, "Out of 30 students in an English course, 16.7 percent were absent from class; but in a chemistry class of 35 students, only 14.3 percent were absent. Therefore the class attendance was better in chemistry." Do you agree with that conclusion?

7 Express the following in the units indicated:

(a) 50 mi/h as feet per minute

(b) 621 blades of grass per square yard as blades of grass per square foot

(c) 2200 atoms per milliliter as atoms per liter

8 If a car is traveling at 35 mi/h, what is its speed in meters per second?

9 "Translate" into SI units the following set of directions for making pancakes (see Appendix A for conversions)

Egg 1	
Milk 1 cup	______ mL
Flour 1½ cups	______ mL
Water ½ cup	______ mL
Sugar 1 tablespoon	______ mL
Baking powder 1 teaspoon	______ mL
Cooking oil 2 tablespoons	______ mL
Cook at 325°F	______ °C
for 5 min	______ s

10 Some stations sell gasoline by the liter. If it costs \$1.11/gal, what is the cost per liter?

11 If it is 9×10^7 mi from the sun to earth and light travels at a rate of 3×10^8 m/s, how long does it take for light to travel from the sun to earth?

Goals

To understand *how* humans arrived at the modern view of matter

To understand the nature of the matter around us

To learn the language by which scientists express atomic information

To be able to calculate the number of each particle in atoms

5

WHAT IS MATTER REALLY LIKE?

Why do different materials behave differently? What makes glass brittle and rubber soft? How is wood different from aluminum? Why does a steel wire bend and conduct electricity, while a quartz crystal is inflexible and does not conduct electricity? Why is gold a solid at room temperature, while oxygen is a gas under those circumstances? Why does iron react with oxygen and rust, while gold does not react with oxygen and does not rust? The answers to these questions will be given in the next several chapters, but let's begin by taking a look at the nature of matter itself.

ANCIENT THOUGHT ON THE NATURE OF MATTER

Human beings have been theorizing for a long time about the nature of the matter which is part of the universe around us. (See Table 5-1 for a summary of major events in this thought.) One of the first to leave detailed written records of serious thought on this subject is the early Greek philosopher Democritus (460–370 B.C.). Later, his Roman student Lucretius (94–55 B.C.) in *The Nature of Things* described matter as consisting of small indestructible entities called *atoms*. He surmised that atoms of different materials differed in some aspect of their nature—perhaps in size, in shape, in number, and in the size of the "hooks" that might be attached to them. While this is surprisingly close to our modern atomic theory, it was *not* a truly scientific theory, because it was not based on systematic observation. On the other hand, it was not totally philosophical either, since certain facts were taken into account and some natural phenomena could be explained by these scholars.

TABLE 5-1

Major events in the development of the atomic theory

Event	Contribution
Leucippus–ca. 400 B.C.	First proposed the idea of matter being composed of invisible, indestructible particles
Democritus—ca. 400 B.C.	First used the word atom to describe these invisible, indestructible particles and left detailed written records
Plato—ca. 400 B.C.	Rejected the idea of atoms on a philosophical basis
Lucretius—ca. 70 B.C.	Revived the atomistic ideas and further explained natural observations on the basis of atoms
Christianity—ca. A.D. 60 to 100	Rejected atomism as a result of discarding the materialistic Epicurian ideas
Alchemy—ca. 300 B.C. to A.D. 1700	Learned to handle and purify matter
Early scientific experiments—ca. A.D. 1200 to 1800	Gradually discovered the laws governing matter
John Dalton—A.D. 1809	Postulated on the basis of these laws that matter was composed of atoms
Structure of atoms elucidated—1860 to 1935	Discovery of electrons, protons, and neutrons and the idea of the nuclear atom (see Table 5-5)

This early view was arrived at by analogy with the grains of sand on a beach. It is said that Democritus' teacher, Leucippus, was walking on a hill above a beach and noticed that although the beach was composed of small particles—sand grains—it looked like a continuous solid, because the grains were too small to be seen from that distance. From his point of view the water near the beach appeared to be continuous also; and he hypothesized that since the beach, which appeared from a distance to be continuous, was really composed of small particles, so the water, which also appeared continuous, was actually composed of invisible particles. We will see that this is essentially the same view that was arrived at many centuries later through many years of intense research. That fact is a great tribute to the minds of the ancient scholars, and indeed to the power of the human mind in general (which all of us possess). Because the idea of atoms was unpalatable to later philosophers—especially Plato, who had a tremendous influence on western thought—it was largely discarded for about two millennia.

It is also important to note another reason why Lucretius' writings were not widely accepted. He was an Epicurian; that is, he believed that no divine being existed and that human happiness and pleasure were the highest good. Jesus Christ was born about 80 years after Lucretius' death. He taught only a short time (2 to 4 years); but because of the rapid spread of Christianity, he had a tremendous impact on western culture—more than any individual before or since. Since the Epicurian views of Lucretius were diametrically opposed to those of Christianity, his writings were neglected. The atomic view of matter was, therefore, discarded along with other views which were objectionable to Christians. In fact, however, Christ had said nothing about atomism, and the early Christian writers had made no comments about matter except to denounce the Stoics' notion that matter is evil.

EARLY USE OF MATERIALS

Since prehistoric times, humans have worked with various materials in practical ways. Wine and other alcoholic beverages were prepared and distilled. Salt was mined or obtained from evaporation of naturally occurring salty water and used as a flavor enhancer and a food preservative. Metals were obtained from their ores and purified by methods which had no doubt been discovered accidentally. These procedures were developed as arts and taught humans that by learning to handle material they could increase their own comfort, conquer their enemies, produce food, and increase the efficiency of daily labor. By 1000 B.C., many of these arts had become highly developed, and in some civilizations—such as China and Egypt—medicine and other scientific activities were apparently at a level not to be surpassed until the 1700s or even later.

During the ensuing centuries, however, many people worked with matter in various special capacities and developed methods of handling, separating, and purifying material to a higher technical level. We refer to these people as *alchemists* (Figure 5-1). They were primarily concerned with bringing about desirable changes in matter. Changing inexpensive materials into valuable

FIGURE 5-1
The Alchemist, by Cornelis Bega. (Courtesy of Dr. Alfred Bader, Aldrich Chemical Company.)

metals such as gold and silver, producing alcoholic beverages, and finding the elixir of life (the source of eternal youth) were among their goals. Their efforts were largely by trial and error—random combinations—and involved little thought about the nature of matter and little if any accurate quantitative measurement. Centuries of this kind of effort, crude as it may seem to a modern scientist, did produce techniques still used by chemists, such as distillation, growing crystals, and use of symbols. Table 5-2 summarizes various features of alchemy; Figure 5-2 shows some of the symbols used by alchemists.

By the seventeenth and eighteenth centuries quantitative methods had been developed with which the first modern scientists could begin to make some real progress. These pioneers recorded their own systematic experiments and shared them with others, who were able to repeat them. Out of this environment of cooperation, recording, systematizing, and quantitative measurements—and a renewed desire by the people involved to really understand matter rather than just bring about desired changes in it—came our present natural sciences of chemistry, physics, and biology. Alchemists, ancient glassmakers, metallurgists, and others showed that many practical, useful changes in matter could be accomplished without really understanding matter. However, it was only after an understanding of the nature of matter had been reached that some of the desired changes could be brought about. In other words, *only with fundamental understanding could the more complex changes be accomplished*. This is a lesson that those who are funding scientific research today should remember. Funding should be provided not only for areas that show promise of an early application (such as curing a disease) but also for areas of basic understanding of a general subject area.

TABLE 5-2

Summary of alchemy

History:
- Alchemy had its beginnings in antiquity in China and India.
- The practice of alchemy became well established in Egypt during the first through third centuries after Christ.
- Alchemical writings were translated into Arabic in the seventh century—Arabia became the center of activity.
- Alchemical writings were translated into Latin in the twelfth century, which introduced alchemy to Europe.
- Rodger Bacon (1214–1292), an English philosopher and alchemist, is considered the first modern scientist.

Goals:
- To change cheap metals into gold—which was believed to be a perfect substance—by using the mythical philosopher's stone
- To perfect all things using the method which would succeed in producing gold
- To discover the elixir of life, which would cure illness and give endless youth and eternal life
- To find the universal solvent—a material that would dissolve anything

View of matter:
- All matter is composed of various combinations of earth, air, fire, water, heat, cold, wet, and dry.

Accomplishments:
- Use of symbols to represent substances
- Development of distillation, crystallization, and other purification methods
- Beginning of modern chemistry

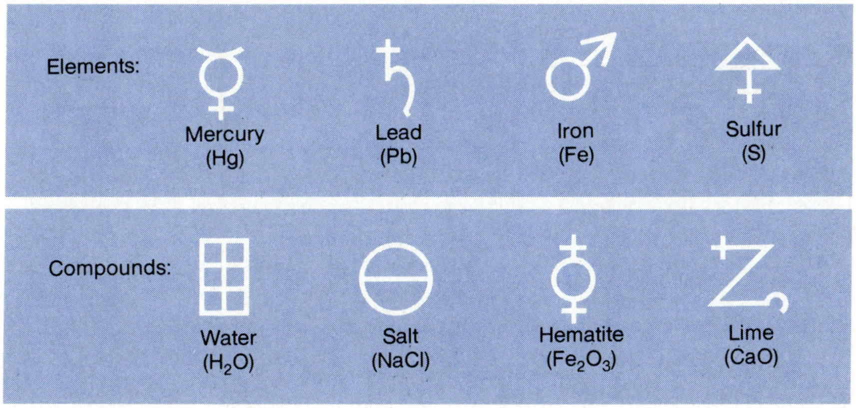

FIGURE 5-2
Symbols used by alchemists. (Information from Rudolph Koch, *The Book of Signs,* Dover, New York, 1955.)

As the alchemical era ended, many people contributed to the development of the modern scientific method. Copernicus (1473–1543), Galileo (1564–1642), Kepler (1571–1630), and Newton (1642–1727), all working in the area of astronomy, described movements of celestial bodies on the basis of observation. Their conclusions were contrary to traditional views and often brought upon them the disapproval or even wrath of their culture, especially the church. At about the same time, Vesalius (1514–1564) and Harvey (1578–1657) made important discoveries in general human anatomy and circulation of blood, which were the beginning of modern medical science. Perhaps even

FIGURE 5-3
Joseph Priestly, who was the first person to prepare oxygen, is known as the "father of modern chemistry." (Photograph from the collection of *Journal of Chemical Education*.)

more important was the establishment of the idea that careful observation and systematizing of observations could greatly expand human knowledge.

When one considers the roots of chemistry in particular, several men stand out for their key accomplishments. Robert Boyle, working in the mid-1600s, described the behavior of gases quantitatively and was the first to establish the relationship between pressure and volume. Joseph Priestly (Figure 5-3) first prepared pure oxygen on August 1, 1774, and noted its value in supporting combustion (the burning of a candle); he also prepared several gaseous compounds for the first time. Priestly is sometimes considered the father of modern chemistry. Lavoisier concluded, in about 1750, that burning (combustion) was the combining of oxygen with the components of the material being burned. He dispelled the then-popular concept of the phlogiston, a substance supposedly "lighter than nothing" whose loss from a compound caused its weight to increase. Some of Lavoisier's predecessors had believed strongly in the phlogiston theory.

BASIC LAWS OF MATTER

During the period which includes roughly the seventeenth and eighteenth centuries, it became clear that matter behaved in certain reliable ways. The most important of these are stated in three laws that are the basis for all chemical work.

The first is the *law of conservation of matter*. This law states that when a chemical or physical change occurs, the total mass after the change is the same as the total mass before the change. If a quantity of ice is carefully weighed and subsequently melted, the liquid water will weigh the same as the ice (Figure 5-4). No change in the *amount* of matter is caused by this physical change—matter is conserved during it. Likewise, if one heats limestone so that it changes to lime and carbon dioxide, careful measurement will show that the total mass of the lime and the carbon dioxide is the same as that of the original limestone. (This must be done in a closed container, of course, so that none of the gaseous carbon dioxide escapes.) In this chemical change (a change in which new substances are produced), the actual amount of matter does not change. *No matter is created or destroyed;* matter has only changed forms. This law holds true for physical and chemical changes, including biological ones. For example, the weight of a person on a given day plus the mass of the food and water ingested and gases breathed in must be equal to the person's weight the next day, plus the gases breathed out and plus any excrement. The mass of all these materials has remained constant; it has only changed location.

Once in a while, one hears it said that a dead person weighs more than a living one (dead weight is heavier), or that a person who dies becomes lighter. The author has even heard the idea once that a person who is thinking negative thoughts (is depressed, for example) will weigh 2 to 3 lb more. All three of these ideas are false: matter cannot be created or destroyed in these ways.

In nuclear chemical reactions (see Chapter 13), a small but measurable

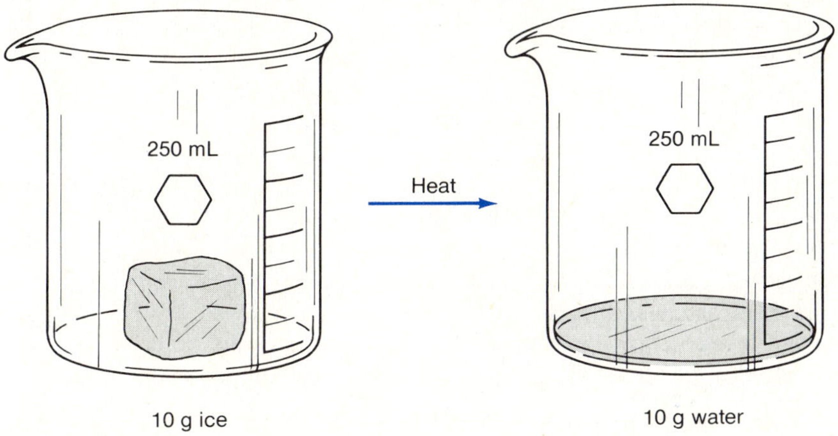

FIGURE 5-4
Conservation of mass during melting.

amount of matter is converted into energy. The law of conservation of matter, as stated above, does hold for most chemical reactions and physical processes. But it would perhaps be better to state it as a law of conservation of matter and energy (taken collectively), as follows: *Matter and energy can be interconverted but cannot be created or destroyed.* We will take up the conversion of matter into energy in Chapter 13. Meanwhile, we can consider that for all practical purposes, the law of conservation of matter does hold.

The second fundamental law about the behavior of matter is the *law of definite proportions* (or definite composition). It has already been pointed out (in Chapter 1) that compounds are substances composed of two or more elements. Every compound has a definite amount of each of its constituent elements. Water, for example, is always composed of 89 percent oxygen and 11 percent hydrogen (by weight). This is true of all samples of water regardless of their origin or history. Water from rain, water distilled from ocean water, water melted from a glacier, or water prepared synthetically in a laboratory—all will contain this same proportion of oxygen and hydrogen. Stated in another way, this law says that two elements always combine in the same proportions. Oxygen and hydrogen will react with each other in a 89 to 11 ratio to produce water. Even if different ratios are mixed, the ratio that will combine to make the water will still be 89 to 11. This can be shown by the numbers in Table 5-3. In reaction 1, 89 g of oxygen and 11 g of hydrogen will react to

TABLE 5-3
Law of definite proportions

	Oxygen	+	Hydrogen	⟶	Water
1	89 g		11 g		100 g
2	178 g		22 g		200 g
3	89 g		25 g (14 g remaining) (11 g used)		100 g
4	300 g (122 g remaining) (178 g used)		22g		200 g

produce 100 g of water. This is a confirmation of the law of conservation of matter and shows that these two elements combine in definite proportions. In reaction 2, there are different amounts of these two elements; but the ratio of 178 to 22 that will react is still 89 to 11 (or about 8 to 1). If, as shown in reaction 3, 89 g of oxygen is mixed with 25 g of hydrogen, only 11 g of the hydrogen reacts with the 89 g of oxygen (the two elements react only in a ratio of 89 to 11); 14 g of hydrogen remains unchanged. Likewise, if one has an excess of oxygen—that is, more than the eight times as much needed to react with the hydrogen—some oxygen stays unchanged. This set of conditions is illustrated in reaction 4, in which 22 g of hydrogen, which requires eight times as much oxygen (178 g) to form water, reacts with only 178 g of the 300 g of oxygen present; 122 g of oxygen does not react. This same phenomenon has been observed for all elements forming compounds.

The third basic law is the *law of multiple proportions*. Sometimes two elements can form two or more compounds. When this happens, the proportions of each element in each compound follow the law of definite proportions and all reactions follow the law of conservation of matter. In addition, it can be observed that the proportion of elements in one compound is some whole-number multiple of that proportion in the other compound. If one compound has a certain ratio of one element to another, the other compound will have exactly one-half or two times or three times that ratio. This is illustrated by the two compounds—water and hydrogen peroxide—formed by oxygen and hydrogen. Water contains oxygen and hydrogen in an 8 to 1 ratio respectively. The other compound, hydrogen peroxide, is 94.1 percent oxygen and 5.9 percent hydrogen, which is a ratio of 16 to 1—exactly twice the 8 to 1 ratio in water. For all compounds composed of the same two elements, the comparative ratios can always be expressed as relatively small whole numbers. This is shown in Table 5-4 for the three compounds of nitrogen and oxygen. Compound 2 has exactly twice as much nitrogen as oxygen compared with compound 1. Compound 3 has four times as much nitrogen as oxygen compared with compound 1. The proportions of nitrogen to oxygen in compounds 2 and 3 are small whole-number multiples (2 and 4, respectively) of that proportion in compound 1.

TABLE 5-4
The law of multiple proportions for compounds of nitrogen (N) and oxygen (O)

COMPOUND NUMBER	COMPOUND NAME	PERCENT N*	PERCENT O*	RATIO N TO O*	RATIO N TO O COMPARED WITH COMPOUND 1
1	Nitrogen dioxide	30.4	69.6	0.44 to 1.00	
2	Nitric oxide	46.7	53.3	0.88 to 1.00	2.0 times
3	Nitrous oxide	63.6	36.4	1.75 to 1.00	4.0 times

*By weight.

DALTON'S ATOMIC THEORY

About 1809, John Dalton (Figure 5-5), having considered the three laws just discussed, proposed that matter is composed of *atoms,* and that atoms have the following characteristics.

First, Dalton proposed that *atoms are indestructible;* they cannot be created or destroyed but can only change the associations they make with other atoms. This is a consequence of the law of conservation of matter. If matter cannot be created or destroyed, then the atoms composing matter likewise must be stable, indestructible, real entities. The law of multiple proportions indicates that matter comes only in discrete units (or packages). Thus one atom of one type could combine with one atom of another kind, or perhaps it could combine with two or three of another type. Half an atom could not exist, however. The fact that the relative ratios of element masses in compounds containing a given pair of elements are always *whole numbers* is indicative that only whole "units" of matter can be involved in chemical processes. These units Dalton called *atoms* (after the name given centuries earlier by Democritus).

FIGURE 5-5
John Dalton, who was the first modern scientist to propose the existence of atoms. (From a drawing by the author.)

Second, *atoms of a given element must all be alike*. For example, all oxygen atoms must be the same as all other oxygen atoms, and all hydrogen atoms must be the same as all other hydrogen atoms. This follows from the law of definite proportions. Since oxygen and hydrogen atoms always combine in the same proportions to make water, then all the atoms of a given element must have the same characteristics.

Third, *atoms of different elements must be different from each other*. Oxygen atoms, for example, must be different from hydrogen atoms. This would explain the different physical and chemical behavior of the various elements.

Dalton believed that the existence of atoms was the only reasonable explanation for the three laws of matter. His view, while not substantially different from that of Democritus, was more detailed, was based on much more observation, and could explain many more facts. Dalton, then, was the first modern scientist to propose atomism. It was, of course, a hypothesis when first stated; but over the next 100 years further evidence was gathered as it was tested by other scientists. Only after about 1900 did the whole scientific community accept atomism. It has been referred to as *atomic theory* since that time. It is a strong theory because it explains all known facts; and it emerged from a cauldron of long, and sometimes intense, controversy. (Remember that a theory is a generally accepted and tested explanation.)

MODERN ATOMIC THEORY

Subatomic particles

Dalton could only guess how atoms of one element differed from those of another. Presumably these differences included shape, size, and "hooks," but scientists of Dalton's era did not have the tools to explore the internal structure of the atom. However, in the years between approximately 1860 and

TABLE 5-5

Discoveries leading to modern atomic theory

SCIENTIST	DATE	DEVICE USED	DISCOVERY
Plucker	1859		Cathode-ray tube
Thomson	1890	Mass spectrometer	Electrons
Goldstein	1886		Canal ray tube
Thomson and Wien	1898	Mass spectrometer	Protons
Millikan	1910		Absolute masses of subatomic particles (protons and electrons)
Rutherford	1911		Nuclear atom
Chadwick	1932		Neutrons

1935 the internal structure of the atom was elucidated. (See Table 5-5 for a summary.)

Let's begin our study of these discoveries by briefly examining two areas of basic physics.

First, let us consider the *laws of electrostatics*. According to these laws, when objects are *charged,* they can have only two kinds of charge: positive (+) and negative (−). When two objects have like charges, either both positive or both negative, they tend to repel (tend to move away from) each other. When their charges are different, one positive and one negative, then they attract (tend to move toward) each other. The exact nature of charge is not understood, but that there are two types of charge and that the laws of electrostatics hold are both well established. These concepts are illustrated in Figure 5-6.

Second, let us examine an instrument that was quite useful in determining atomic structure: the *mass spectrometer*. This is a device that allows one to find out information about the charge and mass of a given particle—a very small object. (*Mass* is a measure of the amount of matter present in a substance. The units of measurement of mass were discussed in Chapter 4.) The

FIGURE 5-6

Electrostatic attraction and repulsion.

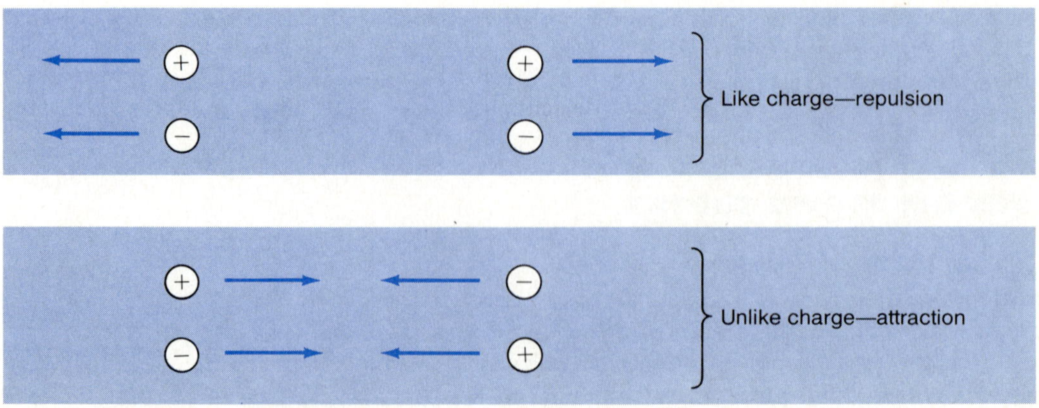

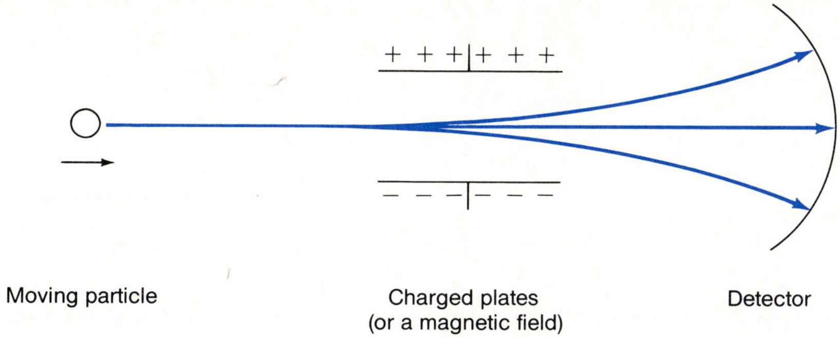

FIGURE 5-7
Diagram of a basic mass spectrometer.

overall design of a mass spectrometer is shown in Figure 5-7. The particles whose characteristics are to be studied are caused to move between two charged plates or magnets. The movement may be by virtue of the natural motion of the particles, or it may be produced by a separate accelerator. In practice, a magnetic field is used in a mass spectrometer, but for purposes of illustration let us assume that charged plates are used. The moving particles are then *deflected* (their path of motion is changed) according to their charge and mass.

A positive particle moves *toward* the negative plates and away from the positive plates, as indicated by the laws of electrostatics. A negative particle is moved *away* from the negative plates and toward the positive plates. The greater the mass of the particle, the less it is moved from its original path. This can be understood by considering two balls being thrown through a wind current created by a strong fan. A golf ball is deflected only a little; a Ping-Pong ball is deflected a great deal. The golf ball is about the same size as the lighter Ping-Pong ball and therefore "feels" about the same force from the wind; but it is not deflected as much, because it has more mass and therefore more inertia. *Inertia* is the tendency for an object to resist movement when a force is brought to bear on it. Thus a heavier particle placed in the charge force field of the mass spectrometer actually moves less than a lighter one experiencing the same force. Finally, the greater the amount of charge possessed by a particle, the more it moves, all other things being equal (such as its own mass, the charge on the plates, or the strength of the magnetic field).

In order to determine if a particle is being deflected—and, if so, in which direction and by how much—a detector is employed. This may be a phosphorescent screen that is read visually or an electronic detector that produces a graph. The scientists who pioneered studies of atomic structure used screens; modern scientists use electronic detectors. These various situations are diagramed in Figure 5-8.

It is possible to determine experimentally only the ratio of mass to charge for a particle. If one assumes that each particle has only one charge (i.e., one positive or one negative charge per particle), then the relative masses of particles can be determined; this is not always a valid assumption, however.

With this basic information, we shall examine several key discoveries that led to our present understanding of atoms.

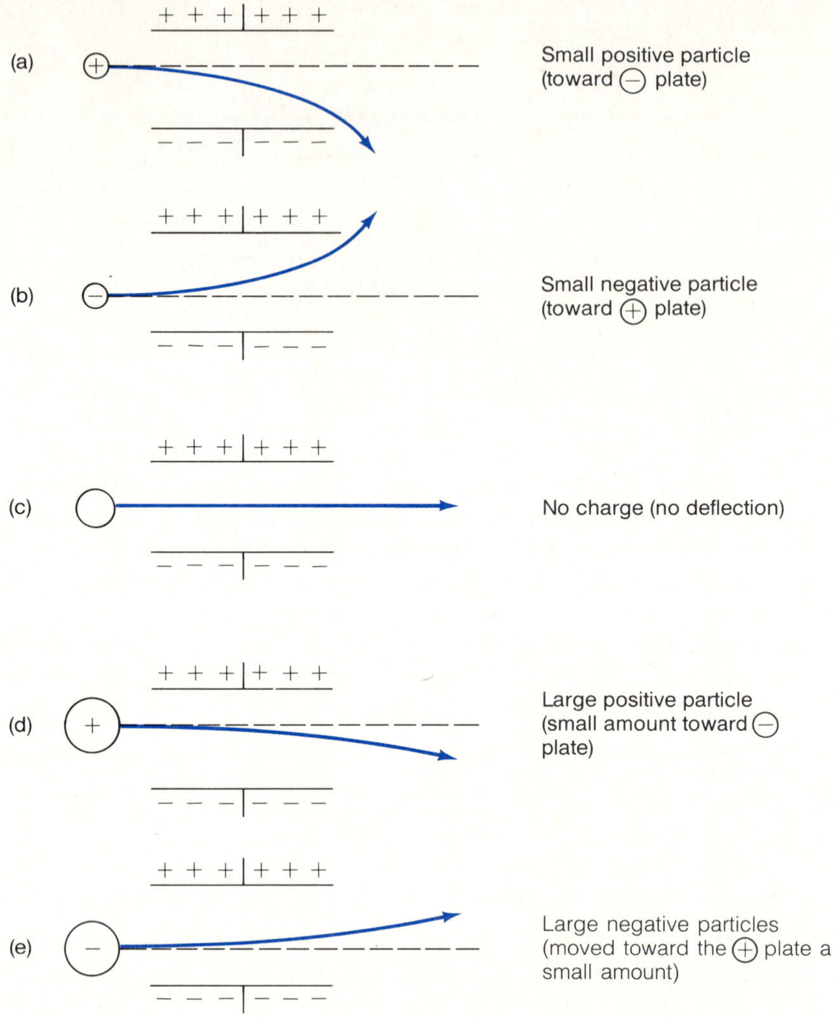

FIGURE 5-8
Behavior of various particles in a mass spectrometer.

In 1859, Plucker developed the cathode-ray tube, which is shown in Figure 5-9. This is a sealed glass tube from which most of the air has been removed and in which two metal plates (electrodes) have been embedded. When a source of electric current is attached to the electrode, a glow (called *cathode rays*) appears between the plates. The video screen and cathode-ray oscilloscope are modern developments of this device.

In the last decade of the nineteenth century, Thomson subjected cathode rays to mass spectrographic analysis (see Figure 5-10). He found that they were deflected strongly toward the positive plate, and drew the following conclusions:

1 These rays were composed of particles rather than of visible light. Light is not deflected either by charged plates or by a magnetic field.

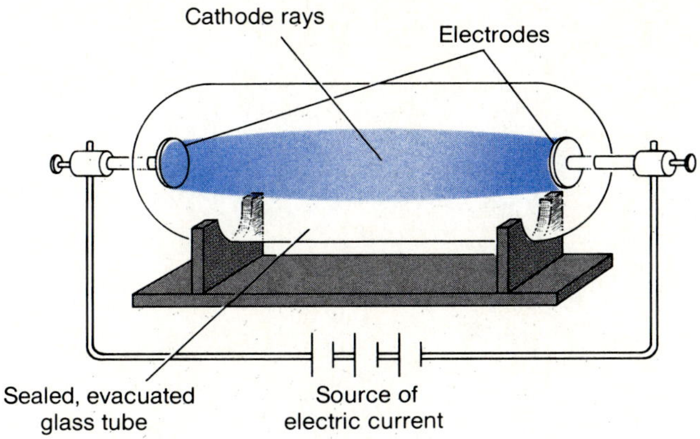

FIGURE 5-9
Cathode-ray tube.

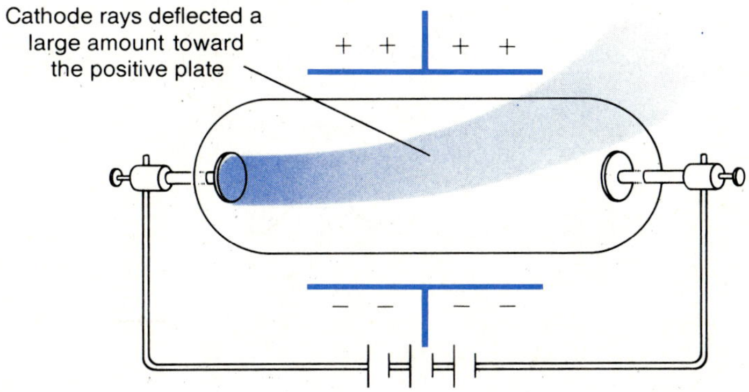

FIGURE 5-10
Cathode rays subjected to mass spectrographic analysis.

2 The particles were *negatively* charged (they were deflected toward the positive plate).

3 They were relatively small particles (they showed a large deflection).

4 Since all electrode materials—iron, silver, etc.—produced rays with identical characteristics, each of these materials must have this same kind of particle within it.

Thomson called these particles *electrons,* symbolized by e^-, and concluded that, since all matter was composed of atoms and all matter had electrons, all atoms must have electrons as component parts. Thus, the electron was a *subatomic particle*.

In 1886, Goldstein discovered canal rays. These are rays emitted from an electrode with holes (canals) in it, as shown in Figure 5-11. Wien and Thomson, just before the turn of the century, placed these rays in a mass spectrometer (see Figure 5-12) and found that they were deflected a small amount toward the negative plates. They concluded that

1 These were positively charged particles.

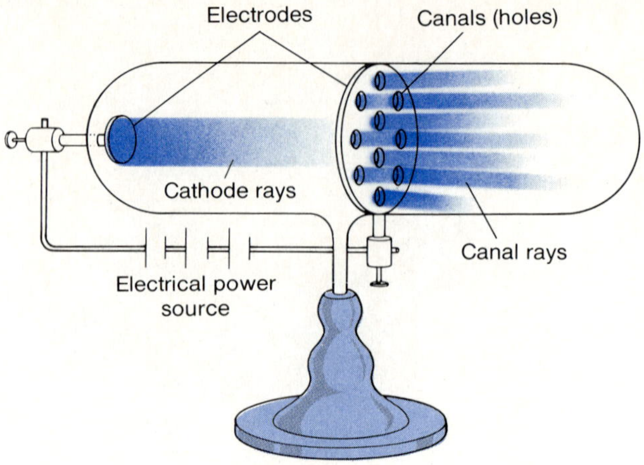

FIGURE 5-11
Canal-ray tube.

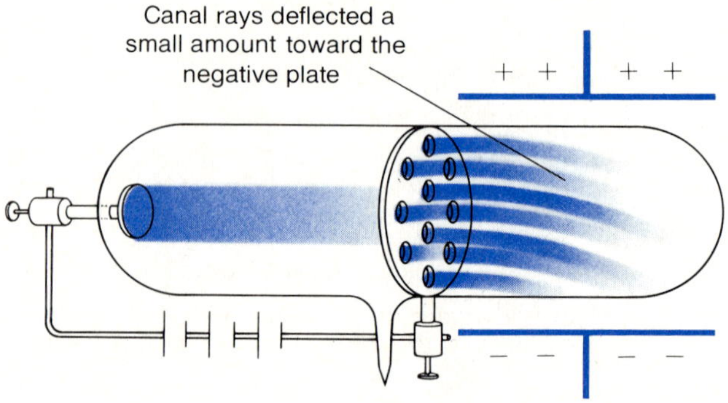

FIGURE 5-12
Canal rays subjected to mass spectrographic analysis.

2 They were large compared with electrons. Calculations showed that those produced by hydrogen has a mass-to-charge ratio 1840 times as large as electrons. If we assume one charge per particle (which turns out be correct in this case), then this particle would be 1840 times as heavy as the electron. This particle had a mass almost identical to that of a hydrogen atom, which was the lightest atom known.

3 All electrode materials produced similar canal rays (all were positive and had large masses). Thus all materials must contain this kind of particle.

Wien and Thomson called these particles *protons*, symbolized by p^+, and concluded that all atoms must contain them. A second subatomic particle had been discovered. Since atoms have no charge and all atoms contain positively charged protons and negatively charged electrons, it was thought, by the year 1900, that all atoms contained *equal numbers* of electrons and protons distributed equally throughout their internal parts, roughly as shown in Figure 5-13. Different atoms simply had different numbers of each type of subatomic particle. (Actually, it was thought that the electrons were embedded in a positive mass—but that there was enough positive charge to neutralize all the electrons.)

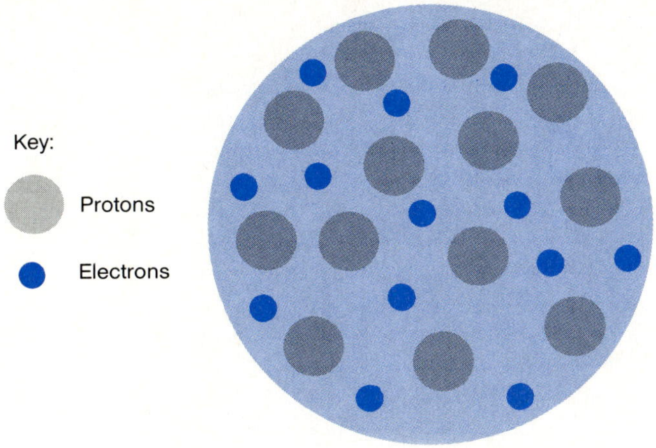

FIGURE 5-13
The Thomson model of the atom.

The nuclear atom

In 1911, Rutherford performed a simple but elegant experiment that caused this view (model) of the atom to be modified. His experiment is diagramed in Figure 5-14. He placed radium, a radioactive element,* in a lead box with an opening at one end. The radium emitted alpha rays (α rays), which came through the hole in the box. These rays had been shown by mass spectrographic analysis to have a positive charge and to be four times as heavy as protons. They were focused on a thin piece of gold foil, and a screen that would detect α rays was placed around the foil. To Rutherford's amazement, most of the α rays penetrated the foil, some were deflected at a slight angle, and some were actually reflected back toward the α-ray source. If nothing heavier than individual p^+ and e^- had been distributed throughout the atom, the α particles, being four times as heavy as protons and thousands of times heavier than electrons, could not have been deflected at a large angle. A large object cannot bounce off smaller ones. It was obvious from Rutherford's experiment that Thomson's view of the atom was incorrect. Since the new evidence was inconsistent with Thomson's hypothesis, Thomson's model of the atom had to be adjusted. Rutherford concluded almost immediately, in a creative insight, that all the heavy protons must be collected together in one place at the center of the atom and that this positive core was much smaller than the atom itself. He called the entire atom the *nuclear atom*, and the dense collection of protons containing most of the mass of the atom the *nucleus*. Figure 5-15 is a diagram of nuclear atoms and shows how the behavior of α particles can be explained. Some of the α particles could travel unhindered through the outer area where the electrons are located. Some would hit the large nucleus at its edge and be deflected at an angle. A few would encounter the nucleus head-on and bounce back.

In 1920, Rutherford calculated the presence of a third subatomic particle; and in 1932, Chadwick presented evidence for it from his studies of radioac-

*An unstable element that gives off penetrating rays. (See Chapter 13.)

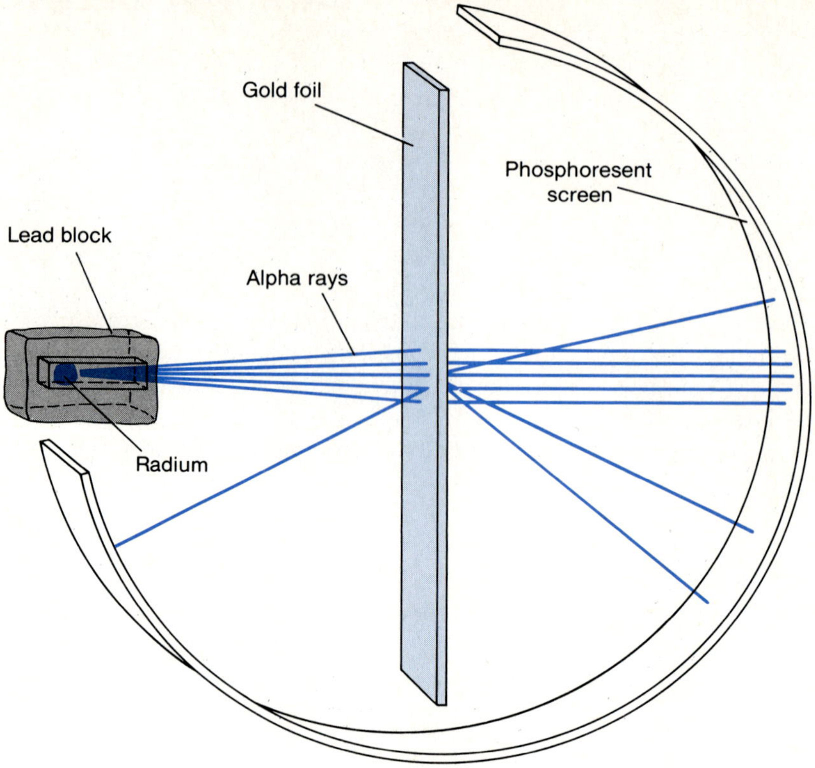

FIGURE 5-14
Rutherford's gold foil experiment.

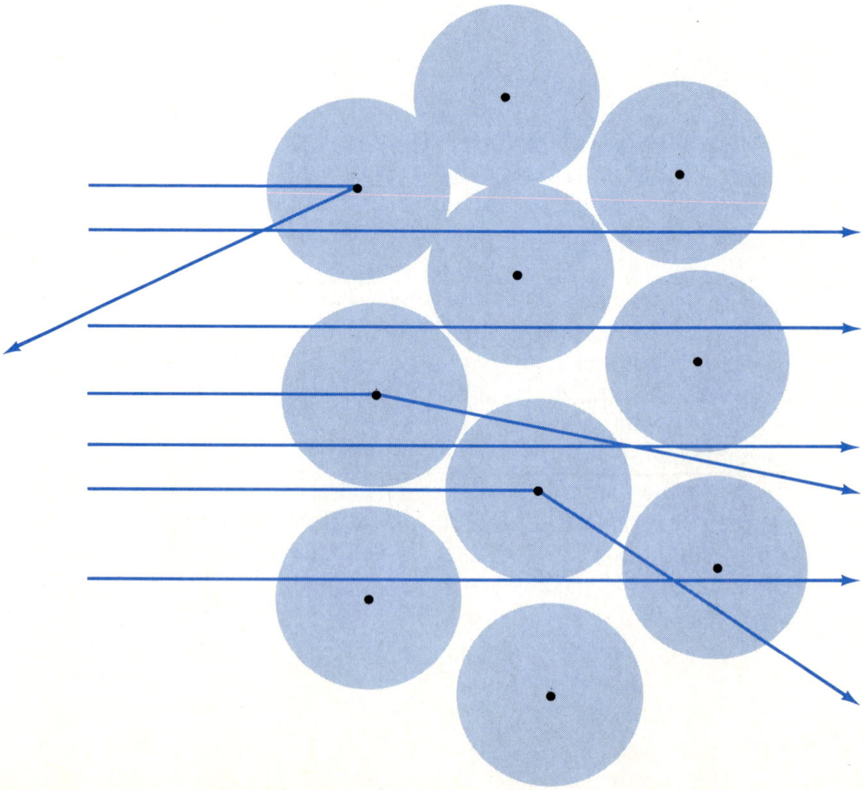

FIGURE 5-15
Penetration of atoms by alpha rays.

tive materials. This particle is called the *neutron* (the symbol for a neutron is n). It was neutral (not charged), had a mass about the same as that of a proton, and was found in the nucleus of the atom along with the protons.

No major change in our view of the atom has occurred since 1932; no evidence has been discovered that is inconsistent with it. This model of the atom is shown in Figure 5-16; it should be our mental picture of an atom during the remainder of this text. Nuclear physicists are continually investigating atomic structure and are refining this model. They have found several small particles involved in holding the nucleus together and now believe that the neutron and proton are each composed of three still smaller particles (quarks). Chemists usually are not concerned with these relatively minor particles, however.

The nucleus contains almost all the mass of the atom, since the protons and neutrons each weigh 1840 times as much as electrons. In most atoms the nucleus contains about 99.98 percent of the mass. The actual size (diameter) of the nucleus is quite small compared with the diameter of the whole atom, however, and thus very little of the space of an atom is occupied. The proportions are as follows: if the diameter of an atom were 100 yards (yd; the length of a football field) the diameter of the nucleus would be about 1 in. This means that only 1 part out of about 50 billion of the volume of the average

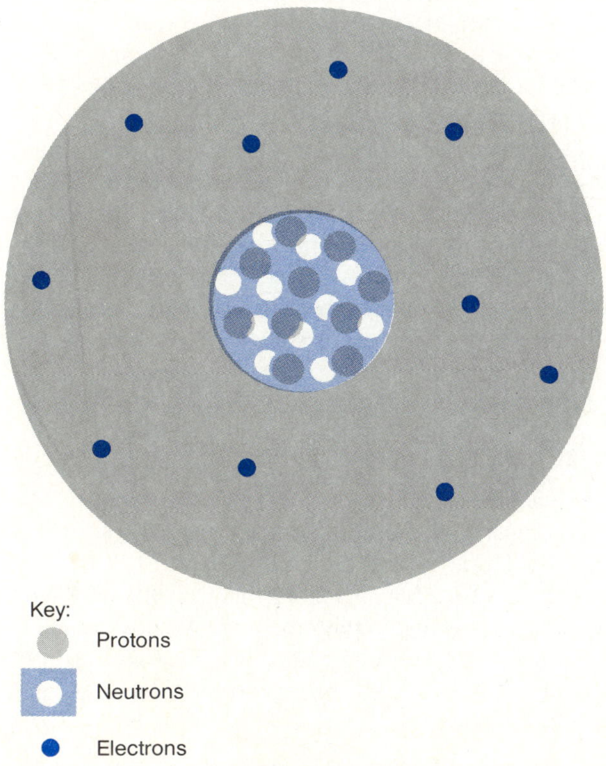

FIGURE 5-16
The nuclear atom.

atom is actually occupied by the nucleus; the rest is mostly empty space in which electrons move.

Another classic experiment (it was not involved in developing this general model of the atom but was important in gaining technical data) was carried out by Millikan. He found that an electron weighed an incredibly small amount: 9.11×10^{-28} g. Protons and neutrons each were calculated to weigh about 1.66×10^{-24} g; this weight, though extremely small, is very large compared with that of the electron.

Terms associated with atomic structure

In order to facilitate communication, chemists use carefully defined terms. Since the masses of all subatomic particles are so small when expressed in grams or pounds (see Chapter 4), a new mass scale was developed, called the *atomic mass unit (amu) scale*. On this arbitrary scale,* protons and neutrons each weigh about 1 amu. (Actually, protons are 1.007 amu and neutrons 1.008 amu, but we will not be concerned with this small difference here.) The masses of whole atoms can be expressed in terms of amu's; and since each neutron and proton weighs 1 amu, the approximate mass of an atom in terms of amu's is simply the number of these particles in the atom.

The *atomic mass* of an atom is the exact mass of an atom on the atomic mass scale. For example, one kind of carbon atom has a mass of 12.00 amu; one kind of oxygen, 15.994 amu; etc.

The *atomic number* of an atom is defined as the number of protons in its nucleus (which is also the number of electrons around this nucleus). An atom has the same number of protons as electrons. Each element is composed of atoms with a unique number of electrons and protons. Neon atoms, for example, have 10 of each; chlorine atoms have 17 of each.

All atoms of a given element have the same number of protons and electrons, but not all have the same weight. For the element chlorine, about 75 percent of the atoms in nature weigh 34.9 amu, while the other 25 percent weigh 36.9 amu. These two kinds of chlorine atoms have the same number of electrons and protons but a different number of neutrons; they are called *isotopes*. Most elements have 2 or 3 isotopes, but some, such as iodine (I), have only 1 naturally occurring isotope (only one kind of iodine atom has been found), and others have more than 3—tin (Sn), for instance, has 10. This is contrary to Dalton's idea that all atoms of a given element are identical.

The term *mass number* refers to the atomic mass of a given isotope to the nearest whole number; it is used to identify each isotope. (It is actually the sum of the neutrons and protons.) For example, the chlorine isotope, which weighs 34.9 amu, is called chlorine-35 and is symbolized by ^{35}Cl.

The average weight of naturally occurring chlorine atoms is the weighted average of 34.9 and 36.9 amu. The value in this case is 35.45 amu, which is closer to 34.9 than 36.9, since there are more atoms of the former type (75 percent). This average atomic mass is called the *atomic weight* of the element.

*The scale was chosen so that one isotope of carbon (^{12}C) weighs exactly 12.00000 amu. All other weights are relative to it.

It is the number used in most chemical calculations, since in most cases chemists use the naturally occurring elements, which are usually mixtures of isotopes.

The atomic number and atomic weight of each element are found both on a list of elements (see Table 1-4) and on the periodic chart (inside back cover). Each is identified by a column heading. In the periodic chart, the atomic number is the whole number printed *above* the element symbol and the atomic weight is the decimal number printed *below* it. Information about each isotope is frequently expressed as follows and is called *isotopic notation;* note that in this method (unlike the periodic chart) the atomic number is on the bottom:

Mass number → $^{12}_{6}C$ ← Element symbol

Atomic number → $^{12}_{6}C$

Numbers of subatomic particles

In order to calculate the number of protons and electrons in an element, one finds the element on a periodic chart or list of elements and then finds its atomic number. Boron (B) has an atomic number of 5; therefore, atoms of boron have 5 electrons and 5 protons. Likewise atoms of uranium (U) have 92 of each.

The mass number of an isotope is the sum of the number of neutrons and protons (the only subatomic particles with appreciable mass—each one weighs about 1 amu). The number of neutrons can be calculated by subtracting the number of protons (atomic number) from the total (mass number):

$$\text{Number of neutrons} = \text{mass number} - \text{atomic number}$$

Carbon, for example, has a mass number of 12 and an atomic number of 6. Since the mass number of 12 indicates 12 particles contributing to the mass (total of the protons plus neutrons), the neutrons would be the total, 12, minus the protons, 6, to give 6 neutrons. Mechanically:

$$\text{Neutrons} = 12 - 6$$

12: Mass number
6: Atomic number (number of protons)

This process is conceptually similar to calculating the number of women in a class if one knows the total enrollment and the number of men. If the class has 50 students and 30 men, since there are only two types of people, the other 20 would have to be women:

$$\text{Number of women} = \text{total students} - \text{number of men}$$

Similarly, since there are only two types of subatomic particles contributing to the mass, subtracting the number of one type from the total will give the number of the other.

TABLE 5-6

Calculations of subatomic particles

SYMBOL	NUMBER OF PROTONS (ATOMIC NUMBER)	NUMBER OF ELECTRONS (ATOMIC NUMBER)	MASS NUMBER	NUMBER OF NEUTRONS
Ca	20	20	40	40 − 20 = 20
H	1	1	1	1 − 1 = 0
He	2	2	4	4 − 2 = 2
Fe	26	26	56	56 − 26 = 30

FIGURE 5-17
Diagrams of atoms of hydrogen, helium and calcium (not to scale).

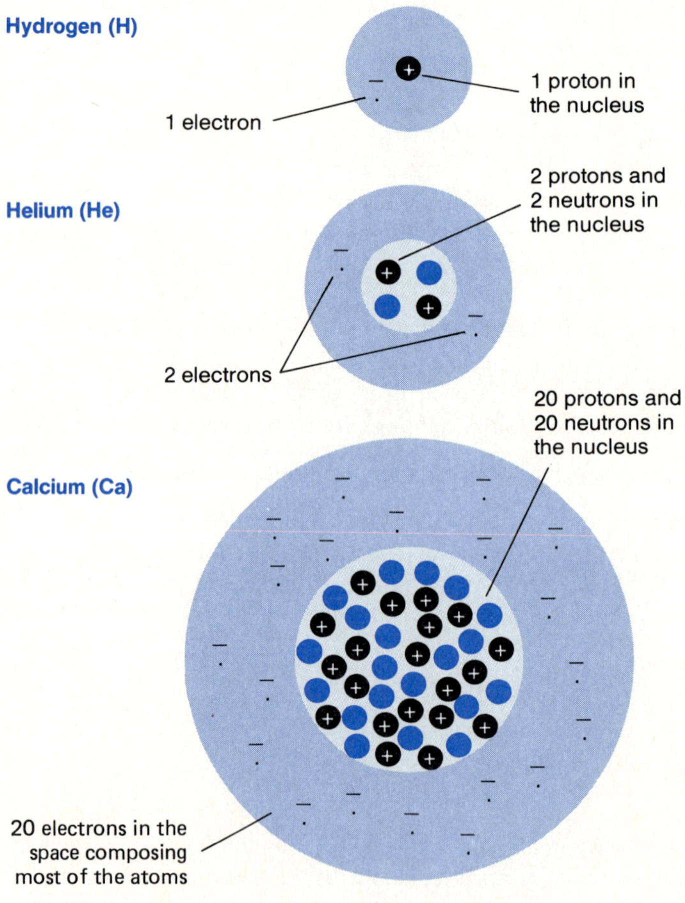

The subatomic particles in atoms of calcium (Ca), hydrogen (H), helium (He), and iron (Fe) can be calculated in a similar way; see Table 5-6. Hydrogen is the only element that has an isotope with no neutrons. For the lower-mass atoms the neutrons and protons are present in about equal numbers, but for the higher-mass ones, the neutrons outnumber the protons.

We are now in a position to construct diagrams of atoms using Rutherford's nuclear model and the number of each subatomic particle as just calculated. Three such diagrams are shown in Figure 5-17. These diagrams give a mental picture of atoms of various elements, although they are not drawn to scale. Remember that the nucleus, which contains most of the mass of atoms, occupies a very tiny volume in the center of the atom. Dalton did not know how atoms of one element differed from those of another, but we now know that the following differences exist:

1 Size (diameter)

2 Mass

3 Number of each subatomic particle

We have come a long way in this chapter. We have followed human discoveries and thought in developing a model of the atom and have formed a mental image of atoms. *Atomic theory*—the term for this view of matter—is perhaps the most basic model used by all modern sciences. You can now actually draw relatively accurate pictures of atoms simply by using a periodic chart. The following program will allow you to develop skill in calculating the number of each particle.

CALCULATING NUMBERS OF SUBATOMIC PARTICLES

Worked example

How many e^-, p^+, and n are in an atom of sodium ($^{23}_{11}Na$)?

1 Read its atomic number as 11.

2 Read its mass number as 23.

3 The number of e^- and p^+ would each be 11 (the same as the atomic number).

4 Calculate the number of neutrons:

$$\text{Number of neutrons} = \text{mass number} - \text{atomic number} = 23 - 11 = 12$$

5 There are, therefore, 11 e^-, 11 p^+, and 12 n in each atom of sodium.

6 You may want to draw a diagram of this atom to aid in developing a mental picture. This may make the idea of atoms a little less abstract.

Worked example

How many e^-, p^+, and n are in the most abundant isotope of titanium ($^{48}_{22}Ti$)?

1 Read its atomic number as 22; therefore each atom has 22 electrons and 22 protons.

2 Read its mass number as 48. This means that atoms of this isotope of Ti have a total mass of 48 amu and a total of 48 nuclear particles.

3 Calculate the number of neutrons:

$$\text{Number of neutrons} = 48 - 22 = 26$$

Problem

Calculate the number of e^-, p^+, and n in an atom of phosphorus ($^{31}_{15}P$).

Solution

1 Since the atomic number is 15, there are 15 e^- and 15 p^+ in each atom.

2 Read the mass number as 31.

3 The number of neutrons = 31 − 15 = 16.

Problem

How many of each subatomic particle would be in this isotope of oxygen: $^{18}_{8}O$?

Solution

1 The protons and electrons would each number eight.

2 The number of neutrons = 18 − 8 = 10.

Note The 18 is different from the mass shown on the periodic chart, since this isotope is present in low abundance in nature. More oxygen atoms are $^{16}_{8}O$.

Problem

Fill in the missing values in the following table. (Answers only are given.)

SYMBOL	NAME	NUMBER OF PROTONS	NUMBER OF ELECTRONS	NUMBER OF NEUTRONS
$^{7}_{3}Li$	______	—	—	—
___	Chromium-52*	—	—	—
$^{108}_{47}Ag$	______	—	—	—
___	______	12	—	12
$^{226}_{88}Ra$	______	—	—	—
___	Copper-64	—	—	—
^{37}Cl	______	—	—	20
___	______	—	16	16
___	______	—	1	0

Solution

SYMBOL	NAME	NUMBER OF PROTONS	NUMBER OF ELECTRONS	NUMBER OF NEUTRONS
Li	Lithium	3	3	4
Cr	Chromium	24	24	28
Ag	Silver	47	47	61
Mg	Magnesium	12	12	12
Ra	Radium	88	88	138
Cu	Copper	29	29	35
Cl	Chlorine	17	17	20
S	Sulfur	16	16	16
H	Hydrogen	1	1	0

STUDY GUIDE

Terms to learn

- Alchemy
- Atom
- Atomic mass
- Atomic number
- Atomic theory
- Atomic weight
- Canal rays
- Canal tube

*This denotes the isotope of Cr with a mass number of 52.

Cathode rays	Law of multiple proportions
Cathode-ray tube	Leucippus
Chadwick	Lucretius
Charge	Mass spectrograph
Dalton	Millikan
Democritus	Neutron
Electron	Plato
Electrostatics	Proton
Goldstein	Rutherford
Isotope	Subatomic particle
Law of conservation of matter	Thomson
Law of definite proportions	Wien

Skill practice problems

1 Define the following terms:

(a) Atomic number of an element
(b) Isotope
(c) A model
(d) Atomic weight of an element
(e) An atom
(f) Electrostatic attraction

2 Determine the number of each subatomic particle and make a drawing of atoms of each of the following types.

(a) $^{28}_{14}Si$ (silicon)
(b) $^{19}_{9}F$ (fluorine)
(c) $^{190}_{76}Os$ (osmium)

3 Complete the following statements or answer these questions.

(a) The first modern scientist to postulate atoms was ____________________.

(b) The fact that an oxide of nitrogen *always* contains 30 percent N and 70 percent O is called the law of ____________________.

(c) A pair of bromine atoms, one weighing 79 amu and the other 81 amu, would be called ____________________.

(d) If electrons were positive, how would they behave in a mass spectrometer?

(e) If protons weighed 6 amu instead of 1 amu, how would canal rays have behaved in a mass spectrometer?

(f) What was the significance of the gold foil experiment?

4 Why did Dalton believe that the concept of the atom explained the behavior of matter?

5 Why was the atomic view of matter discarded after Democritus postulated it? Why was it discarded again after Lucretius revived it?

6 What contribution did alchemy make toward the development of modern chemistry?

7 Why is atomic theory considered to be especially strong?

Goals

To understand the orderliness of matter

To understand the nature of light and color

To grasp the concept of energy levels within atoms

To understand the nature of atomic electronic orbitals and the standard way of expressing them

To understand the relationship between the periodic chart and atomic structure

To learn more about the nature of matter and to understand how common experiences (such as color) originate in atomic processes

6

ELECTRONIC STRUCTURE OF ATOMS

In Chapter 5 the existence and overall internal structure of atoms were examined, but nothing was said about the arrangement of the electrons within the atom. All the evidence suggests that it is this very thing—how the electrons are arranged—which determines the chemical behavior of all elements and the resultant nature of the compounds formed by the elements.

THE PERIODIC LAW AND THE PERIODIC CHART

Even before the structure of atoms had been elucidated, it had been observed that there were repeating sequences of chemical and physical properties among the elements. If one arranges the elements in order of their atomic numbers, one finds that every so often an element has properties quite similar to a previous one. If one continues along the series, a third element is found with properties similar to the first two.

Consider the elements with atomic numbers 3 (Li) through 10 (Ne). Each of them has distinctly different chemical and physical properties. The element with the atomic number 11 (Na) has properties similar to that with atomic number 3; the element with atomic number 12 (Mg) is quite similar to that with atomic number 4; the element with atomic number 13 (Al) is quite similar to that with atomic number 5; and so on. This pattern is shown in Table 6-1. If we continue still farther, we find that the element with atomic number 19 (K) is very much like the elements with atomic numbers 3 and 11 in its properties.

This situation is called the *periodic law*. It is stated concisely as follows:

> Chemical properties of the elements are a periodic (or repeating-sequence) function (result) of their atomic numbers.

TABLE 6-1

Comparison of some properties of elements with atomic numbers 3 to 18

Sequence 1		Sequence 2		
ATOMIC NUMBER	ELEMENT	ATOMIC NUMBER	ELEMENT	PROPERTIES
3	Li	11	Na	Metal, very reactive with water
4	Be	12	Mg	Metallic
5	B	13	Al	
6	C	14	Si	Forms chains of atoms
7	N	15	P	Each atom bonds with three H atoms.
8	O	16	S	Each atom bonds with two H atoms.
9	F	17	Cl	Poisonous, corrosive gas
10	Ne	18	Ar	Totally inert—no reaction with other elements

The recognition of this repeating-sequence relationship led to the organization of the elements into what is called the *periodic table* or *periodic chart*. This table has already been presented (see Figure 1-15). It is the most useful way of listing the elements because it gives a vast amount of chemical information.

The columns in the periodic table are called *groups* or *families* of elements. The members of families have similar chemical (and often physical) properties. These groups are designated according to the Roman numeral–letter combination at the top of the chart and also, in some cases, by names as well. Figure 6-1 gives these group names. For example, group IA, sometimes called the *alkali metals,* contains the elements Li, Na, K, Rb, Cs, and Fr, all of which are reactive, soft metals. The names of the families should be memorized. The horizontal groups are called *periods* and are designated by the numbers on the left of the chart. Elements with atomic numbers 19 to 36 are in period 4, for example.

The periodic chart also separates the elements into three general group-

FIGURE 6-1
Group names on the periodic chart.

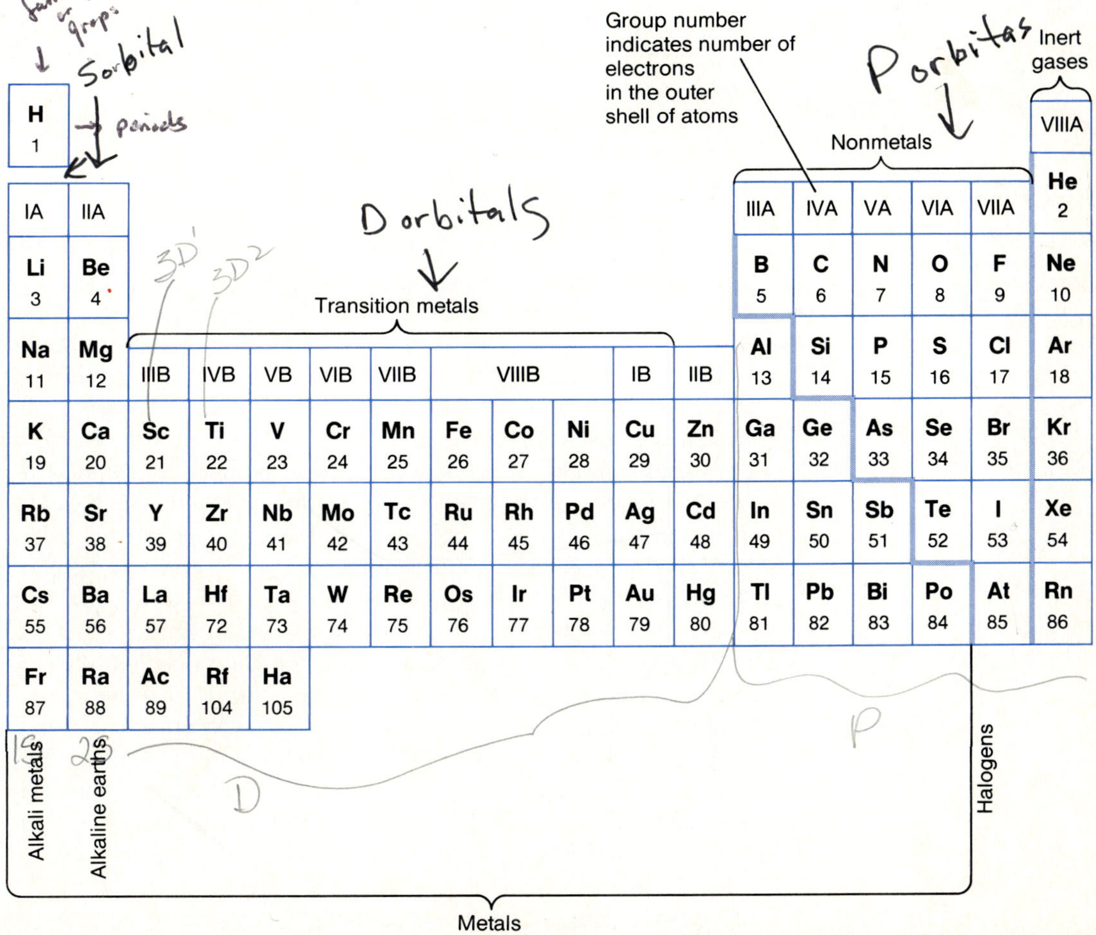

ings—metals, nonmetals, and inert gases—each of which has different characteristics. These divisions are also shown in Figure 6-1 and are indicated by the heavy dark lines. The shaded zigzag line that separates metals from nonmetals is called the *zintl border*. Metals are usually solids with a "metallic" shine (except mercury, Hg, which is a metallic liquid). They conduct electricity, can be shaped by hammering (i.e., are malleable), and can be drawn into wires without breaking (i.e., are ductile). Nonmetals are, in most cases, gases or brittle solids and do not conduct electricity.

You can, therefore, tell a good deal about an element just by its position on the periodic chart without having to memorize individual cases. The periodic chart will be referred to frequently throughout this text. Since it is so valuable, you might want to post a copy of it on your wall above your desk, get a wallet-sized one to keep with you at all times, and perhaps even get an adhesive one for the dashboard of your car! We will return to the study of the periodic chart and examine the various trends in it at the end of this chapter.

LIGHT AND COLOR

In order to understand the periodic chart and modern ideas about chemical behavior, one must first be familiar with some observations involving light and color. The explanation for these observations led to an understanding of the arrangement of electrons within atoms, which in turn allowed scientists to explain why each element has a unique set of chemical and physical properties. As has just been discussed, these properties are reflected in the organization of the periodic chart.

In the early 1900s, Niels Bohr, the Danish physicist, proposed an explanation for certain interesting experiments which had been carried out earlier. It had been found that when solutions of compounds of various elements are heated in an ordinary gas flame, each element imparts its own characteristic color to the flame. Focusing each element's color onto a prism divides that color into its component colors. The instrument used, basically a flame and a prism, is called a *spectroscope;* it is shown in Figure 6-2. The display of the separated component colors of an element is called an *atomic line spectrum* and is different for each element. Figure 6-3 shows the line spectra for several common elements, and Table 6-2 lists some flame colors.

The line spectrum or even the flame color observed by the naked eye can be used to identify a given element once the colors of the common elements are learned. For example, if an unknown solution gives a violet color when a wire dipped in it is placed in a flame, it contains potassium. Fireplace flames can be given beautiful (and romantic!) colors by pouring a solution of a lithium or strontium compound (purchased at a pharmacy or chemical supply company) on the wood and allowing the wood to dry before burning it. In a chemical laboratory the unknown solution can be heated and the line spectrum obtained. This spectrum can then be compared with the known line spectra of various elements, and the elemental composition of the unknown can be identified by the similarity of its spectrum to these known spectra.

This same emission of colored light from elements can be observed when

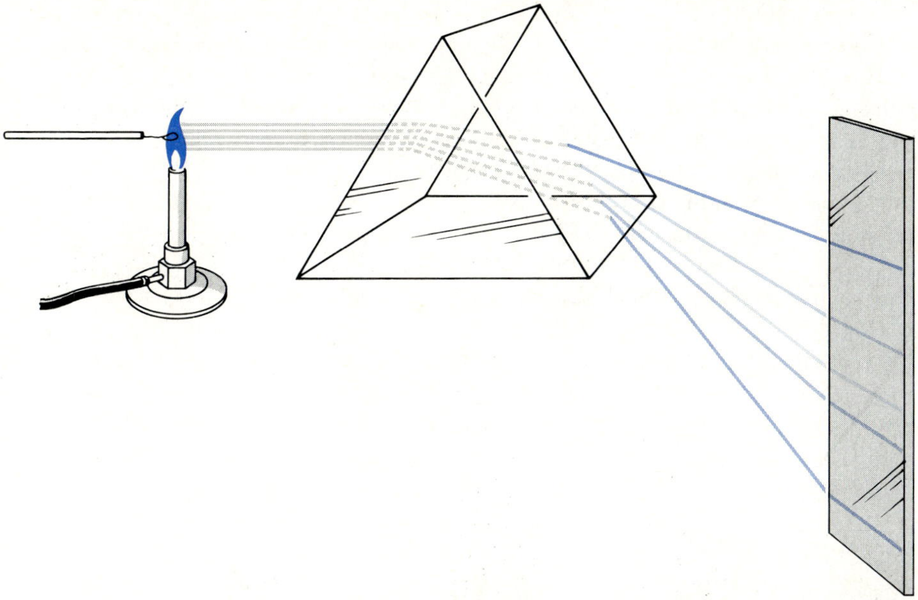

FIGURE 6-2
Spectroscope.

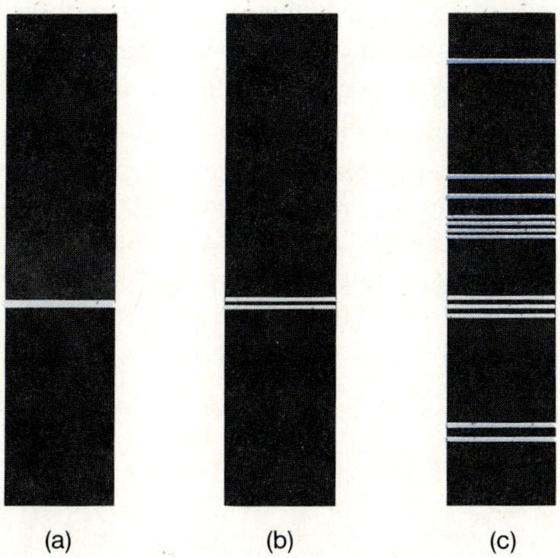

FIGURE 6-3
Atomic line spectra. (a) Sodium (crude). (b) Sodium (refined). (c) Potassium (refined).

TABLE 6-2
Flame colors of selected elements

ELEMENT	COMPOUND USED	COLOR
Sodium (Na)	Sodium chloride (table salt)	Orange-yellow
Lithium (Li)	Lithium nitrate	Brilliant red
Barium (Ba)	Barium nitrate	Yellow-green
Calcium (Ca)	Calcium nitrate	Orange-red
Potassium (K)	Potassium nitrate	Violet
Strontium (Sr)	Strontium nitrate	Crimson-red
Copper (Cu)	Cupric nitrate	Green
Boron (B)	Borax	Green

an electric spark is passed through a gaseous element. ''Neon'' lights represent a common application of this process. When the gaseous element neon is placed in a glass tube from which air has been removed, and an electric current is passed through this tube, the neon gives off a bright red glow, just as it would if placed in a flame. Other elements, such as argon and xenon, can be used to obtain lights of different colors.

In order to understand this phenomenon, and Bohr's interpretation of it, we must first understand the nature of light. For most purposes, light can be thought of as a traveling electrical field alternating from positive to negative values. In other words, it consists of waves of positiveness and negativeness (*not* positive and negative particles) moving through space. Humans experience light and color when the retina, optic nerve, and brain receptors respond to these oscillations of charge. Light can be characterized by two of its properties: wavelength, symbolized by the Greek letter lambda (λ); and frequency, symbolized by the Greek letter nu (ν). All light, regardless of its color, travels through space at a constant speed of 186,000 miles (mi), or 300,000,000 meters (m)—300,000 kilometers (km)—per second. The wavelength (λ) is the *distance* between crests of waves, and the frequency is the *number of times* each second a new wave moves past a fixed point or hits the optic nerve.

In order to understand this better, let's take a brief look at waves in water. Figure 6-4 shows a beach. The distance between wave crests is 15 ft, which would be the wavelength (λ). The waves are coming to the beach at a rate of five each minute, which is the frequency (ν) of the waves. Figure 6-5 illustrates this same idea for waves of light. At one instant, a wave of positiveness is interacting with the eye of the observer. At the next instant, a wave of negativeness comes into the eye. The wavelength (λ) is shown as the distance between crests, and the frequency is the number of waves that hit the eye per unit of time (usually each second).

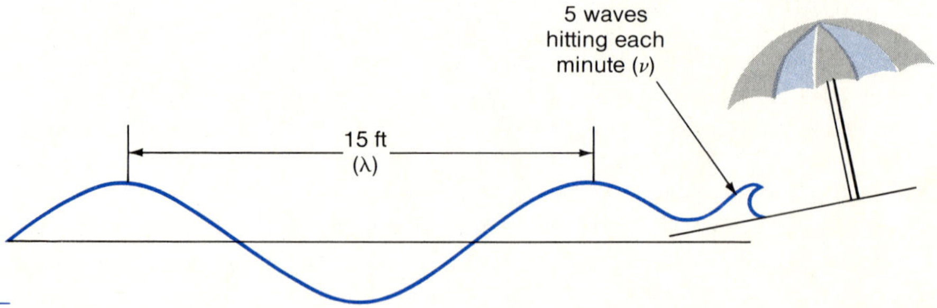

FIGURE 6-4
Water waves.

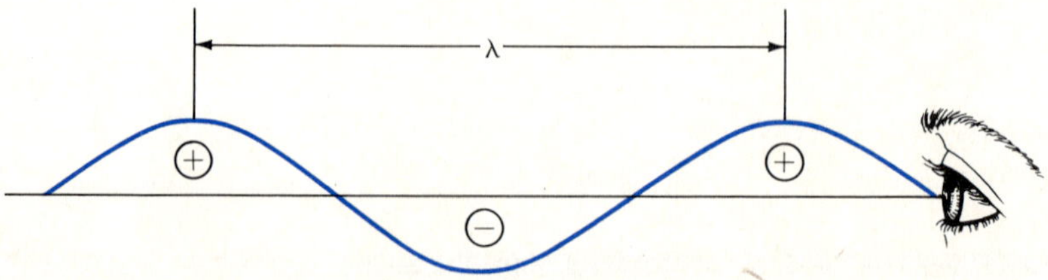

FIGURE 6-5
Light waves.

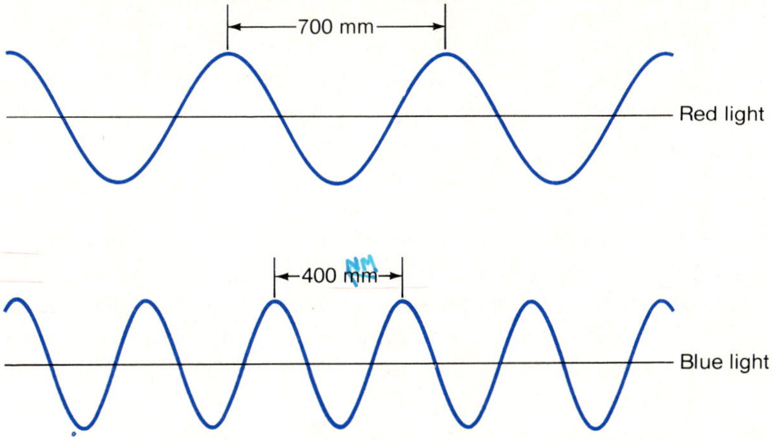

FIGURE 6-6
Difference in wavelength between red and blue light.

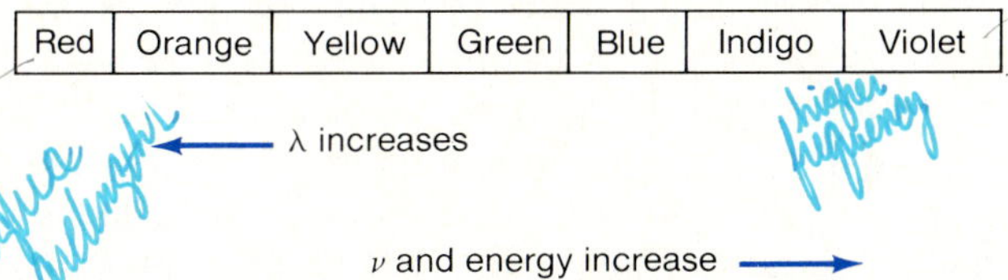

FIGURE 6-7
The visible spectrum.

Each color of the visible spectrum (i.e., the colors humans can see) has its own wavelength and frequency. Since all colors travel at the same speed, if the wavelength is greater, then the frequency must be smaller, and vice versa. This can be illustrated by two people walking side by side. If one takes large steps and the other small, in order for them to walk at the same speed, the one with the shorter steps must walk faster (have a higher frequency) than the other. Likewise, if a given kind of light has shorter distances between waves (lower λ) than another kind of light, then it must have more waves per second (higher ν) for their speeds to be the same. As shown in Figure 6-6, red light has a greater λ and lower ν compared with blue light, which has a shorter λ and higher ν. The red light (with a long wavelength) is hitting the eye at a relatively low frequency. The blue light (with a shorter wavelength) is hitting with a higher frequency. Color, then, is the human optical system's response to different frequencies of light. When one looks at a red sweater, the light coming from it hits the eye at low frequency and the brain interprets that as ''red.'' Each color in the spectrum, which is shown in Figure 6-7, has its own λ and ν. As one goes toward the violet end, the λ becomes shorter and the ν higher.

Another concept that is of considerable importance is the *energy* of various colors. It can be shown that violet light has more energy per photon* than red light. As the frequency increases, so does the energy of that light.

* See the end of this section for a description of a photon of light.

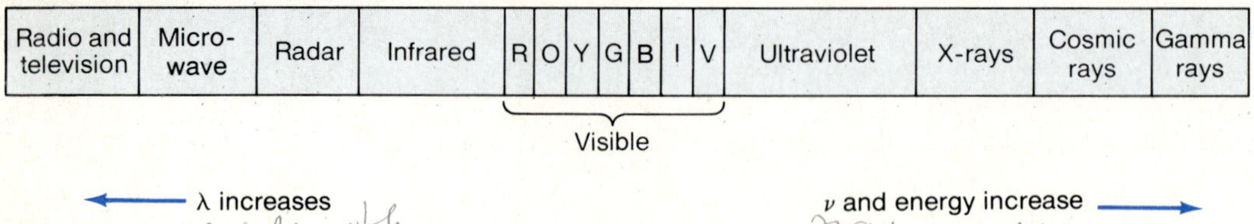

FIGURE 6-8
The entire electromagnetic radiation spectrum.

Visible light is part of a larger spectrum called the *electromagnetic radiation spectrum*. There are wavelengths shorter than the wavelength of violet light and longer than that of red light. Humans cannot see these kinds of light, but they do exist and can interact with matter and cause various responses, even though the human retina is not sensitive to them. Figure 6-8 shows the entire electromagnetic radiation spectrum, along with the changes in wavelength, frequency, and energy within this spectrum.

White light is a mixture of approximately equal intensities of all wavelengths of visible light. When white light shines on an object and the materials in that object *absorb* certain of the wavelengths but not others, that object *reflects* the color complementary to the one being absorbed. Figure 6-9 shows a ''color wheel'' in which the complementary colors are shown opposite each other. If white light from the sun or an electric light bulb shines on a sweater and the dye absorbs only the red light, then the green (its complement) is reflected, the eye experiences the wavelength for green, and the sweater looks green. The intensity of the color is dependent on the amount of dye and thus on the proportion of the red waves absorbed. If the object selectively absorbs no one wavelength, it reflects the whole spectrum and appears white. If all the wavelengths are absorbed equally, the object looks black or gray.

Other areas of the spectrum are of interest also. Sunlight is composed mainly of infrared, visible, and ultraviolet rays. The infrared light is invisible

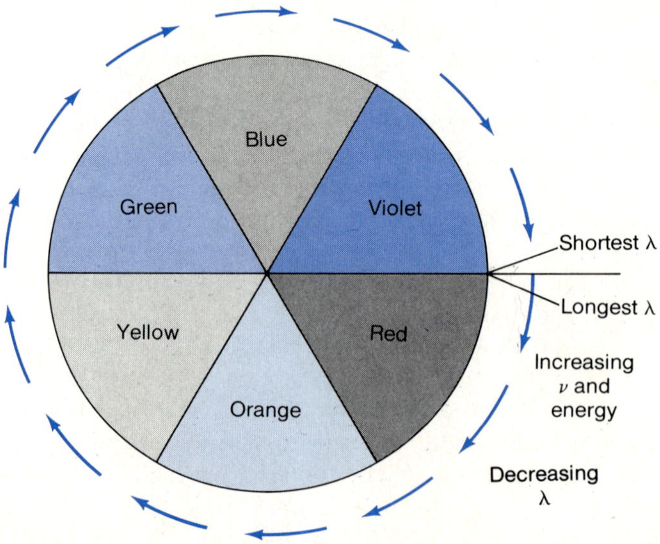

FIGURE 6-9
Position of complementary colors on the color wheel.

to the eye but is experienced (or "seen," if you will) by the skin as *heat*. The ultraviolet rays are invisible to the eye, but because of their relatively high energy they can cause the skin to change color (tan). If they are received in too large a quantity, they can cause a severe sunburn. (See Chapter 16 for more on this subject.)

In a few cases light acts as though it were composed of massless particles (an apparent contradiction in terms). These are called *photons*. That is, light sometimes behaves like small *packages* (quanta) rather than continuous waves. Putting these two ideas together, we arrive at a paradox called the *wave-particle duality theory of light*. Most of the time, light behaves like a wave, but sometimes it behaves like particles; perhaps, then, it has a dual nature which is quite difficult to conceptualize. Everything else we experience is either a wave (for example, water or sound) or an object (for example, balls, books, or people). Nevertheless, the evidence demands that we accept this theory of duality even if we cannot readily understand it.

ELECTRONIC ENERGY LEVELS IN ATOMS

Now that we understand something about light and color, let's return to Bohr's flame colors. Remember that different atoms give *different colors* which have *different energies*. Bohr reasoned that if various atoms are able to give off different colors of light having different energies, then they must have different *energy levels* within them. He further postulated that the electrons are in those energy levels and could be only at certain discrete locations within the atom. In the space around the nucleus, he reasoned, there must be paths (orbits); the electrons would circle the nucleus, like planets around the sun, within these orbits.

This is called the *quantum concept:* the location of the electrons, and thus their energy, is quantitized. This means that they cannot float around anywhere but must have only certain limited positions. This idea is not as foreign to everyday experience as it may seem at first. Books on a bookshelf can have only certain locations. One can "place" a book halfway between two shelves, but it will immediately fall to the lower one and stay there. Just as the book cannot rest between shelves but must be at one level or another, so the electron cannot be *between* orbits but must be *in* one of them. (See Figures 6-10 and 6-11.)

Bohr postulated that when the atoms of a particular element are placed in a flame, the heat of the flame causes the electrons to jump to a larger (higher-energy) orbit. When this *excited* atom (one having excess energy) comes to the edge of the flame, it cools down and the electron drops back to its normal orbit again (its *ground state* or normal state). When the electron drops down, it gives off the energy equal to the difference between the two orbits but gives it off as *light* rather than as heat. This is diagramed in Figure 6-11. In Figure 6-11*b*, the difference in distances, and thus in energy, is greater than in the atom shown in Figure 6-11*a*. When the electron in the excited state in Figure 6-11*b* drops back to its normal orbit, it gives off *more* energy than the electron in Figure 6-11*a*. Since this energy is emitted as light, the light will have a

FIGURE 6-10

Right: Quantitized energies of books on a bookshelf.

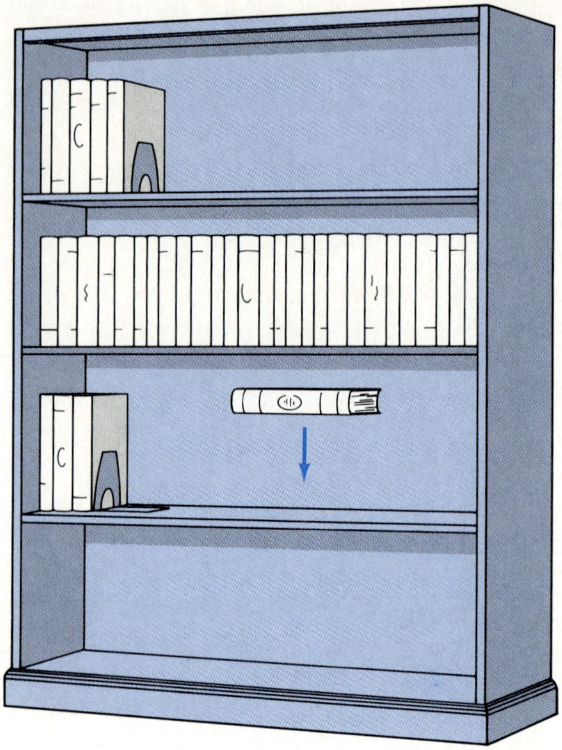

FIGURE 6-11

Below: Atomic source of flame colors.

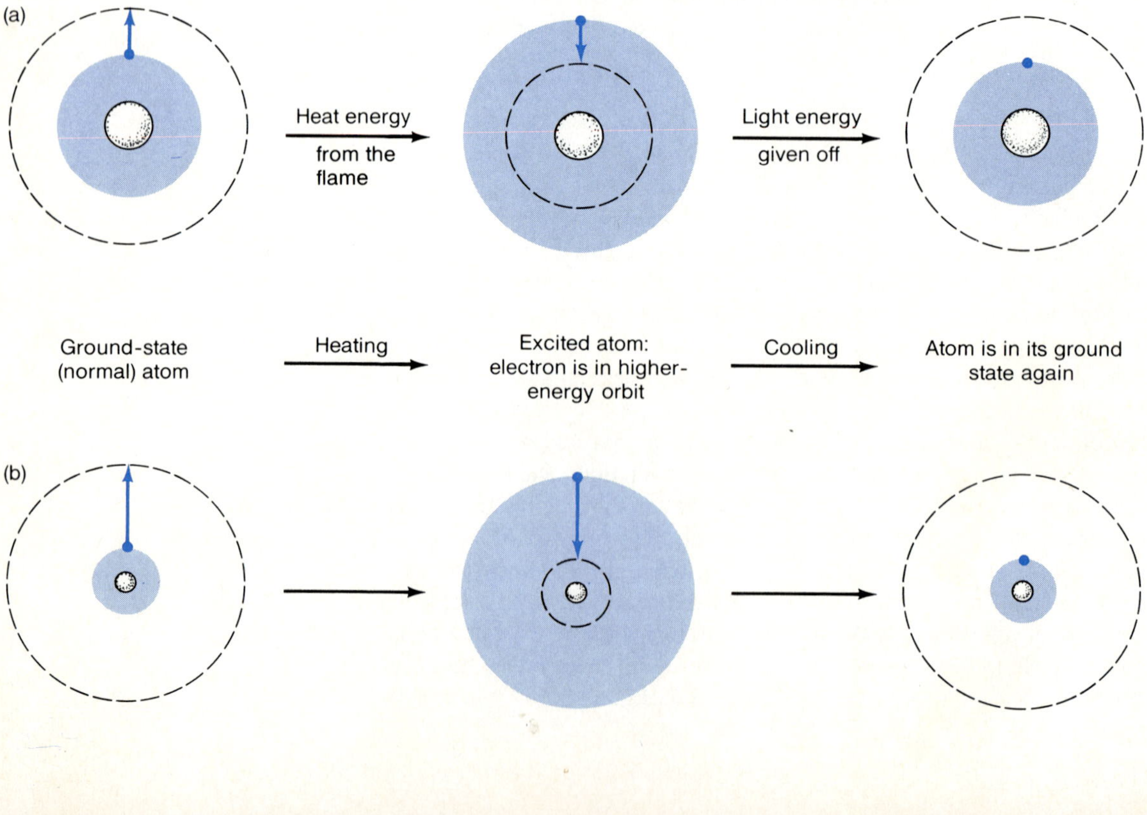

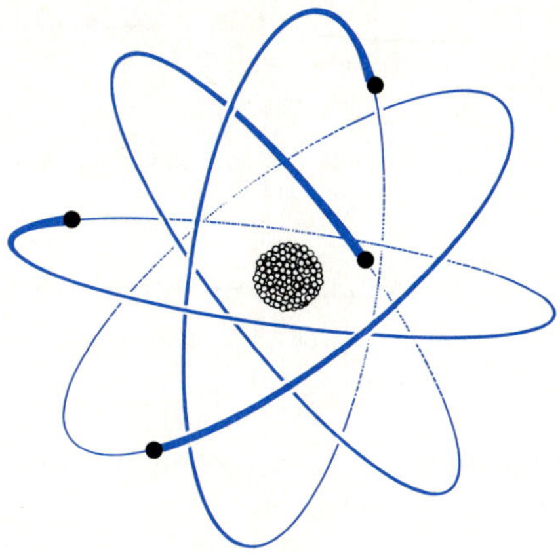

FIGURE 6-12
The Bohr model of the atom.

higher energy and a shorter wavelength (higher frequency) and be a different *color* than the light emitted by the atom in Figure 6-11*a*.

It was concluded that since each element had its own unique atomic line spectrum, indicating unique transitions of the electrons between energy levels, atoms of each element must have a unique set of electronic orbits or electronic energy levels. Later scientists worked with more sensitive spectroscopes and discovered component lines in the atomic line spectra. Where Bohr had seen a wide, colored spot, others saw a series of closely spaced lines and therefore proposed that there must be sublevels within the major orbits. The familiar symbol for nuclear energy, shown in Figure 6-12, is a result of these ideas. This is called the *Bohr model of the atom*. As we will see later, the Bohr model is not entirely correct; but it does give a crude mental picture of the idea of quantitized energy levels. (Refer back to Figure 6-3 for a comparison of the two types of line spectra.)

ATOMIC ORBITALS

Subsequent work by chemists, physicists, and mathematicians has led to a better understanding of the distribution of electrons within atoms. The reasoning involved in writing detailed electronic structures and the skills required are covered in Appendix D. In this chapter, only the basic ideas, the results of the process, and the conclusions that will be used later in the text are presented.

Electrons can be visualized as existing in three-dimensional volumes around the nucleus. These volumes are called *orbitals* and should be understood as different from orbits (which are circular lines). Each orbital has a different size, shape, and orientation within the atom, and each is given a number-letter designation. One can simply think of the empty space around the nucleus of an atom as having within it volumes which can be occupied by electrons. These are somewhat like roads or streets which can be occupied by cars. While there are a large number of such orbitals, real atoms under ordinary conditions have a relatively small number of them occupied by electrons.

A list of the standard designations of the first 18 atomic orbitals (i.e., the 18 closest to the nucleus) follows:

$1s$	$2s\ \ 2p2p2p$	$3s\ \ 3p3p3p\ \ 3d3d3d3d3d$	$4s\ 4p4p4p$
First level	Second level	Third level	Fourth level

The number indicates the size (i.e., the diameter) of the orbital, and the letter indicates the shape. The *s* orbitals are spherical, and the *p* orbitals are dumbbell-shaped, as shown in Figure 6-13. The $1s$ is small and spherical, the $2s$ is large but still spherical, and the $2p$ orbitals are dumbbell-shaped and the same size as the $2s$. The *d* orbitals are more complex. These are shown diagramatically in Figure 6-13.

The group of orbitals of a given size is called a major *level* or *shell*. There is only one orbital, the $1s$, at the first size, or level. The second size, or level, has four orbitals, and the third has nine. Each orbital can hold one or two electrons. Notice that there are three *p* orbitals of a given size and five *d* orbitals in the third level. *Each* can hold up to two electrons. Electrons occupy the orbitals to the left in the list (smaller size and simpler shape) first, and fill the higher orbitals only after the others are filled. The ease of filling an orbital decreases as one goes to the right in the list. In other words, for electrons to be in orbitals to the left in the list they need *lower energy* than electrons in orbitals to the right in the list, which need *higher energy*.

Here, we will not be concerned with how one determines the location of electrons in each atom. (See Appendix D for that if you are interested or are required to do so.) But we shall look at several examples. These examples use a standard way of indicating electronic structures (or arrangements); that is, a superscript is used to indicate how many electrons are in that orbital:

Be $1s^2,\ 2s^2$

C $1s^2\ 2s^2\ 2p^1 2p^1$

Fe $1s^2\ 2s^2\ 2p^2 2p^2 2p^2\ 3s^2\ 3p^2 3p^2 3p^2\ 3d^2 3d^1 3d^1 3d^1 3d^1\ 4s^2$
or $1s^2\ 2s^2\ 2p^{6*}\ 3s^2\ 3p^{6*} 3d^{6*} 4s^2$

Beryllium atoms, for example, have two electrons in the $1s$ orbital and two in the $2s$, as shown above. Carbon atoms have two electrons in the $1s$ orbital, two in the $2s$ orbital, and one in each of two $2p$ orbitals. The larger atoms, such as Fe, have more electrons and a more complex electron structure. The electron structure of three members each of groups IA and VIIA is shown in Figure 6-14.

*The *p* orbitals and *d* orbitals can be designated collectively.

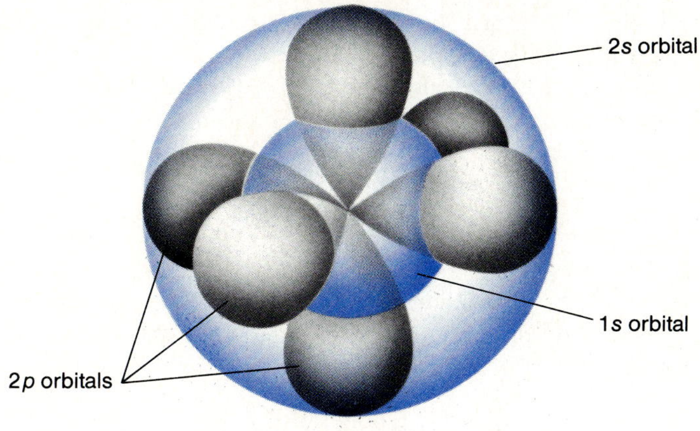

FIGURE 6-13
1s, 2s, and 2p atomic orbitals. (There also exist "f" orbitals, but this text will not treat them.)

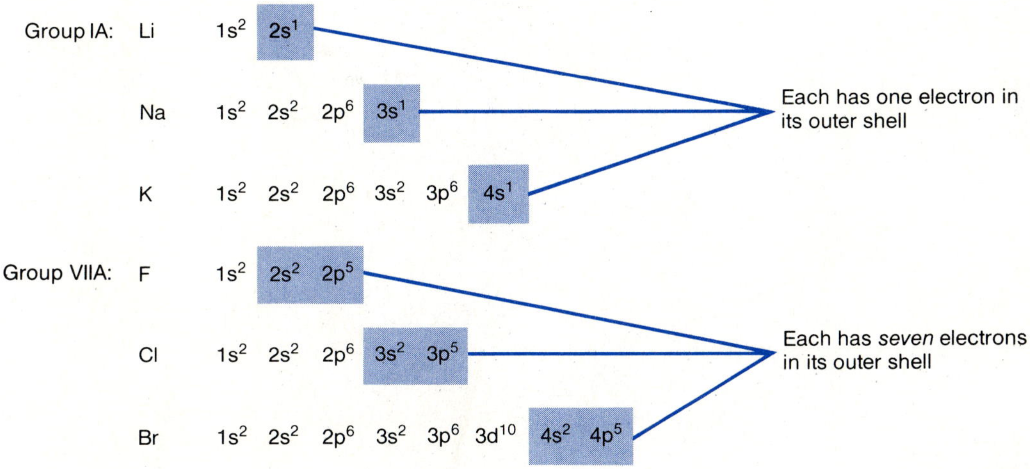

FIGURE 6-14
Electron structure of three members of group IA and three members of group VIIA.

ELECTRONIC STRUCTURE AND THE PERIODIC CHART

From the structures given in the previous section, one can see that the electron configuration of the *outermost* levels of atoms of K, Li, and Na is the same: each has one electron in an *s* orbital. These three elements are all in the *same* family (group IA) of the periodic chart. The outer-shell configurations of Cl, Br, and F are also the same: two electrons in an *s* orbital and five in *p* orbitals. These are also in the same family (group VIIA). *All members of each family have identical outer-shell configurations.* It is for this reason that they have similar chemical properties. *The chemical properties of any element are a direct result of the electron structure of the atoms of that element.* This is the most fundamental principle of modern chemistry, and it explains the organization of the periodic chart. Li, K, and Na are all soft, silvery metals that react violently with water. They combine with chlorine to form compounds that taste salty. F, Cl, and Br are all nonmetals that react with hydrogen to form compounds that taste sour (these compounds are acids—see Chapter 10). The *group numbers* on the periodic chart indicate the *number of electrons* in the outer shell of atoms of the elements in that family, and the period numbers indicate the highest occupied shell.* For example, Na and Cl are both in period 3, and thus in each the third shell is the largest one occupied with electrons.

There are two other kinds of information stored in the organizational features of the periodic chart. The first type indicates the relative size (diameter) of atoms. As one moves *down* a given family, a whole new shell is occupied by electrons. For example, in group IA, Na has its last (outermost) electron in the 3*s* orbital, K in the 4*s*, Rb in the 5*s*, etc. Since the 4*s* orbital is larger than the 3*s*, K atoms are larger than Na atoms. Thus, as one goes *down* a family, the atomic diameter always *increases*.

As one goes from *left to right* in a given period, the atomic weight of course increases, but in most cases the atomic size *decreases*. In period 2, for example, neon atoms are smaller than those of fluorine, which in turn are smaller than oxygen atoms. There are a number of exceptions to this trend, but it generally holds except in the central portion of the chart. The explanation for this pattern is that as one goes to the right in a period, atoms have more protons in the nucleus (for example, oxygen has eight protons and fluorine has nine). The nine positively charged protons of fluorine are pulling the nine negatively charged electrons toward the nucleus more effectively than the eight protons of oxygen. *Each* electron of oxygen feels the full force of fewer protons than *each* electron in fluorine feels. These trends in atomic size are shown in Figure 6-15. As one goes *down* and to the *left,* diameter *increases*. This means that calcium atoms are smaller than those of potassium and that selenium atoms have a greater diameter than those of oxygen.

The second type of information indicates the relative *electronegativity* of atoms, that is, the attraction of atoms for electrons. The atoms of inert gases do not tend to gain or lose electrons, while those of metals lose electrons and nonmetals gain electrons. As one goes *up* and to the *right* in the periodic chart, electronegativity *increases*. (The last group—inert gases—are not included in this trend, however.) Fluorine (F) is thus the *most* electronegative element, and francium (Fr) is the *least* electronegative. This trend is also shown by the arrows in Figure 6-15. Using this rule one can determine, for example, that oxygen (O) is more electronegative than sulfur (S) and sodium (Na) is less electronegative than magnesium (Mg). This means that oxygen gains electrons from some other source more easily than sulfur does and that sodium gives up its electrons more easily than magnesium does (magnesium holds onto its electrons more tightly). We will be using these concepts in Chapter 7, when we consider reactions of elements with each other.

---

*Strictly speaking, this applies only to the groups designated with the letter A: IIA, VIIA, etc.

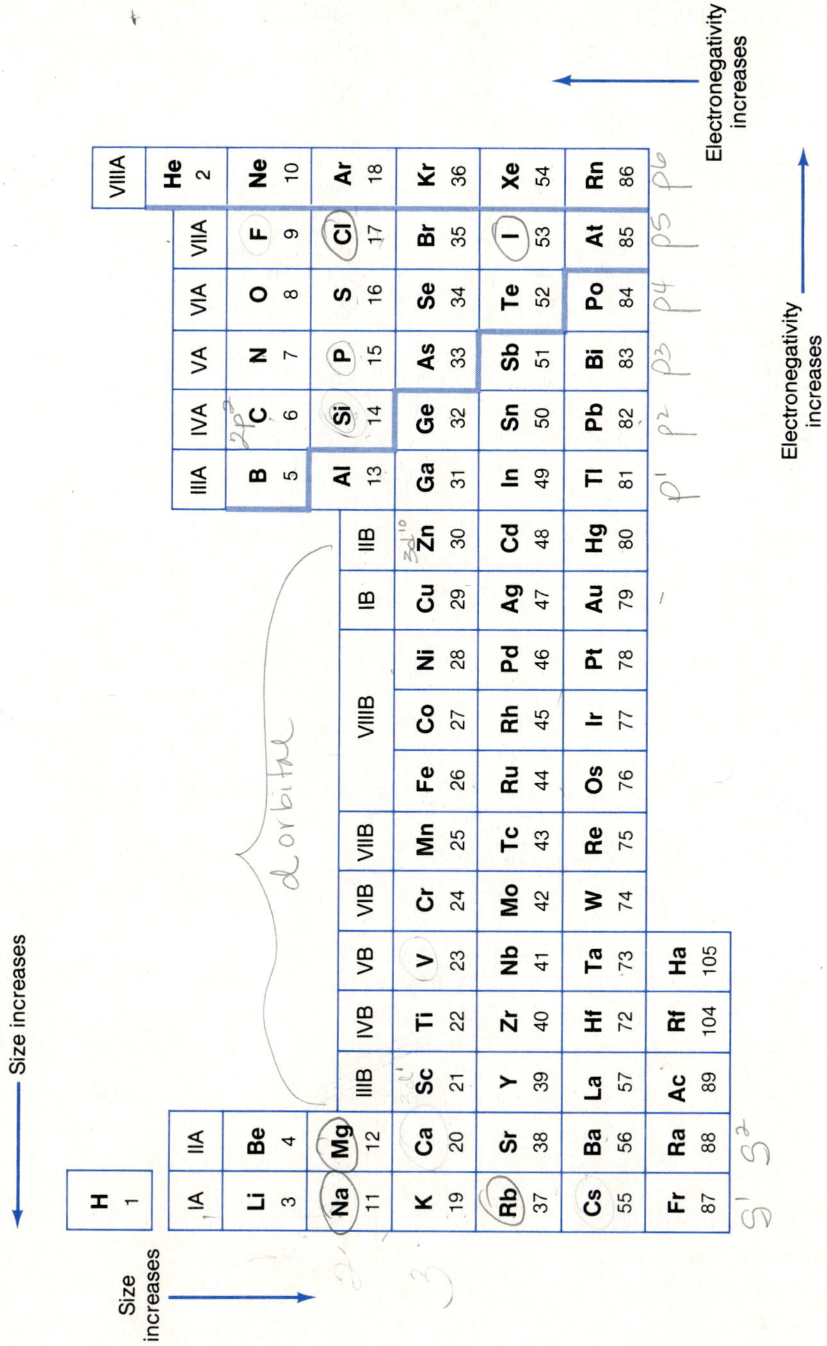

FIGURE 6-15
Summary of trends on the periodic chart.

PRODUCTION OF COLORED LIGHT

As was mentioned previously, when the various elements are heated, they give off different colors of light. Most flames are yellow because of the presence of the ubiquitous element sodium, or because of glowing particles of carbon. In order to produce a truly beautiful red flame one can purchase a lithium compound (such as lithium nitrate), dissolve it in water, and pour the solution on wood or other combustible material that is to be burned later in the fireplace. After the wood or other combustible material has dried, burning it will produce a flame that is red. A light-green flame can be produced by soaking wood in a solution of household borax and allowing the wood to dry before burning it.

The brilliant colors seen in fireworks displays are the result of mixing salts of various elements with the explosive materials. The explosion itself—which produces the heat and noise—is usually produced by black powder, the same material used throughout history for explosives and fireworks. Most gunpowder and commercial explosives now use safer mixtures, but for fireworks displays the best explosive is still black powder. Black powder is a mixture of potassium nitrate (KNO_3), sulfur (S), and carbon (C) from charcoal. Because this material is highly susceptible to detonation by sparks, it is dangerous to manufacture and to prepare (at least by inexperienced people).

If one wants a display with brilliant colors, inorganic salts are added to the black powder or new mixtures are prepared with similar ingredients. Sr is used for red (in the form of $SrCO_3$); Ba, in the form of $Ba(NO_3)_2$, is used for green; and $CuCO_3$ is used when a blue color is desired. Brilliant yellow is produced by using small pieces of iron or carbon, which burn like a shower of meteors as they fall through the air. The bright white light in some fireworks is produced by pieces of aluminum (Al) or magnesium (Mg). These are ignited by the powder explosion and burn with an exceptional brightness as they arch through the sky.

This information is given to show the atomic origin of the colors we enjoy. *It should not be used by anyone who is not trained in the use of fireworks and explosives*. Explosives themselves are dangerous, and the addition of small amounts of foreign materials to explosives can lead to disastrous results.

STUDY GUIDE

Terms to learn

- Atomic line spectrum
- Bohr
- Color
- *d*-type orbital
- Electromagnetic spectrum
- Electronegativity
- Electronic structure (arrangement)
- Energy
- Energy level
- Family of elements
- Flame colors
- Frequency
- Group of elements
- Infrared

Light
Orbit
Orbital
p-type orbital
Periodic chart
Periodic law
Period of elements
Photon
Quantitized
s-type orbital
Shell
Spectroscope
Ultraviolet
Waves
Wavelength
Wave-particle duality

Skill practice problems

1 Which of the following is *larger?*

(a) Ca or Sr
(b) Cu or Au
(c) Al or Cl
(d) Ga or P

2 Arrange the following series in *increasing* order of size (smallest first):

Si, F, Ca, V, Cs

3 Decide which would *give up* its electrons the most *easily*.

(a) N or As
(b) Sn or I
(c) N or Al
(d) B or F
(e) Li or Ba

4 Arrange the following in order of *decreasing* electronegativity:

I, Na, Cl, Mg, Rb

5 Complete the following statements or answer the questions.

(a) In the Bohr model, the electron is considered to be a (wave) (particle) moving in an (orbit) (orbital). (Circle each correct choice.)

(b) Lithium compounds give off what color when heated?

(c) If the dye in the cover of this book absorbed yellow light, what color would the book be?

(d) Which of the following are metals?

Li, S, Xe, Pb, C

(e) All the following, except one, have similar chemical properties. Which one is the exception?

Ca, Mg, K, Be, Sr

(f) The element iodine (I) belongs to what family?

Number designation: ____________ Name: ____________

(g) Who first postulated electronic energy levels in atoms?

(h) Which has the *shortest wavelength:* infrared, x-ray, orange, or radio waves?

(i) Which has the *lowest frequency:* infrared, x-ray, orange, or radio waves?

(j) Which has the *highest energy:* infrared, x-ray, orange, or radio waves?

(k) How many electrons are in the outer shell (level) of atoms of P, Si, and B? (Refer to the periodic chart.)

6 Give the group and period in the periodic chart in which each of the following elements is found:

(a) Carbon (C)
(b) Chromium (Cr)
(c) Silver (Ag)
(d) Arsenic (As)
(e) Argon (Ar)

7 Briefly explain *why* Bohr postulated discrete electronic orbitals in atoms.

8 Brilliant white fireworks are produced by what material?

Goals

To develop skill in reading chemical symbols

To understand why elements combine to form compounds

To understand how physical properties are a result of types of bonding

7

CHEMICAL CHANGE

In Chapter 6, we studied the chemical elements and the atoms of which they are composed. This chapter will be devoted to the study of the ways in which atoms of these elements interact with each other to form *compounds*. It will also present the symbols that are used by chemists to express these chemical changes.

THE LANGUAGE OF CHEMISTRY

Chemical equations

The elements magnesium (a shining metal) and oxygen (a gas) react with each other when exposed to a lighted match or a spark, and a new substance is formed. The new substance is a white powder and has physical, chemical, and physiological properties different from those of either the magnesium or the oxygen. It is said to be a *compound,* since it is not a mixture of the two elements composing it but rather a new substance. A compound is any pure material composed of two or more elements. It cannot be separated into the elements again by ordinary physical methods, but can, in theory at least, be changed back by chemical methods (for example, the white powder could, in theory, be changed back into magnesium and oxygen).

The standard way of expressing such a change is as follows:

$$2Mg + O_2 \longrightarrow 2MgO$$

This is called a *chemical equation*. The various parts of this expression are shown in Figure 7-1. The materials to the left of the arrow are called *reactants* (since they react to form the new materials); the materials on the right are called *products* (since they are the product, or the result, of the chemical change).

The physical picture of what is happening to the atoms is shown below. It indicates approximately what one would see as the reaction occurred, if atoms, molecules, or ions were visible. *Molecules* are units with two or more atoms linked together, and *ions* are charged atoms or groups of atoms. In this example, ions are formed:

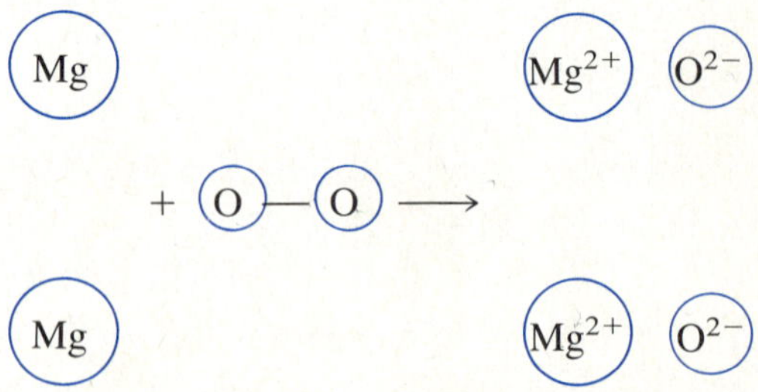

What one observes in the laboratory is shown in Figure 7-2.

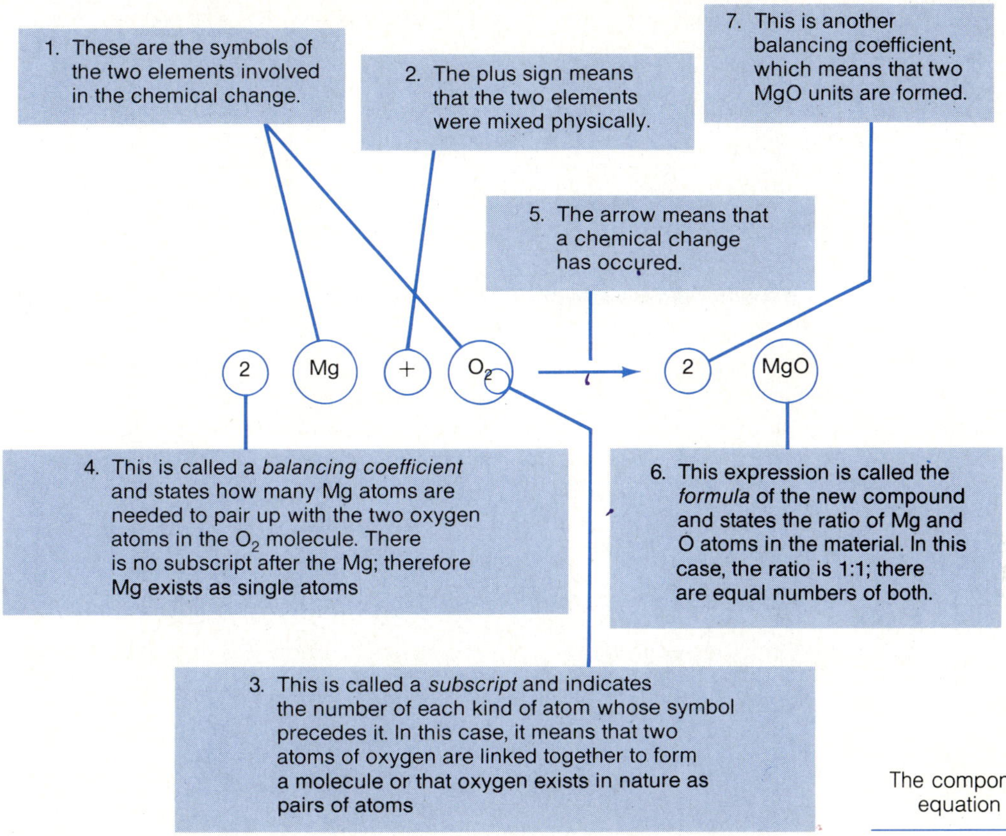

FIGURE 7-1
The components of a chemical equation and their meaning.

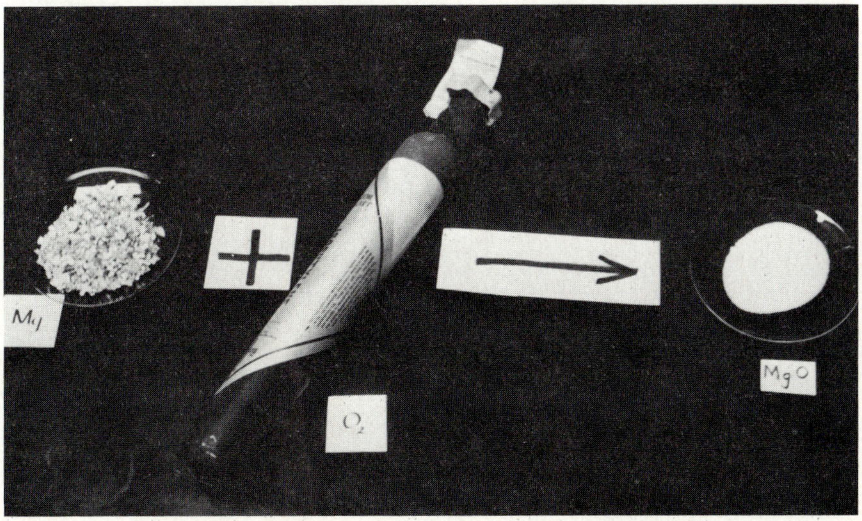

FIGURE 7-2
Magnesium metal (Mg) and oxygen gas (O_2) combine to form a white powder, magnesium oxide (MgO).

TABLE 7-1

Elements that exist in diatomic form under ordinary conditions

ELEMENT NAME	ATOMIC SYMBOL	MOLECULE SYMBOL (CHEMICAL FORMULA)
Hydrogen	H	H_2
Oxygen	O	O_2
Nitrogen	N	N_2
Fluorine	F	F_2
Chlorine	Cl	Cl_2
Bromine	Br	Br_2
Iodine	I	I_2

TABLE 7-2

Elements that exist as polyatomic molecules

ELEMENT NAME	ATOMIC SYMBOL	MOLECULE SYMBOL (CHEMICAL FORMULA)
Phosphorus	P	P_4 (tetraatomic)
Arsenic	As	As_4 (tetraatomic)
Sulfur	S	S_8 (octaatomic)

IA	IIA	IIIB	IVB	VB	VIB	VIIB	VIIIB			IB	IIB	IIIA	IVA	VA	VIA	VIIA	0
1 H_2																	2 He
3 Li	4 Be											5 B	6 C	7 N_2	8 O_2	9 F_2	10 Ne
11 Na	12 Mg											13 Al	14 Si	15 P_4	16 S_8	17 Cl_2	18 Ar
19 K	20 Ca	21 Sc	22 Ti	23 V	24 Cr	25 Mn	26 Fe	27 Co	28 Ni	29 Cu	30 Zn	31 Ga	32 Ge	33 As_4	34 Se	35 Br_2	36 Kr
37 Rb	38 Sr	39 Y	40 Zr	41 Nb	42 Mo	43 Tc	44 Ru	45 Rh	46 Pd	47 Ag	48 Cd	49 In	50 Sn	51 Sb	52 Te	53 I_2	54 Xe
55 Cs	56 Ba	57 La	72 Hf	73 Ta	74 W	75 Re	76 Os	77 Ir	78 Pt	79 Au	80 Hg	81 Ti	82 Pb	83 Bi	84 Po	85 At	86 Rn
87 Fr	88 Ra	89 Ac	104														

58 Ce	59 Pr	60 Nd	61 Pm	62 Sm	63 Eu	64 Gd	65 Tb	66 Dy	67 Ho	68 Er	69 Tm	70 Yb	71 Lu
90 Th	91 Pa	92 U	93 Np	94 Pu	95 Am	96 Cm	97 Bk	98 Cf	99 Es	100 Fm	101 Md	102 No	103 Lr

FIGURE 7-3
Periodic chart with the diatomic elements and polyatomic elements indicated.

Chemical formulas

Let's look at each aspect of the chemical equation in more detail. Oxygen exists as pairs of atoms called *molecules*. Since there are two atoms in each molecule, oxygen molecules are said to be *diatomic* (*di* = 2). There are seven elements that exist in diatomic form under ordinary conditions. These are listed in Table 7-1. You should note that, except for hydrogen, all these appear in an angular array in the upper right part of the periodic chart—a fact which can be used as a memory aid.

Of the elements commonly encountered, there are three that exist as molecules containing more than two atoms (*polyatomic* molecules). These are phosphorus, arsenic, and sulfur; see Table 7-2.

The ten elements listed in Tables 7-1 and 7-2 should be memorized and, when used in a chemical equation, expressed in the proper way. Most other elements should be expressed as single atoms. See Figure 7-3.

The *formula of a compound* expresses the number and type of atoms in the compound. The *empirical formula* is the simplest ratio of atoms. The *molecular formula* is the actual number of each type of atom in the molecule. For example, the molecular formula of benzene is C_6H_6, which means that there are six atoms of carbon and six atoms of hydrogen in a molecule of benzene. The ratio of carbon atoms to hydrogen atoms is 1 to 1; therefore, the empirical formula of benzene is simply CH. Other examples are shown in Table 7-3.

TABLE 7-3

Examples of molecular and empirical formulas

MOLECULAR FORMULA	DIAGRAM OF MOLECULE	EMPIRICAL FORMULA
H_2O	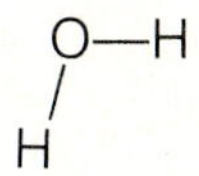	H_2O*
C_2H_6	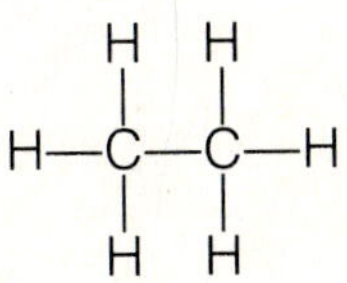	CH_3
CH_4O	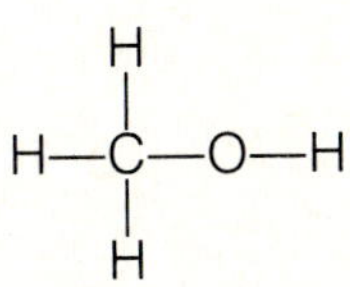	CH_4O
$C_2H_6O_2$	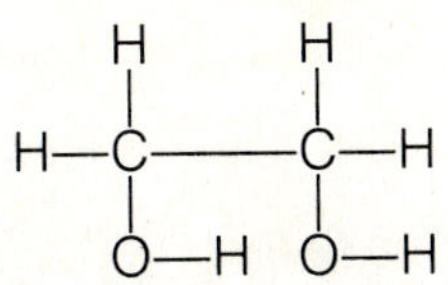	CH_3O

*Two hydrogen atoms to one oxygen atom is the simplest ratio possible for this molecule; therefore, the empirical and molecular formulas are the same.

Balancing equations

Once we know the molecular or atomic forms of the reactants and products and are given, have memorized, or have determined (we will see how this is done in Chapter 10) the formulas of the compounds, an equation describing the chemical reaction can be written. At this point, however, the equation is said to be *unbalanced*. Below is the unbalanced chemical equation for the reaction of magnesium with oxygen:

$$Mg + O_2 \longrightarrow MgO$$

The formulas of the reactants and products are expressed correctly, but the number of atoms of each type is *not* the same on both sides of the arrow. Therefore the equation written in this form is not a totally correct expression of what is occurring. *Balancing coefficients* must be placed in front of the chemical symbols in the equation so that the same number of each type of atomic symbol appears on both sides of the equation. We can, in the process of balancing the equation, place numbers in front of any formula, but we *cannot* change any of the subscripts already in the formulas. That is to say, we cannot change the ratios of atoms in a compound as expressed by the formula. This is controlled by nature alone. The balancing coefficients are also controlled by nature; and in balancing an equation, we are simply determining what these coefficients are. While more sophisticated methods are needed in some cases, most equations can be balanced by inspection and use of a little reasoning. The following activities will aid in learning this skill.

BALANCING CHEMICAL EQUATIONS

Worked example

Balance the following equation:

$$Mg + O_2 \longrightarrow MgO$$

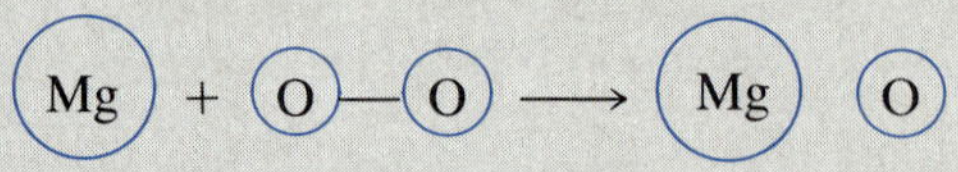

(Diagram of relationship of atoms)

1 There is one Mg atom on the left and one Mg atom on the right.

2 There is a molecule of O_2 on the left (i.e., two oxygen atoms total), but on the right there is only one oxygen atom.

3 To get two oxygens on the right, we must have two units of MgO. Remember that we cannot change the 1 : 1 ratio of Mg to O. We therefore place a 2 in front of the MgO.

$$Mg + O_2 \longrightarrow 2MgO$$

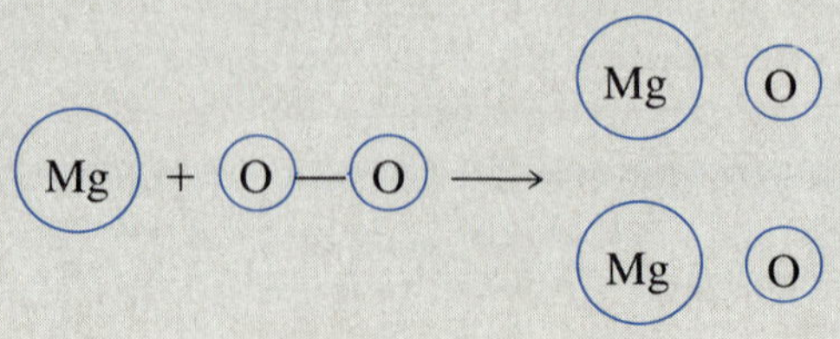

4 There are now two oxygens on each side, but we now have two Mg on the right and only one on the left.

5 By placing a 2 in front of the Mg on the left, we have completed the balancing process.

$$2Mg + O_2 \longrightarrow 2MgO$$

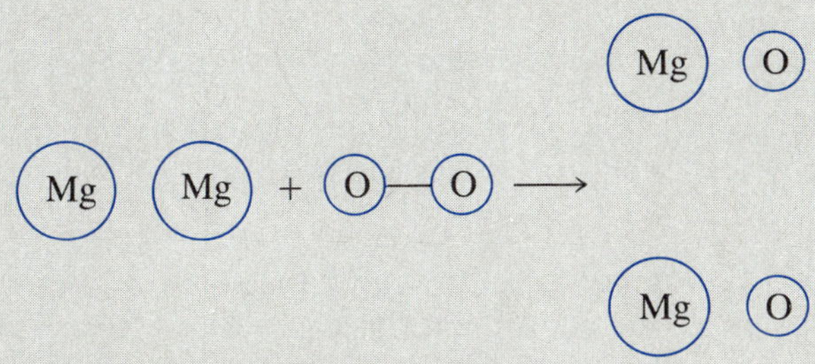

6 We should check all elements just to be sure. There are now two Mg on both sides and two oxygens on both sides.

Worked example

Balance the following equation (see Figure 7-4):

$$C_2H_4 + Br_2 \longrightarrow C_2H_4Br_2$$

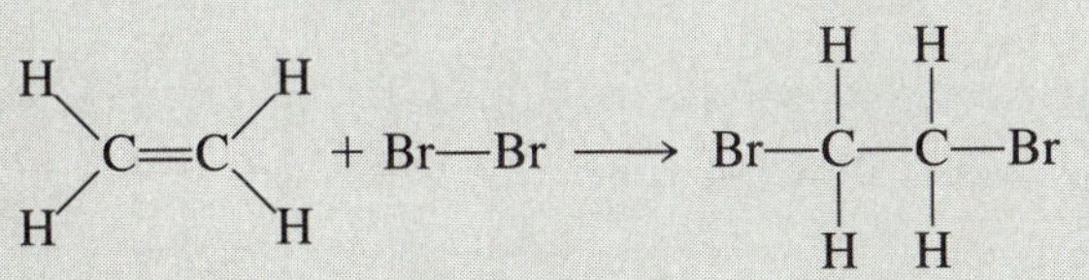

1 There are two carbons, four hydrogens, and two bromines on both sides.

2 Therefore, the equation is balanced as written, and no balancing coefficients are needed.

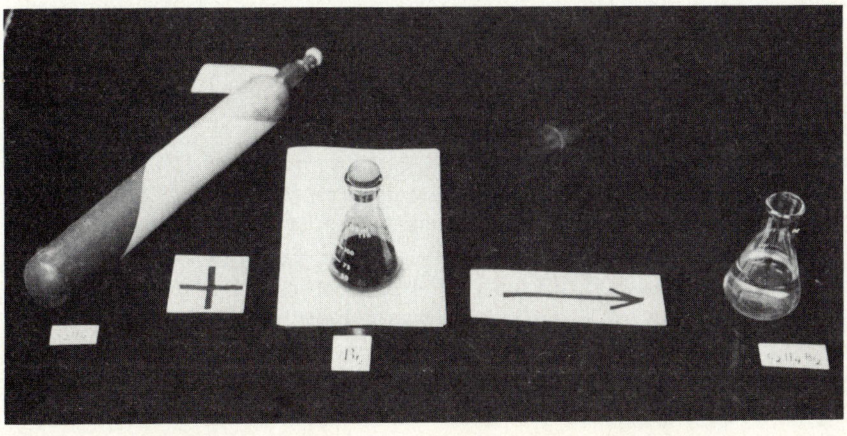

FIGURE 7-4
Gaseous ethylene (C_2H_4) and red liquid bromine (Br_2) react to form the colorless liquid ethylene bromide ($C_2H_4Br_2$).

Worked example

Balance the following equation (see Figure 7-5):

$$Al + Cl_2 \longrightarrow AlCl_3$$

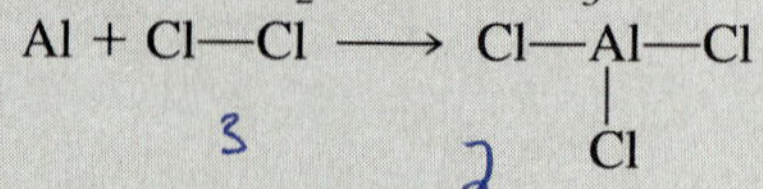

1 There are two Cl on the left and three on the right.

2 Since we cannot change either subscript, the only way to get the same number and type of atoms on both sides is to find the lowest common denominator, which for 2 and 3 is 6.

3 We place a 3 in front of the Cl_2 to give a total of six Cl atoms on the left and then place a 2 in front of the $AlCl_3$ to give a total of six Cl atoms on the right.

$$Al + 3Cl_2 \longrightarrow 2AlCl_3$$

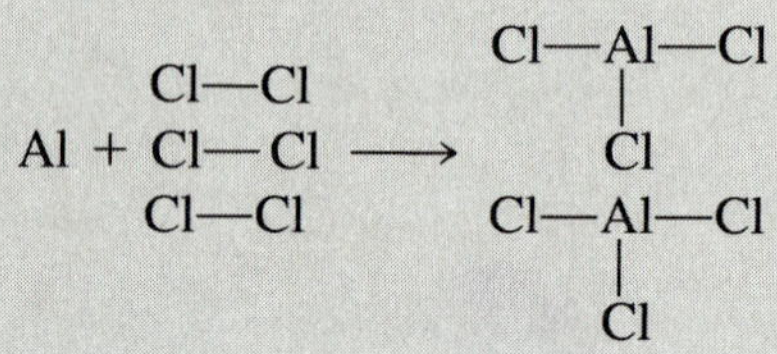

4 There are two Al on the right, so we place a 2 in front of the Al on the left. The equation is now balanced.

$$2Al + 3Cl_2 \longrightarrow 2AlCl_3$$

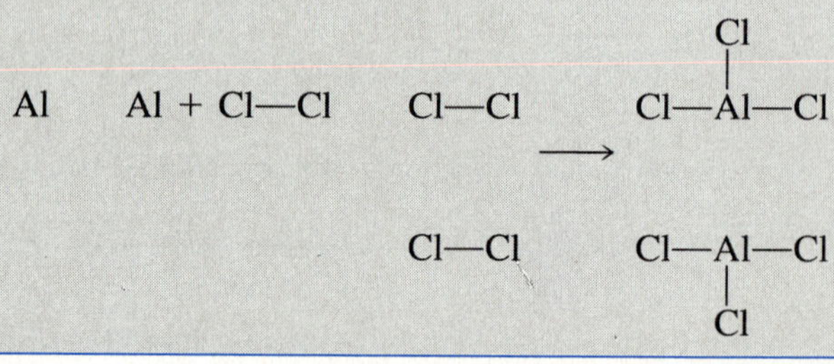

Problem

Balance the following equation:

$$H_2 + O_2 \longrightarrow H_2O$$

After attempting to balance this equation, look at the correct method and compare it with yours. The order in which the balancing coefficients are placed is not important; it is only the result that matters. If you are having difficulty, however, it would be wise to study the solution in order to develop a feel for the order that will most easily give the correct answer.

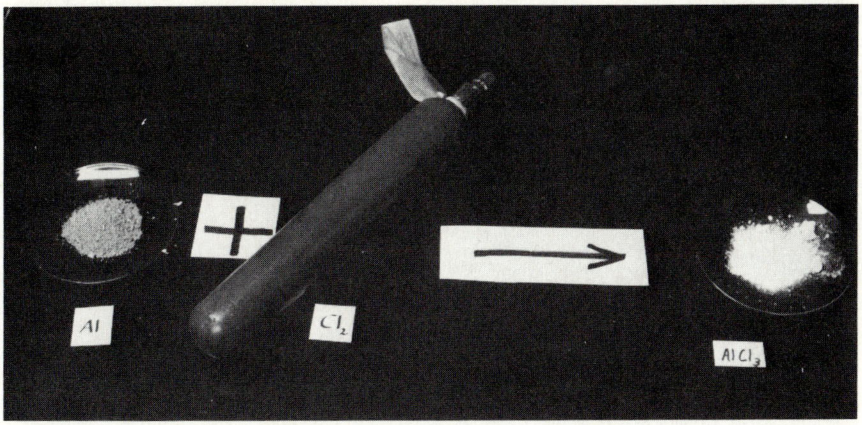

FIGURE 7-5
Aluminum powder (Al) and gaseous chlorine (Cl_2) react to form a white solid, aluminum chloride ($AlCl_3$).

Solution

$$\mathrm{H{-}H + O{-}O \longrightarrow \; H{-}O{-}H}^{*}$$

1 There are two H on both sides.

2 There are two O on the left and only one on the right.

3 Place a 2 in front of the H_2O.

$$H_2 + O_2 \longrightarrow 2H_2O$$

$$\mathrm{H{-}H + O{-}O \longrightarrow H{-}O{-}H \quad H{-}O{-}H}$$

4 There are now four H on the right and only two on the left.

5 Place a 2 in front of the H_2 to give a total of four H on the left also. The balancing is complete.

$$2H_2 + O_2 \longrightarrow 2H_2O$$

$$\mathrm{H{-}H \quad H{-}H + O{-}O \longrightarrow H{-}O{-}H \quad H{-}O{-}H}$$

* The water molecule is angular, although the student would not have been expected to know that before this point.

Problem

Balance the following equation:

$$Na_2O + H_2O \longrightarrow NaOH$$

Refer to the solution below *after* you have attempted this problem.

Solution

1 There are two Na on the left and only one on the right.

2 Place a 2 in front of the NaOH.

$$Na_2O + H_2O \longrightarrow 2NaOH$$

3 There are two H on the left and two on the right. Therefore, no coefficient is needed, except the 1 which is understood to be there when nothing is written.

4 There are two O on the left (one in each compound) and two on the right. Thus the equation is balanced.

Problem

Balance the following equation:

$$Si_2I_6 + H_2O \longrightarrow H_2Si_2O_4 + HI$$

Solution

1 There are two Si on each side. No balancing coefficient is required.

2 There are six I on the left and one on the right.

3 Place a 6 in front of HI.

$$Si_2I_6 + H_2O \longrightarrow H_2Si_2O_4 + 6HI$$

4 There are eight H on the right (two in $H_2Si_2O_4$ and six in HI).

5 Place a 4 in front of the H_2O to get eight H on the left.

$$Si_2I_6 + 4H_2O \longrightarrow H_2Si_2O_4 + 6HI$$

6 There are four O on both sides. Balancing is complete.

Problems

Now attempt to balance the following equations. The answers only are given.

a $H_2 + Br_2 \longrightarrow HBr$

b $N_2 + H_2 \longrightarrow NH_3$

c $NI_3 \longrightarrow N_2 + I_2$

d $SiO_2 + Al \longrightarrow Si + Al_2O_3$

e $SnO_2 + HCl \longrightarrow H_2O + SnCl_4$

Answers

a $H_2 + Br_2 \longrightarrow 2HBr$

b $N_2 + 3H_2 \longrightarrow 2NH_3$

c $2NI_3 \longrightarrow N_2 + 3I_2$

d $3SiO_2 + 4Al \longrightarrow 3Si + 2Al_2O_3$

e $SnO_2 + 4HCl \longrightarrow 2H_2O + SnCl_4$

BONDING

The octet rule: Why bonding occurs

The driving force behind atoms combining with each other is the tendency for atoms to *gain, lose,* or *share* electrons so as to fill their outermost electronic energy levels. For the elements hydrogen and helium, the outermost shell is the $1s$ shell, which requires two electrons to be full. Remember that the first shell has no p orbital. For atoms in period 2, the outside shell has one s orbital and three p orbitals, which means that eight electrons are needed to fill it (a total of four orbitals, each with a capacity of two electrons). Atoms tend to be especially stable (i.e., in a low energy state, which is always preferred) when the outer shell is full. Thus atoms tend to do whatever is necessary to get the ''magic'' eight electrons in their outer levels—except for the smaller atoms, which tend to get two electrons in the outer shell. This is called the *octet rule*. The octet rule explains why the inert gases (group VIII) such as He, Ne, and Ar do not undergo chemical changes under most circumstances. The outer shells *are already full;* hence there is no need to interact with other elements to gain, lose, or share electrons. Elements in periods 3 through 7 also usually obey the octet rule (i.e., they tend to fill the outer shell s and p orbitals even though d or f orbitals may be present):

He $\boxed{1s^2}$ ⟵ 2 electrons

Ne $1s^2\ \boxed{2s^2\ 2p^6}$ ⟵ 8 electrons

Ar $1s^2\ 2s^2\ 2p^6\ \boxed{3s^2\ 3p^6}$ ⟵ (8 electrons)

Ionic bonding

There are two basic ways in which atoms fulfill the octet rule. The first of these is called *ionic bonding*. (The second is *covalent bonding;* it is discussed in the next section.) In ionic bonding, atoms of metallic elements *lose* electrons when combining, and atoms of nonmetals *gain* electrons. For instance, consider the electronic configurations of atoms of sodium (Na) and fluorine (F); see Figure 7-6:

Na $1s^2\ 2s^2\ 2p^6\ \boxed{3s^1}$ ⟵ 1 electron in outer shell

F $1s^2\ \boxed{2s^2\ 2p^5}$ ⟵ 7 electrons in outer shell

Na has one electron in its outer shell and F has seven. If Na gives its one $3s$ electron to F, as follows,

Na $1s^2\ 2s^2\ 2p^6\ 3s^{①}$

F $1s^2\ 2s^2\ 2p^5$ ⟵

the octet conditions of *both* will be satisfied. F will have eight electrons in its second shell (which is the outermost), and Na will have eight in what is now its outer shell—the second shell.

Na $1s^2\ \boxed{2s^2\ 2p^6}$ ⟵

F $1s^2\ \boxed{2s^2\ 2p^6}$ ⟵ 8 electrons

The Na still has 11 protons in the nucleus, but it now has a total of only 10 electrons. This means that it has one more positively charged particle than negatively charged particles, which gives the whole ''atom'' a net charge of $+1$. The F still has 9 protons, but it now has 10 electrons, which gives it a net charge of -1.

These charged ''atoms,'' which have gained or lost one or more electrons, are called *ions*. *Positively* charged ions are called *cations*. (This term is easily remembered because the letter *t* in it is similar to the plus sign.) *Negatively* charged ions are called *anions*.

The negatively and positively charged ions which result from a transfer of electrons from one atom to another are held together by electrostatic attraction; that is, entities of opposite charge attract each other. This attraction of ions of different charge for each other is called *ionic bonding*. It occurs when the compound contains a metal and one or more nonmetals. The following compounds fulfill this criterion and are thus ionic in nature:

$MgCl_2$	FeI_2	CuF
Na_2O	Na_2SO_4	$Zn(C_2H_3O_2)_2$

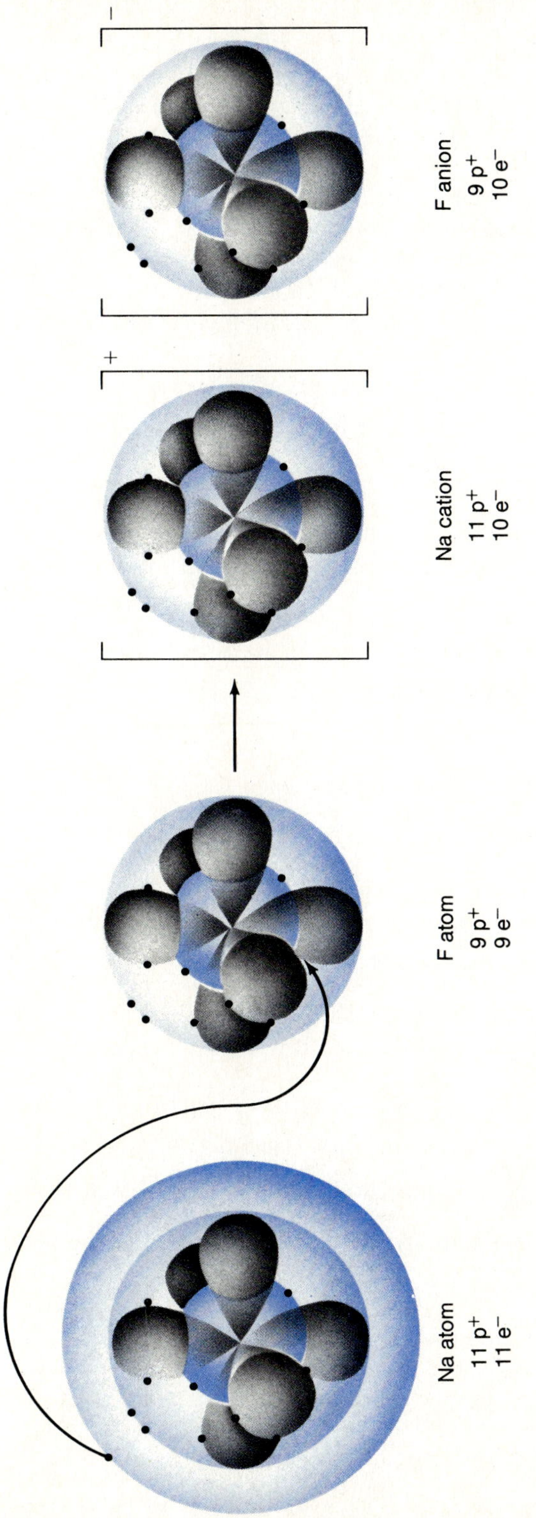

FIGURE 7-6

Transfer of an electron during the reaction of sodium (Na) with fluorine (F) to form an ionic bond.

Another way of showing this same transfer process is as follows:

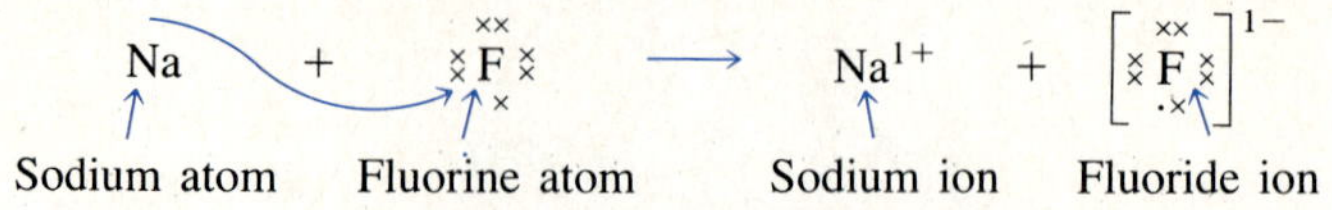

In this method, the *outer electrons only* (called *valence electrons*) are represented by symbols such as dots, circles, or x's. The electrons are then moved so that the octet rules of all atoms are satisfied, and the ions are formed. These representations are called *Lewis dot diagrams* after G. N. Lewis, one of the pioneers, along with Linus Pauling, in the understanding of chemical bonding.

For reaction of magnesium with fluorine, the following process occurs:

Mg $1s^2\ 2s^2\ 2p^6\ \boxed{3s^2}$ ⟵ 2 electrons in outer shell

F $1s^2\ \boxed{2s^2\ 2p^5}$ ⟵ 7 electrons in outer shell

Since the Mg needs to lose two electrons and the F needs to accept only one electron, two F atoms are needed to accept the two electrons from one Mg:

Mg $1s^2\ 2s^2\ 2p^6\ 3s$:
F $1s^2\ 2s^2\ 2p^5$ ←
F $1s^2\ 2s^2\ 2p^5$ ←

This gives:

Mg $1s^2\ \boxed{2s^2\ 2p^6}$ ⟵
2F $1s^2\ \boxed{2s^2\ 2p^6}$ ⟵ 8 electrons in outer shell

The Mg now has a charge of +2, and each F has a charge of −1. An ion develops *one* plus charge for each electron lost and *one* negative charge for *each* electron gained. Using Lewis dot diagrams:

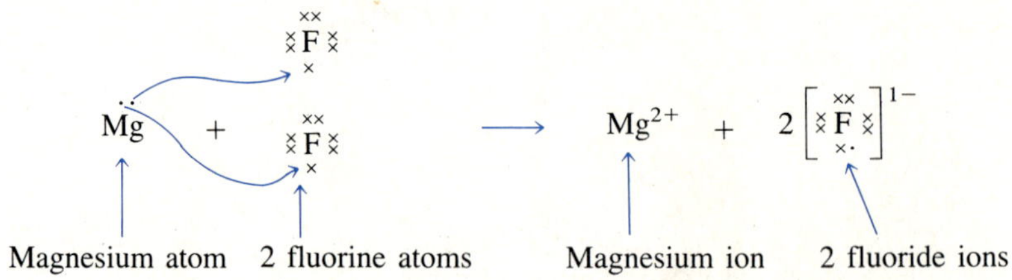

The standard way of expressing this process in a chemical equation is:

$$Mg + F_2 \longrightarrow MgF_2$$

The atom diagram is shown below:

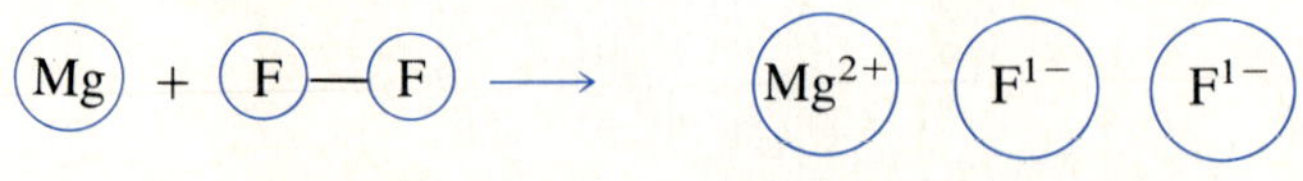

Not linked individually but attracted by the positive and negative charges

A transfer of electrons from atoms of metals to atoms of nonmetals to form ions is possible because metals have a *low electronegativity* (low ability to hold onto electrons); thus electrons can be easily lost. The nonmetals, on the other hand, have a *high electronegativity* and can therefore attract electrons from metals.

Because of the strong attraction of ions for each other, ionic materials have a high melting point and are solids at room temperature. The melting points of ionic materials mentioned so far in this section are shown below:

FORMULA	MELTING POINT, °C
CuF	908
FeI_2	592
MgF_2	1396
$MgCl_2$	708
NaF	985
NaCl	801

Covalent bonding

When two nonmetals react chemically, both atoms involved in the bonding have high electronegativity and a transfer of electrons is not possible. Neither one is willing to give up electrons. In this case, a sharing of the electrons is necessary in order to fulfull the octet rule.

For example, consider the interaction of atoms of hydrogen and chlorine:

H $\boxed{1s^1}$ ⟵ 1 electron in outer shell

Cl $1s^2\ 2s^2\ 2p^6\ \boxed{3s^2\ 3p^5}$ ⟵ 7 electrons in outer shell

The H and Cl atoms both need one more electron to fill the outer shells. Using the Lewis dot method, if both atoms come into physical proximity and allow one electron from each to be *shared* by both, then the octet conditions of both will be met:

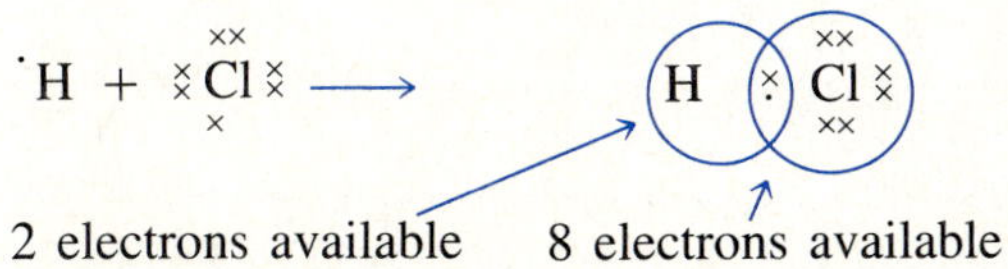

The orbital ''picture'' for this process is illustrated in Figure 7-7.

The shared electrons now belong to *both* atoms (a situation similar to a joint bank account). Hydrogen now has two electrons available to it (its original electron plus the one shared with chlorine), and chlorine has eight (its original seven plus the one shared with hydrogen). These two shared electrons cause the H and Cl atoms to be physically linked to each other (bonded); this link is called a *covalent bond*. The resulting unit is called a *molecule*.

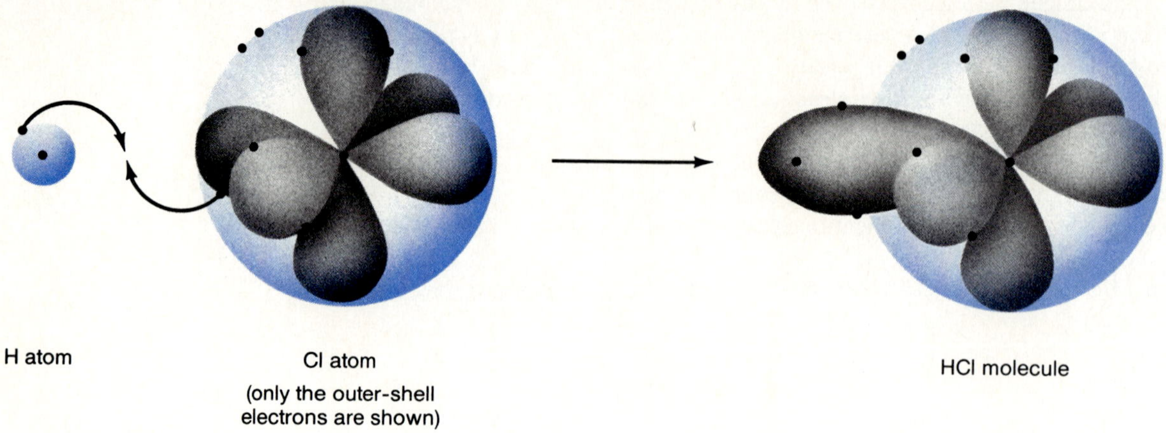

FIGURE 7-7
Covalent bonding in an HCl molecule.

Covalent bonding, as has been noted, occurs between *two nonmetals*. It is usually represented by a straight line:

$$\mathrm{H\overset{\times}{\bullet}Cl} \qquad \text{which equals} \qquad \mathrm{H{-}Cl}$$

Covalent bonding can occur between two identical nonmetal atoms. For example:

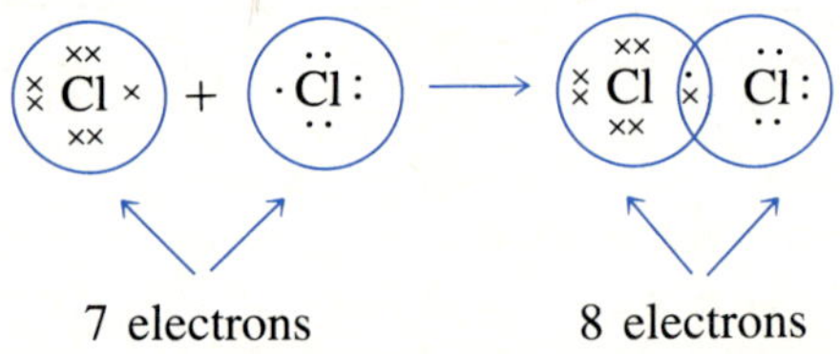

This explains the diatomic nature of chlorine (Cl_2). The individual atoms, by pairing up and sharing two electrons, can fulfill the octet rules of both.

Oxygen illustrates a second aspect of covalent bonding:

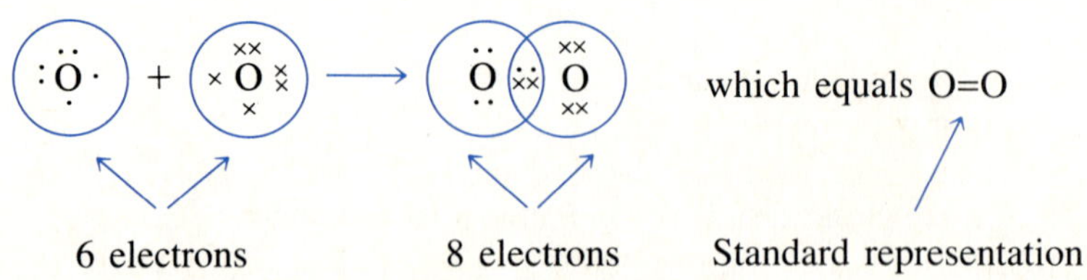

Since each oxygen atom has only six electrons, and thus needs two more to obtain an octet, it has to share *two* pairs of electrons. This is called a *double covalent bond* and is commonly represented by two straight lines. We have, thus, also explained the diatomic nature of oxygen. By pairing and sharing two pairs of electrons, the oxygen atoms have octets of electrons and are, therefore, more stable than in the single-atom state.

TABLE 7-4

Examples of covalent materials and their melting points

NAME	FORMULA	MELTING POINT , °C
Chlorine	Cl_2	−103
Nitrogen	N_2	−210
Oxygen	O_2	−218
Hydrogen chloride	HCl	−112
Aspirin	$C_9H_8O_4$	143
Water	H_2O	0
Ethyl alcohol	C_2H_6O	−117
Benzene	C_6H_6	5

A third way atoms can bond covalently is called a *triple bond*. Such a bond is found, for instance, in molecules of nitrogen (N_2). Atoms of nitrogen have five electrons in the valence shell, and so three more are needed to fulfill the octet rule. These can be obtained by sharing *three* pairs of electrons as shown below:

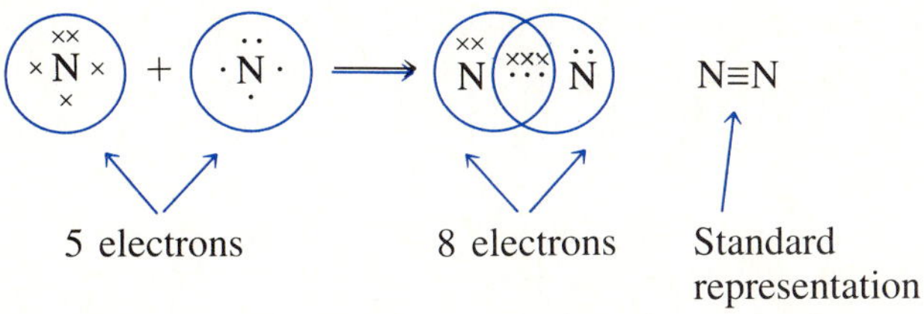

Covalent materials are usually gases, liquids, or relatively low-melting solids (an exception, among others, would be diamond, which melts above 3500°C). Several covalent materials along with their melting points are listed in Table 7-4. Notice that they all are composed of nonmetals only. A melting-point apparatus is shown in Figure 7-8.

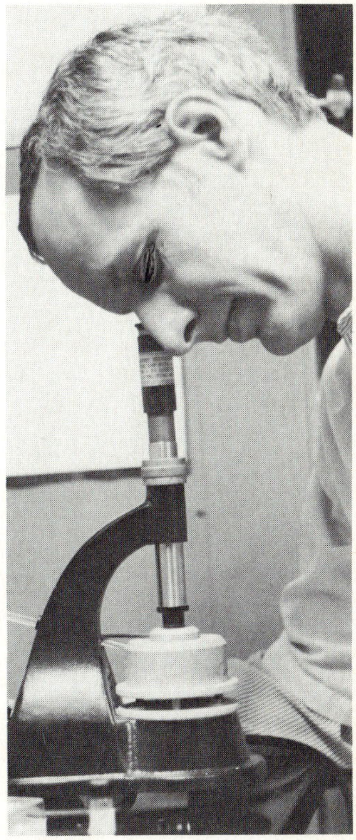

FIGURE 7-8
Melting-point apparatus, used to determine the melting point of solid materials in order to identify substances or estimate their purity.

Polar covalent bonding

In most covalent bonds, the two atoms sharing the electrons have somewhat different electronegativities and the electrons in the bond are *not* being shared equally. They spend more of their time closer to one of the atoms than to the other. In HCl, for example, the two electrons in the bond spend more time near the Cl atom because it is more electronegative than the H atom. The difference in electron attracting abilities is not great enough to allow a complete transfer of electrons and can be thought of as intermediate between an ionic (complete transfer) and pure covalent (equal sharing) bond:

$$\delta^{+}\text{H} \quad \dot{\times}\,\overset{\times\times}{\underset{\times\times}{\text{Cl}}}\,\overset{\times}{\times} \quad \delta^{-}$$

The Greek letter δ (delta) is used as a symbol for "partial."

The Cl end of the molecule has a partial negative charge, and the H end has a partial positive charge. This is called a *polar covalent bond,* and the molecule is said to be *polar*—that is, it has a negative end and a positive end.

As a general rule, only identical atoms bonded to each other exhibit pure covalent bonding (that is, equally shared nonpolar bonding). In most other cases there is a small difference in electronegativities, and a slightly polar bond results. Therefore, the H—H and C—C bonds are pure covalent bonds, and bonds such as C═O, C—Cl, and O—H are polar. When a molecule has nonpolar bonds, the molecule is nonpolar also. If the bonds are polar, the shape of the molecule determines if the molecule is polar or not. Carbon dioxide is a linear (straight) molecule (bond angles of 180°); and while each bond is polar, the molecule as a whole is not polar:

$$\delta^- \quad \overset{\delta^+}{\mathrm{O{=}C{=}O}} \quad \delta^-$$

Both ends of the molecule are negative, and the center is positive because of the relative electronegativities of carbon and oxygen; but since the two *ends* do not have opposite charges, it is *not* a polar molecule. The polarities of the two bonds cancel each other.

Water molecules, on the other hand, are bent, or angular, so that the bond polarities do not cancel. The angle formed by the two H—O bonds is 105°. The oxygen "end" is negative and the hydrogen "end" is positive, resulting in a polar molecule:

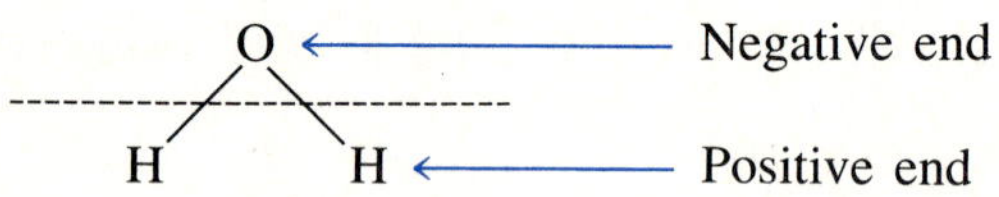

Metallic bonding

Metals, since they have a low electronegativity, give up their valence-shell electrons easily. Metal atoms can be thought of as atomic nuclei with their inner-shell electrons "floating" in a cloud of valence-shell electrons. Iron can be roughly pictured as shown in Figure 7-9. In this representation of the atomic structure of metals electrons are shared in all directions, and thus bonded in all directions, which helps to explain the great strength of most metals. All metals except mercury (Hg) (which is a liquid) are solids at room temperature.

Table 7-5 summarizes bonding in general; Table 7-6 summarizes bonding in some common substances. Figure 7-10 and Figure 7-11 show two important figures in the development of thought about bonding: G. N. Lewis and Linus Pauling.

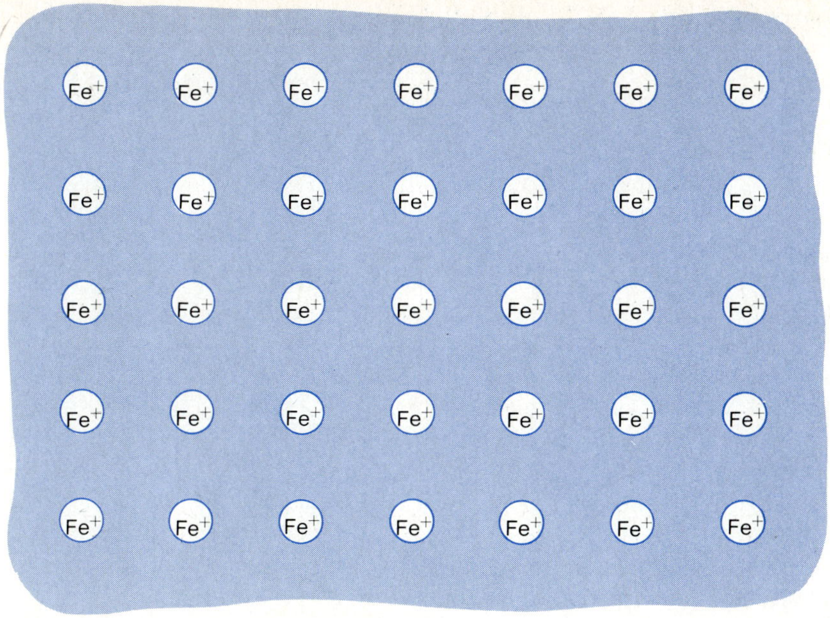

FIGURE 7-9

Metallic bonding in iron (Fe).

TABLE 7-5

Summary of bonding

TYPE OF BOND	TYPE OF ATOMS	EXAMPLES
Ionic	Metal and nonmetal	ZnF_2 CsCl
Covalent	Two nonmetals	
Pure covalent	Identical nonmetal atoms	Cl_2 N_2
Polar covalent	Two nonmetals with different electronegativities	HI CO
Metallic	Metals only	Na Mn

TABLE 7-6

Bonding in common substances

MATERIAL	FORMULA	TYPE OF BONDING	STATE (AT ROOM TEMPERATURE)
Metals	Fe (iron) Cu (copper) Au (gold) Al (aluminum)	Metallic	Solid
Calcium chloride	$CaCl_2$	Ionic	Solid
Salt	NaCl	Ionic	Solid
Boric acid	H_3BO_3	Covalent	Solid
Gasoline	C_8H_{18}	Covalent	Liquid
Alcohol	C_2H_6O	Covalent	Liquid
Aspirin	$C_9H_8O_4$	Covalent	Solid
Oxygen	O_2	Covalent	Gas
Ammonia	NH_3	Covalent	Gas
Stannous fluoride	SnF_2	Ionic	Solid

FIGURE 7-10
G. N. Lewis, one of the founders of modern ideas about chemical bonding. (From the collection of *Journal of Chemical Education*.)

FIGURE 7-11
Linus Pauling, a pioneer in developing current ideas about chemical bonding. (Smithsonian Institution.)

The following program will allow you to develop skill in recognizing types of bonding.

DETERMINING TYPES OF BONDS

Worked example

What kind of bonding occurs in the Br_2 molecule?

1 Both atoms are nonmetals (Br is to the right of the zigzag line on the periodic chart that separates metals from nonmetals).

2 Therefore the bonding is covalent (see Table 7-5).

3 Since both atoms are the same, the bonding is *pure covalent*.

Worked example

What kind of bonding do you expect in Rb_2S?

1 Rb is a metal (to the left on the periodic chart), and S is a nonmetal (to the right on the periodic chart).

2 Therefore the bonding is *ionic* (see Table 7-5).

Problem

What kind of bonding do you expect between atoms of carbon and fluorine in the compound CF_4? Refer to the solution *after* you have tried this problem on your own.

Solution

1 Both are nonmetals (to the right on the periodic chart).

2 Therefore covalent bonding is expected.

3 Since the two elements have different electronegativities, *polar covalent* bonding is expected.

Problem

What type of bonding occurs between atoms of copper in a piece of copper metal?

Solution

1 There is only one kind of atom present, since copper is an element.

2 Since all the atoms are metals, *metallic* bonding holds the atoms together.

SOLUBILITY: HOW TO DISSOLVE MATERIALS

The concepts of ionic and polar covalent bonds have already been discussed. These concepts are used to predict the solubilities of various materials. The general rule is that ionic and polar substances dissolve in each other and nonpolar materials dissolve in each other. This is sometimes stated as "like dissolves like." It was shown earlier in this chapter that water is polar and NaCl is ionic. Using the above rule one would predict that the ionic NaCl (table salt) will dissolve in the water, and indeed it does. On the other hand, gasoline (C_8H_{18}) is a nonpolar material and, therefore, does not dissolve in water. Sugar molecules contain the polar —C—O—H group (oxygen has a high electronegativity), and so sugar dissolves in water but not in gasoline.

One does not always know the molecular formulas of everyday substances, much less their molecular shapes. However, in the course of everyday life one has probably encountered many commonly used materials that do not dissolve in water but do dissolve in alcohol (such as rubbing alcohol or shellac thinner), which is somewhat less polar than water, or in a nonpolar solvent such as gasoline, charcoal lighter fluid, or ethylene dichloride. If a material such as floor wax, tar, or crayon is not removed by water, it can usually be dissolved in one of the nonpolar solvents listed above. The use of nonpolar solvents such as benzene or carbon tetrachloride is discouraged because of demonstrated health hazards.

If the job can be done outdoors, that is preferable; and since there is plenty of ventilation outdoors, there is little if any health or safety hazard presented by the organic (nonpolar) solvents. If the task must be done indoors or in an area with little ventilation, ethylene dichloride, which is nonflammable and relatively nontoxic, should be used. Even with ethylene dichloride, the area should be well ventilated or the exposure time should be kept short, since prolonged breathing of the fumes can be harmful. If the exposure time is kept to within 5 to 10 minutes (min) and only small amounts of ethylene dichloride are used, this solvent is relatively safe to use indoors. One should *never* use gasoline in a confined area, since even such a small (and easily performed) operation as turning on a light switch can provide enough of a spark to ignite the flammable mixture and cause a violent explosion. For more information on dissolving unwanted materials and stains, refer to Chapter 17.

PHYSIOLOGICAL PROPERTIES OF ELEMENTS AND COMPOUNDS

The author once read a letter to the editor of a newspaper discussing the dangers of using sodium fluoride (NaF) in public water supplies to reduce tooth decay. It has been shown that NaF, when placed in drinking water in trace amounts, does greatly improve the hardness of tooth enamel. The writer of the letter, however, had read that sodium is highly reactive (which is correct) and that fluorine is highly poisonous (which is also correct), and concluded that fluoridation is a dangerous practice. The conclusion was not necessarily correct, since—as we have seen—compounds usually have properties quite different from those of the elements from which they are composed. This can be shown by a few examples.

Salt (NaCl) is required for normal physiological functions, but chlorine (Cl) is very toxic. Water (H_2O) does not burn, even though it is composed of hydrogen (which is highly flammable) and oxygen (which supports combustion). Sodium fluoride will prevent tooth decay, but neither sodium (Na) nor fluorine (F_2) will prevent tooth decay. Mercury metal (Hg), which is an element, is somewhat toxic, but its compounds are much more harmful. The following toxicities have been found in humans (these amounts are $L.D._{50}$ values—that is, the amount needed to cause death in 50 percent of the population):

Hg (oral)	1429 mg per kilogram of body weight
$HgCl_2$ (oral)	29 mg per kilogram of body weight
$Hg(CH_3)_2$ (breathing)	0.1 mg per cubic meter of air breathed

$HgCl_2$ is about 50 times as toxic as mercury itself; and—although the numbers cannot be compared, since the units are different—$Hg(CH_3)_2$ is much more dangerous than either of the others.

To return to the letter about fluoridation: NaF is actually quite toxic (4 mg per kilogram of body weight), but not because of the toxicity of F_2. And in any case the extremely small amount used in water supplies (usually 1 mg per liter of water) is not harmful but beneficial. As consumers, we have to be careful not to think of compounds as simply mixtures of their component elements and not to draw erroneous conclusions about consumer issues. The actual experimentally determined toxicities of each material must be used in making decisions. Even similarities in chemical composition cannot be used. For example, NaF has a toxicity of 4 mg per kilogram of body weight; NaCl, which might be expected to have a similar toxicity, has a value of 8200 mg per kilogram of body weight. This higher value indicates a much lower toxicity—it would take 560 g or 1.2 pounds (lb), of NaCl to cause poisoning in a 150-lb person.

STUDY GUIDE

Terms to learn

Anion	Metallic bonding
Balancing	Molecular formula
Bonding	Molecule
Cation	Octet rule
Chemical equation	Polar covalent bond
Covalent	Pure covalent bond
Diatomic	Polyatomic
Double bond	Sharing electrons
Empirical formula	Single bond
Formula	Transfer of electrons
Ion	Triple bond
Ionic bond	Valence electrons
Lewis dot diagram	

Skill practice problems

1 Match the following terms with the example in the equation given.

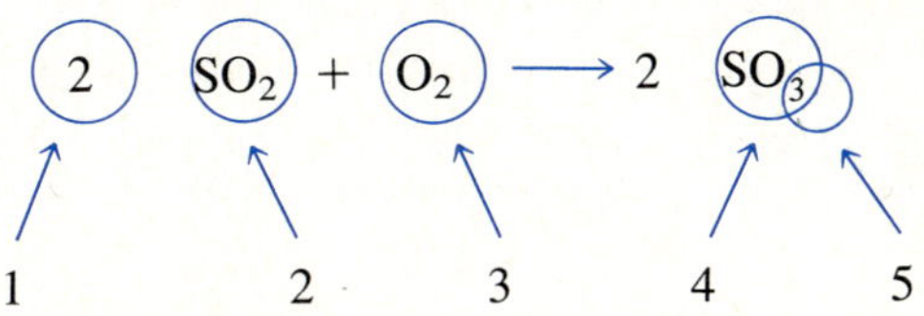

(a) ______ Formula of a product

(b) ______ A diatomic molecule

(c) ______ Formula of a reactant

(d) ______ A balancing coefficient
(e) ______ A subscript

2 Balance the following equations.
(a) $CH_4 + Cl_2 \longrightarrow HCl + CCl_4$
(b) $C_2H_4 + O_2 \longrightarrow CO_2 + H_2O$
(c) $NH_3 + O_2 \longrightarrow H_2O + N_2$
(d) $KCN + As_2O_3 \longrightarrow KOCN + As_4$
(e) $Na_2S_2O_5 + H_2O \longrightarrow NaHSO_3$

3 Write the Lewis dot diagram for selenium (Se) and indicate what has to happen for the octet rule to be satisfied. Does it have to gain or lose electrons? Repeat for carbon (C) and for aluminum (Al). Repeat for calcium (Ca).

4 Indicate the type of bonding you would expect between each of the following pairs of atoms:
(a) C—O
(b) I—I
(c) Cs—F
(d) N—N
(e) H—N
(f) Li— Br
(g) Zn— Cu
(h) Ca—F
(i) Na—Na
(j) O—N

5 Would you expect floor wax to be more soluble in water or in ethylene dichloride?

6 Are crayons composed of polar or nonpolar materials? Would you use water or paint thinner to remove crayon from a skirt?

7 Why should gasoline *never* be used as a cleaner indoors?

8 Classify each ion formed in problem 3 as an anion or a cation.

9 Would you expect NaF or NaCl to be toxic because they are both prepared from highly reactive Na and either toxic F_2 or toxic Cl_2?

10 If NaF has a toxicity in humans of 4 mg per kilogram of body weight, how much of it would a 150-lb human have to ingest at one time to feel ill effects?

11 If typical water supplies have 1 mg NaF per liter, how many liters would one have to drink to receive a toxic dose? Assume that all the NaF is taken up by the body. How many gallons is this equal to?

Goals

To gain insight into the kind of thinking used by chemists

To understand how the actual number of atoms or molecules can be found

To develop skill in doing simple calculations

8

THE MOLE CONCEPT: COUNTING ATOMS AND MOLECULES

In the chemical laboratory, the experimenter can determine a quantity of material by measuring its weight (or volume); see Figures 8-1 and 8-2. The laboratory balance is the main tool used to find the amount of a substance to be used in a reaction or the amount of material produced from a chemical change. On the other hand, chemists must think about numbers of atoms or at least ratios of numbers of atoms (i.e., the relative number of atoms).

Since matter exists as atoms and molecules, and it is these entities that interact during chemical processes, a chemist must think about their relative numbers.

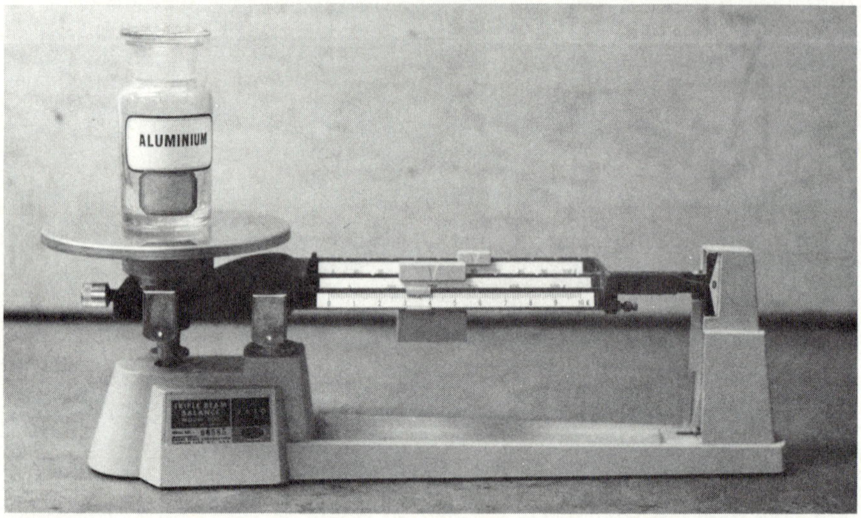

FIGURE 8-1

Platform balance used by chemists to measure the weight (mass) of a material to the nearest 0.01 g.

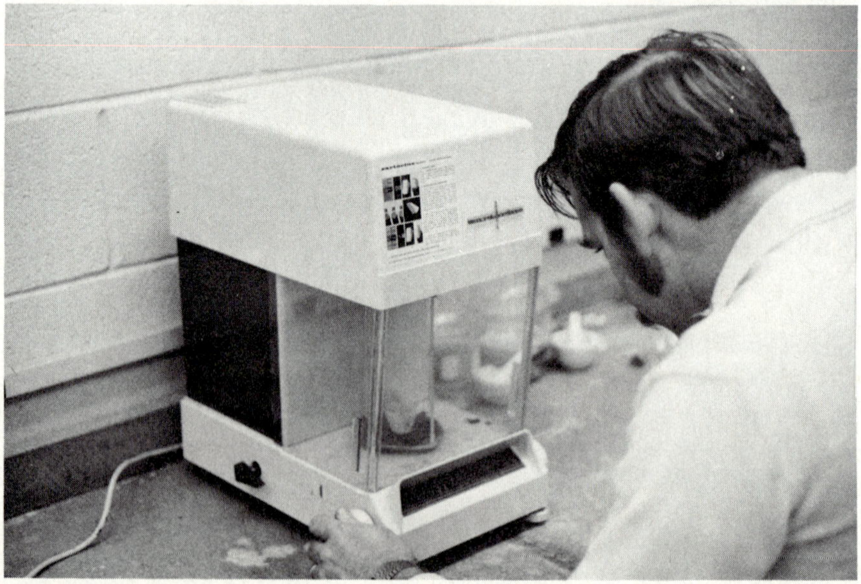

FIGURE 8-2

The analytical balance can be used to find the mass of a quantity of material or an object to ±0.0001 g.

In the example shown by the equation below,

$$2Mg + O_2 \longrightarrow 2MgO$$

two atoms of magnesium combine with one oxygen molecule. If a chemist wants to prepare MgO with maximum efficiency and a minimum of starting materials (and a minimum of cost!), the starting mixture should contain twice as many atoms of Mg as molecules of O_2. However, atoms and molecules cannot be counted, because they are so small that they are invisible, even with the best of tools. Even if they could be seen, the time needed to count them would be prohibitive. To put it succinctly, the chemist must work with weights of material but has to *think about* numbers of atoms or molecules. The *mole concept* is the connection between the two:

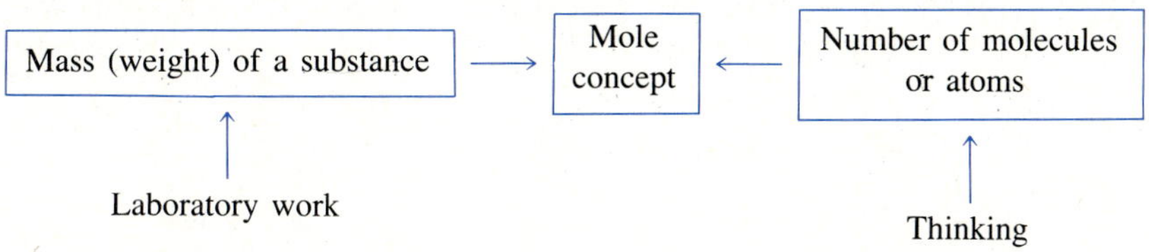

THE MOLE CONCEPT

Consider two groups of balls, one of low mass (for example, each ball weighing 1 lb) and one of larger mass (for example, each ball weighing 10 lb)—baseballs and bowling balls, for instance. Consider equal numbers of each, as shown in Figures 8-3 and 8-4. Notice that no matter what number of balls we take, *as long as the number of each type is the same,* the mass of the whole group of bowling balls is 10 times the mass of the group of baseballs. This is because the mass of *each* bowling ball is 10 times the mass of *each* baseball. For an equal number of balls, the relative masses of the groups is the same as the relative masses of the individual balls.

If one counts equal numbers of baseballs and bowling balls, one always gets a ratio of masses of 10 to 1. Therefore, if one takes 10 times the mass of bowling balls, one is assured of having *equal numbers* of them. For example, let's suppose that there are 10,000 lb of baseballs and 100,000 lb of bowling balls. Since the weight of the bowling balls is 10 times the weight of the baseballs, there must be the *same* number of each. One knows this by using the proportionality idea just stated, but one can confirm it by calculating the actual number of each:

$$\frac{10{,}000 \text{ lb}}{1 \text{ lb per ball}} = 10{,}000 \text{ baseballs}$$

$$\frac{100{,}000 \text{ lb}}{10 \text{ lb per ball}} = 10{,}000 \text{ bowling balls}$$

FIGURE 8-3
The 10 bowling balls weigh about ten times as much as the 10 baseballs.

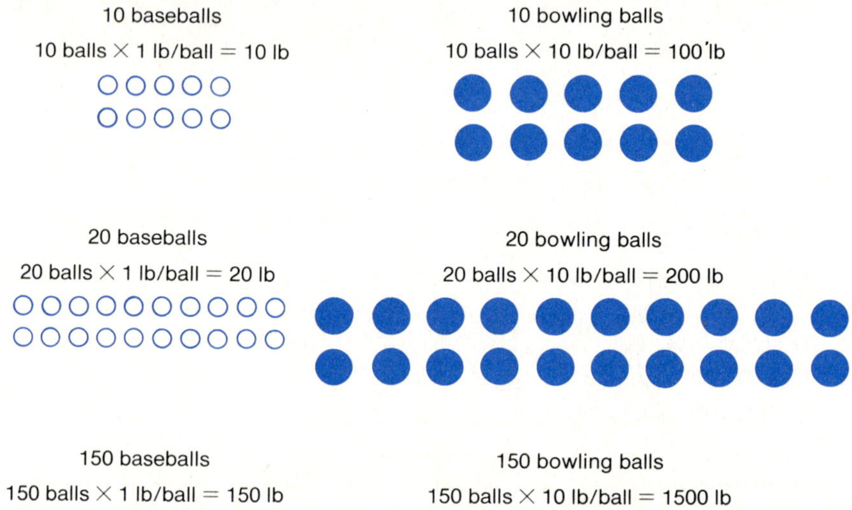

FIGURE 8-4
Low-mass and high-mass groups.

Atoms and molecules are real physical entities, and the concept we have just been considering applies to them as well. Consider the two elements carbon and magnesium:

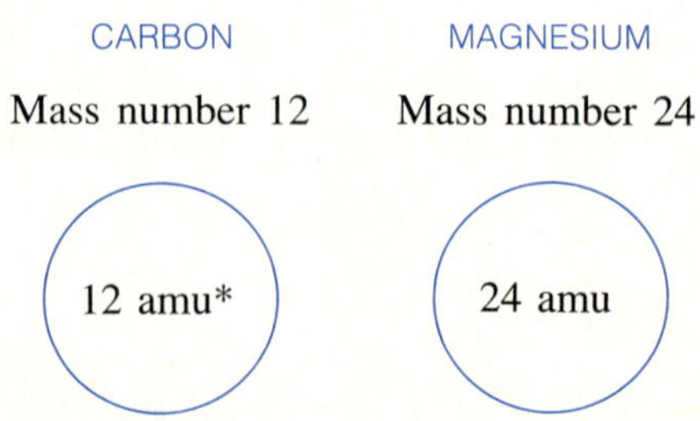

*The abbreviation amu stands for atomic mass unit—see Chapter 5.

TABLE 8-1

Relationship between mass and number of atoms of carbon and magnesium

	CARBON		MAGNESIUM	
PART	NUMBER OF ATOMS	MASS	NUMBER OF ATOMS	MASS
a	1	12 amu	1	24 amu
	2	24 amu	2	48 amu
	200	2400 amu	200	4800 amu
b	*X**	12 g	*X*	24 g
	10*X*	120 g	10*X*	240 g

* *X* represents an unknown number.

Atoms of magnesium have twice the mass of those of carbon. Therefore, if we take *twice the weight* of magnesium as carbon, we will have *equal numbers of atoms* (see Table 8-1).

From Table 8-1, part *a,* one can see that as long as the mass of magnesium one takes is twice that of carbon, the number of atoms of both elements is equal. It therefore follows that for *any mass scale* (grams, pounds, tons, etc.), as long as the mass of magnesium one takes is twice that of carbon, the number of atoms of each will be equal—although at this point one does not know what that number is. From part *a* one observes that the 24 amu of magnesium is twice the 12 amu of carbon. Each is one atom. Likewise, 4800 amu of magnesium is twice the 2400 amu of carbon, yet each has 200 atoms. In part *b,* the 24 grams (g) of magnesium is twice the 12 g of carbon; therefore, each must contain equal numbers of atoms (*X* represents this unknown number). The 120-g sample of carbon has 10 times as many atoms of carbon as the 12-g sample, since the weight is 10 times as great. Thus one can see that *weight ratios* can be used to determine *ratios of atoms*.

Originally, chemists worked with ratios of weights and ratios of numbers of atoms. If they wanted equal numbers of carbon and magnesium atoms, they simply weighed out twice as much magnesium as carbon. Or if they wanted twice as many atoms of carbon as magnesium, they weighed out equal amounts of each (twice as much carbon, relative to magnesium, as would be required for an equal number of atoms).

Since this procedure was a bit cumbersome, chemists decided to choose standard amounts of each material that would contain the same number of atoms. It seemed convenient to work with weights equal to the atomic and molecular weights, although this choice was purely arbitrary. Thus the standard weight of carbon was taken as 12 g, which is numerically identical to its atomic weight of 12 amu. Likewise, 24 g was taken as the standard amount of magnesium. These standard quantities were called *moles*—12 g of carbon was referred to as 1 mole (mol), 24 g of carbon as 2 mol, etc. If chemists weighed equal numbers of moles, they could be assured of equal numbers of atoms. The absolute number of atoms in a given number of moles was a matter of curiosity, but it was not essential to doing chemical calculations. We now know that a mole of a substance contains 6.02×10^{23} units of that substance.

The term *mole* should now be considered similar to other words that stand for certain numbers (see Table 8-2 for examples). It is equal to 6.02×10^{23} of *anything*, and it is called *Avogadro's number* (in honor of the Italian chemist Amedeo Avogadro, 1776–1856, who was a pioneer in the development of the mole concept, although the numerical value of the number was determined by Albert Einstein in the late nineteenth century). Avogadro's number, which when written out in standard notation is 602,000,000,000,000,000,000,000, is too large for us to readily conceptualize. Several examples may help, however. If one had a mole of seconds it would be equal to 1.9×10^{16} years; this is 4 million times the supposed age of the earth (which is assumed to be 4.5 billion years). It is thus obvious that one could never count a mole of atoms even if they were visible. A mole of baseballs would occupy a cube almost 4000 miles (mi) on a side. A mole of sand grains would be a pile of sand as large as the United States and about 20 feet (ft) deep, which is a lot of grains of sand! However, as we will see, a mole of atoms easily fits into the palm of one's hand.

TABLE 8-2

Some words that stand for numbers

WORD	NUMBER
Dozen	12
Gross	144
Pair	2
Ream	500
Mole	6.02×10^{23}

CALCULATIONS USING THE MOLE CONCEPT

The number of grams of a monoatomic element numerically equal to the atomic weight of that substance is called the *gram atomic weight*. It contains 1 mol of that substance (see Table 8-3). Since the atomic weight of carbon is 12 amu, 12 g of carbon is one gram atomic weight or *1 mol*. A more proper way to state the same thing is to say that 12 g of carbon contains 1 mol of carbon atoms or that 12 g of carbon is composed of 6.02×10^{23} carbon atoms. Likewise, 24 g of magnesium contains 1 mol of magnesium atoms (since each atom weighs 24 amu) or 24 g of magnesium is composed of 6.02×10^{23} magnesium atoms. Thus, 24 g of magnesium contains the same number of atom as 12 g of carbon.

TABLE 8-3

Comparison of amu values with mole weight* values

MATERIAL	AMU VALUES	AMOUNT CONSTITUTING 1 MOL (g/mol)
Na	23.0 amu per atom	23.0
Cl_2	71.0 amu per molecule	71.0
CO_2	44.0 amu per molecule	44.0
Zn	65.4 amu per atom	65.4
Na_2CO_3	106 amu per unit	106
C_8H_{18}	114 amu per molecule	114
$C_{10}I_{10}$	1389 amu per molecule	1389
O	16.0 amu per atom	16.0
O_2 (normal form of oxygen)	32.0 amu per molecule	32.0

*The weight of 1 mol.

It is thus possible to calculate both the number of moles of such elements and the number of atoms. The procedure for this type of calculation is shown below, and the program which follows will allow the student to develop skills in this area.

METHOD

1 *Number of moles from mass.* One determines the number of moles of a substance from its mass as follows:

(a) For monoatomic elements:

$$\text{Number of moles of atoms} = \frac{\text{mass (g)}}{\text{gram atomic weight (g/mol)*}}$$

(b) For ionic substances:

$$\text{Number of moles of units} = \frac{\text{mass (g)}}{\text{gram formula weight (g/mol)}}$$

Note The gram formula weight is the *sum* of all the atomic weights in the formula.

(c) For covalently bound substances:

$$\text{Number of moles of molecules} = \frac{\text{mass (g)}}{\text{gram molecular weight (g/mol)}}$$

Note The gram molecular weight is the sum of the weights of all the atoms in the molecule.

Note The three terms *gram atomic weight, gram formula weight,* and *gram molecular weight* can all be simply referred to as the *particle weight* or *mole weight* (abbreviated MW) of the substances. Each is expressed as grams per mole.

2 *Number of atoms, ions, or molecules from number of moles:*

$$\boxed{\begin{array}{c}\text{Number of atoms}\\ \text{or}\\ \text{molecules}\end{array}} = \text{number of moles} \times (6.02 \times 10^{23})$$

$$\text{Number of ions} = \text{number of moles} \times (6.02 \times 10^{23}) \times \text{ions per unit}$$

Note To determine if a material is composed of atoms, ions, or molecules, refer to Chapter 7.

---

*This is read as ''grams per mole.''

CALCULATING NUMBER OF MOLES AND MOLECULES OR ATOMS OF A SUBSTANCE

Worked example

How many moles of atoms does one have in 48 g of Mg? (See Figure 8-5.) How many atoms does one have?

1 Since this is a monoatomic element (*not* one of those listed as diatomic in Chapter 7), it exists as single atoms.

2 One therefore uses the following relationship:

$$\text{Number of moles} = \frac{\text{mass}}{\text{MW}}$$

3 The mass is given as 48 g, and the mass of one Mg atom (from the periodic chart) is 24 amu. Therefore one MW is 24 g/mol.

4 Putting these data into the above relationship:

$$\text{Number of moles} = \frac{48\ \text{g}}{24\ \text{g/mol}} = 2\ \text{mol}$$

Conceptually, since each mole contains 24 g, 48 g is twice this (2 mol). *Mechanically,* one simply places the numbers in the correct relationship and divides 48 by 24 to obtain *2*.

5 The number of atoms can then be calculated by using

$$\text{Number of atoms} = \text{number of moles}\ (6.02 \times 10^{23})$$

The number of moles is calculated to be 2. Therefore:

$$\begin{aligned}\text{Number of atoms} &= 2\ \text{mol}\ (6.02 \times 10^{23}\ \text{atoms per mole})\\ &= 12 \times 10^{23}\ \text{atoms}\end{aligned}$$

Or, expressed more properly,

$$1.2 \times 10^{24}\ \text{atoms}$$

FIGURE 8-5
1 mol (24 g) of magnesium and 1 mol (12 g) of carbon.

Worked example

How many moles of molecules are in 200 g of benzene (C_6H_6)? (See Figure 8-6.) How many molecules does this amount contain?

1 This is a covalently bound substance; therefore, the relationship that is useful is:

$$\text{Number of moles} = \frac{\text{mass}}{\text{MW}}$$

2 The mass is given as 200 g. The gram molecular weight is the sum of all the atomic weights, which for C_6H_6 is 78 g/mol. Each carbon weighs 12 amu, and each hydrogen weighs 1 amu $[(6 \times 12) + (6 \times 1)] = 78$.

3 Putting these values into the above formula:

$$\text{Moles} = \frac{200 \text{ g}}{78 \text{ g/mol}} = 2.56 \text{ mol}$$

4 Since there are 6.02×10^{23} molecules in 1 mol, the number of molecules in the 200 g is:

$$\begin{aligned}\text{Number of molecules} &= \text{number of moles} \times (6.02 \times 10^{23}) \\ &= 2.56(6.02 \times 10^{23}) \\ &= 1.54 \times 10^{24}\end{aligned}$$

This means that in the 200 g of C_6H_6, there are 2.56 mol (or more properly stated, there are 2.56 mol of molecules) or that there are 1.54×10^{24} molecules present.

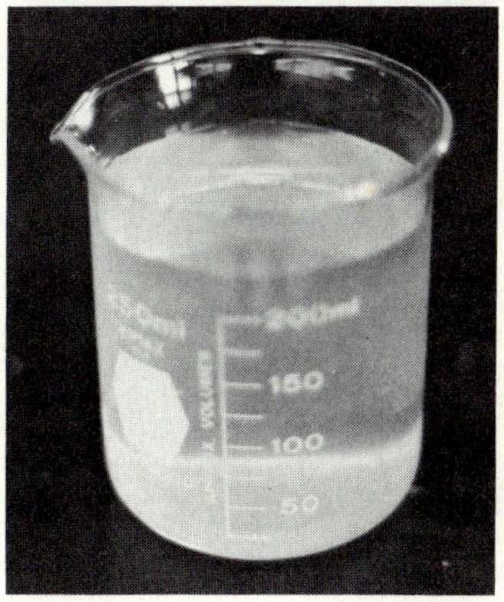

FIGURE 8-6
This container has 200 g (2.56 mol) of benzene.

Worked example

In 85 g of $CaCl_2$, how many moles of compound are there? How many ions are present?

1 This is an *ionic* material (that is, a metal and nonmetal). Therefore it exists as *ions* (one Ca^{2+} and two Cl^-, for a total of three ions per formula).

2 The gram formula weight of $CaCl_2$ is $40 + (2 \times 35.5) = 111$ g/mol.

3 $\frac{85 \text{ g}}{111 \text{ g/mol}} = 0.77$ mol of $CaCl_2$

4 $(6.02 \times 10^{23})(0.77 \text{ mol}) = 4.6 \times 10^{23}$ $CaCl_2$ units

5 4.6×10^{23} $CaCl_2$ units (3 ions per unit)
$= 1.4 \times 10^{24}$ ions in the 85 g

Worked in two steps:

(a) $\frac{85 \text{ g}}{111 \text{ g/mol}} = 0.77$ mol

(b) (0.77 mol)(6.02×10^{23} unit per mole)(3 ions per unit) = 1.4×10^{24} ions

Problem

How many moles are in 50 g of ethylene (C_2H_6)? How many molecules are there? (First attempt to work this; *then* look at the solution below.)

Solution

1 C_2H_6 would have a gram molecular weight of 30 g/mol

$$(2 \times 12) + (6 \times 1) = 30$$

2 Use the following relationship:

$$\text{Number of moles} = \frac{\text{mass}}{\text{MW}}$$

$$= \frac{50\ \text{g}}{30\ \text{g/mol}} = 1.7\ \text{mol}$$

Therefore there are 1.7 mol of C_2H_6 in 50 g.

3 To calculate the number of molecules we use:

$$\begin{aligned}\text{Number of molecules} &= \text{number of moles} \times (6.02 \times 10^{23}) \\ &= 1.7(6.02 \times 10^{23}) \\ &= 1.02 \times 10^{24}\end{aligned}$$

Therefore there are 1.02×10^{24} molecules of C_2H_6 in 50 g of this material.

Problem

How many moles of $Ca(OH)_2$ are in 0.17 g of this material? How many oxygen atoms are there?

Solution

1 The gram formula weight of $Ca(OH)_2$ is 74 g/mol.

2 The number of moles is:

$$\frac{0.17\ \text{g}}{74\ \text{g/mol}} = 0.0023\ \text{mol}$$

3 There are $0.0023(6.02 \times 10^{23}) = 1.4 \times 10^{21}$ $Ca(OH)_2$ units in the 0.17 g.

4 There are two oxygen atoms for each molecule of $Ca(OH)_2$. Therefore there are twice as many oxygen atoms as $Ca(OH)_2$ units. The number of oxygens is:

$$2 \times (1.4 \times 10^{21}) = 2.8 \times 10^{21}$$

APPLICATIONS OF THE MOLE CONCEPT

Although the mole concept is included in this book, that is not because you will necessarily be using it in your future course work or profession. It will be referred to a few times in the remainder of this book, however, and its usefulness will become apparent. The main reason for presenting it is to expose you to one of the most important thought processes in the field of chemistry. No chemistry course for any group of students would be complete without at least mentioning it. The types of calculations a chemist can do with it are covered in some detail in Appendix E (depending on the nature of your course, your instructor may refer you there at this point or later).

In all chemical processes—including synthesis and qualitative and quantitative analysis—the ratio of atoms or molecules is usually important to the process. For example, in Chapter 11 the synthesis of a sulfa drug is presented. The exact ratios of molecules are important in each step, and the number of moles is proportional to the number of molecules. Since chemists cannot count molecules, the mole concept gives them the tool they need to do the job. The chemist can mix equal moles of the two reactants and be assured of having an equal number of molecules. The weight needed for equal moles can be calculated as outlined in this chapter. That amount can then simply be weighed on a laboratory balance.

If a physician is trying to find the correct amount of an antidote to give to a victim of poisoning, once the blood or other body fluid has been analyzed, the mole ratio of poison to antidote can be used to calculate the amount of antidote quickly. Even if the physician does not do the actual calculating, but refers to a chart or to memory, the information was originally determined using the mole concept. The measurement of acidity, which is discussed in Chapter 10, is based on the number of moles of hydrogen ion in each liter of solution.

The mole concept is one of the most important ideas used by the chemist, second only to basic atomic theory, which is a description of matter, and the electronic orbital theory, which accounts for the chemical behavior of all elements. In practice, most chemical calculations are based on the mole, and chemists in both industrial and academic laboratories as well as chemical engineers, biologists, physicists, and people in the health sciences use it when doing quantitative work.

STUDY GUIDE

Terms to learn

Avogadro's number	Mole
Gram atomic weight	Mole weight
Gram formula weight	Ratio
Gram molecular weight	

Skill practice problems

1 Why do chemists use the mole concept?

2 What does the term *mole* mean?

3 How does one determine how many moles of a substance one has?

4 How many moles of Fe atoms are in 168 g of Fe?

5 How many atoms of Fe are in the 168 g in problem 4?

6 How many moles of NaCl are in 10 g?

7 How many Na^+ ions are present in the 10 g in problem 6?

8 If one wanted to weigh out 1 mol each of $MgCl_2$ and $ZnCl_2$ for a chemical reaction, how many grams would one weigh?

9 List several practical ways chemists use the mole concept.

10 How many years would it take to count the atoms in a tiny speck of iron (0.001) g)? Assume that one can count one atom each second.

11 What is the *ratio* of number of atoms in 32 g of sulfur and 39 g of potassium?

12 From your answer to problem 11, why did chemists not need to know the absolute number of atoms in a mole to deal with atom-molecule ratios?

13 How many moles are in 280 g of H_2SO_4?

14 What is the ratio of oxygen atoms to Fe_2O_3 units?

15 How many molecules of naphthalene ($C_{10}H_8$) are in one naphthalene mothball (each mothball weighs 3.0 g)?

16 If a mothball (3.0 g) disappears (sublimes—goes from a solid to a vapor) in 1 week, how many molecules are escaping from its surface each minute?

Goals

To gain a better understanding of the behavior of commonly used materials

To understand the reason for certain procedures used in everyday life

To understand the meaning of chemical terms used in the news media and in later chapters of this book

9

THE STATES OF MATTER

As you will remember from Chapter 1, matter can exist in three states, or phases: gas, liquid, and solid. We shall examine each of these states from a molecular point of view and consider four characteristics: attractive forces between molecules, distance between molecules, motion of the molecules, and orderliness of the arrangement of molecules. Much of our understanding of molecular characteristics comes from everyday observation or from simple laboratory experiments rather than from sophisticated ones.

KINETIC MOLECULAR THEORY

Attractive forces between molecules

First, consider attractive forces. One cannot walk through a solid material (such as a brick wall), but one can walk through a liquid (water) with some effort and through a gas (air) with ease. This means that since it takes great effort to push aside the molecules of a solid, they must be attracted strongly to each other (there must be strong attractive forces). Liquid takes some effort to part, and so there must be some attraction between liquid molecules—but much less than for a solid. Gaseous molecules must have essentially no attraction for each other, since little effort is needed to pass through a material in this phase.

Distance between molecules

Solids and liquids cannot be significantly compressed (made smaller in volume) no matter how much pressure is put on them. Therefore, their molecules must be touching each other (there must be no space between them). On the other hand, gases can be compressed; this indicates that there is a large space between molecules in this state.

Motion of molecules

If one places a black solid object or substance on a white one, no mixing of the two is observed. This means that molecules of solids are not spontaneously moving (moving of their own accord) from one place to another (there is no translational motion).

When one places a crystal of colored dye in a beaker of water and leaves the system undisturbed (does not stir the system), the color gradually moves throughout the entire volume of the water. This means that the molecules are moving within the liquid. A classical experiment that can be repeated fairly easily is to place a fat-soluble dye in homogenized milk. The microscopic globules of butterfat absorb the dye and become visible under a microscope. When one watches the globules, one sees each globule moving in a random manner as though it were being bombarded from all sides by the invisible molecules of the liquid. This is called *Brownian motion* (see Figure 9-1); it was among the evidence used by early scientists to postulate the continuous, random motion of molecules in the liquid state.

A similar conclusion can be drawn about the motion of molecules in the gaseous state. By looking at smoke particles under a microscope, one also

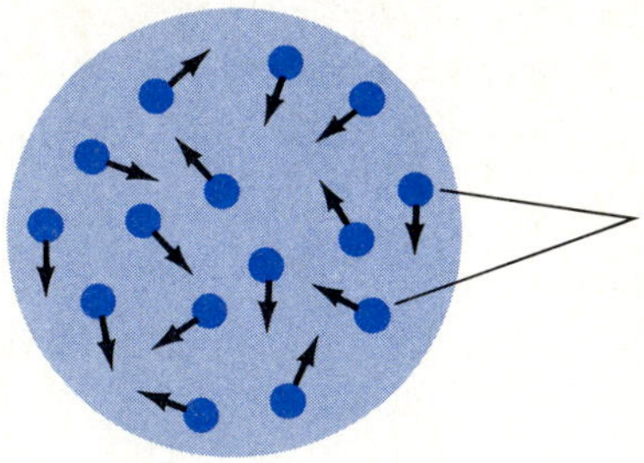

FIGURE 9-1
Brownian motion: view through a 100 X microscope of homogenized milk stained with Sudan IV (a red dye).

observes random movement of the particles as a result of bombardment by constantly moving gas molecules. If one places a quantity of colored gas in a closed flask, it quickly colors the air in the whole container. Gaseous molecules must, therefore, be moving from one place to another even more rapidly than those in liquids.

One can also observe that for both the gaseous and liquid states, this moving of a colored material (a process called *diffusion*) is faster at higher temperature. Remember that higher temperature means higher molecular kinetic energy or faster movement of molecules.

Ordering of molecules

The ordering of molecules cannot be demonstrated by the same simple types of experiments and observations that have just been discussed, but it is known from other experiments that molecules in most solids are highly ordered. This means that the molecules in most solids exist in rows and layers, or in other regular geometric patterns; solids that have this characteristic are called *crystalline*. A few solid materials such as glass and rubber have much less order and are called *amorphous*. These latter materials are sometimes actually very slow-flowing liquids in one respect. If one places a quartz crystal (a crystalline solid) in a particular location and returns years later, it will not have changed its shape or size. In contrast, it is reported that if one examines a window glass from a very old house, it will be thicker at the bottom and thinner at the top than it was originally. The glass has been flowing like very thick syrup during those years. Some have questioned this observation; but if one heats glass, this slow flowing can be observed within a short period of time.

It has also been determined that gases have no orderliness at all and that liquids have very little.

Table 9-1 summarizes the three states of matter. One can draw macroscopic analogies to the molecular behavior of these three phases. Gas molecules behave much like the constantly flying bees that are contained in a jar; liquids behave much like marbles that fill only half a jar; and solids behave much like an orderly stack of magnets. Figure 9-2 shows drawings of molecules in the three states. This view of matter, which is consistent with all known observations, is called the *kinetic molecular theory* (a theory of matter based on the motion of molecules).

TABLE 9-1

Summary of the three phases of matter according to the kinetic molecular theory

	GAS	LIQUID	SOLID CRYSTALLINE	SOLID AMORPHOUS
Attractive forces between molecules	None	Small	Great	Great
Space between molecules	Great	None	None	None
Translational motion of the molecules	Great	Some	None*	None*
Orderliness of the molecules	None	Very little	High	Low

*In the solid phase, molecules vibrate, even at very low temperature, but do not move from one place to another to any appreciable extent.

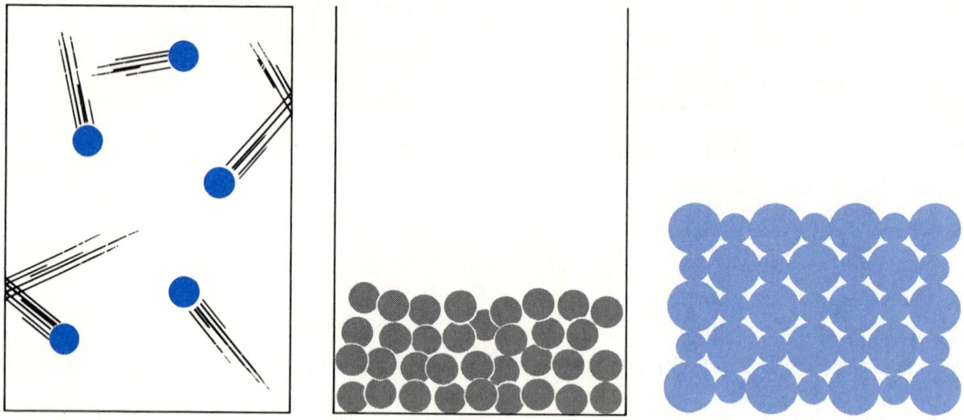

FIGURE 9-2

The three states of matter on the atomic or molecular level.

PHASE CHANGES

Most substances can exist in all three phases and change from a solid to a liquid (melting) and finally to a gas (evaporation) as the temperature is raised. This process can be reversed by cooling the substance; gaseous materials can be changed to liquids (condensation) which, in turn, change to solids (freezing) as the temperature is lowered still further. It is also possible under certain conditions of pressure to change most materials directly from the solid to the gaseous state (sublimation) and vice versa (condensation). These processes are diagramed in Figure 9-3.

Several examples will help to clarify these processes. Water, which at ordinary temperature is a liquid, can be changed to a solid (ice) by lowering the temperature to 0°C or to a gas (steam) by raising the temperature to 100°C (or 212°F); see Figure 9-4. Each material has its own boiling and melting

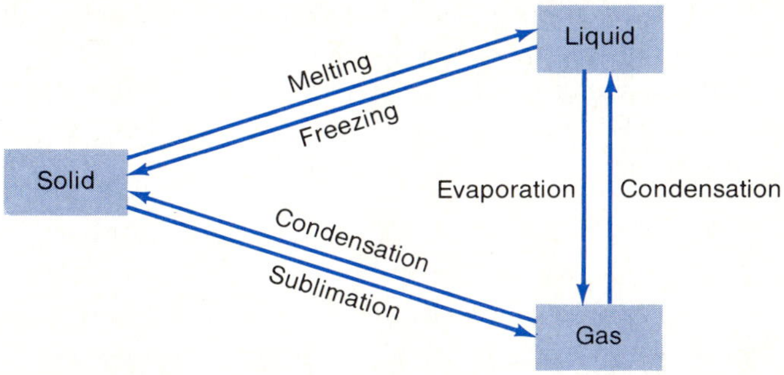

FIGURE 9-3
Interconversions between the phases.

FIGURE 9-4
Ice, water, and steam—the three states of matter.

points. Nitrogen, which is normally encountered as a gas, can be cooled until it condenses to a liquid (at −196°C) and then cooled still further until it, too, changes to a solid (at −210°C). Iron, which is usually a solid, can be heated until it melts (at 1535°C) or, if an extremely high temperature is used, until it boils (at 3000°C). All these phase changes are reversible. There are many solid materials that decompose before any melting occurs, and thus they do not exist as liquids or gases. Limestone ($CaCO_3$) decomposes to CaO and CO_2 before any melting occurs, and wood burns in air or decomposes to methyl alcohol (among other substances) in the absence of air before it can reach a high enough temperature to melt. The decomposition of wood in the absence of air to methyl alcohol is called *destructive distillation* and was formerly used as a way of making methyl alcohol (wood alcohol). Three common examples of sublimation are (1) the evaporation of ice from a sidewalk or the drying of frozen wet laundry even when the temperature is not above the freezing point of water, (2) the evaporation of moth balls, and (3) the disappearing of a piece of dry ice (solid CO_2).

These phase changes are brought about by a change in temperature, which is in turn caused by the addition or removal of heat. Temperature has already been defined as a measure of the average kinetic energy of the molecules of a substance. *Heat* can be defined as the total thermal energy of a quantity of a substance; it is the energy possessed by a material as a result of the movement or vibration of its molecules. Although temperature and heat are related, one is *not* a measure of the other. For example, the water in a cup containing boiling water is at the same temperature (100°C) as the water in a bucket containing boiling water (100°C), yet the total heat content of the water in the bucket is higher than the total heat content of the water in the cup. The average molecule in the cup is moving just as fast (has the same kinetic energy) as the average molecule in the bucket; but because there are more molecules in the bucket, the total energy of the water molecules in the bucket is greater.

Each material has its own characteristic boiling and melting temperatures and its own *heat capacity*. Heat capacity is the *amount of heat* needed to change a material's temperature by a given amount. The units of heat measurement are the calorie* (abbreviated as cal) and the joule (abbreviated as J). A calorie is the amount of heat needed to raise one gram of water one degree Celsius, and a calorie is equal to 4.184 joule. If one adds 1 cal of heat (from a flame or electric heater, for example) to 1 g of water, the temperature of the water will rise only 1°C. On the other hand, if one adds 1 cal to 1 g of copper metal, the temperature of the copper will rise 11°C. Put another way, it takes only 1⁄11 or 0.09 cal to raise 1 g of copper 1°C. This quantity of heat, the amount needed to increase the temperature of one gram of a substance one degree Celsius, is called its *specific heat*.

Each phase of a material has its own specific heat. The specific heat of ice and steam are only 0.5 cal/g · degree (read as "calories per gram per degree"), while the specific heat of liquid water is 1 cal/g · degree. This means that ice changes temperature twice as fast for a given rate of heating as water does, since only half as much heat energy is needed to bring about a given temperature change.

Likewise, each substance has its own characteristic *heat of fusion,* the amount of heat needed to *melt* one gram of it, and *heat of vaporization,* the energy needed to boil or evaporate one gram of it. It requires 80 cal to melt 1 g of ice to water (heat of fusion) and 540 cal to evaporate 1 g of water to steam (heat of vaporization). These same quantities of heat are given off when the steam is condensed or the water is frozen. The change in heat content is signified by the symbol ΔH [Δ (the Greek letter delta) = change; H = heat] and is sometimes called *enthalpy*.

The melting and boiling processes are *not* accompanied by a temperature change. When water reaches 100°C, it begins to boil and changes to steam, which *remains at 100°C*. The additional energy put into the water is used to separate the water molecules from each other, and no increase in the *average kinetic energy* of the molecules occurs until all the water has been changed to steam. Thus steam at 100°C contains a lot more heat than water at 100°C (540 cal/g more), but the temperature of each is the same. Likewise, when ice reaches 0°C, it begins to melt and changes to water, which *remains at 0°C* until all the ice has changed to water. The temperature stays at 100°C during

*The Calorie, which is often used in discussions of diet and nutrition, is 1000 cal, or 1 kcal. The kilocalorie is sometimes referred to as a *big calorie*.

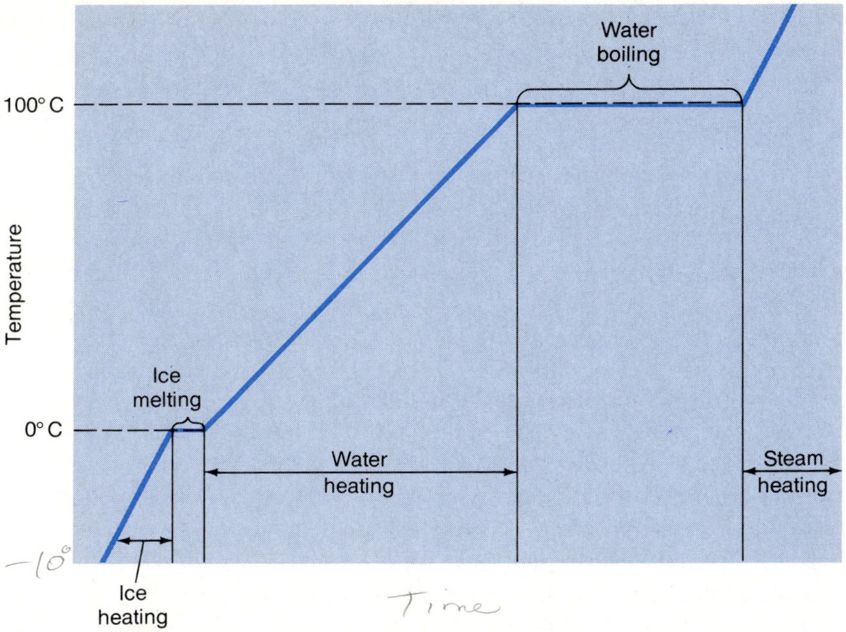

FIGURE 9-5
Heating curve for water.

the entire boiling process and at 0°C during the melting process. Boiling and melting are said to be *isothermal* (same-temperature) changes.

Figure 9-5 shows what happens to ice as it is heated. This type of graphic representation is called a *heating curve* and is a plot of temperature versus time. Figure 9-5 depicts the changes in the temperature with time of a block of ice at −10°C that is placed on a heater. First the ice temperature rises to 0°C. While the ice is changing to water, the temperature of the ice-water mixture remains at 0°C. Only after all the ice melts does the water temperature rise slowly to 100°C. Then, the temperature again stops rising until all the water has been changed to steam. Finally, if this were done in a closed container and the steam were trapped, the temperature would again begin to rise as heating continued.

Water, which is the solvent in all living systems, is unusual in many ways. It has one of the highest heats of fusion, heats of vaporization, and specific heats of any common substance. Thus more energy is needed to bring about a temperature or phase change in water than in any other commonly encountered material. Or, expressed another way, water resists temperature or phase changes more than most other materials. This makes living systems more stable—less apt to change temperature or state—when they are placed in a hot or cold environment.

We can do a few simple calculations to help understand some of the energy change phenomena we experience in everyday life. The useful relationships are as follows.

FOR TEMPERATURE CHANGES:

$$\text{Heat} = \text{specific heat} \times \text{amount of material} \times \text{temperature change}$$

Or, put in more compact mathematical terms,

$$\Delta H = (\text{Sp H})(\text{weight})(\Delta T)$$

FOR FREEZING:

$$\begin{aligned} \text{Heat} &= \text{heat of fusion} \times \text{amount of material} \\ \Delta H &= (\text{H}_{\text{fus}})(\text{weight}) \end{aligned}$$

FOR EVAPORATION:

$$\begin{aligned} \text{Heat} &= \text{heat of vaporization} \times \text{amount of material} \\ \Delta H &= (\text{H}_{\text{vap}})(\text{weight}) \end{aligned}$$

Let's consider the amount of heat given off by 0.5 g of steam condensing. The heat *given off* when the steam condenses is the *same* as the amount of heat needed to *cause evaporation*. Since condensation is just the reverse physical process of evaporation, energy goes in the opposite direction. Just as the heat content of the water has to be increased to cause evaporation, the heat content of the steam has to be reduced by the same amount to cause condensation. Heat (and all other forms of energy) cannot be created or destroyed. It can change forms (electrical to thermal, etc.) or it can change from kinetic to potential energy, or vice versa, but the total amount of energy in a system and its surroundings must be the same at all times. This is called the *law of conservation of energy*.

Let's get back to our problem of calculating the amount of heat given off by 0.5 g of steam condensing.

$$\begin{aligned} \Delta H &= (\text{H}_{\text{vap}})(\text{weight}) \\ &= (540\ \text{cal}/\cancel{\text{g}})(0.5\ \cancel{\text{g}}) \\ &= 270\ \text{cal} \end{aligned}$$

This means that if one placed one's hand in a jet of steam long enough for 0.5 g to condense on it, the 270 cal of heat would be transferred to the hand. One should not attempt this, however, since it can cause severe burns.

Now let's consider what would happen if 0.5 g of boiling water were placed on a person's hand. The water would cool from 100°C to 37°C (body temperature), which is a change of 63°C. Since this is a temperature change, not a phase change, the applicable equation is

$$\begin{aligned} \Delta H &= (\text{Sp H})(\text{weight})\Delta T) \\ &= (1\ \text{cal}/\cancel{\text{g}} \cdot \cancel{\text{degree}})(0.5\ \cancel{\text{g}})(63\ \cancel{\text{degrees}}) \\ &= 31.5\ \text{cal} \end{aligned}$$

This means that if 0.5 g of boiling water were placed on a hand, only about 31.5 cal would be transferred to that hand. If the 0.5 g of steam were condensed on a hand (see previous problem), the condensed water, at 100°C, would then cool to 37°C. So the total energy transfer to the hand would be 270 cal + 31.5 cal = 301.5 cal. It can be seen that about 10 times as much heat is transferred when steam is used (301.5 cal) than when boiling water is used (31.5 cal). This difference in energy transfer is the reason steam causes a more severe burn than an equal amount of boiling water.

Water is also the most efficient cooling agent. When 1 g of water evaporates from the skin during perspiration, over 500 cal is transferred *from* the body in the process. This is more than would be transferred by the evaporation of any other material, since the heat of vaporization of water is higher than that of any other material. The high heat of vaporization of water also explains why a person who is wet quickly feels cold when a breeze hits him or her. Since the wind causes evaporation of the water from the skin, a great deal of heat is transferred from the body as the water molecules change to the gaseous state.

Let us now take a brief look at the three states of matter and see how the behavior of matter can be explained on the basis of the order, attractive forces, movement, and distance of atoms or molecules.

SOLIDS

Solids are formally defined as those materials that have a definite shape and size of their own. In the first part of this chapter, we saw that in solids, the atoms, molecules, or ions are very close to each other; that they are arranged in an orderly manner; that there is no transitional motion; and that there are strong attraction forces. This last characteristic is what gives solids their definite shape. The units hold on to each other so tightly that gravity cannot force them into a different shape.

The most characteristic solids are those of *ionic* crystals. Figure 9-6*a* shows a crystal of NaCl before and after it has been broken. The small crystals

FIGURE 9-6
Sodium chloride crystals. (a) *Left,* whole single crystal; *right,* broken crystal. (b) Natural salt crystals (halite). (From the collection of Dr. Kurt Niedenzu.)

(a)

(b)

that result from the breaking are similar in shape to the larger one (they are all cubes). This *cleavage* pattern is the result of the arrangement of the Na^+ and Cl^- ions in the solid. Figure 9-7 shows a model of the arrangement of the two types of ions. Notice the high degree of order and the cubic nature of this arrangement. Figure 9-8 shows a crystal of calcite ($CaCO_3$), along with a model of the ionic arrangement. Notice the angular array in the model which is manifested in the actual crystal shape. Breaking a large calcite crystal produces smaller similarly shaped pieces.

FIGURE 9-7
Model of NaCl crystal lattice.

FIGURE 9-8
Left, model of crystal lattice of calcium carbonate; *right,* calcium carbonate crystal. (The crystal, a form of calcite called *Icelandic spar,* is from the collection of Dr. Kurt Niedenzu.)

Materials such as quartz are also highly crystalline and have a definite shape—clear hexagonal needles. These are shown in Figure 9-9 and are composed of *chains* of SiO_2 units. The bonding here is covalent rather than ionic (Si and O are both nonmetals); and solids of this kind are called *covalent* or *extended-lattice,* solids, since the molecules are long chains extended indefinitely. Wood is composed of long chains of cellulose; plastics also contain long molecules. Wood and plastic are both classified as extended-lattice solids although neither wood nor plastic is highly crystalline in nature. Diamond (pure carbon) is the ultimate extended-lattice solid. Figure 9-10 shows a model of the arrangement of carbon atoms in a diamond. The lattice extends in all directions, so that a diamond gemstone can be considered to be one very large molecule.

Crystals of Al_2O_3 are used as jewels. If they are colored blue (by impurities of cobalt), they are sapphires; if they are colored red (by chromium), they

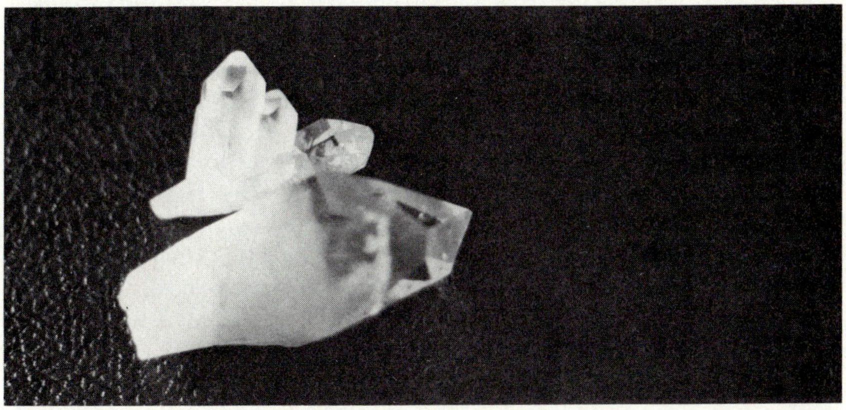

FIGURE 9-9
Quartz crystals. (From the collection of Dr. Kurt Niedenzu.)

FIGURE 9-10
Model of the molecular structure of diamond.

are rubies. Another form of Al_2O_3 is called *corundum;* this is the hard gray material used for grinding stones. The exact nature of the arrangement of atoms, ions, and molecules determines the nature of the material. Al_2O_3, when forming a solid along with water molecules, exists as the very soft, crumbly *bauxite* used as an aluminum ore. Among all natural materials, the purer forms (sapphire and ruby) are second only to diamond in hardness (9 on a scale of 10).

A class of materials called *amorphous solids, or glasses,* have, for all practical purposes, a definite shape (they are rigid materials). They differ from true solids, however, in that the molecules are not highly ordered. Indeed glass is actually a very slow-flowing liquid (as was mentioned earlier in this chapter). Some glasses are harder, more heat-resistant, clearer, etc., than others. Table 9-2 gives the name, composition, and special properties of various types of glass.

The basic solid nature of glasses is provided by the SiO_2 (sand—small pieces of quartz). Other materials in the mixture help to break up the orderliness of the natural SiO_2 and cause the amorphous character. This in turn lowers the melting point from 1400°C for pure SiO_2 to between 900°C and 1200°C, depending on the exact nature of the glass. This is a commonly observed effect—*adding impurities usually lower the melting point of a solid.* This principle is used to lower the temperature of ice-water mixtures. Salt can be added to ice-water mixtures, which lowers the melting point of the ice to below the normal 0°C. It is lowered 3.7°C for each 58 g (1 mol) of salt in a liter of ice water. Salt (NaCl) or $CaCl_2$ (calcium chloride) placed on icy steps and walkways causes the ice to melt at a lower temperature. This is a commonly used safety precaution.

One can change the properties of other materials by adding relatively small amounts of another material. For example, paraffin wax is a fairly soft

TABLE 9-2

Composition of glasses

NAME	APPROXIMATE COMPOSITION	PROPERTIES	USES
Soft	75% SiO_2 15% Na_2O 5% CaO 2% Al_2O_3 3% MgO	Inexpensive	Windows, bottles
Flint	45% SiO_2 45% PbO 10% K_2O or Na_2CO_3	Heavy, optically clear	Crystal goblets, cut glass, eyeglasses
Borosilicate (Pyrex)	80% SiO_2 12% B_2O_3 4% Na_2O 4% Al_2O_3	Chemically inert; little expansion when heated; does not break easily when exposed to rapid temperature changes	Cooking ware, laboratory glassware
Lead	20% SiO_2 80% PbO		Shielding to block radioactive emissions

solid that melts at a low temperature and, when formed into candles, tends to drip while the candle is burning. If a small amount of stearic acid is added, the paraffin becomes harder and has a higher melting point. When paraffin mixed with stearic acid is used in candles, the candles do not drip (they are dripless candles). This rise in melting point is an exception to the general rule and produces an effect that is opposite in direction to that observed when other materials are added, for example, to SiO_2. (The author is unaware of a good explanation for this anomaly.)

Molecular solids are those composed of covalently bonded individual molecules (not long chains of atoms). Most materials composed of nonmetals fall into this category. Dry ice (CO_2), sugar, and aspirin are of this type. The exceptions are those few known specifically to have an extended lattice, such as quartz (see above).

LIQUIDS

Liquids are those materials that take the shape of their container but do not necessarily fill it. On a molecular level, liquids have fairly rapidly moving particles that are close to each other and are mildly attracted to neighboring molecules. There is little if any discernible orderliness in the liquid state.

Liquids, along with gases, are called *fluids*. Materials are sometimes changed into the liquid state so that they can be handled more easily. Plastics and synthetic fibers are frequently processed at high temperatures and placed in molds or forced through small holes in metal pieces called *spinnerettes*. Once the thin filaments of liquid are extruded, they cool and harden into fibers that can be woven into cloth. Figure 9-11 shows a spinnerette in operation.

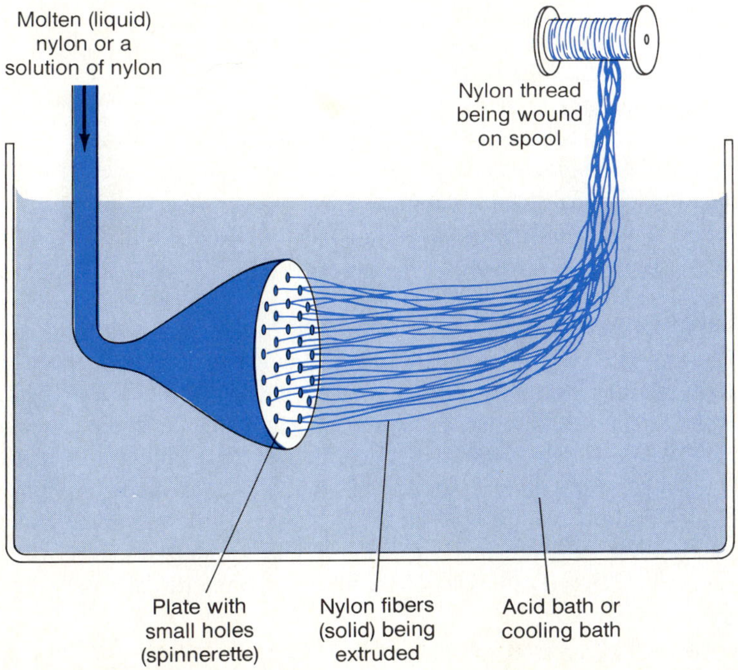

FIGURE 9-11
Nylon being changed from the liquid state to solid fibers.

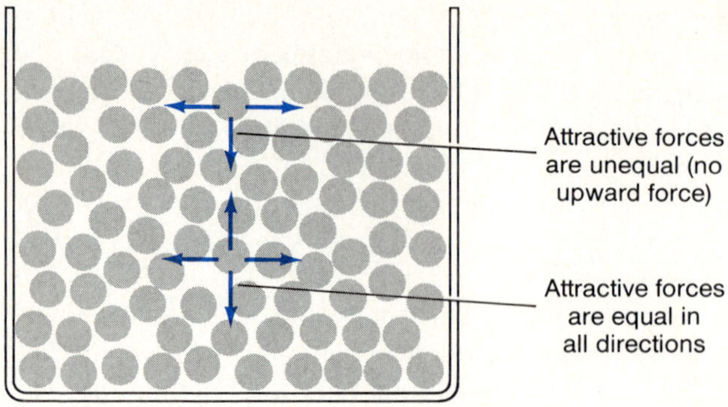

FIGURE 9-12
Attractive forces in liquids.

FIGURE 9-13
Curved meniscus of water in a graduated cylinder.

Surface tension is a phenomenon that is unique to the liquid state and is responsible for part of the behavior of liquids. Figure 9-12 shows the attractive forces in liquids. Molecules on the surface experience attraction forces only on the bottom and sides rather than from all directions. This produces a ''layer'' of molecules that is harder to penetrate than the molecules in the center of the liquid mass. The force holding the surface molecules together is called *surface tension* and accounts for the fact that water, for example, can be ''piled up'' beyond the top of a glass and tends to climb the walls of a container. This curved surface, shown in Figure 9-13, is called a *meniscus*.

If the diameter of the container holding water is reduced to several millimeters, the *attractive forces* of the liquid molecules for the sides of the glass container cause the liquid to literally ''climb the walls'' so that the liquid level rises in the narrow tube (capillary). This phenomenon is referred to as *capillary action*. It is part of the mechanism by which water rises in the capillaries of plants to great heights with no outside pressure (actually, there is no complete explanation for this).

Surface tension also causes drops of oil in water (or vice versa) to assume spherical shapes. The polar water molecules are more attracted to each other than to the nonpolar oil molecules. Thus they clump together and assume a roughly spherical shape. Materials called *surfactants* (see Chapter 17 for more detail) cause this surface tension to break down and allow water and oil to mix in a homogeneous manner.

Liquid crystals—A special class

There is a class of liquids in which the molecules (usually rather large ones) are naturally lined up parallel to each other. Since this constitutes a regular ordered pattern, these are called *liquid crystals* (LC). There are several ways they can be ordered. One is like toothpicks in a holder and another is like layers on pieces of glass turned at angles to each other (see Figure 9-14). Liquid crystals can exist in different phases. Both temperature changes and electric voltage can cause one phase to change into another or can cause the material to lose all ordering. These phase changes are used as the basis for two consumer products.

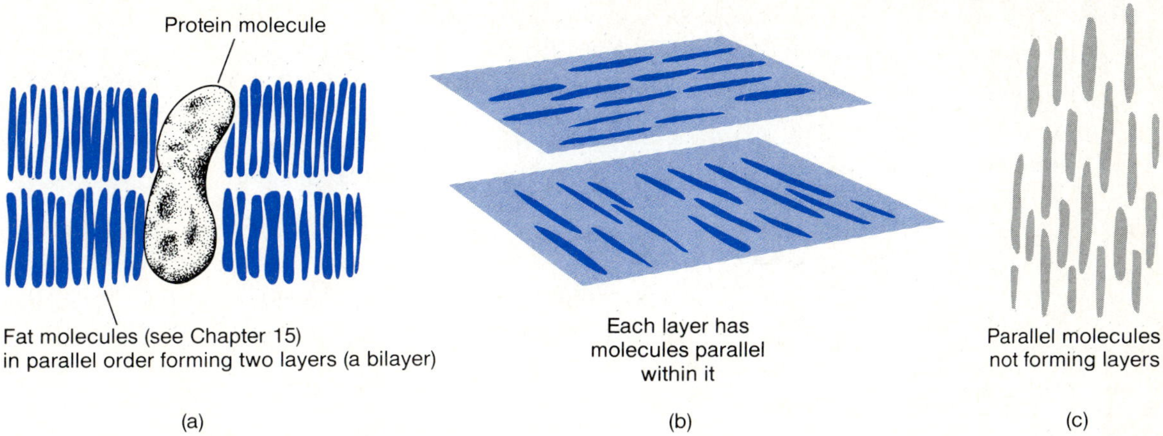

FIGURE 9-14
Liquid crystal arrangements. (a) Cross section of the lipid bilayer in a cell membrane. (b, c) Other ways in which molecules can be ordered under the influence of heat or an electric current.

The first product is a convenient way of measuring temperature. Since the various phases interact with light in different ways, the different phases are colored differently. Sheets of plastic-containing LC material can be placed on the forehead of a small child. If the child has a fever, the plastic will be one color (red, say); if the child's temperature is the normal 98.6°F (37°C), the plastic will be another color (such as blue). Sheets of a similar material can be placed on the hands of people with circulatory problems. Where blood flow is normal, more heat is dissipated and the plastic will be a different color. Thus, closed blood vessels can be observed without surgery or expensive diagnostic procedures. (See Figure 9-15.)

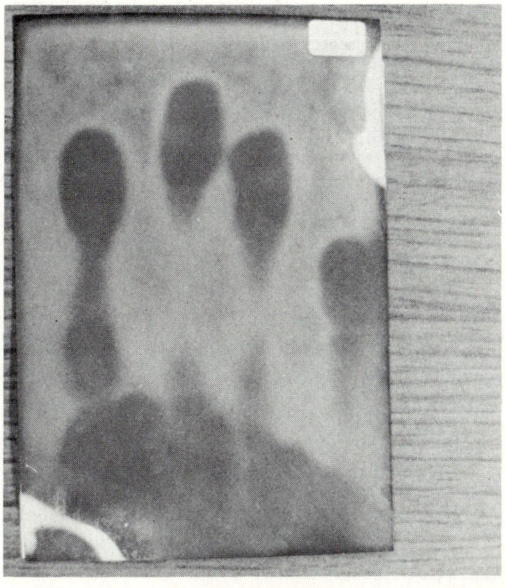

FIGURE 9-15
Liquid crystal sheet on which a pattern has appeared as a result of color change caused by the heat of a hand.

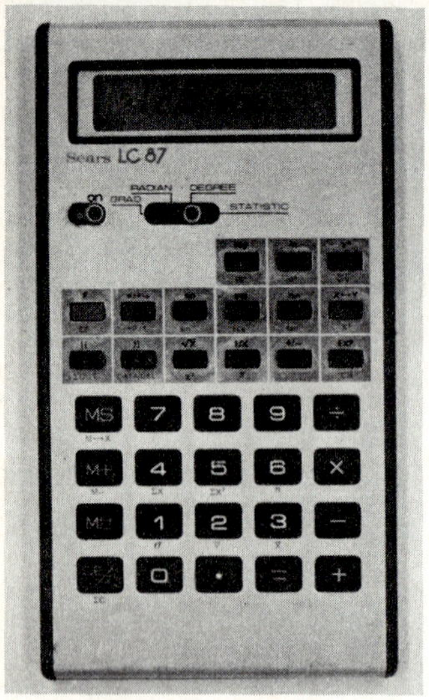

FIGURE 9-16
LC display on a pocket calculator.

The second common application is in LC displays on pocket calculators (see Figure 9-16). Normally the LC is transparent; but when certain sections of the display are activated by electric voltages from the computer, various number-shaped patterns appear. This application can also be used for large billboard-type displays and even for colored pictures, since different voltages produce different phases, causing a variation in color.

GASES

Gases are fluids that take the shape of their container and fill it. The molecules of a gas are in constant motion, do not attract each other, and have large distances between them. There is no order in the gaseous state. Because there is constant motion, the molecules are bombarding the sides of the container. This causes an outward push which we call *pressure*. Pressure is defined as the force per unit area.

Gaseous materials are used to pressurize spray containers of paint, deodorant, and hair spray (see Figure 9-17) and to inflate vehicle tires. Gases produced chemically in baking cause breads and cakes to be light and fluffy (see Chapter 17), and gaseous CO_2 is used to produce the bubbles (or ''fizz'') in soft drinks, beer, and champagne.

Air

The most common gas we encounter in our everyday experience is air. Air is a mixture of 78 percent N_2, 20 percent O_2, 1 percent Ar, and smaller amounts of H_2, CO_2, He, H_2O, and several other gases. This mixture covers the earth to a depth of about 1000 mi (99 percent of it is within 100 mi), and because of the constant bombardment of objects near the earth's surface, produces a constant pressure on them. This force is commonly called *atmospheric pressure*. It is measured in several different units.

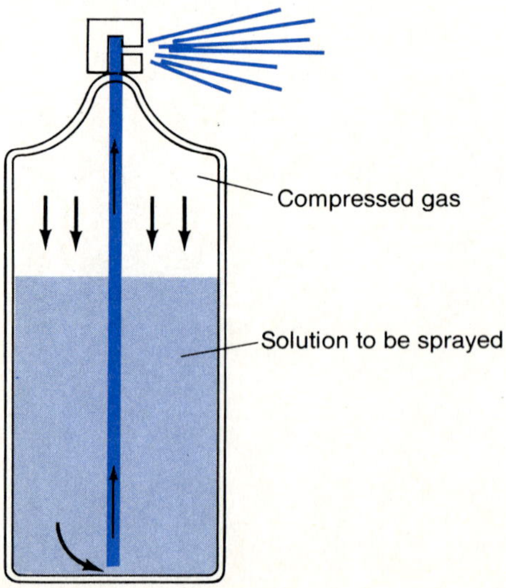

FIGURE 9-17
Aerosol can.

Pressure measurement of air The two devices used to measure air pressure are the *mercury manometer* (sometimes also called a *mercury barometer*) and the *aneroid* (dry) *barometer*. Both are illustrated in Figure 9-18.

In the manometer a U-shaped glass tube is sealed at one end, and mercury, the only metal which is a liquid at ordinary temperatures, is placed in the tube. In the area above the mercury at the sealed end, there is no air. There are no molecules bombarding the surface of the mercury. Such an area—containing no matter of any kind—is called a *vacuum*. It is simply completely

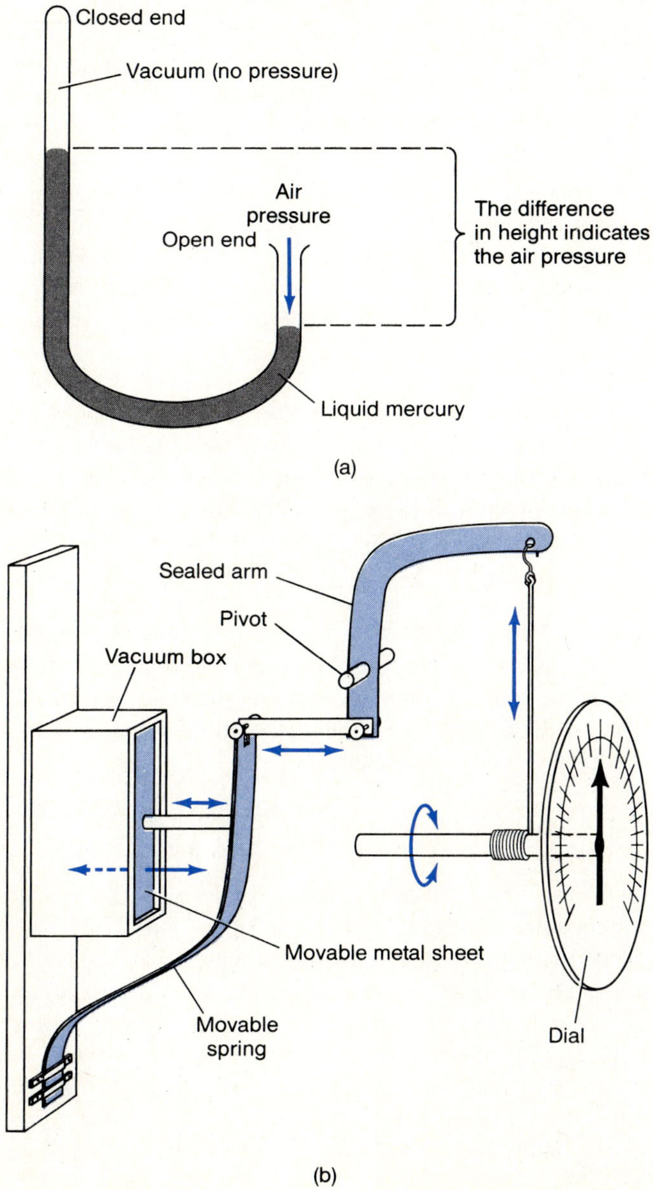

FIGURE 9-18
Devices for measuring the pressure of a gas. (a) Mercury manometer. (b) Aneroid barometer.

empty space. On the open end of the tube, air molecules hit the mercury surface, forcing it downward in the tube. This, in turn, forces the vacuum end to rise, causing a difference in the level of mercury in the two arms of the manometer. The vacuum end stops rising when the *weight* of the mercury pushing back down equals the *force* of the gas on the open end. The greater the pressure, the greater the difference is.

The average pressure of the atmosphere at sea level produces a difference of 760 mm or 29.92 in. Atmospheric pressure (also referred to as *barometric pressure*) is used as an indicator of weather conditions and is given on local weather reports. Low pressures (28 to 29.5 in) indicate that rain or snow is expected, and high-pressure areas (29.5 to 31 in) are usually clear. Very low pressures (<28 in) occur during thunderstorms or tornadoes.

The aneroid barometer works as shown in Figure 9-18 and is a more convenient device, since it contains no mercury. Household atmospheric pressure indicators are usually this type. Such devices use a metal vacuum box that expands and contracts with atmospheric pressure changes. These changes in pressure are indicated on a dial.

While weather forecasters use inches of mercury to indicate pressure, the units millimeters of mercury or torr can also be used and are usually the choice of scientists. Thus, a pressure can be given as 760 mm of Hg or 760 torr. The two are synonymous. Likewise, one can use pounds per square inch (lb/in^2). Ordinary atmospheric pressure is 14.6 lb/in^2, which means that each person has 14.6 lb of force on each square inch of his or her body. This same pressure is both inside and outside of most objects so that no force is actually felt. In those rare cases such as tornadoes, which are very low pressure areas, the 14.6-lb pressure inside of buildings causes a tremendous force outward (up to 1 lb/in^2), which causes the walls to literally explode outward.

Relative humidity of air Air contains water vapor, and the amount of water present has a significant impact on living systems. The term *relative humidity* is used to indicate the amount of water in the air. This term means the amount of water dissolved in air compared with the maximum amount that could be dissolved at that temperature:

$$\text{Relative humidity} = \frac{\text{amount } H_2O \text{ dissolved in air}}{\text{maximum amount}} \times 100$$

Thus, if the air contains half as much as is possible, it is said to have 50 percent relative humidity. If the relative humidity is too low, most house plants do not grow well, human skin dries out and itches, and one feels colder at a given temperature. The relative humidity is usually low during the winter when buildings are being heated with hot, dry air, and it is necessary to add moisture to the air to raise its water content. This can be done by mechanical devices that spray droplets of water into the air, by placing a water tray on radiators or inside hot-air ducts, or by having a large number of well-watered house plants. The ideal relative humidity for skin health and comfort is 50 to 70 percent.

If the humidity is very high (greater than 90 percent), the air feels damp and mold and mildew can grow. This happens in low-lying areas during the summer and in basement living areas. A dehumidifier, which is a device that removes water vapor from the air, can be used to lower the relative humidity. A dehumidifier is basically a small refrigeration unit that causes coils of tubing to cool and condenses the water vapor into the liquid state. The liquid water drips off the coils and is discarded or used for any source where distilled or demineralized water is needed, such as in humidifiers in the winter or steam irons. It can be collected in the summer, stored in plastic containers, such as clean gallon milk containers, and used later.

It should be pointed out that as the temperature of the air increases, the air can dissolve more water. For example, a given sample of air containing 1 g of water per liter at 60°F has a relative humidity of 50 percent (since 2 g of water can be dissolved in air at 60°F). If the air is heated to 80°F, the *relative* humidity goes down. The 80°F air can now dissolve 5 g of water, and the 1 g it actually has is now only one-fifth of the maximum, or 20 percent relative humidity. At the 20 percent level, water evaporates from skin and plant leaves at a higher rate.

Behavior of gases

A given sample of a gas can be characterized by several measurements (parameters); see Figure 9-19. These parameters include:

Pressure—the force per unit area exerted by the moving molecules bombarding the side of the container

Volume—the amount of space occupied by the sample

Temperature—the average kinetic energy of the molecules

Quantity of material—the weight (mass) of the actual gas molecules or the number of moles of the material

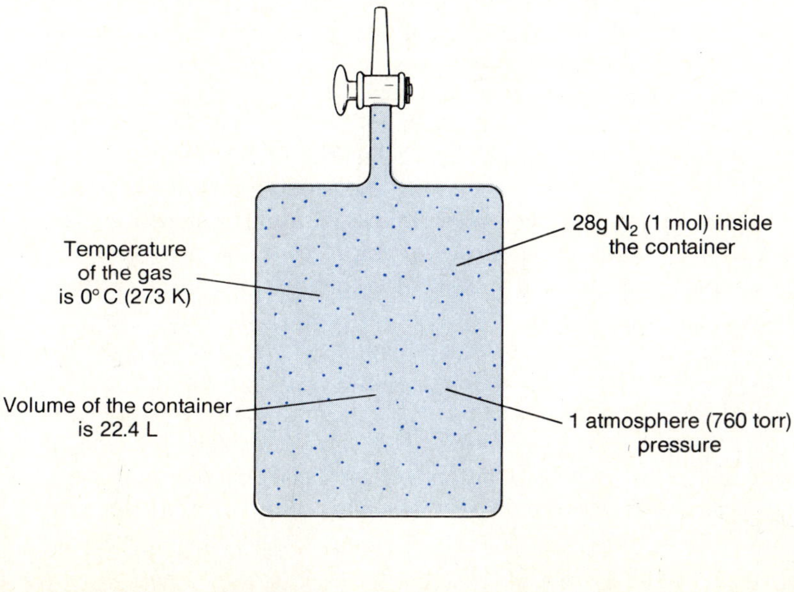

FIGURE 9-19
Parameters of a sample of a gaseous material.

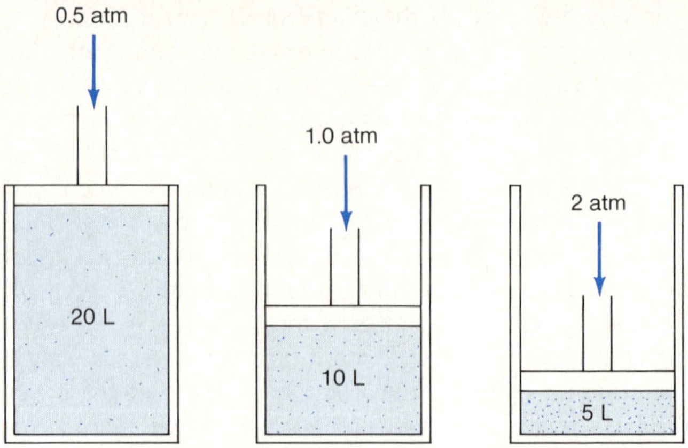

FIGURE 9-20
Boyle's law: As pressure is increased on a sealed gas sample, the volume decreases proportionally.

As one (or more) of these parameters changes, it causes others to change also. As the pressure on a sample of gas is *increased,* the molecules are pressed closer together and the volume *decreases*. This effect of the pressure of a gas on its volume is called an *inverse relationship;* that is, a change in the pressure causes a change in the opposite direction in the volume. As the pressure on the gas is *doubled,* the volume is cut to *half* its original value and vice versa. Figure 9-20 illustrates this principle, which is called *Boyle's law*. This relationship holds as long as the temperature of the gas does not change.

The temperature also affects the volume or the pressure of the gas. If the temperature is doubled, the volume doubles (as long as the pressure is not changed). This is shown in Figure 9-21*a* and is called *Charles's law*. If the volume is held constant (as in a rigid container), an increase in temperature causes a corresponding increase in pressure, as shown in Figure 9-21*b*. This is *Gay-Lussac's law*. It is important to note that the temperature must be measured on the Kelvin (K) scale (also called the *absolute scale*), since that is the only scale on which the numerical readings are proportional to the actual kinetic energy of the molecules. (See the discussion of temperature scales near the end of Chapter 4.)

As the absolute temperature doubles from 200 K to 400 K, the volume doubles from 6 L to 12 L. You will notice, however, that the reading on the Celsius scale does not double. (See Figure 9-21*a*.)

Gay-Lussac's law explains why one should not heat a closed container. The pressure inside the container will build up and eventually cause an explosion, which can cause destruction of property and personal injury. If one heats an unopened can of beans over an open campfire (or some other heat source), the can can explode, causing hot beans and pieces of metal to spray over—and burn—people, tents, and sleeping bags. The water evaporates, and the gaseous steam builds up pressure. While canning fruits and vegetables, one should leave the lids loose enough for steam to escape so that the jars do not break under the pressure buildup. This law also explains why the pressure in automobile tires increases during summer heat and decreases during winter cold, and why one should keep aerosol cans away from sources of heat.

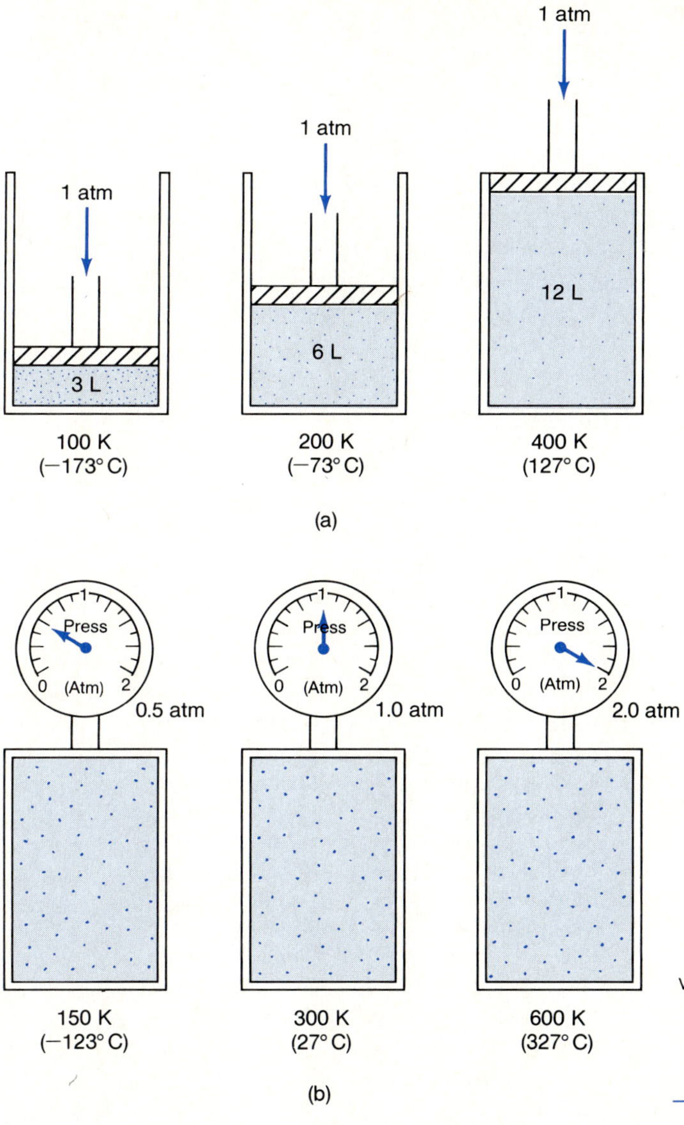

FIGURE 9-21
(a) Charles's law: As the temperature is raised, the volume increases proportionally. (b) Gay-Lussac's law: As temperature is raised in a sealed container, the pressure increases proportionally.

Chemists have chosen a standard or reference set of conditions. They are 1 atmosphere (atm) of pressure and a temperature of 0°C (273 K). These values are called *standard temperature and pressure,* or STP. If one has *1 mol* of *any ideal gas* (for example, 28 g N_2, 32 g O_2, 20 g Ne, or 4 g He), it will occupy 22.4 L at STP. This 22.4 L is called the *molar volume,* or *volume of one mole*.

When one pumps air into a bicycle tire, one is using muscular force to increase the pressure on the gas trapped in the pump. This increase in pressure causes a decrease in volume (Boyle's law), forcing the molecules into a smaller volume. Since there are more molecules in less volume, they exert more force on the walls of the pump (more pressure). This forces the air into the tire. The more molecules one forces into the relatively rigid (constant) volume of the tire, the greater the pressure in the tire.

When gaseous CO_2 is generated chemically during baking (see Chapter 17 for the chemistry involved), the pressure within a given CO_2 bubble forces the dough, etc., to expand. This creates a cavity and makes the bread or cookies light and fluffy. Omelets are prepared by beating bubbles of air into the eggs. When the eggs are heated, this air expands and some of the natural water turns to steam. This causes the cooked eggs to be lighter and have a larger volume than plain scrambled eggs.

STUDY GUIDE

Terms to learn

Amorphous
Atmospheric pressure
Attractive forces
Barometer
Boiling
Boyle's law
Brownian motion
Calorie
calorie
Capillary action
Charles's law
Condense
Crystalline
Evaporate
Extended lattice
Fluid
Gas
Gay-Lussac's law
Glass
Heat
Heat capacity
Heat of fusion
Heat of vaporization
kcal
Kinetic molecular theory
Law of conservation of energy
Liquid crystal
Liquid
Manometer
Molecular solid
Pressure
Relative humidity
Specific heat
Solid
Sublime
Surface tension
Temperature
Torr

Skill practice problems

1 How many calories are needed to melt 15 g of ice at 0°C to water at 0°C? The heat of fusion (H_{fus}) is 80 cal/g.

2 How much heat is needed to change 500 g of water used to boil eggs from 20°C to its boiling temperature of 100°C?

3 How much heat is needed to heat 500 g of water from 20°C to 100°C and then boil away (evaporate) 100 g of it? The heat of vaporization (H_{vap}) is 540 cal/g.

4 One can prepare hard-boiled eggs by heating the water to boiling and *then turning the heat off* and letting the eggs sit in the hot water for 10 min. The alternative is to heat the eggs in boiling water for about 8 min so that some of the water evaporates (assume that about one-fifth of it evaporates). Refer to your answers to problems 2 and 3 and decide which method uses the *least* energy.

5 About how much heat is produced from burning a 20-gal (64-kg) tank of gasoline (C_8H_{18})? Gasoline produces 10.3 kcal/g.

6 List several everyday examples other than those given in the text that model molecules in the gas, liquid, and solid states.

7 Why should one not incinerate (i.e., heat to high temperature) aerosol cans along with paper trash?

8 Explain the term *relative* in *relative humidity*.

9 If a sample of gas originally at a temperature of 200 K, a pressure of 4 atm, and a volume of 8 L has its pressure changed to 2 atm while the temperature stays the same, what will its new volume be?

10 If the pressure in a tire sitting in a warm garage (30°C) is 35 lb/in^2, what will the pressure change to when the tire stands outside at 0°C? (Hint: do not forget to use kelvin temperature.)

11 An aerosol can is designed to withstand 4 atm pressure before it will burst. If it has enough gas in it to produce a pressure of 2 atm at 25°C (298 K—about room temperature), will it burst if heated to 350°C in a paper fire?

12 Whose law was used to solve problem 9? Whose law was used to solve problem 10?

13 Does the metal vacuum box in an aneroid barometer expand or contract when the atmosphere pressure is rising? Why?

14 Will water rise higher or less high than mercury if it is placed in a manometer? Water weighs 1 g/mL and mercury weighs 13.5 g/mL.

15 If mercury weighs 13.5 times as much as water, how far will water rise in a manometer on an average day? The average atmospheric pressure is 760 torr.

16 If one drinks 1000 g (1 L) of ice water (0°C), how many Calories (kilocalories) does the body use to bring the water to 37°C (body temperature)?

Goals

To develop skill in converting from chemical names to chemical formulas and from formulas to names, which will facilitate discussion of chemical substances in later chapters

To learn the chemical nature of some commonly used inorganic chemicals

To examine the concept of acidity and to understand how it is measured and how it affects common human experiences

10

INORGANIC CHEMISTRY

Compounds are designated by *names* as well as by *formulas,* and the relationship between the two is called a *system of nomenclature*. Because there are so many compounds which carbon can form with other elements, carbon-containing compounds (organic compounds) are named using a self-contained system of *organic nomenclature;* all compounds of all other elements are named using a system of *inorganic nomenclature*. We will examine both systems: inorganic nomenclature in this chapter and organic in Chapter 11.

The International Union of Pure and Applied Chemistry (IUPAC) is the international organization that has determined the official system of naming to be used in professional journals and other chemical literature. Before the development of the IUPAC system, compounds were named using a variety of systems. Some compounds had what are now called *common names*. Whole systems (now called *classical systems*) for naming different kinds of compounds had been developed and were widely used by chemists. These non-IUPAC names are still used by many chemists and are still found on labels and in older books; they will therefore be presented along with the IUPAC names.

In addition to inorganic nomenclature, in this chapter we will examine the concept of acidity and the pH scale, several common inorganic materials, and "hard" water.

INORGANIC NOMENCLATURE

For purposes of naming, inorganic materials—those not containing carbon bonded covalently, but including compounds with the carbonate (CO_3^{2-}) and acetate ($C_2H_3O_2^-$) ions—have been divided into several groups. These groups and their distinguishing characteristics are:

1 *Binary compounds*—compounds containing only two elements. More than one atom of each element can be present, however.

2 *Ternary compounds*—compounds containing three or more elements. **(a)** *Acids*—compounds in which the letter H is the first letter in the formula. (Figure 10-1 shows examples of acids.)

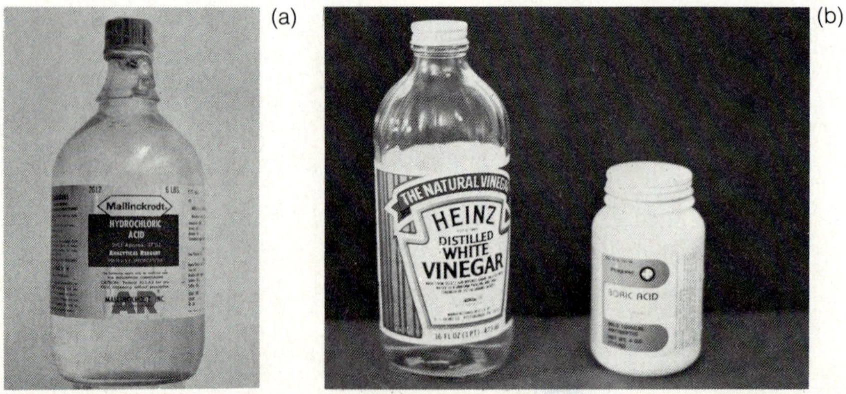

FIGURE 10-1

Commonly encountered acidic materials. (a) Hydrochloric acid (HCl) in concentrated form, used in chemical laboratories. (b) *Left,* acetic acid ($HC_2H_3O_2$) in a 5% solution, used as household vinegar. *Right,* boric acid (H_3BO_3), available at drugstores for household use.

(b) *Bases*—compounds containing hydroxide (OH^-) group along with a metal ion or ammonium ion (NH_4^+). (Figure 10-2 shows examples of bases.)

(c) *Salts*—compounds containing a metal or ammonium ion and one or more nonmetals (other than the OH^- group). This nonmetal portion is called an *anion*. (Figure 10-3 shows examples of salts.)

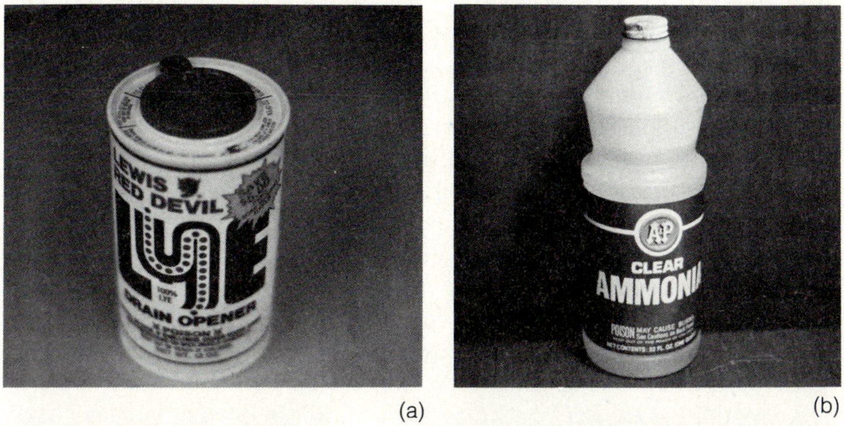

FIGURE 10-2
Common bases. (a) Lye (NaOH). (b) Household ammonia (NH_4OH), used for cleaning.

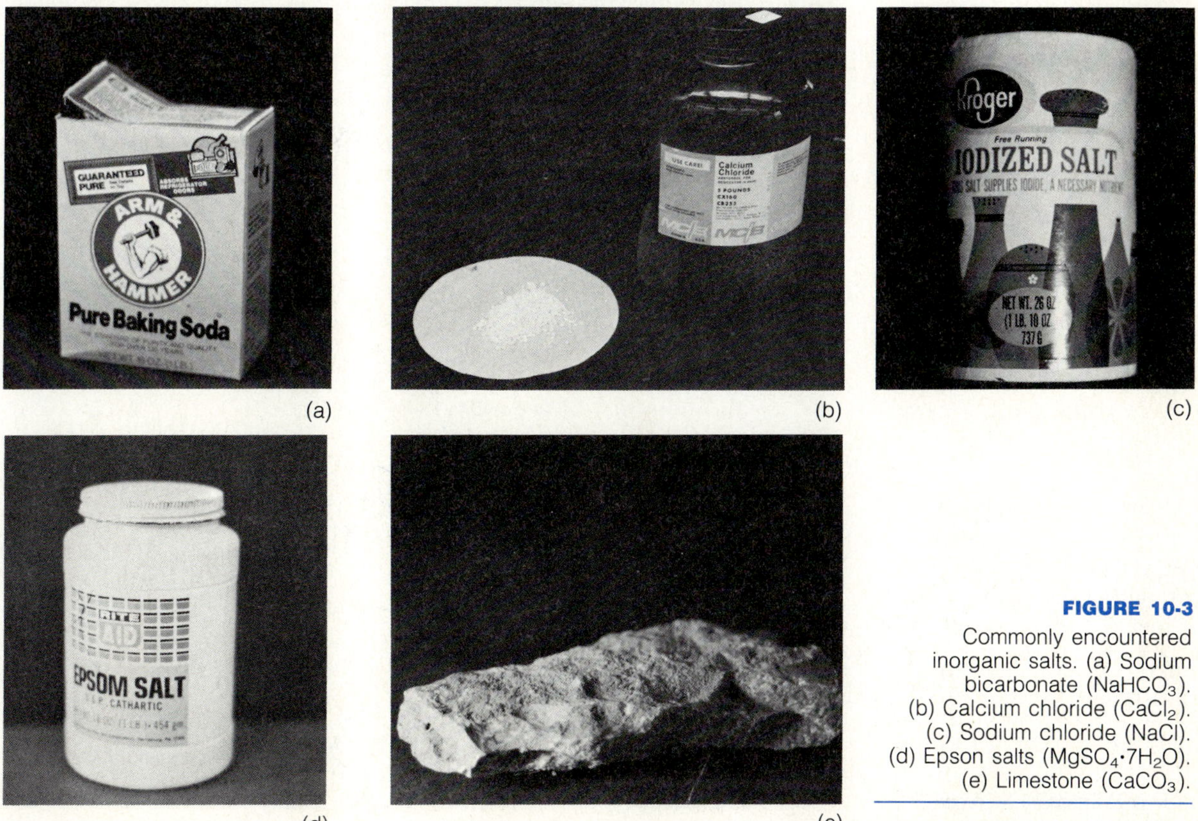

FIGURE 10-3
Commonly encountered inorganic salts. (a) Sodium bicarbonate ($NaHCO_3$). (b) Calcium chloride ($CaCl_2$). (c) Sodium chloride (NaCl). (d) Epson salts ($MgSO_4 \cdot 7H_2O$). (e) Limestone ($CaCO_3$).

A given compound can be classified in more than one way and can therefore have more than one name. For example, HCl is both a binary compound (it has two elements) and an acid (the letter H is written first).

In order to be able to name inorganic compounds without difficulty, certain chemical symbols and formulas should be memorized. The names of certain elements (Table 10-1), acids (Table 10-2), anions (Table 10-3), and cations (Table 10-4) are presented; these should be *memorized*. The system for each class of compounds will be presented next, along with the rules for constructing names from the memorized entities. After that, there is a program for developing skill in naming, as well as a game which can be played in or out of class as a drill and as practice in using the system.

It is suggested that each of the substances to be learned be written on small cards (chemical name on one side and chemical formula on the other) that can be easily carried in a pocket or handbag and used for drill at opportune times (while one is waiting in line for a meal, for a date, for an instructor to arrive, etc.). Using short but frequent blocks of time will probably be more effective than spending a long time at one sitting.

The following is a presentation in outline form of the inorganic naming system. *The boxes contain the method of actually constructing the name* for each type of compound and are followed by examples. The outline is followed by a program which can be used to develop skill in naming inorganic compounds. But neither the outline nor the program will be of much value until Tables 10-1 to 10-4 have been memorized.

TABLE 10-1
Common elements whose name-symbol relationships should be learned

NAME	SYMBOL
Aluminum	Al
Barium	Ba
Bromine	Br
Calcium	Ca
Carbon	C
Chlorine	Cl
Copper	Cu
Fluorine	F
Gold	Au
Hydrogen	H
Iodine	I
Iron	Fe
Lead	Pb
Lithium	Li
Magnesium	Mg
Mercury	Hg
Nitrogen	N
Oxygen	O
Phosphorus	P
Platinum	Pt
Potassium	K
Silicon	Si
Silver	Ag
Sodium	Na
Sulfur	S
Tin	Sn
Uranium	U
Zinc	Zn

TABLE 10-2
Common acids which should be memorized

NAME	FORMULA
Acetic acid	$HC_2H_3O_2$ (abbreviated HOAc)
Boric acid	H_3BO_3
Carbonic acid	H_2CO_3
Chloric acid	$HClO_3$
Chlorous acid	$HClO_2$
Hydrochloric acid	HCl
Hypochlorous acid	HClO
Nitric acid	HNO_3
Nitrous acid	HNO_2
Perchloric acid	$HClO_4$
Phosphoric acid	H_3PO_4
Sulfuric acid	H_2SO_4
Sulfurous acid	H_2SO_3

TABLE 10-3
Common ions with negative charge (anions)

NAME	FORMULA AND CHARGE
Acetate	$C_2H_3O_2^-$
Bicarbonate	HCO_3^-
Bromide	Br^-
Carbonate	CO_3^{2-}
Chloride	Cl^-
Chlorate	ClO_3^-
Chlorite	ClO_2^-
Cyanide	CN^-
Fluoride	F^-
Hydroxide	OH^-
Hypochlorite	ClO^-
Iodide	I^-
Nitrate	NO_3^-
Nitrite	NO_2^-
Oxide	O^{2-}
Perchlorate	ClO_4^-
Phosphate	PO_4^{3-}
Sulfate	SO_4^{2-}
Sulfide	S^{2-}
Sulfite	SO_3^{2-}

Common ions with positive charge (cations)

CATEGORY	NAME: CLASSICAL	NAME: IUPAC	SYMBOL
Nonmetal	Ammonium		NH_4^+
Metals with more than one oxidation number*	Cupric	Copper(II)	Cu^{2+}
	Cuprous	Copper(I)	Cu^+
	Ferric	Iron(III)	Fe^{3+}
	Ferrous	Iron(II)	Fe^{2+}
	Mercuric	Mercury(II)	Hg^{2+}
	Mercurous	Mercury(I)	Hg_2^{2+}
	Stannic	Tin(IV)	Sn^{4+}
	Stannous	Tin(II)	Sn^{2+}
Metals with only one oxidation number*	All group IA metals		M^+†
	All group IIA metals		M^{2+}†
	Aluminum		Al^{3+}
	Zinc		Zn^{2+}
	Silver		Ag^+

* *Oxidation number* (*oxidation state*) is the term chemists use to designate the charge on an atom in a compound. The list given here is a partial one appropriate for this text.
†M is used here for convenience as an abbreviation for either group IA or group IIA metals.

An outline of the inorganic system

1 *Acids.* Acids contain an ionizable hydrogen (*ionizable* means that a compound can break up into its component ions). One can recognize acids by the presence of the letter H as the *first letter* in the formula.

Memorize name-formula relationships from Table 10-2.

EXAMPLES:

H_3PO_4	Phosphoric acid
HCl	Hydrochloric acid

Note Binary acids can be named as *either* acids *or* binary compounds. For example, HCl can be named *hydrochloric acid* (acid name from Table 10-2) or *hydrogen chloride* (binary compound name—see section 4 of this outline).

2 *Bases.* Bases contain a hydroxide (OH^-) group along with a metal or ammonium ion.

Metal name + hydroxide	or	Ammonium + hydroxide

EXAMPLES:

$NaOH$	Sodium hydroxide
$Zn(OH)_2$	Zinc hydroxide
NH_4OH	Ammonium hydroxide

Note If the metal is one of those that can have more than one oxidation number, refer to Table 10-4.

3 *Salts and oxides*. Salts contain a metal (or ammonium) ion and an anion—a single atom or group of atoms with a negative charge (see Table 10-3). Oxides contain a metal and the oxide ion.

EXAMPLES:

(a) *Metals with only one oxidation number* (see Table 10-4).

Metal name + name of anion

EXAMPLES:

MgS	Magnesium sulfide
$BaSO_4$	Barium sulfate
Li_2CO_3	Lithium carbonate
$(NH_4)_3PO_4$	Ammonium phosphate
Al_2O_3	Aluminum oxide

(b) *Metals with variable oxidation states* (see Table 10-4).

Cation name + name of anion

EXAMPLES:

	IUPAC Name	Classical Name
$FeCO_3$	Iron(II) carbonate	Ferrous carbonate
$CuCl$	Copper(I) chloride	Cuprous chloride
Hg_2SO_4	Mercury(I) sulfate	Mercurous sulfate
SnO_2	Tin(IV) oxide	Stannic oxide

4 *Binary compounds of two nonmetals*. Many nonmetals have variable oxidation states; therefore, the naming system is only for variable states.

(a) *IUPAC system:*

SO_3	sulfur(VI) oxide
NO_2	nitrogen(IV) oxide

This system requires mastery of the idea of oxidation numbers and is therefore beyond the scope of this book. We will present the classical system only.

(b) *Classical system:*

____name of the least electronegative element + ____anion name

Note The blank spaces indicated by a low line are for Latin prefixes that designate the number of that kind of atom in the formula.

NUMBER OF ATOMS	PREFIX
1	mono
2	di
3	tri
4	tetra
5	penta

EXAMPLES:

SO_3	Sulfur *tri*oxide
P_2O_5	*di*phosphorous-*pent*oxide (or phosphorous *pent*oxide)
NO_2	Nitrogen *di*oxide

Note Some binary compounds such as NO have been named using another classical system and incorporate the *-ic* and *-ous* endings. For instance, NO is called *nitric oxide* and N_2O is called *nitrous oxide*. This part of the classical system is beyond the scope of this book and will not be covered. (Aren't you glad?)

5 *Common names*. Some compounds are so commercially useful or naturally abundant that special names are used for them that have nothing to do with the system just outlined. These names must simply be memorized.

EXAMPLES:

H_2O	Water
NH_3	Ammonia

See Table 10-8 (page 221) for more examples.

NAMING INORGANIC COMPOUNDS

Writing names for compounds whose formulas are known

Worked example

Name Al_2O_3.

1 The compound contains a metal (Al) and a nonmetal (O).

2 Since the nonmetal is oxygen, the compound is an oxide.

3 Al has only one oxidation state. We are therefore led to section 3, part *a*, of the outline, where we find that we should construct the name by writing the metal name followed by the anion name.

4 The metal name is *aluminum* and the anion name is *oxide*.

5 The name of Al_2O_3 is *aluminum oxide*.

Worked example

Name NO_2.

1 This is a binary compound composed of two nonmetals.

2 We can also refer to section 4, part *b*, for the classical name.

Least electronegative element name + prefix + anion name

Nitrogen dioxide

Note Prefixes are usually used when there is more than one atom of an element in the molecule. Thus in nitrogen dioxide, oxygen has the prefix *di* but nitrogen does *not* have the prefix *mono*.

Worked example

What is the name of $Ca(OH)_2$?

1 This compound has a metal and the OH^- group.

2 We recognize the OH^- group as that which classifies a compound as a *base*. We therefore refer to section 2, where we see that bases are named by writing the name of the metal followed by the word *hydroxide*.

3 Thus $Ca(OH)_2$ is named *calcium hydroxide*.

Problem

Name KBr. (Try this before looking at the solution below.)

Solution

1 This compound is a salt (metal and anion). The metal is in group IA and therefore has only one oxidation state (+1). See Table 10-4.

2 Refer to section 3, part *a*, of the outline. Metals with only one oxidation number are named by using the name of the metal plus the name of the anion.

3 The name is *potassium bromide*.

Problem

Name SnO.

Solution

1 This compound contains a metal and oxygen. The metal has variable oxidation states (see Table 10-4).

2 The IUPAC name is constructed as follows from section 3, part *b*, in the outline:

Cation name + anion name

↘ ↙

Tin(II) oxide

Note Since the charge of the oxide ion is −2 (see Table 10-3), and since the total charge on the compound as a whole must be zero, the charge of the Sn must be +2. By learning the charges on the anions in Table 10-3, one can determine the charge on the metal ion (if it is one of the variable ones from Table 10-4) and thus the name to be used.

3 Using the classical method (see section 3, part *b*, and Table 10-4):

Cation name + anion name

↘ ↙

Stannous oxide

Problem

What is the name of H_2CO_3?

Solution

1 Since the H is the first letter written, the compound is an acid.

2 From memory (or from Table 10-2), the name is *carbonic acid*.

Problem

What is the name of $HgSO_4$?

Solution

1 This compound contains a metal and a cation and is therefore a salt.

2 The name of the SO_4^{2-} group is sulfate.

3 Since the sulfate group has a charge of -2, the Hg ion has to have a $+2$ charge. Hg^{2+} is the mercuric ion or the mercury(II) ion (see Table 10-4).

4 The IUPAC name is mercury(II) sulfate, and the classical name is mercuric sulfate.

Writing formulas for compounds whose names are known

Mechanically one goes through the following steps in writing the correct formula when the name is known:

1 The symbol of the metal or ammonium or of the *least* electronegative element is written down *first*.

2 This is followed by the formula of the anion or the symbol of the *most* electronegative element.

3 The correct charges are placed over each one.* See Tables 10-3 and 10-4.

Note The metal or ammonium is always positive, and the anion is always negative.

4 The absolute value of the charge (the number without the charge) on the right entity is placed as a subscript after the left entity and *vice versa*.

5 These subscripts are reduced to the simplest whole number.

Worked example

What is the formula for potassium nitrite?

1 The symbol for potassium is K (see Table 10-1), and its oxidation number is $+1$ (it is in group IA).

2 The formula for nitrite is NO_2^- (see Table 10-3).

* If the compound is binary with only nonmetals, simply refer to the prefixes (*di*, *tri*, etc.) to write the subscripts and ignore steps 4 and 5.

3 Write the metal first, then the anion formula (each with its charge).

$$K^+ \qquad NO_2^-$$

4 Since each one has a charge of 1, they occur in a 1 to 1 ratio. Therefore, simply write:

$$KNO_2$$

Mechanically one goes through the following steps:

(a) Each ion with its charge is written.

$$K^+ \qquad NO_2^-$$

(b) The absolute value (the number without the sign) of each charge is moved to the subscript positions on the *opposite ion*.

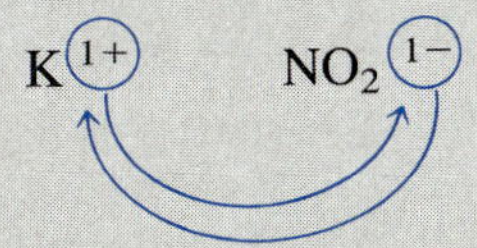

This gives $K_1(NO_2)_1$.

(c) Any subscript with a value of 1 is ignored (if nothing is written, one understands the subscript to be 1).

$$KNO_2$$

Worked example

What is the formula for barium acetate?

1 Barium (a metal) is in group IIA, has the symbol Ba, and has a charge of +2 (as have all metals in group IIA).

2 The acetate anion is $C_2H_3O_2^-$.

3 Write this information down putting the metal first:

$$Ba^{2+} \qquad C_2H_3O_2^-$$

4 Exchange the absolute charge values to obtain:

$$Ba_1(C_2H_3O_2)_2$$

Note When a subscript is used with a complex anion, the anion formula is enclosed in parentheses.

5 Ignoring the subscript 1 (for purely arbitrary reasons), one writes the formula as $Ba(C_2H_3O_2)_2$.

Note Since the *sum of all charges must be zero,* and since the charge on the acetate is -1, it takes two acetates to neutralize the $+2$ charge on the barium. This is the rationale, but the mechanical exchanging of absolute charge values to give subscripts always works.

Worked example

What is the formula for cupric bicarbonate?

1 Cupric ion is Cu^{2+}.

2 Bicarbonate is HCO_3^-.

3 Write:

$$Cu^{2+} \qquad HCO_3^-$$

4 Exchange absolute charge values:

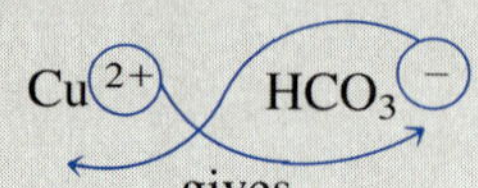

gives

$$Cu_1\ (HCO_3)_2$$

or

$$Cu(HCO_3)_2$$

Problem

What is the formula for iron(III) hydroxide?

Solution

1 Iron(III) is Fe^{3+}.

2 Hydroxide is OH^-.

3 Write:

$$Fe^{3+} \qquad OH^-$$

4 Exchanging absolute charge values and ignoring the subscript 1 gives $Fe(OH)_3$.

Problem

What is the formula of boric acid?

Solution

1 Since this compound is an acid, we call on our memory or go to Table 10-2.

2 There we find boric acid as H_3BO_3.

Note As mentioned at the beginning of this chapter, memorization of the information in Tables 10-1 through 10-4 is necessary in order to name inorganic compounds.

Problem

What is the formula of tin(II) phosphate?

Solution

1 Tin(II) is Sn^{2+}.

2 Phosphate is PO_4^{3-}.

3 Write:

$$Sn^{2+} \qquad PO_4^{3-}$$

4 Exchanging the absolute charge values gives $Sn_3(PO_4)_2$.

Problem

Write the formula for tin(IV) sulfate.

Solution

1 Tin(IV) is Sn^{4+}.

2 Sulfate is SO_4^{2-}.

3 Write:

$$Sn^{4+} \qquad SO_4^{2-}$$

4 After exchanging absolute charge values, one gets $Sn_2(SO_4)_4$.

5 In this case the ratio can be reduced. The 2 to 4 becomes 1 to 2. Therefore the correct way to express the formula is $Sn(SO_4)_2$.

Problems

The solutions only are given.

a Name the following compounds.

(1) SO_3
(2) $Ba(OH)_2$
(3) $CaSO_4$
(4) H_2CO_3
(5) $AlPO_4$
(6) $Zn(C_2H_3O_2)_2$
(7) $FeSO_4$

b Write formulas for the following compounds.

(1) Silicon tetrachloride	**(5)** Carbon dioxide
(2) Sodium hypochlorite	**(6)** Ammonium hydroxide
(3) Copper(I) cyanide	**(7)** Silver phosphate
(4) Perchloric acid	

Answers

a **(1)** Sulfur trioxide or sulfur(VI) oxide
(2) Barium hydroxide
(3) Calcium sulfate
(4) Carbonic acid
(5) Aluminum phosphate
(6) Zinc acetate
(7) Iron(II) sulfate or ferrous sulfate

b

(1) $SiCl_4$	**(5)** CO_2
(2) NaClO	**(6)** NH_4OH
(3) CuCn	**(7)** Ag_3PO_4
(4) $HClO_4$	

Chem-Deck*

Card games similar to rummy or poker that are very effective in teaching inorganic nomenclature were published in the *Journal of Chemical Education* by Alan and Sharan Sherman, who teach at Middlesex County College (New Jersey). Like other card games, these involve both chance and skill; they give practice in naming compounds, writing their formulas, and calculating their formula weights (a skill that will be needed for the chemical calculations in Appendix E). The deck can be made by the students, or the instructor can have several decks prepared for general class use. (See Figure 10-4.)

"The objective of the game is to give students practice in forming chemical compounds. This is done by combining cation (positive ion) cards with anion (negative ion) cards and the proper number (subscript) cards to insure electroneutrality of the compounds. By playing the various card games with CHEM-DECK, students will learn the charges of the various ions which will help them learn to write the chemical formulas of many compounds. Of course, there exists the element of chance in this game; a person needs to obtain the proper cards to make compounds. However, by using their chemical knowledge, students should be able to manipulate the cards to form the best chemical compounds. (Best in this game means the compounds worth the most points—those with high molecular masses.)

"Before we discuss the rules for the various games, let us review some information about the CHEM-DECK. The deck consists of 88 cards. There are 18 cation cards, 25 anion cards, and 45 subscript cards. You can make your own deck by using heavy cardboard to cut out 88 cards of playing card

* A. Sherman and S. Sherman, *Journal of Chemical Education*, **57**:503 (1980).

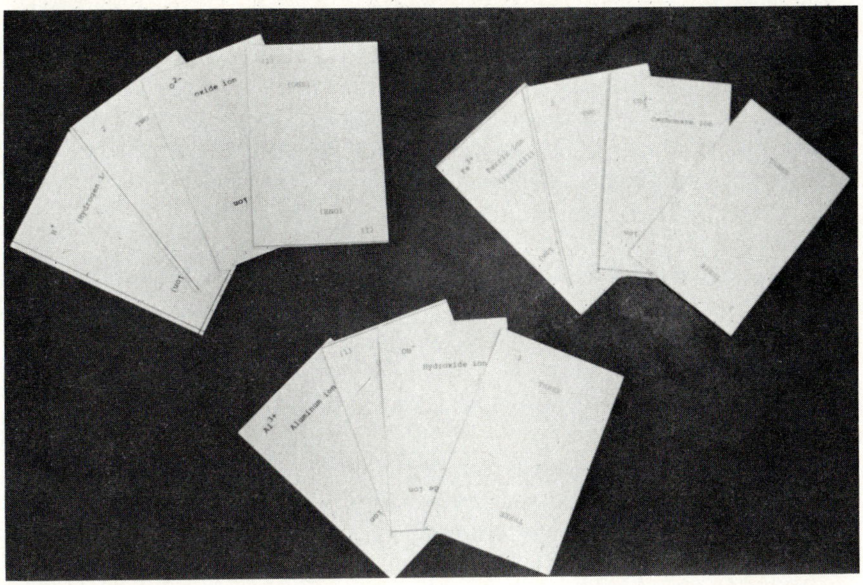

FIGURE 10-4
Chem-Deck hands.

size. Or you can simply take 88 cards from two decks of standard playing cards and paste the chemical symbols on them. The breakdown of the deck is as follows; one of each of the following cards, unless otherwise indicated.

Na^{1+}, NH_4^{1+}, Ag^{1+}, K^{1+}, Li^{1+}, Cu^{1+}, Sn^{2+}, Ca^{2+}, Zn^{2+}, Cu^{2+},
Mg^{2+}, Hg^{2+}, Ba^{2+}, Fe^{2+}, Pb^{2+}, Al^{3+}, Fe^{3+} (two of these),
H^{1-}, OH^{1-}, Cl^{1-}, $C_2H_3O_2^{1-}$, HCO_3^{1-}, ClO_3^{1-}, $H_2PO_4^{1-}$,
NO_2^{1-}, I^{1-}, Br^{1-}, HSO_4^{1-}, NO_3^{1-}, F^{1-}, SO_3^{2-}, CO_3^{2-},
O^{2-}, S^{2-}, SO_4^{2+}, HPO_4^{2-}, CrO_4^{2-}, AsO_4^{3-}, PO_4^{3-},
P^{3-}, N^{3-} (two of these)

1 (15 of these)
2 (15 of these)
3 (15 of these)

We chose ions that are frequently used in introductory chemistry courses. You can, of course, add more ions of your own. However, if you increase the ions in the deck, you should also increase the number of subscript cards in the deck by an equal amount.

"There are three basic card games students can play with CHEM-DECK: rummy, five-card draw poker, and seven-card stud poker. The rules are slightly different from those of Hoyle and will be explained below. Also, a table of atomic masses is necessary for scoring.

Chem-Deck rummy "The object of the game is to obtain three compounds. It takes four cards to make a compound (anion card, cation card, and two subscript cards). Each player is dealt 12 cards. The regular rules of rummy are now followed as each player tries to obtain three neutral compounds.

"Scoring in the game is very important and adds to the strategy of the players. The first player who obtains three compounds wins the hand and receives 50 points. This person lays out his or her hand, as in regular rummy, and adds up the molecular mass of each compound. The sum of the molecular masses is then added to that player's score. The other players add up the molecular masses of their compounds (if they have any) and add that to their score. The player who obtains 1000 points wins the game.

Chem-Deck five-card draw poker "The object of this game is to obtain one compound, preferably of high molecular mass. Each player receives five cards. Players may then exchange up to four cards if they wish. The player with the highest molecular mass compound wins the hand. If no player has a compound, the player having the heaviest ion wins the hand.

Chem-Deck seven-card stud poker "The object of this game is to obtain one compound. The game is played exactly like ordinary seven-card stud. Scoring is the same as in five-card draw.

Examples and rules for forming compounds "A chemical compound is formed from a positively charged species (cation) and a negatively charged species (anion). These species must be balanced to form an electrically neutral compound. The charges are balanced using subscripts.

"Example 1—formation of sodium carbonate:

Na^{1+} cation, CO_3^{2-} anion forms Na_2^{1+} $(CO_3)_1^{2-}$ or using CHEM-DECK

Na^{1+}	2	$CO_3{}^{2-}$	1
Sodium ion	Subscript card	Carbonate ion	Subscript card

"Example 2—formation of potassium chloride:

K^{1+} cation, Cl^{1-} anion forms $K_1^{1+}Cl_1^{1-}$ or using CHEM-DECK"

K^{1+}	1	Cl^{1-}	1
Potassium ion	Subscript card	Chloride ion	Subscript card

Chemantics

Another chemical formula card has been developed by Albert K. Sawyer, who is at the University of New Hampshire. It is called *Chemantics* and is described in the *Journal of Chemical Education*.*

ACIDITY AND THE pH SCALE

An *acid* is a material that gives up a hydrogen ion (H^+) when the acid is added to water. Such materials can be recognized, as already mentioned, by the presence of the letter H as the *first* letter (to the left) in the chemical formula. An acid solution is one containing a large quantity of H^+ ions. *Bases* are those materials that contain hydroxide ions (OH^-), which they give up when the base is added to water. They are easily recognized by the presence of the OH^- ion in the chemical formula in combination with a metal ion or NH_4^+ ion.

The following processes occur when acids and bases are added to water:

$$\text{Acids:}\quad \underset{\text{An acid}}{HCl} \longrightarrow \underset{\text{Hydrogen ion}}{H^+} + Cl^-$$

$$\text{Bases:}\quad \underset{\text{A base}}{NaOH} \longrightarrow Na^+ + \underset{\text{Hydroxide ion}}{OH^-}$$

These processes are called *ionization*. A few acids ionize almost completely and are called *strong* acids (HCl, HNO_3, $HClO_4$). Most others ionize only partially and are called *weak* acids.

Water is an example of a material that is *amphoteric*—it can act as both an acid and a base. This is because it automatically decomposes (or dissociates) to a *very small* extent to produce both acid (H^+) and base (OH^-) according to the following reaction:

$$H_2O \longrightarrow H^+ + OH^-$$

Since an equal, although small, amount of H^+ and OH^- is produced, pure water has an excess of neither H^+ or OH^- and is called a *neutral compound* (it is neither acidic nor basic). At 25°C, 6×10^{16} molecules from 1 L of water dissociate out of a total of 3.3×10^{25} molecules. This is only 1 molecule out of 5×10^8 molecules, or 1 molecule out of 500 million, or 1×10^{-7} mol in each liter.

Water solutions containing acids have an excess of H^+ over OH^- and are called *acidic*. In solutions containing bases, the OH^- are present in excess and the solution is said to be *basic* (or *alkaline*).

A scale—called the *pH scale*—has been developed to help quantify the amount of acidity or basicity of water solutions. Table 10-5 shows the pH scale. The numbers on the pH scale originate from the exponent of 10 when the number of moles of H^+ in each liter of solution is expressed in scientific notation. Since pure water has 1×10^{-7} mol of H^+ in each liter, the pH, which is the absolute value (the number without the plus or minus sign) of the

Journal of Chemical Education,* **53:780 (1976).

TABLE 10-5

The pH scale used to measure acidity

	pH	COMMON MATERIAL	CHARACTERISTIC
Acidic	1	Hydrochloric acid*	Strongly acidic
	2	Lime juice; lemon juice	
	3	Vinegar; carbonated beverages	Mildly acidic
	4	Apple juice	
	5	Boric acid solution	Very mildly acidic
	6		
Neutral	7	Drinking water; milk; pure water	Neutral
Basic	8	Dilute sodium bicarbonate solution	Very mildly basic
	9	Soaps; detergents	Very mildly basic
	10		
	11	Household ammonia	Mildly basic
	12		
	13		
	14	Lye* (sodium hydroxide)	Strongly basic

*Although the pH values of very dilute or very concentrated solutions of these materials differ, hydrochloric acid and sodium hydroxide are strongly acidic and basic, respectively.

exponent, is 7. A solution with 1×10^{-6} mol of H^+ has a pH of 6. Remember that 10^{-6} is 10 times *larger* than 10^{-7}.

Pure water has a pH of 7, which is neutral; solutions with a pH of less than 7 are acidic; solutions with a pH greater than 7 are basic. Each unit on the pH scale is 10 times the next one. This means that a solution with a pH of 4 has 10 times the acidity of one with a pH of 5 (the *stronger* the acid, the *lower* the pH). Likewise, a solution with a pH of 9 has 10 times the OH^- ions of one with a pH of 8 (the *higher* the pH, the *greater* the concentration of OH^- ions).

The importance of pH in everyday life and in the chemist's laboratory can hardly be overemphasized. Most chemical processes are highly pH-dependent. This means that they will work (or work better) at one pH but will not work (or work less well) at a different pH. The form of many materials can be changed by varying acidity. At a pH of 6 to 7, for example, the proteins of milk remain suspended; but if the pH is lowered to 4 or less by adding vinegar or some other acidic material, the proteins coagulate, or curdle. When milk "sours" on its own, it is because bacteria change milk sugar (lactose) into lactic acid. The lactic acid lowers the pH of the milk, which causes the proteins to curdle. This milk also tastes sour. All acidic materials—such as lemon juice, spoiled milk, and vinegar—taste sour. Basic materials such as $NaHCO_3$ taste bitter; and when they are used in cooking, they give a bitter taste to the food if an excess is used. Although $NaHCO_3$ does not contain an OH^- ion when it is added to water, an excess of OH^- ions is formed—thus it is a base.

Almost all biochemical processes take place in the narrow pH range between 6 and 8, and it is essential that the body correctly regulate the blood and cellular pH.

The pH of consumer products can also be important. For instance, shampoos with a high or low pH sometimes burn the user's eyes; shampoos with a pH of about 6 to 7 do not. This is an important consideration when chemists are formulating shampoos. (See Chapter 17 for more on shampoos and pH.)

Farming and gardening require close attention to the pH of soils, since a plant's health, longevity, and ability to produce fruit or blossoms depend on the soil acidity. Table 10-6 shows various common plants and the pH at which they grow best. The adjustment of soil pH can be accomplished in several ways.

If the soil is too acidic, because of rotting plant material or other natural sources, ground limestone ($CaCO_3$), lime (CaO), or hydrated lime [$Ca(OH)_2$] can be added at a ratio of about 50 to 100 lb per 1000 ft^2. The following reactions exemplify those that occur:

$$\begin{array}{rcl} 2CaCO_3 + 2H^+ & \longrightarrow & Ca(HCO_3)_2 \\ \text{Excess acidity} & \longrightarrow & \text{very mildly basic} \end{array}$$

$$\begin{array}{rcl} CaO + H^+ & \longrightarrow & Ca^{2+} + H_2O \\ \text{Acid} & \longrightarrow & \text{neutral} \end{array}$$

If the soil is very acidic, one can add wood ashes (which contain NaOH and KOH) or large amounts of commercial $Ca(OH)_2$, CaO, or $CaCO_3$.

If the soil is too basic, several materials can be used to make it more acidic. Sawdust or leaves work slowly, since the acid they produce comes from rotting (which is a slow process) and the production of CO_2, as shown below:

$$\text{Leaves} \xrightarrow[\text{Slow rotting}]{O_2} CO_2 \xrightarrow{H_2O} \underset{\text{An acid}}{H_2CO_3}$$

Powdered sulfur will also tend to lower the pH of soil, as shown by the reaction below:

$$S + O_2 \xrightarrow[\text{Slow oxidation}]{} SO_3 \xrightarrow{H_2O} \underset{\text{An acid}}{H_2SO_4}$$

A faster-acting material is aluminum sulfate, $Al_2(SO_4)_3$, which also produces acid and can be applied at a ratio of about 20 lb per 1000 ft^2.

Most states have laboratories that will test soil for acidity and give you a pH value. You can then add the proper material if the soil is not the proper pH

TABLE 10-6

Optimum soil pH values for the growth of selected plants

Strongly acidic, pH 4–5	Blueberries, cranberries, ferns, holly trees, orchids, laurel, and rhododendron
Acidic, pH 5–6	Most berries and shrubs, dogwood, hemlock, and spruce trees
Neutral, pH 6–8	Most vegetables, flowers, trees, and grass
Basic, pH 8–9	Hay and soybeans

TABLE 10-7

pH color indicators

MATERIAL	RELATIONSHIP BETWEEN pH AND COLOR	AVAILABILITY
Litmus	Red below pH 7 Blue above pH 7	Paper-impregnated form available at drugstores
Phenolphthalein	Red above pH 9 Colorless below 8	Exlax laxative preparation mashed up with alcohol and filtered through a piece of paper towel
Red cabbage	Red below pH 7 Blue above pH 7	Grocery store
pH paper	Different colors at each step on the pH scale	Chemical supply houses, pet shops, drugstores, garden centers
Grape juice	Violet ("normal" grape color) below pH 7 Blue above pH 7	Grocery store

for the plants to be grown. In a laboratory, pH meters are used (see Chapter 12), but some simple color pH indicators can be used at home. Table 10-7 lists several such materials. A small amount of the indicator material is placed in the solution whose pH you wish to determine. The color of the indicator tells you the region of the pH scale in which the solution falls. The most convenient form of indicator is pH paper. An indicator is adsorbed on the paper, and the pH is measured by matching the color to a color chart printed on the container.

COMMON INORGANIC MATERIALS

Commonly encountered inorganic materials (a list appears in Table 10-8) include fertilizers, some household and industrial materials, and paint pigments.

Fertilizers

Among the inorganic materials most frequently used commercially are fertilizers. They are mixtures of salts that contain the nutrients needed by plants. Those nutrients needed in relatively large amounts include phosphorus (P), nitrogen (N), calcium (Ca), magnesium (Mg), and potassium (K); and those needed in smaller amounts are iron (Fe), copper (Cu), manganese (Mn), zinc (Zn), and boron (B). Most solid commercial fertilizers are composed of ammonium nitrate (NH_4NO_3), sodium nitrate ($NaNO_3$), potassium carbonate (potash of soda, K_2CO_3), potassium phosphate (K_3PO_4), potassium chloride (KCl), or calcium phosphate [$Ca_3(PO_4)_2$]. Calcium carbonate (limestone, $CaCO_3$) or lime (CaO) is also used to provide Ca; and since most limestone also contains a significant amount of dolomite ($MgCO_3$), it provides Mg as well.

TABLE 10-8

Common inorganic materials

NAME	FORMULA	USE OR COMMON NAME
Sodium chloride	$NaCl$	Table salt
Sodium bicarbonate	$NaHCO_3$	Baking soda
Sodium carbonate	Na_2CO_3	Washing soda
Magnesium sulfate	$MgSO_4 \cdot 7H_2O$	Epsom salts
Calcium carbonate	$CaCO_3$	Chalk, limestone, marble
Calcium chloride	$CaCl_2$	Used to melt ice
Sodium hydroxide	$NaOH$	Lye; used to make homemade soap
Ammonium hydroxide	NH_4OH	Household ammonia
Hydrochloric acid	HCl	Muriatic acid; used to clean rust from iron and to wash mortar from brick
Boric acid	H_3BO_3	3% solution used for eyewash
Sulfuric acid	H_2SO_4	Battery acid, also used in many industrial processes
Borax	$Na_2B_4O_7 \cdot 10H_2O$	Washing aid
Lime	CaO	Soil nutrient
Slaked lime (hydrated lime)	$Ca(OH)_2$	Whitewash
Iron oxide	Fe_2O_3	Paint pigment, rouge, iron rust
Magnesium oxide	MgO	Milk of magnesia; antacid
Plaster of paris	$CaSO_4 \cdot \frac{1}{2}H_2O$	Taking casts of tracks, making casts to set bones
Gypsum	$CaSO_4 \cdot 2H_2O$	Plasterboard or "dry wall"
Sodium hypochlorite	$NaOCl$	Household chlorine bleach
Stannous fluoride	SnF_2	Enamel hardener in toothpaste
Titanium dioxide	TiO_2	Pigment in white paints
Zinc oxide	ZnO	Pigment in artist colors and in certain skin lotions
Ammonium nitrate	NH_4NO_3	Fertilizer
Ammonia	NH_3	Household ammonia, fertilizer
Ferric ferrocyanide	$Fe_4[Fe(CN)_6]_3$	Blue pigment in paints and inks

As was mentioned in the previous section, wood ashes can be used to raise the pH (reduce the acidity) of soils. These ashes are rich in all the metal ions mentioned above and also contain some phosphorus. Nitrogen is the only nutrient which is essentially lacking in ashes. Therefore, even if the soil is not too acid, ashes make good fertilizers; and the alkalinity they produce can be counteracted by adding aluminum sulfate, $Al_2(SO_4)_3$, or sulfur, S, to the soil at the same time.

Nitrogen is sometimes provided by adding ammonia (NH_3) in liquid form. Ammonia has a boiling point of $-35°C$ and is a gas at ordinary temperature and pressure. It can be cooled to $-35°C$, where it condenses to a very cold liquid. It can be stored at ordinary temperatures in a sealed container under pressure and applied under the ground by means of tubes attached to plows. Liquid ammonia is commonly referred to as *liquid nitrogen,* but this is a misnomer from the standpoint of a chemical nomenclature, and the use of this term is discouraged.

Originally most nitrogen for the manufacture of fertilizer was supplied from deposits of NH_4NO_3 in Chile. Because Germany was cut off from these deposits during World War I, the German chemist Fritz Haber was given the challenge of finding a way of making NH_3 synthetically. He succeeded in finding a catalyst, a mixture of iron and aluminum oxide, that allows the direct combination of N_2 and H_2 to produce NH_3:

$$N_2 + 3H_2 \xrightarrow[550°C]{Fe—Al_2O_3} 2NH_3$$

A *catalyst* is a material which allows an otherwise difficult chemical reaction to proceed with relative ease. It is not actually used up in the reaction and can be used over again. Haber's process allowed the Germans to produce fertilizer and explosives; it is used now to produce the liquid ammonia used by farmers. The NH_3 is, in turn, used to produce the nitrates used in solid fertilizers.

Industrial and household inorganic compounds

In addition to fertilizers, there are a number of important inorganic materials used in manufacturing and by consumers in everyday life.

Sulfuric acid (H_2SO_4) is one of the most important. It is a viscous (thick) liquid in concentrated form, used, for example, in lead storage batteries in automobiles. It is a highly toxic and corrosive material if brought into contact with biosystems, but it is valuable in the manufacture of other products. It is produced by the *contact process:*

$$S + O_2 \xrightarrow{\text{catalyst}} SO_3 \xrightarrow{H_2O} H_2SO_4$$

Sodium bicarbonate ($NaHCO_3$, baking soda) is used in baking, in controlling acidity during chemical reactions (it is slightly basic), and in fire extinguishers. It is manufactured by the *Solvay process:*

$$NH_3 + NaCl + CO_2 \longrightarrow NaHCO_3 + NH_4Cl$$

When it is placed on a hot object or in a flame, the following reaction occurs:

$$2NaHCO_3 \xrightarrow{\text{heat}} Na_2CO_3 + H_2O + CO_2$$

Thus, $NaHCO_3$ can be used to put out small fires in the kitchen, especially grease fires, by sprinkling it on the fire. The carbon dioxide released forms a blanket of gas that is denser than air and prevents O_2 from getting to the flame, thereby extinguishing it.

Other inorganic compounds are also in general use. Calcium chloride ($CaCl_2$), which is used to melt snow and ice, is used in large amounts by highway crews during the winter. Sodium chloride (NaCl, table salt) is, of course, used by consumers for many purposes, including flavoring and preserving foods, melting ice, and lowering the temperature of a salt-ice mixture while making homemade ice cream. Sodium hypochlorite (NaOCl) is used as the chief ingredient in household bleaches. Inorganic materials such as S,

$Ca(OH)_2$, and $CuSO_4$ are sometimes sprayed on plants as pesticides to rid the plant of insects, fungi, etc.

Paint pigments

Paints for covering walls and other structural parts of buildings as well as artists' paints are colored using inorganic materials (pigments); see Figure 10-5. (The coatings themselves are organic polymers and are discussed in Chapter 11.) White is usually produced in exterior (house) paint and interior (wall) paint by using titanium dioxide (TiO_2), and in artists' paints by zinc oxide (ZnO) or sometimes by TiO_2. Barium sulfate ($BaSO_4$) is also a white material and is frequently used to lighten other colors. For example, cadmium sulfide (CdS) is used as a yellow pigment; but if the paint is to be a light yellow (say, cadmium yellow-light), some $BaSO_4$ is also added. Green is produced from chromium hydroxide $[Cr(OH)_3]$ or chromium oxide (Cr_2O_3). The latter is more an earth green (less brilliant); the former produces a bright color. Brown can be produced with iron oxide (Fe_2O_3), and red with a mixture of CdS and SeS. Prussian blue is pigmented with the complex ferric ferrocyanide, $Fe_4[Fe(CN)_6]_3$; ultramarine blue is a mixture of sulfur and sodium and aluminum silicates. Mixtures of pigments can be purchased to give many different hues. These can all be used in either oil or acrylic (latex) paint bases. Formerly, the compounds PbO, $Pb(C_2H_3O_2)_2$, or $PbCO_3$ ("white lead") and Pb_3O_4 ("red lead") were used in paints; but because of the toxic nature of most lead compounds, they are no longer used. But it is still possible for a

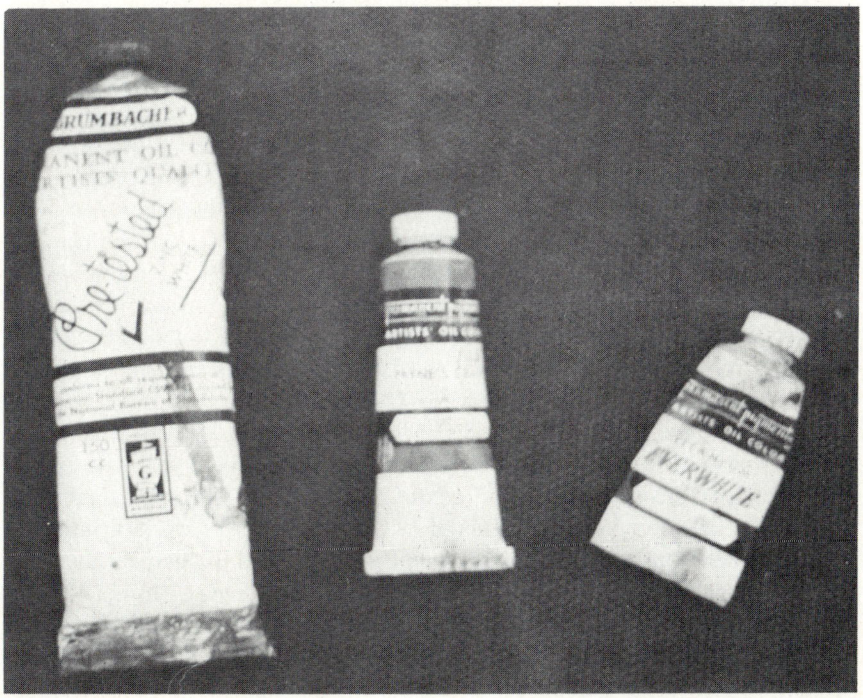

FIGURE 10-5
Artists' paints containing inorganic chemical pigments.

child to chew on a piece of painted wood in an older house that has lead paint under several coats of non-lead-based paint; this can cause lead poisoning.

HARD WATER

Hard water is defined as water containing ions of the metals Ca, Mg, or Fe. If these ions are present along with the bicarbonate ion, the water is said to be *temporarily hard* because heating metal bicarbonates causes carbonates to form. These metal carbonates are fairly insoluble in water and precipitate out of the water when it is heated:

$$\underset{\text{Calcium bicarbonate in solution}}{Ca^{2+} + 2HCO_3^-} \xrightarrow{\text{heat}} \underset{\text{Hard solid}}{CaCO_3\downarrow} + H_2O + CO_2$$

After heating, the metal content of the water is reduced and the water is then said to be *soft*. If the anions present in the water along with the metal ions are not bicarbonate, heating does not remove the metal and the water stays hard. This kind of water is said to be *permanently hard*.

Hard water causes two major problems. The first is the inactivation of soap and the formation of calcium (or magnesium) stearate, which causes scum to form (see the sections on soaps, shampoos, and bathtub cleaners in Chapter 17). Not only is the presence of calcium stearate undesirable in itself; it also makes more soap necessary for a given job: part of the soap is used to react with the metal ions before any cleaning can occur.

The second problem is scale buildup in water heaters and hot water pipes (see Figures 10-6 and 10-7). The reaction above shows what happens when temporarily hard water is heated. Most hard water is of this type, and the $CaCO_3$ that is formed is deposited as a solid on the inside of the heating vessel; this solid deposit is called *scale*. Scale can cause very serious problems in water heaters. The layer of $CaCO_3$ acts as an insulator and prevents the heat from the gas flame or electric heating elements from getting to the water. Not only is energy being lost; the electric heating elements also burn out faster. This scale also tends to clog hot water pipes so that the flow of water is restricted. It appears as a white deposit in teakettles and other pans used to boil hard water repeatedly. This deposit can be removed from teakettles, etc., by filling them with a normal concentration of a toilet bowl cleaner ($NaHSO_4$ or HCl) and letting them sit for a while. This will not work for aluminum vessels, since aluminum reacts with the acid.

Many municipal water supplies have water that is only slightly hard, which is not a serious problem. But in some locations, where the home or city water comes from a well, the Ca^{2+} content can be quite high and scale deposits rapidly. This rapid accumulation of $CaCO_3$ is a result of the large amount of time water spends underground, which gives it time to become saturated with $Ca(HCO_3)_2$.

Water hardness is measured as parts per million (ppm) of $CaCO_3$, which is equal to the milligrams of $CaCO_3$ formed per liter of water when the water is heated. If a water sample has a hardness of 10 ppm, it contains enough

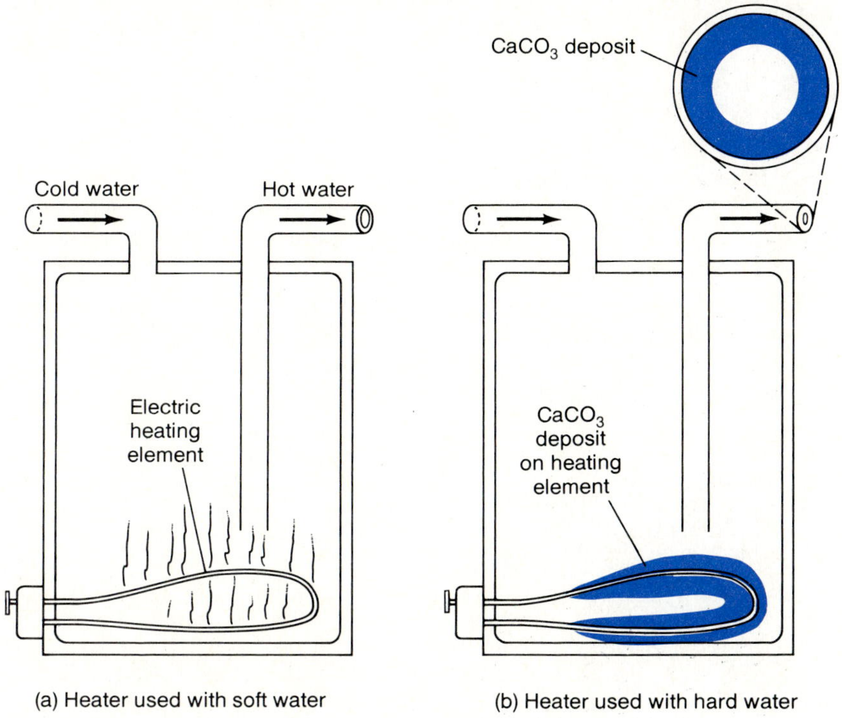

FIGURE 10-6
Scale buildup during water heating.

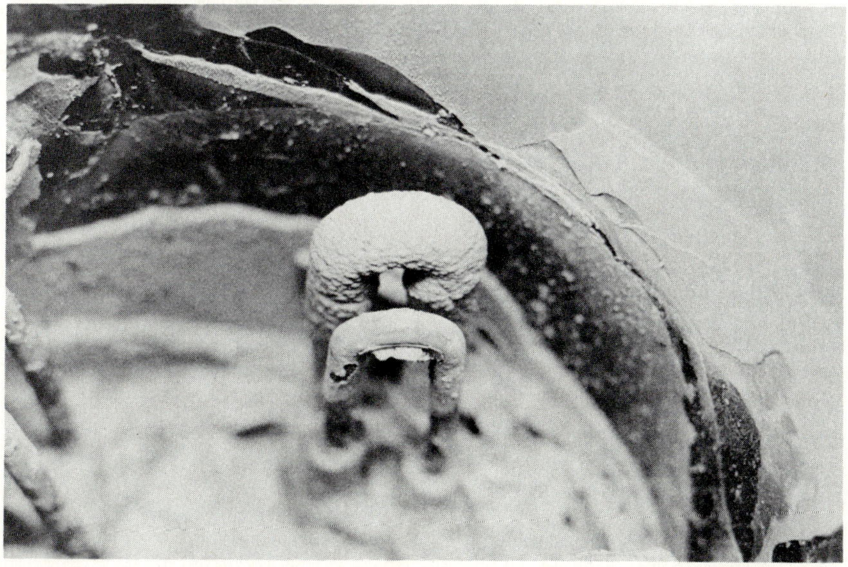

FIGURE 10-7
Scale buildup in a hot water system.

$Ca(HCO_3)_2$ to form 10 mg of $CaCO_3$ in each liter of water when the water is heated. $CaCO_3$ is slightly soluble in water; 6.7 mg of $CaCO_3$ can be dissolved in 1 L of water at room temperature. Thus, of the 10 mg formed, only $10.0 - 6.7 = 3.3$ mg precipitates out and forms scale. At a hardness of 20 ppm, 13.3 mg of scale forms per liter; and at 100 ppm, 93.3 mg forms per liter. Up to a hardness level of 6.7 ppm, of course, *no scale forms*.

Because the hardness of most water supplies is under 20 ppm, water hardness is not a serious problem, but some deep wells within limestone strata can have as much as 1000 ppm, or even more. Almost 1 g of $CaCO_3$ (993 mg) forms when 1 L of this water is heated; thus, it should not be used in hot water systems or even in cooking, since it may impart strange flavors to foods. It is fine for drinking and has a unique taste that is preferred by some, and is also appropriate for watering lawns and gardens as well as for flushing toilets. If iron is present along with the calcium (as it frequently is), a reddish solid forms on the inside of all containers in which the water is used.

Table 10-9 gives some information about water hardness.

Since most hard water is of the temporary type containing $Ca(HCO_3)_2$, we will focus on that. The reactions shown below illustrate how water becomes hard. Decomposing organic matter (containing CO_2) becomes wet, which produces carbonic acid (H_2CO_3). This carbonic acid solution soaks into the ground, where it reacts with the limestone ($CaCO_3$), producing $Ca(HCO_3)_2$, which is soluble in water:

$$\underset{\text{From decomposing leaves}}{CO_2} + \underset{\text{From rain}}{H_2O} \longrightarrow \underset{\text{Weak solution of carbonic acid}}{H_2CO_3}$$

$$H_2CO_3 + \underset{\text{Limestone}}{CaCO_3} \longrightarrow \underset{\text{Temporary hardness}}{Ca(HCO_3)_2}$$

This process causes the limestone to dissolve and produces caves and caverns in limestone strata. Figure 10-8 diagrams the aquatic cycle and the source of water hardness.

TABLE 10-9

Summary of water hardness

HARDNESS LEVEL, mg $CaCO_3$/L	RELATIVE HARDNESS	NOTES
0–1	Soft	Rainwater falls into this category.
1–7	Slightly hard	No scale is formed on heating.
7–20	Moderately hard	Most city water is in this range. Problems with scale are minor. It is quite usable with soap.
20–100	Hard water	Springwater and some well water fall into this category. Scale builds up rapidly.
100–1000	Very hard water	It is not usable with soap or in hot water systems.

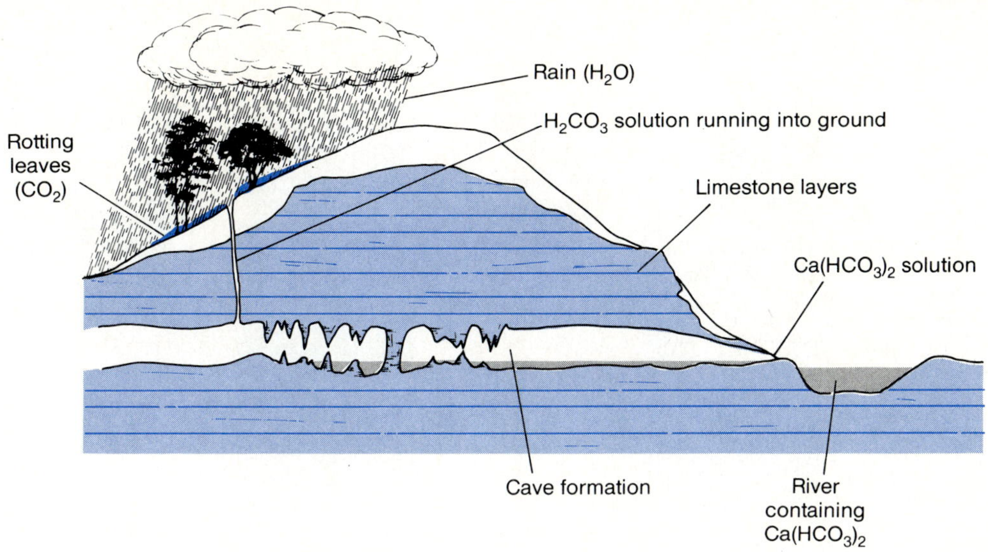

FIGURE 10-8
The aquatic cycle and the source of water hardness.

Softening hard water

There are two procedures that can be used to get rid of $Ca(HCO_3)_2$ and produce *soft* water. The first, called *temporary softening* (the Ca is not actually removed but only inactivated), involves making the water basic, which causes $CaCO_3$ to form as a precipitate:

$$Ca(HCO_3)_2 \xrightarrow{\text{base}} \underset{\text{Powder}}{CaCO_3 \downarrow}$$

This reaction removes the Ca^{2+} ions from solution, much as heating the water does, but the product is a powder rather than a scale. The borax added to laundry detergents is a base (as is Na_2CO_3); in addition to neutralizing the acidic odor-causing materials, it causes temporary softening. Phosphates are added to laundry detergents to help keep the Ca^{2+} complexed (or "tied up") so that it cannot cause the problems noted above:

$$3Ca^{2+} + Na_6P_6O_{18} \longrightarrow Ca_3P_6O_{18} + 6Na^{1+}$$

The second method, called *permanent softening,* uses a material called an *ion exchange resin* to replace the Ca^{2+} ions with Na^+ ions. Ion exchange resins are used in commercial household water softeners to remove the Ca^{2+} from the water. The process is diagramed in Figure 10-9. The zeolite is a natural plastic resin with small pores or cavities. Initially these pores are filled with Na^+ ions. As the water containing Ca^{2+} ions is passed over the resin, the Ca^{2+} ions enter the pores and Na^+ ions are released into the water. The following reaction occurs; *two* Na^+ ions are released for every Ca^{2+} ion absorbed:

$$2Na^+ \ldots \text{zeolite} + \underset{\text{Hard water}}{Ca^{2+}} \longrightarrow \underset{\text{Soft water}}{2Na^+} + Ca^{2+} \ldots \text{zeolite}$$

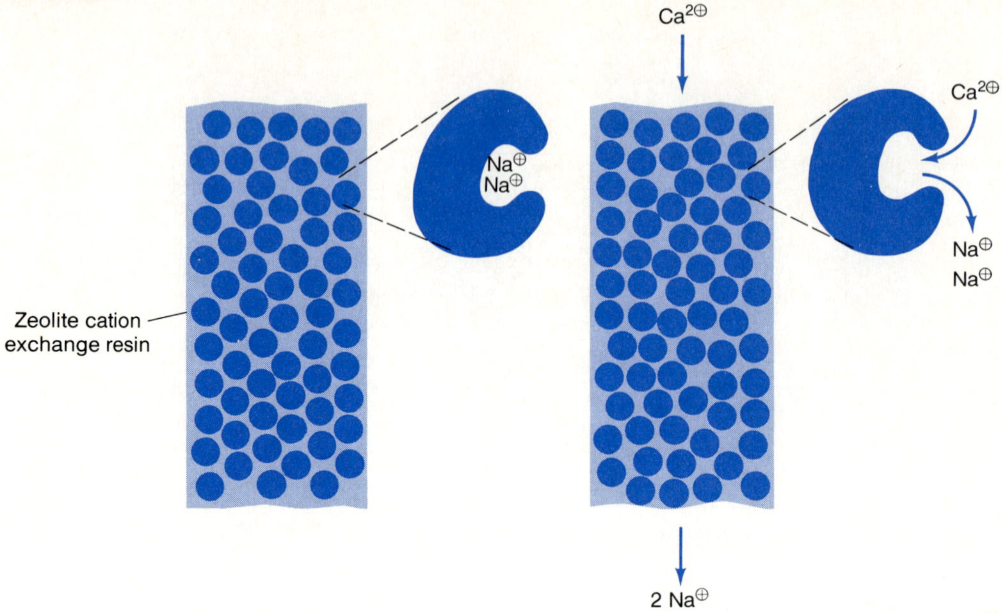

FIGURE 10-9
Operating principle of a permanent water softener.

(a) Water softener containing zeolite saturated with Na⊕

(b) Operating softener

The water coming out of the softener contains two Na^+ ions for every Ca^{2+} ion entering it. The Na^+ ions cause none of the problems associated with hard water—no scale buildup and no soap deactivation. However, the Na content of the drinking water is raised. This is a problem for people who are on restricted-sodium diets because of high blood pressure, or hypertension.

The amount of Na produced if 1 mg of $CaCO_3$ is placed into 1 L of water (hardness of 1 ppm) in a softener is calculated below (see Chapter 8 for a review).

1 1 mg = 0.001 g $CaCO_3$

2 $$\frac{0.001 \text{ g}}{100 \text{ g } CaCO_3 \text{ per mole}} = 0.00001 \text{ mol } CaCO_3$$

3 0.00001 mol of $CaCO_3$ gives 0.00001 mol of Ca^{2+} ion.

4 This amount of Ca^{2+} ion causes twice as many moles of Na^+ ions to be released.

5 $$0.00001 \text{ mol } Ca^{2+} \left(\frac{2 \text{ mol } Na^+}{1 \text{ mol } Ca^{2+}}\right) = 0.00002 \text{ mol } Na^+$$

6 0.00002 mol Na^+ (23 g/mol) = 0.00046 g Na^+

7 0.00046 = 0.46 mg Na^+ produced in each liter of water for every ppm hardness

If one is using water that has a hardness of 20 ppm $CaCO_3$ (20 mg $CaCO_3$ per liter of water), after softening it would contain 20(0.46) = 9.2 mg Na^+ per liter. This level of Na^+ ion does not present a serious problem; but if the softened water originally had a hardness of 100 ppm, it would now have about 50 mg Na^+ per liter and would taste slightly salty. Very hard well water containing 1000 ppm $CaCO_3$ that is softened to remove Ca^{2+} has 500 mg (0.5 g) Na^+ per liter, tastes distinctly salty (unpleasantly so), and is not helpful for people on a sodium-restricted diet.

Eventually all the Na^+ ions in the resin are replaced by Ca^{2+} ions, and the water passes through without being changed. This is detected electronically, and the softener then goes through a recharging cycle. During recharging the water flow is stopped and the resin is soaked in a concentrated solution of sodium salt (brine). This replaces the Ca^{2+} in the resin with Na^+. The calcium is thus "flushed" out of the resin and discarded. This recharging is usually done at night when the demand for water is low. By morning the system is ready to start softening again.

Hard water: Implications for consumers

In this chapter and Chapter 9 we have spent some time on consumer education and on learning the chemistry of some consumer products. As consumers, we must also be able to recognize misleading (or even downright false) advertising claims. Let's use the water-softening industry as an example.

The industry has generally adopted a figure of 20 ppm as "hard" water, with the implication that a water softener is needed at that point. In practice, almost everyone would need a softener if that were the case. A water softener is seldom needed until the level is significantly above 20 ppm (see Table 10-9). The author has seen hot water pipes that have been used for 20 years with unsoftened water containing hardness levels of 20 ppm that have only about ⅛ in of scale deposit.

Another example of deceptive practices is home water-testing demonstrations. The salesperson who comes to give a "free water-hardness analysis" will put several chemicals into a test tube containing a sample of your tap water. A large amount of precipitate forms in the test tube, and the customer is told that this precipitation is caused either by hardness or by pollution—either of which the product being promoted promises to remove. But pollution is always present in trace amounts, and no simple test will reveal it anyway. Moreover, even the hardest water (1000 ppm, containing 1 g $CaCO_3$/L) would form only 0.01 g of solid even if all the $CaCO_3$ in a test tube full of it (about 10 mL) could be induced to form a solid. This would be only 10 mg, which is barely enough to see. This "test" is actually the adding of one chemical (such as $BaCl_2$) to another (such as Na_2SO_4). These two react together to form a voluminous solid (in this case, $BaSO_4$). Since this reaction has nothing to do with hardness or pollution, the technique is fraudulent. This is not to say that all companies use dishonest procedures; nor does it necessarily imply that an individual salesperson is aware of any fraud (salespeople do what they are trained to do). And it does not necessarily mean that you do not need a water softener (some water is indeed quite hard). It does mean that you must be very careful and very critical of sales techniques.

STUDY GUIDE

Terms to learn

Acid	Liquid nitrogen
Acidic solution	Nutrient
Amphoteric	Oxidation number
Base	Oxidation state
Basic solution	Oxide
Baking soda	Permanent water hardness
Binary compound	Permanent water softening
Catalyst	pH
Contact process	ppm
Fertilizer	Potash of soda
Haber process	Salt
Hard water	Scale
Hydrated lime	Softening of water
Inorganic	Solvay process
Ion exchange resin	Temporary water hardness
IUPAC	Temporary water softening
Lime	Ternary compound
Limestone	Zeolite

Skill practice problems

1 Write the formulas for the following compounds.

(a) Lithium sulfate
(b) Sulfurous acid
(c) Sodium oxide
(d) Cuprous phosphate
(e) Carbon tetrabromide
(f) Aluminum cyanide
(g) Silver nitrate
(h) Iron(III) iodide
(i) Carbon disulfide
(j) Potassium sulfite

2 Write the names of the following compounds.

(a) $Co(NO_3)_2$
(b) $Fe(ClO_4)_2$
(c) $AlCl_3$
(d) MnO_2
(e) $ZnSO_4$
(f) H_2CrO_4
(g) UF_6
(h) $Sn(OH)_4$
(i) $Mg(CN)_2$
(j) SiI_4

3 What is the ratio of acidity (i.e., H^+ concentration) between vinegar and lime juice? (See Table 10-5.)

4 How many milligrams of Na per liter would one obtain by passing hard water containing 80 ppm $CaCO_3$ through an ion exchange water softener?

5 If $CaCO_3$ dissolves in water to the extent of 6.7 mg per liter of water, how many grams of $CaCO_3$ precipitate from 1 gal of hard water containing 50 ppm $CaCO_3$?

6 What is the chemical reaction that causes caves to form in limestone strata?

7 What is the pH of a solution with a H^+ concentration of 1×10^{-2}?

8 List three reasons why control of pH is important.

9 What color would phenolphthalein be if it were placed in a glass of household ammonia?

10 What is the material commonly called *liquid nitrogen?* Is this the proper terminology to use?

11 How does baking soda put out grease fires?

Goals

- To learn about commonly used organic materials
- To learn the systematic way chemists name complex molecules
- To gain the ability to recognize various kinds of organic materials by their functional groups
- To understand how chemistry is involved in producing the materials we use
- To gain insight into how a chemist approaches a problem—how research is done

11

ORGANIC CHEMISTRY

Chapter 10 discussed the naming system for inorganic compounds. The majority of these compounds do not contain carbon. In this chapter we will study the chemistry of carbon-containing compounds, or *organic chemistry*.

ORGANIC COMPOUNDS

Carbon, as you will remember, has four electrons in its outer shell. When it combines with other nonmetals (such as hydrogen, oxygen, nitrogen, and the halogens), it needs four more electrons to complete its octet (see Chapter 7). Carbon therefore always forms four covalent bonds.

Because a carbon atom can readily form covalent bonds with other atoms of carbon as well as with other nonmetals, *chains* of carbon atoms can be formed. The number of carbon atoms that can be in a molecule varies; and for a given number of carbon atoms, several different spatial and bonding arrangements are possible. There are literally millions of organic compounds. By comparison, for example, sodium and bromine form only one compound, NaBr.

Let's consider the number of compounds containing only four carbon atoms, one oxygen atom, and ten hydrogen atoms. All possible ways of arranging these atoms are shown on the opposite page. Each of these compounds has the formula $C_4H_{10}O$, but every one is a different compound with different physical, chemical, and physiological properties.

Such compounds—compounds which have the same formula but whose atoms are arranged differently—are called *isomers*. It may not be obvious to the beginning student that each of these is different, and it is beyond the scope of this text to teach the necessary skills for recognizing this or for drawing all isomers of a given formula. However, the chemical structures shown above do illustrate the concept of isomerism and suggest why so many organic compounds are possible. This is further suggested by the fact that for the formula C_6H_{12} there are 5 possible isomers, for $C_{10}H_{22}$ there are 75, and for $C_{20}H_{42}$ there are (almost unbelievably) 366,319. These figures give us an idea why there are millions of known organic compounds and why even people who study organic chemistry for a lifetime are not able to learn all this vast field. This chapter will present as clearly as possible a few basic aspects of organic chemistry and emphasize those which have implications for consumers.

Let's begin by taking a closer look at isomers. Methyl ether and ethyl alcohol have the same molecular formula, but they are quite different materials with different physical, chemical, and physiological properties. (See Table 11-1 and Figure 11-1.) The differences in properties and structure between these two compounds are an excellent illustration of the *structure-function principle,* which states that *all* properties of a compound (function) are a consequence of the kind of atoms in molecules of that substance and the arrangement of those atoms (structure). Some isomer pairs have even more subtle differences. They have the same kinds of bonds, but the atoms are arranged differently.

ALL POSSIBLE WAYS OF ARRANGING FOUR CARBONS, ONE OXYGEN, AND TEN HYDROGENS

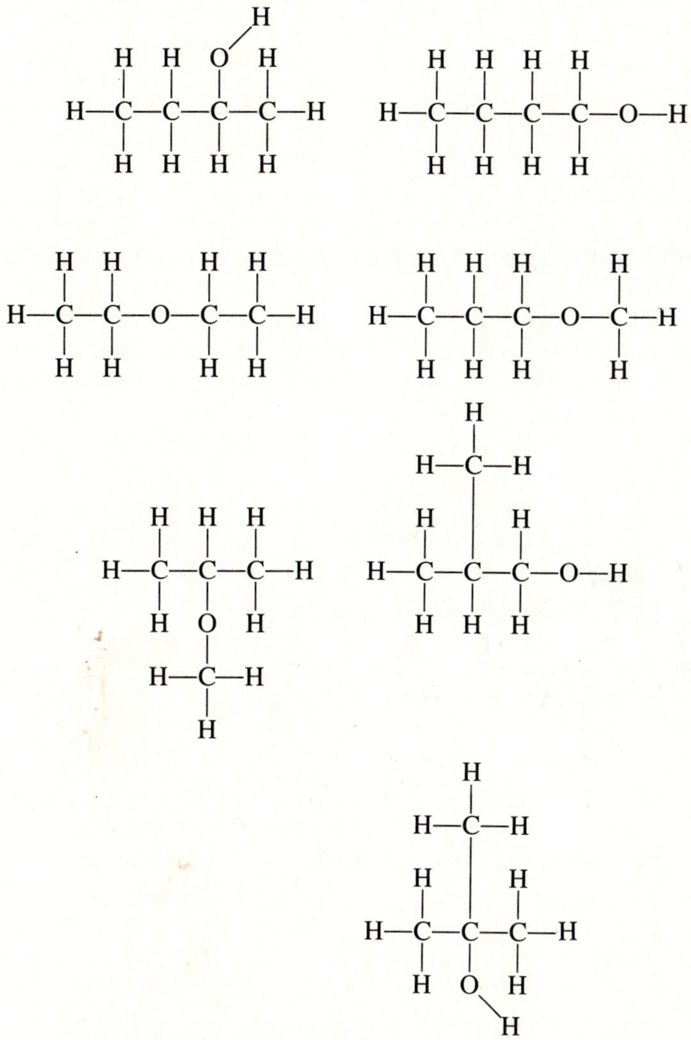

TABLE 11-1

Summary of various differences between ethyl alcohol and methyl ether

NAME	ARRANGEMENTS	KINDS OF BONDS	MOLECULAR FORMULA	PROPERTIES
Methyl ether	$CH_3—O—CH_3$	O—C	C_2H_6O	Gas at room temperature
		C—H		Chemically inert
				An anesthetic
Ethyl alcohol	$CH_3—CH_2—OH$	C—C C—H	C_2H_6O	Liquid at room temperature
		O—C		Chemically reactive
		O—H		An intoxicating beverage

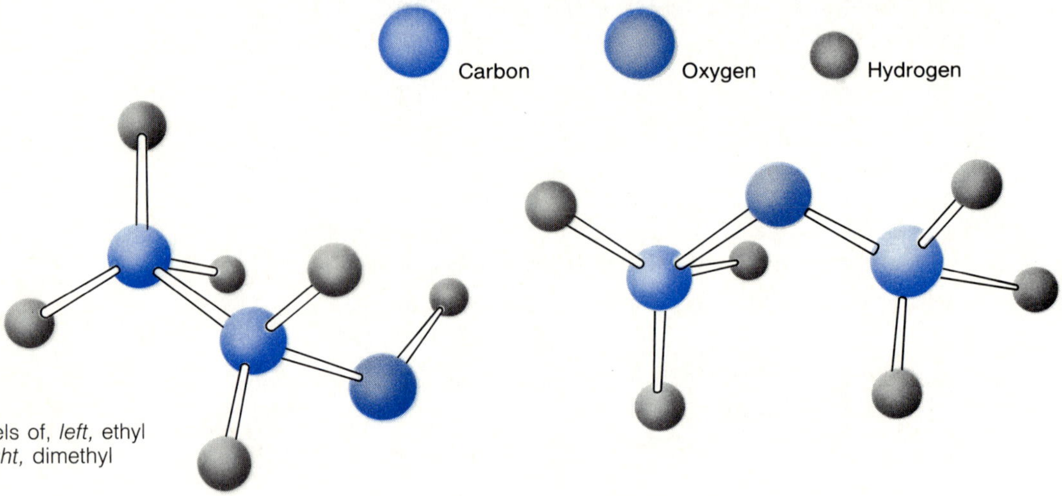

FIGURE 11-1
Molecular models of, *left,* ethyl alcohol and, *right,* dimethyl ether.

The following two pairs of molecules are examples:

Formula = C_4H_{10} $CH_3—CH_2—CH_2—CH_3$

and

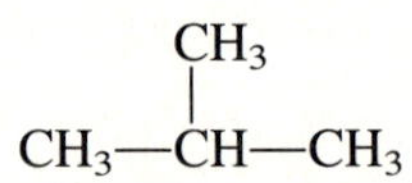

Different arrangement of carbons

Formula = $C_2H_2Cl_2$

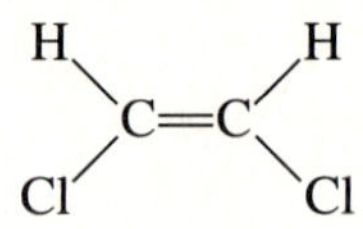

and

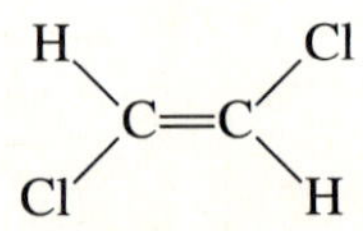

Different relationship of chlorine atoms to each other

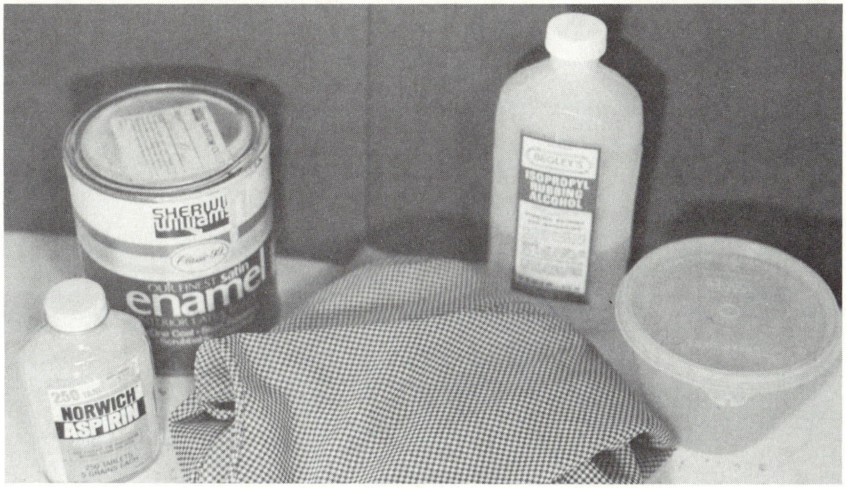

FIGURE 11-2
Some of the consumer products made by organic chemists.

Many, if not most, of the products produced by the chemical industry for direct consumer use are organic chemicals (see Figure 11-2). Plastics, fibers, fuels, and drugs are among those produced in the greatest quantities. We will look at some of these later in this chapter (and later in the text) in relation to specific areas of interest to consumers.

CLASSIFICATION OF ORGANIC COMPOUNDS: FUNCTIONAL GROUPS

Covalent bonding between atoms produces molecules with definite *shapes* and *arrangements* of atoms. For purposes of naming, organic compounds have been divided according to the various arrangements of atoms. We will examine the more common of these here. These unique arrangements, called *functional groups*, determine how each material will act chemically and what its physical properties are.

Alkanes and substituted alkanes

Alkanes Alkanes are compounds containing only carbon and hydrogen and having only *single* bonds. Examples follow:

NAME	MOLECULAR FORMULA	STRUCTURAL FORMULA	USE
Methane	CH_4	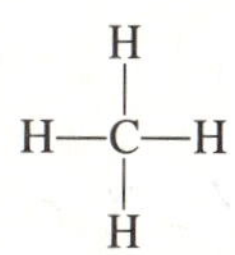	Natural gas
Butane	C_4H_{10}	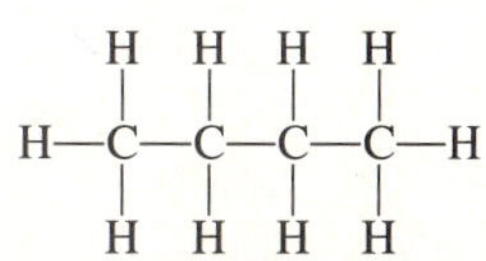	Fuel for portable stoves

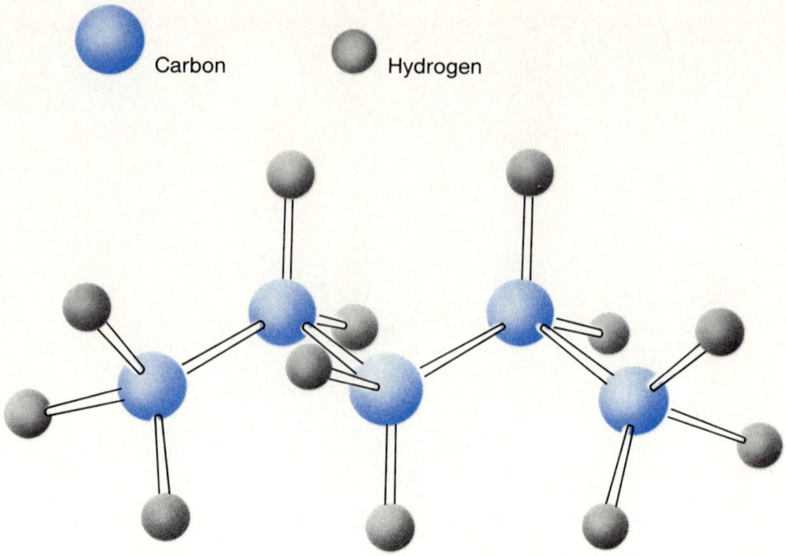

FIGURE 11-3
Molecular model of pentane.

NAME	MOLECULAR FORMULA	STRUCTURAL FORMULA	USE
Pentane (see Figure 11-3)	C_5H_{12}	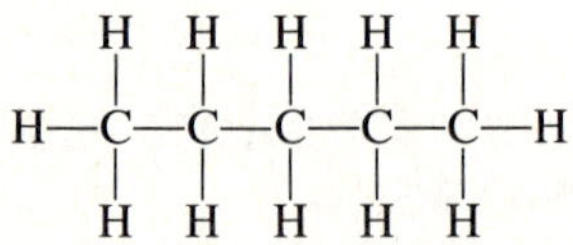	Cigarette lighter fluid
Octane	C_8H_{18}	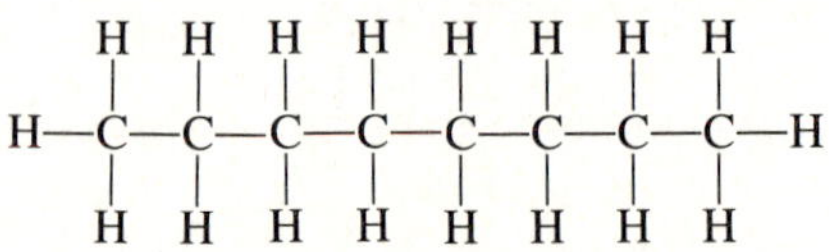 or $CH_3(CH_2)_6CH_3$	Major component of gasoline
Paraffin	$C_{20}H_{42}$		Wax for candles

Substituted alkanes In substituted alkanes, one or more of the hydrogens have been replaced by a halogen (F, Cl, Br, I). Examples follow:

NAME	MOLECULAR FORMULA	USE
Chloroform (see Figure 11-4)	$CHCl_3$	Solvent
Carbon tetrachloride	CCl_4	Used in some types of fire extinguishers
Dichloroethane (ethylene chloride)	$CH_2Cl—CH_2Cl$	Dry-cleaning solvent, chewing gum solvent

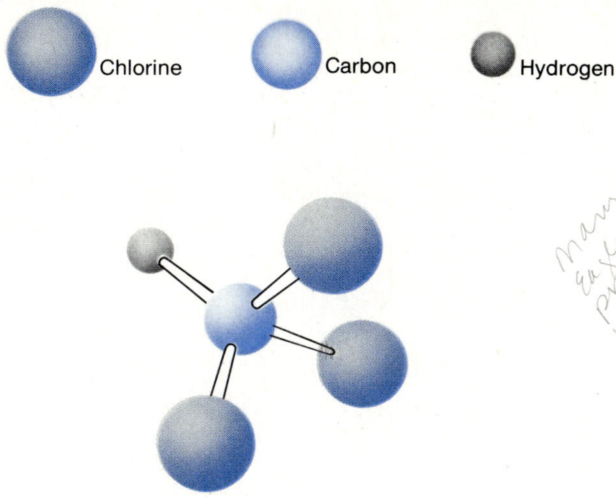

FIGURE 11-4
Molecular model of chloroform.

NAMING ALKANES AND SUBSTITUTED ALKANES

The purpose of this program is to teach the IUPAC naming (nomenclature) system for hydrocarbons and substituted hydrocarbons. After finishing this unit, a student should be able to write the name for any compound of these types whose structure is given and draw the structure for any whose name is given (the latter is the easier for most students).

Latin prefixes are used to designate the number of carbons in a given molecule or part of the molecule. The more common of these prefixes are listed in Table 11-2.

Structure to name

Rules for constructing names are as follows.

1 Straight-chain alkanes (no branching):

(Prefix for number of carbons + *ane*

Note All alkanes end in *-ane*.

2 Branched alkanes.

(a) Find the longest continuous chain of carbons.

(b) Decide what groups are attached and name them:

(Prefix for number of carbons) + y*l*

TABLE 11-2
Common prefixes

NUMBER OF CARBONS	PREFIX
1	meth-
2	eth-
3	prop-
4	but-
5	pent-
6	hex-
7	hept-
8	oct-
9	non-
10	dec-

(c) Assign numbers to the carbons in the longest chain so that the branches are at the lowest numbers:

> (Position where group is attached to longest chain)
> + (prefix for number of carbons in group + y*l*)
> + (prefix for number carbons in longest chain + ane).

Note When more than one group is attached to the longest chain, list the groups *alphabetically*.

Worked example

Name the compound whose structure is shown below:

$$CH_3—CH_2—CH_2—CH_2—CH_2—CH_3$$

1 Since there are no branches, simply count the number of carbons in the molecule (see rule 1 above):

There are six carbon atoms.

2 Find the prefix that designates 6:

hex-

3 Combine hex- and -ane into one word:

Hexane

Worked example

Name the compound whose structure is written below:

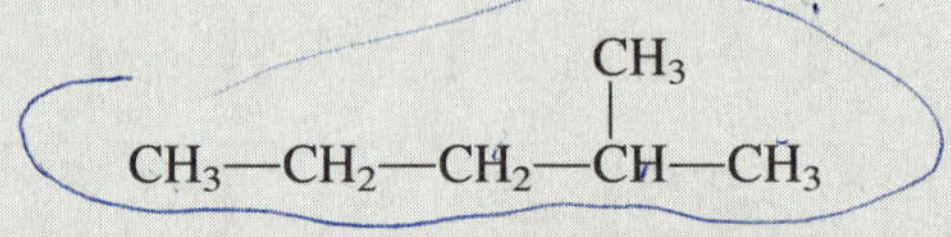

1 Find the longest carbon sequence and draw a "box" around it:

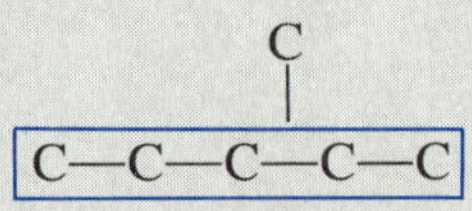 ⟵ 5 carbons in sequence

Note Since this alkane is branched, we use rule 2 above.

2 Write the "base" name by using the prefix for five carbons:

pent-

3 Add an ending to designate only single bonds. This is the name of the longest chain:

Pent*ane*

4 Draw a circle or circles around any groups *attached* to this longest sequence of carbons.

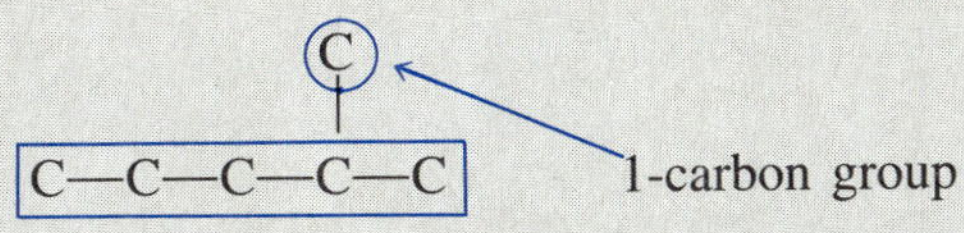

5 First write the *prefix* for the number of carbons in this group. Next add the ending -yl (to designate that it is a group added to the longest chain):

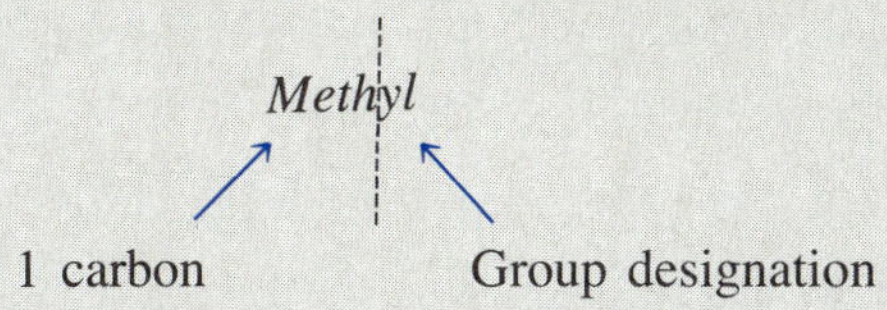

6 Place numbers next to each carbon in the sequence (longest carbon chain) chosen in step 1. Start at one end and continue sequentially to the other end. Start at the end *closest* to the most attached groups (this gives the position of the attached group the lowest number possible):

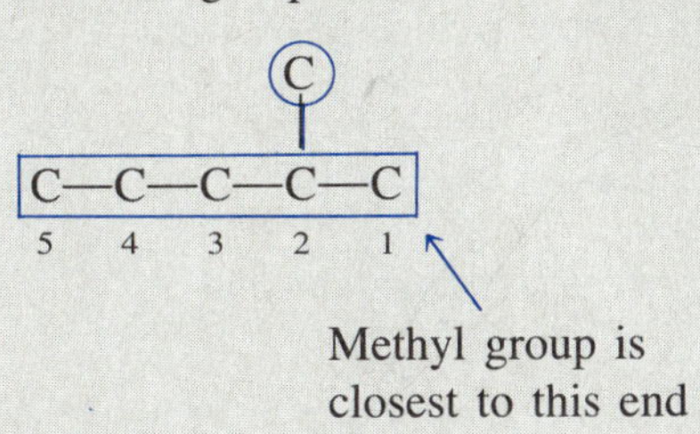

7 Construct the name of this compound as follows using rule 2 above:

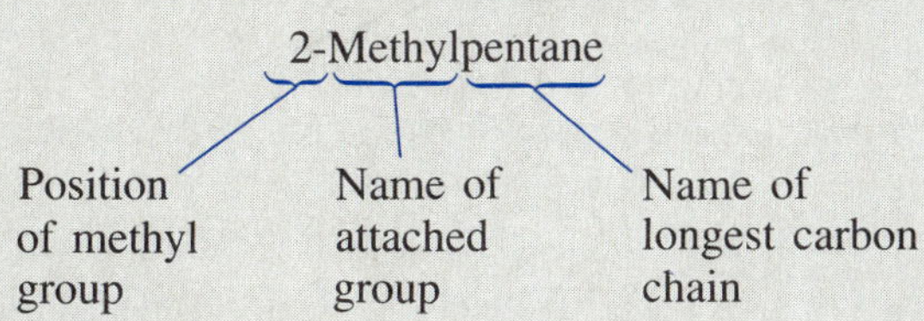

Worked example

Name the compound whose structure is written below:

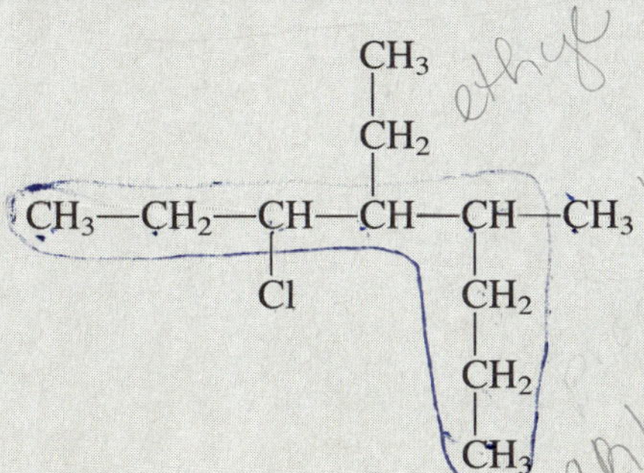

1 Find the longest carbon chain and name it:

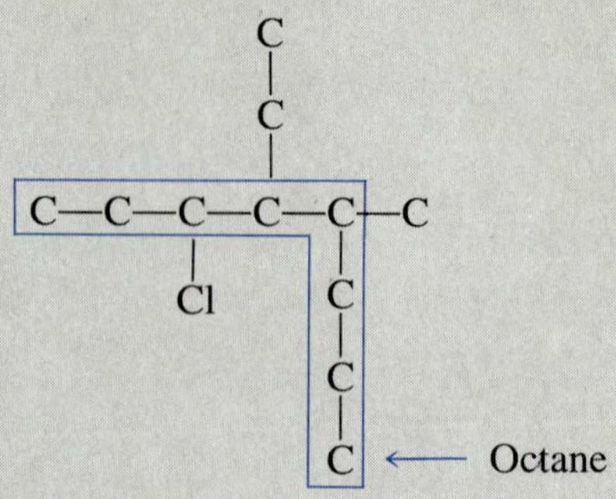

Note The student should be aware that the longest chain is *not* always written in a straight line.

2 Circle the groups attached to the longest chain and name them:

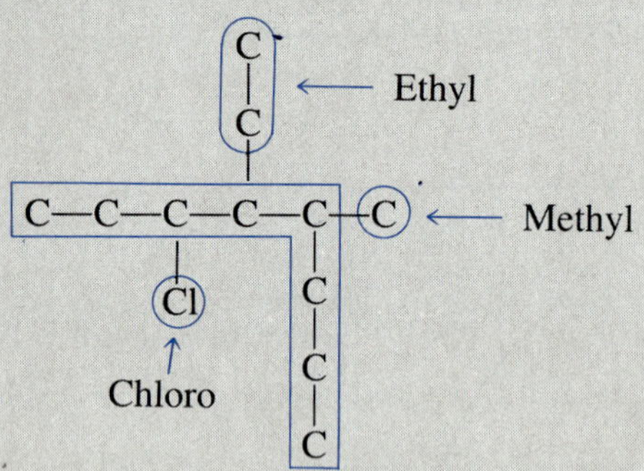

Note The halogens all end in *-o*; thus: fluoro-, chloro-, bromo-, iodo-.

3 Assign numbers to the longest chain (in this case the groups are closest to the left end, and so we start numbering from the left end):

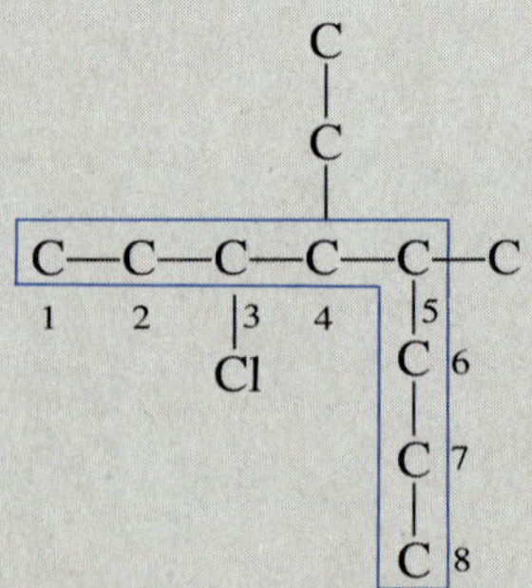

4 Write the name of the longest chain + *ane*:

Octane

5 List the attached groups before the word *octane* (alphabetically).

*C*hloro *e*thyl *m*ethyloctane

6 Place numbers before each group name to show where on the longest chain it is attached:

The chloro group is attached at carbon 3.

The ethyl group is attached at carbon 4.

The methyl group is attached at carbon 5.

3-Chloro-4-ethyl-5-methyloctane

Note One always places a hyphen between numbers and letters in chemical names. This is the *complete name*.

Worked example

Name the compound whose structure is written below:

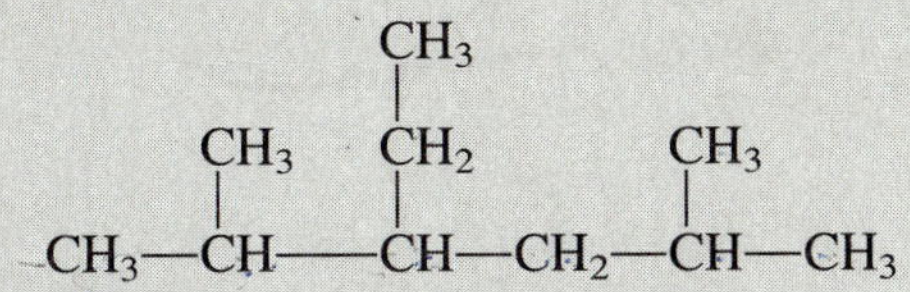

1 Find the longest carbon chain and name it. When there is more than one chain of the same length, choose the one that leaves the groups with the simplest structure attached to it:

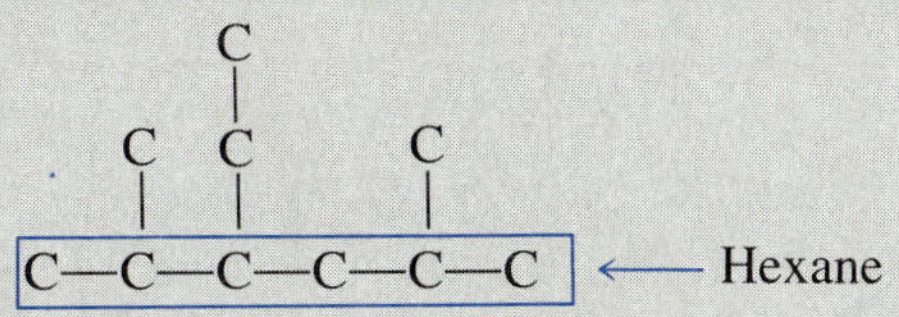

2 Circle the groups and name them:

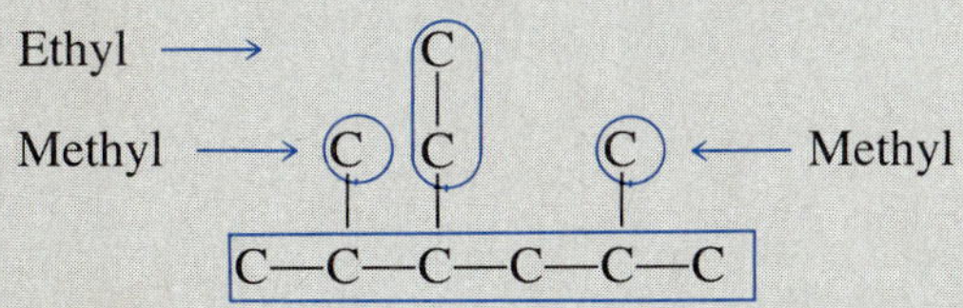

3 Number the longest chain:

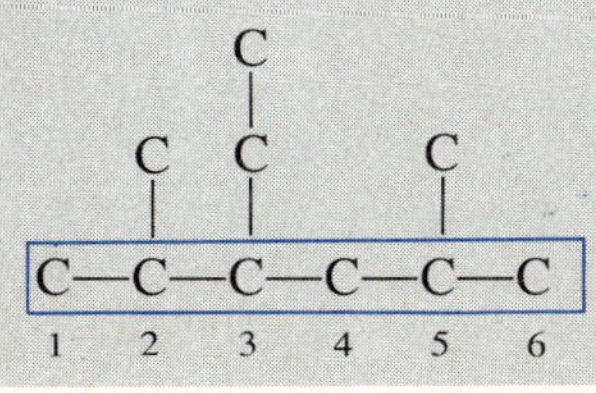

TABLE 11-3
Commonly used prefixes

NUMBER OF GROUPS OF A GIVEN TYPE	PREFIX
2	di-
3	tri-
4	tetra-
5	penta-
6	hexa-

Note This numbering system places the groups on carbons 2, 3, and 5. Starting at the other end place the groups on carbons 2, 4, 5, which are larger numbers.

4 Write the name of the longest chain:

Hexane

5 List the attached groups before the word *hexane* alphabetically. Since there are two methyl groups, rather than listing one twice use a prefix to indicate how many are present (*di*methyl). Commonly used prefixes are shown in Table 11-3.

Note In alphabetizing, the prefixes designating the number of groups are not considered. For instance, in di*m*ethyl the m is used, *not* the d.

6 Place numbers before the group names to show where they are attached to the longest chain. When there are 2 or more of the same group, there should be a number designation for each one:

3-Ethyl-2,5-dimethylhexane

Showing where *both* methyl groups are attached

Note A comma is placed between two adjacent numbers in a chemical name.

Problem

Attempt to name the following compound. *After* you have tried, examine the solution to see if you reasoned correctly:

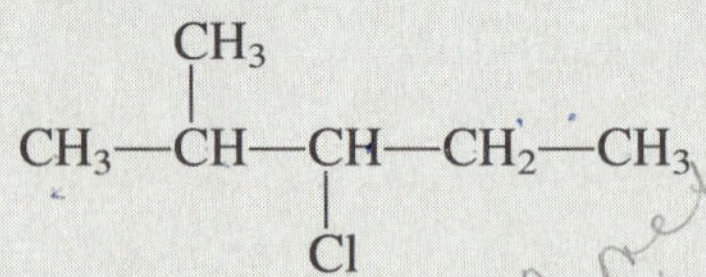

Solution

1 Find the longest chain and name it:

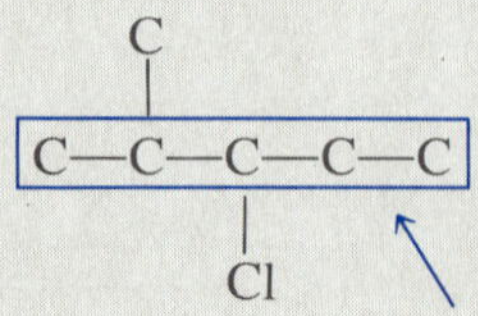

Pentane

2 Circle and name each group:

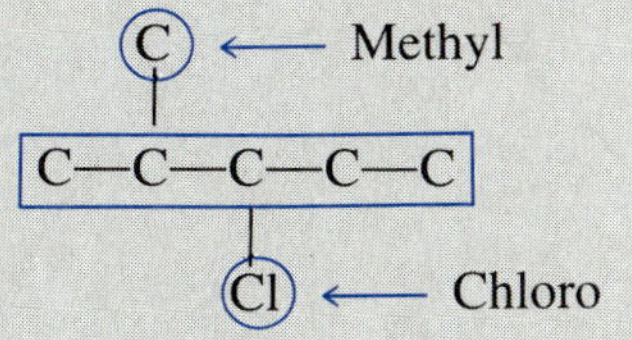

3 Number the longest chain from the left end:

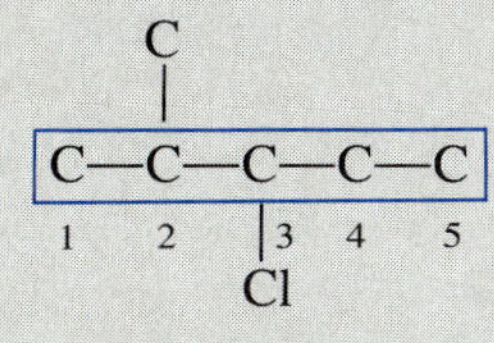

4 Write out the name of the compound:

3-Chloro-2-methylpentane

Problem

Name the following compound:

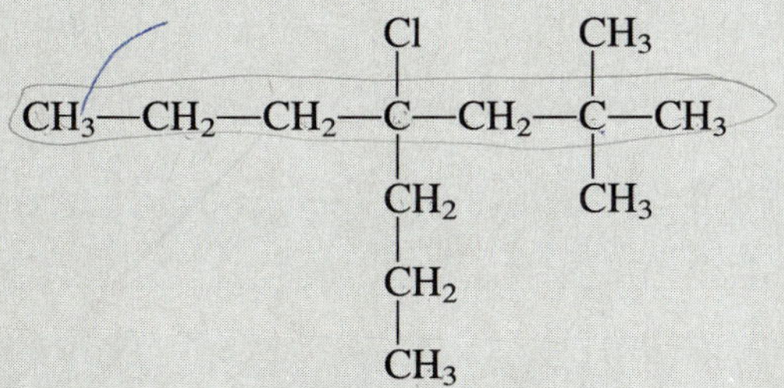

Solution

1 Find the longest chain and name it:

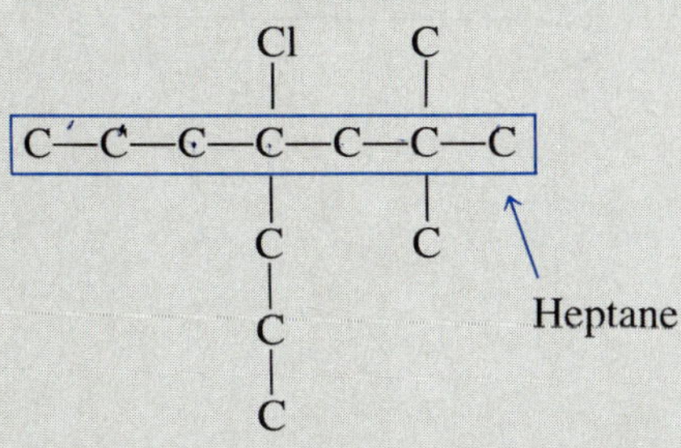

2 Circle and name all groups:

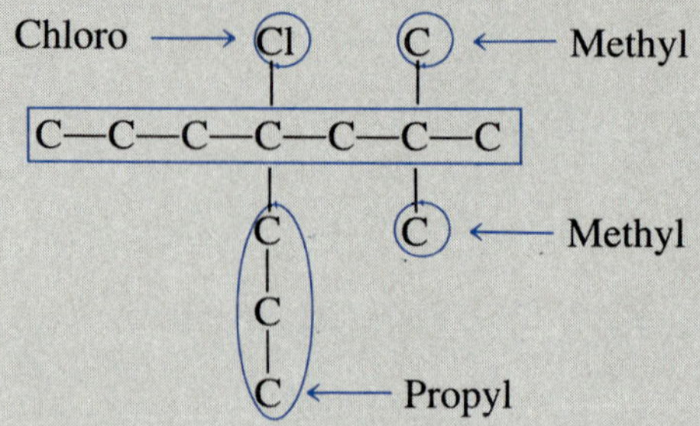

3 Number the longest chain from the right end:

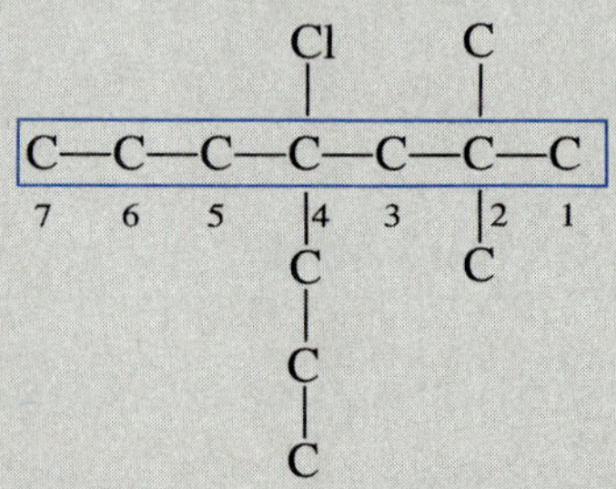

4 Name the compound:

4-Chloro-2,2-di*m*ethyl-4-*p*ropylheptane

Note Even though both methyl groups are on the same carbon, one must write the 2 twice (hence, -2,2-dimethyl).

Problem

Name the following compound:

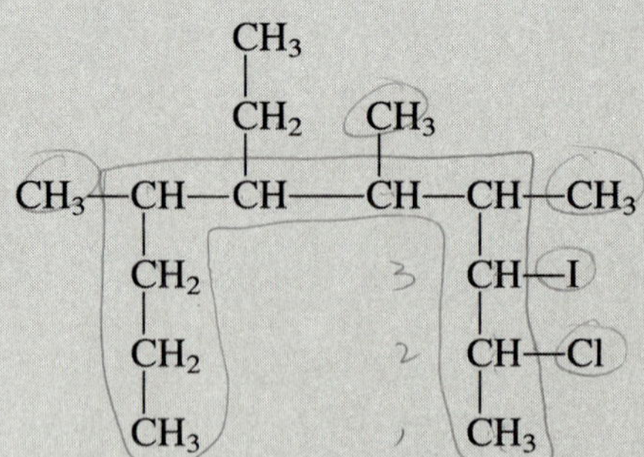

Solution

1 Find the longest chain and name it:

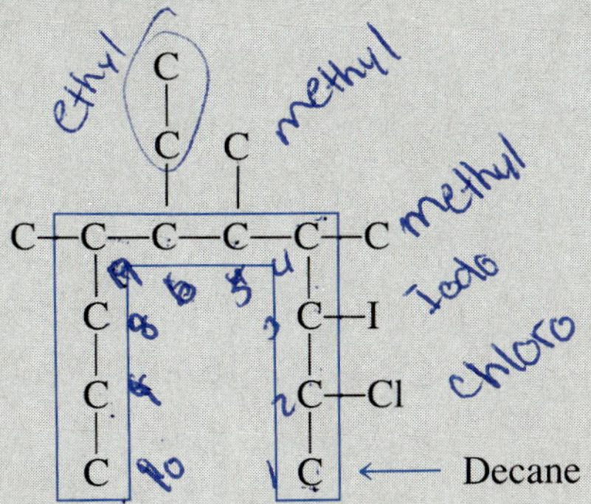

2 Circle and name each group:

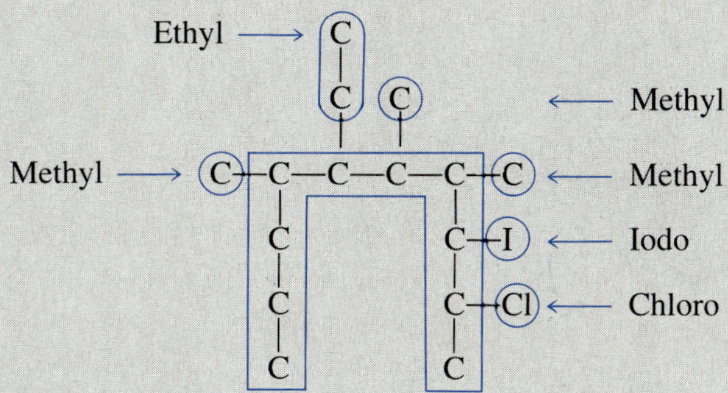

3 Number the longest chain from the right end:

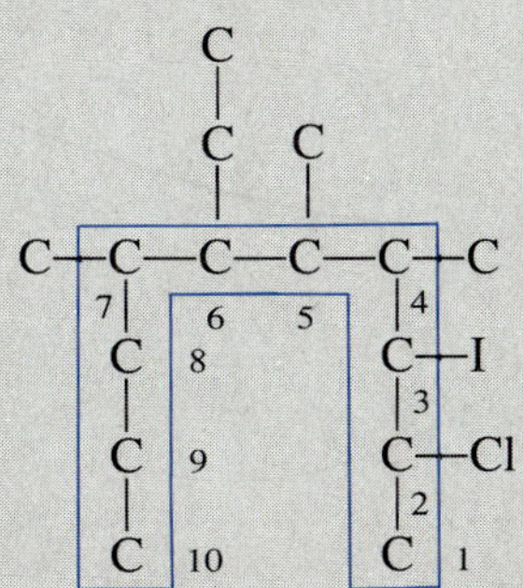

4 Write out the complete name:

2-*C*hloro-6-*e*thyl-3-*i*odo-4,5,7-tri*m*ethyldecane

That is a terrible-looking name. But when a systematic approach is used, it is really not too difficult to decipher.

Problem

Try to name each of the following (names only are given as answers):

a

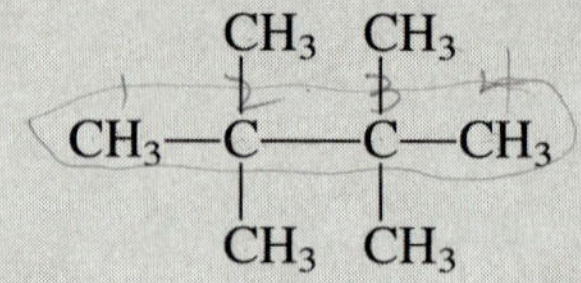

b

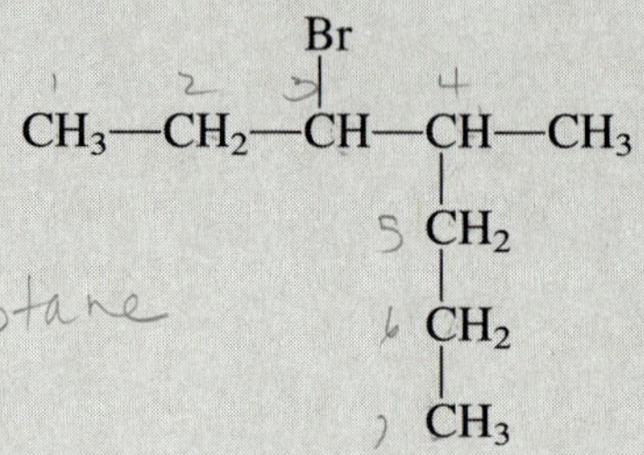

c

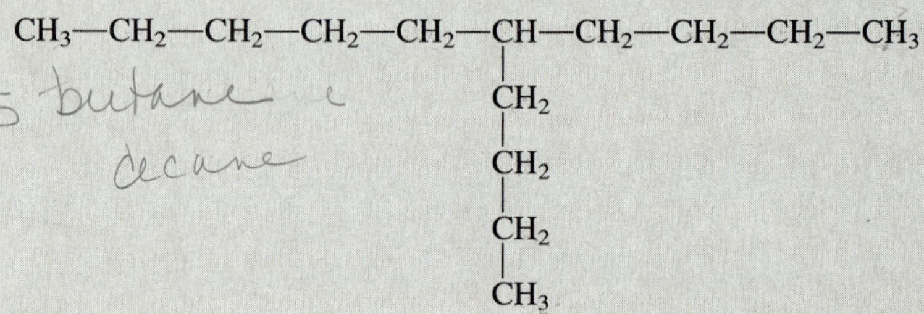

Answers

a 2,2,3,3-Tetramethylbutane

b 3-Bromo-4-methylheptane

c 5-Butyldecane

Name to structure

Worked example

Draw out the structure of 2,2-dimethylpropane.

1 Link carbons together to make a chain of the length indicated by the last part of the name (*propane*, in this case). The prefix *prop-* indicates three carbons.

C—C—C

2 Number this chain from *either* end:

C—C—C
1 2 3

3 Attach the indicated groups at the designated carbons. In this case, attach two methyl groups to carbon 2:

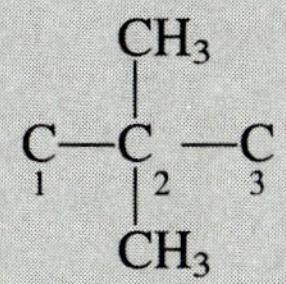

4 Attach hydrogen atoms to any carbons that do not have four bonds so that each carbon has four bonds:

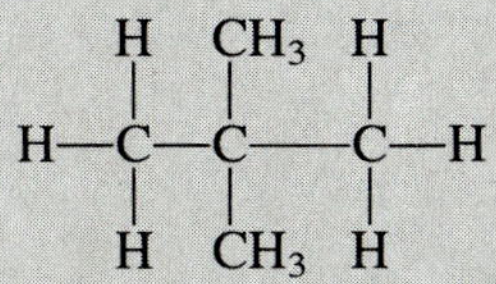

5 Write out the formula in standard (condensed) form:

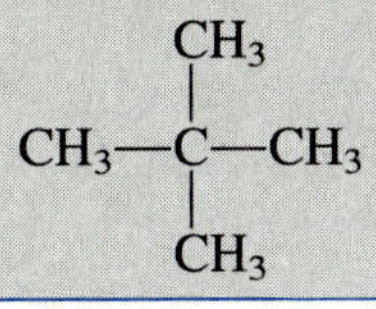

Problem

Now, you try one. Draw the structure of 3,3-dibromo-1-chloro-5-ethylheptane.

Solution

1 Link seven carbons in a row:

C—C—C—C—C—C—C

2 Number them:

C—C—C—C—C—C—C
1 2 3 4 5 6 7

3 Attach the indicated groups:

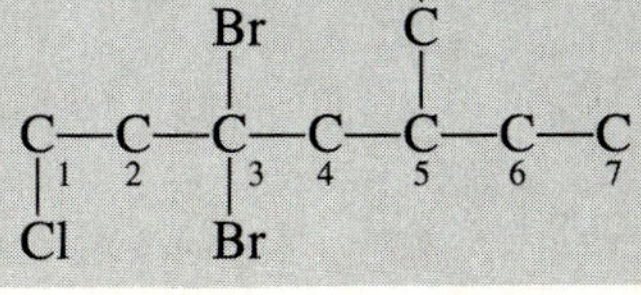

4 Attach hydrogens to any carbons that do not have four bonds:

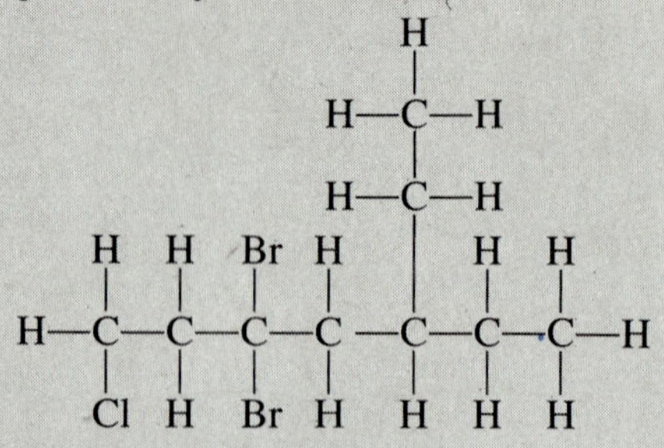

5 Write the structure in condensed form:

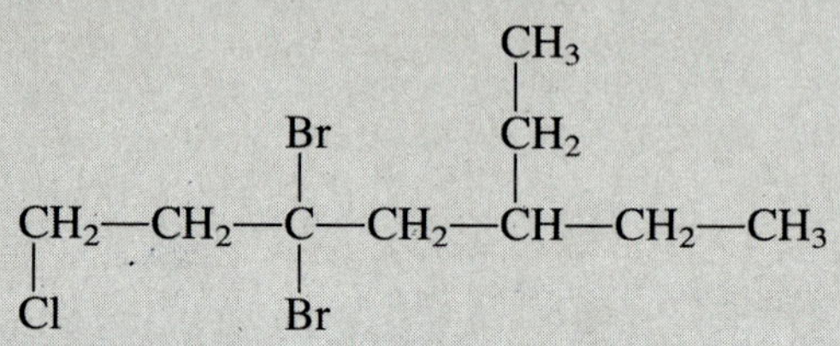

Other functional groups

We will now take a briefer look at several other common functional groups. We will consider the characteristic features of each, common materials of each type that are of general interest, and (only briefly) the names of some of them. The names are based on the system used for alkanes and therefore require little further study for mastery.

Cycloalkanes Cycloalkanes are compounds containing only carbon and hydrogen (all single bonds) but also containing a *ring* of carbon atoms. Examples follow:

NAME	STRUCTURE	USE
Cyclopropane	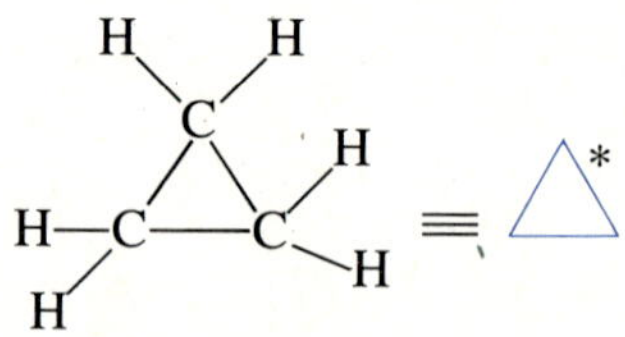	General anesthesia
Cyclopentane	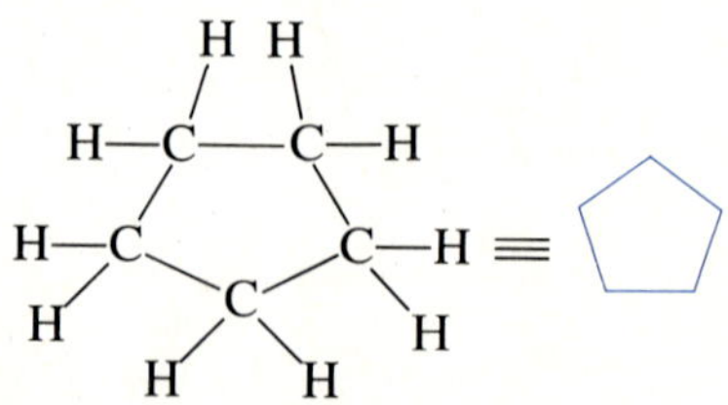	Solvent

*These are often abbreviated by simply drawing a geometrical figure with the same number of points as the carbon atoms in the ring.

These compounds are named using a prefix to designate the number of carbons followed by *-ane*, just as in alkanes. This name is simply preceded by the prefix *cyclo:*

Cyclopropane

Indicates a cyclic structure	Prefix for three carbons	Ending indicative of alkanes

Alkenes Alkenes are compounds that contain only carbon and hydrogen atoms and have at least one carbon-carbon *double* bond.

Examples follow:

NAME	STRUCTURE	USE
Common: Ethylene (see Figure 11-5) IUPAC: Ethene	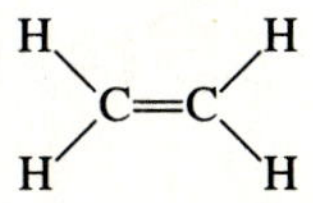	Production of polyethylene
1,3-Butadiene	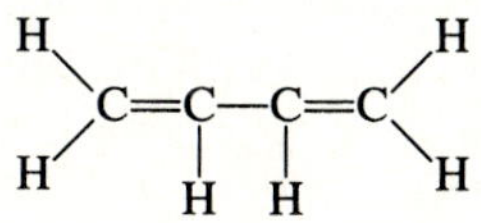	Production of butyl rubber

Note that, as with inorganic nomenclature, there are sometimes common (or classical) names as well as the IUPAC names.

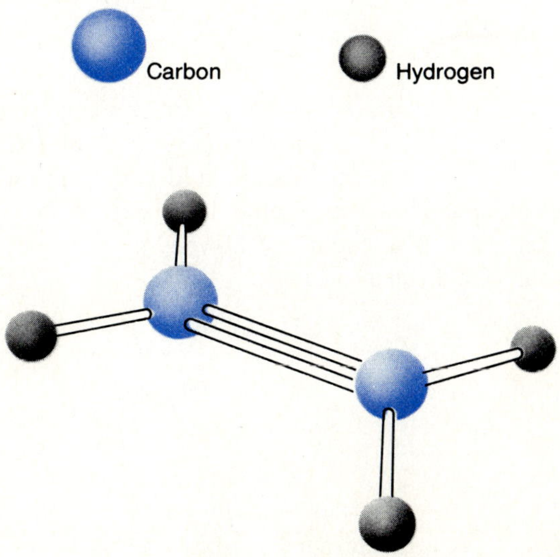

FIGURE 11-5
Molecular model of ethylene (ethene).

The naming of alkenes is quite similar to that for alkanes except that the *-ane* ending is changed to *-ene* and the location of the double bond must be indicated in compounds with four carbons or more by identifying the first carbon involved in the double bond:

$$CH_3—CH{=}CH—CH_3$$

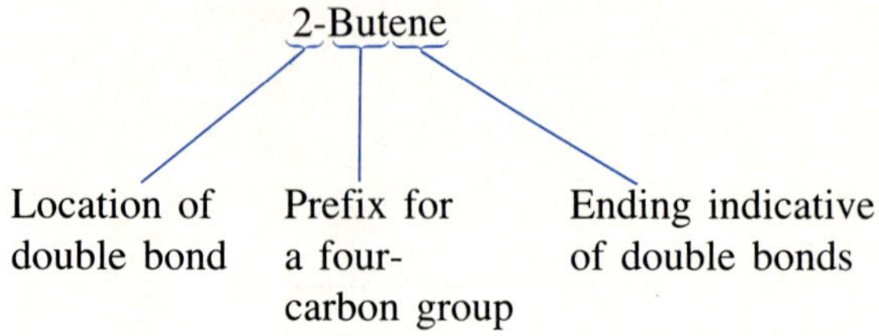

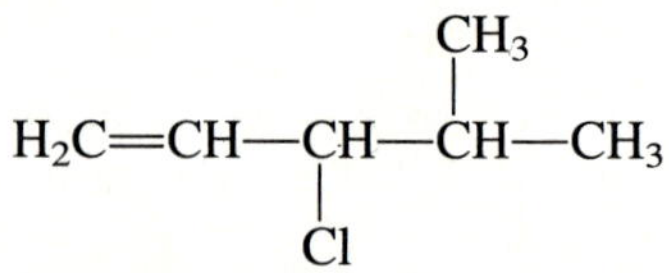

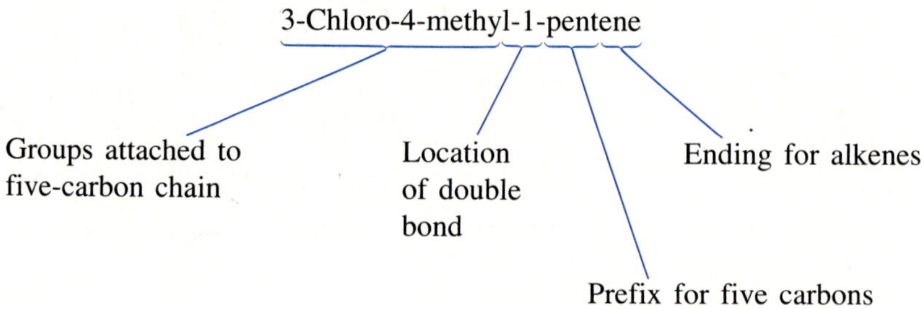

Alkynes Alkynes contain only carbon and hydrogen and have one carbon-carbon *triple* bond. Acetylene (Figure 11-6), CH≡CH, whose IUPAC name is *ethyne,* is the only alkyne used widely in industry. It is used as fuel for welding torches (see Figure 11-7) and in the synthesis of other industrial chemicals. The ending of alkynes is *-yne,* and the triple-bond location is usually indicated. Thus:

$$CH_3—CH_2—CH_2—C{\equiv}C—CH_3$$

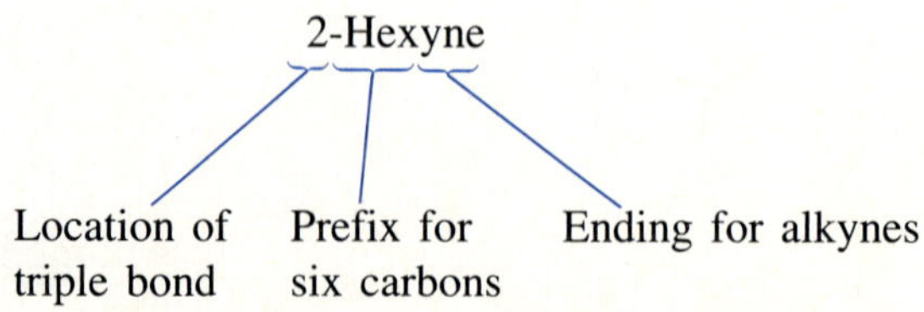

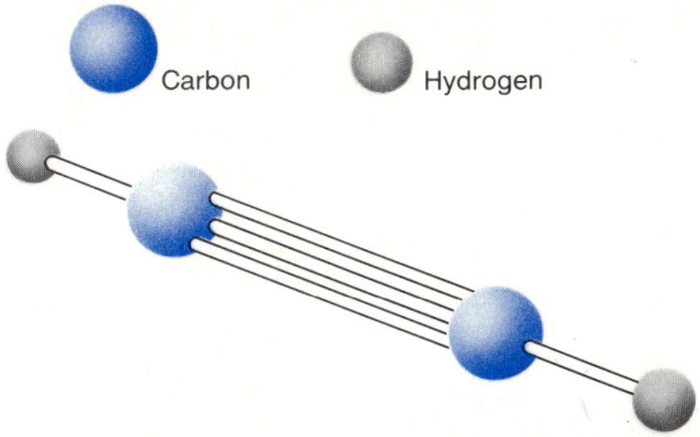

FIGURE 11-6
Molecular model of acetylene.

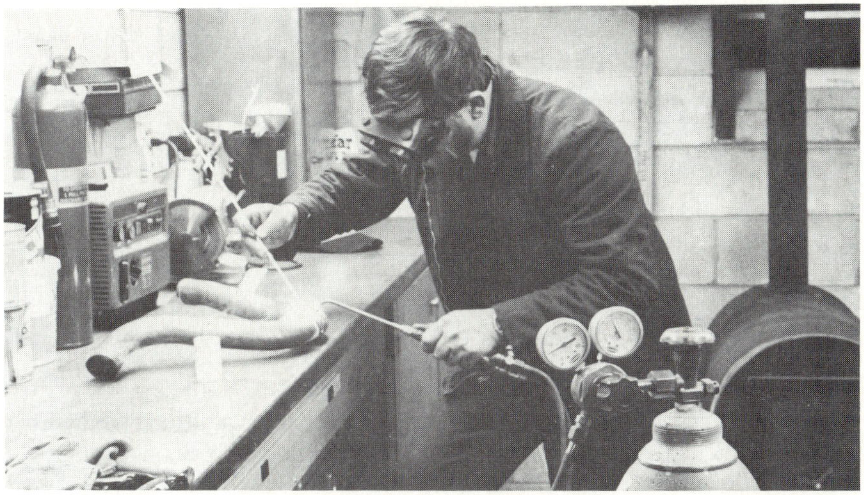

FIGURE 11-7
Welding operation using acetylene fuel (in tank).

Aromatics Aromatics are compounds containing *rings* of carbon which are sometimes written with alternating double and single bonds in this ring. The rings are usually six-membered. These double bonds are not localized, however. The six electrons in the three bonds are spread equally around the ring. This delocalization of the electrons around the ring accounts for the unusual stability of aromatic compounds and, in effect, produces a 1½ bond between each two carbons. This is shown in Figure 11-8 as rings of electron density above and below the carbon atoms. Examples of aromatics follow:

NAME	STRUCTURE	USE
Benzene (see Figure 11-8)	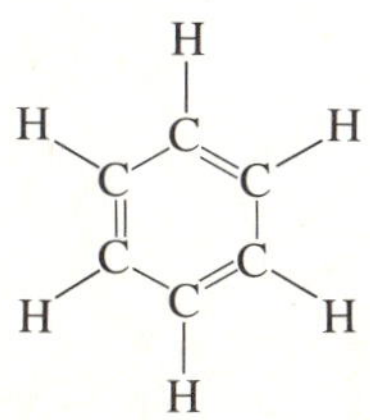	Solvent

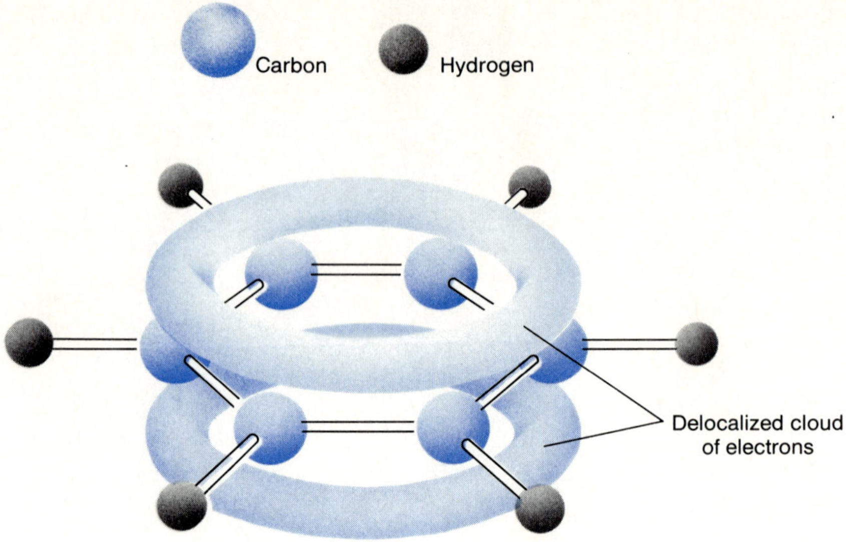

FIGURE 11-8
Molecular model of benzene.

Because the benzene ring occurs so often in compounds, it is abbreviated

where each ''point'' on the hexagon symbolizes a carbon atom and the hydrogens are understood to be present. Also, usually the double bonds are designated by drawing a circle to indicate the delocalization of electrons. Thus:

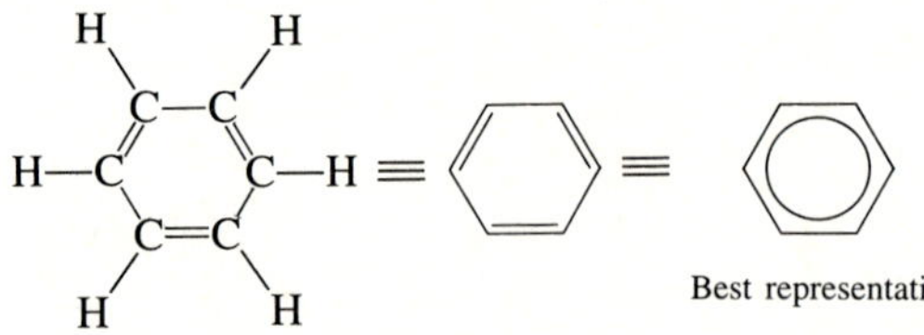

Benzene is used less now than it was in the past because of its (very weak) carcinogenicity.

NAME	STRUCTURE	USE OR OCCURRENCE
Toluene 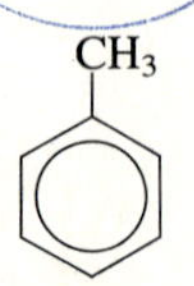	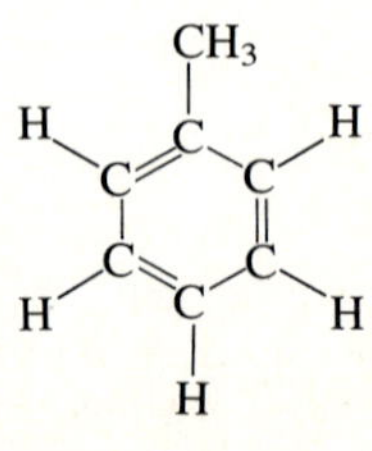	Solvent in glues

NAME	STRUCTURE	USE OR OCCURRENCE
Naphthalene 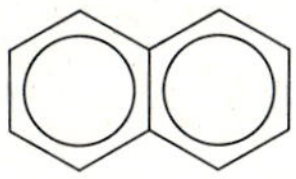	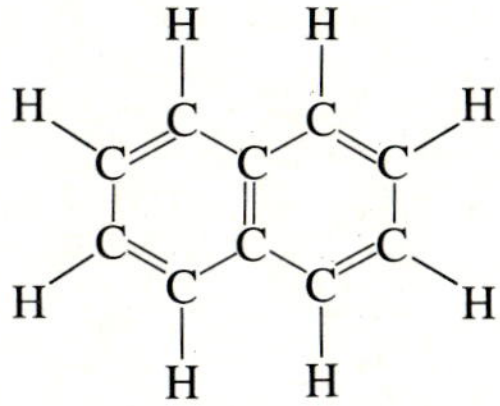	Mothballs
Benzpyrene* 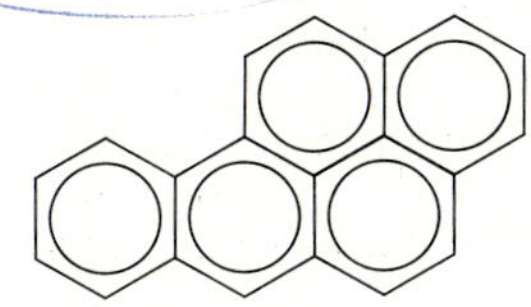		Formed when most materials burn. Highly carcinogenic. Found in tobacco smoke in relatively large amounts.
para-Dichlorobenzene† 	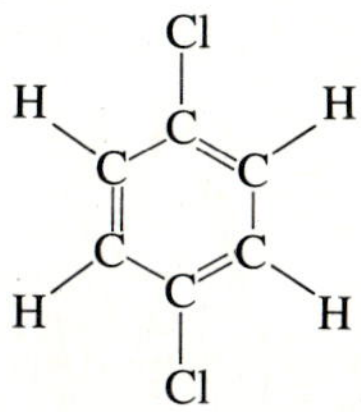	Mothballs can be made of this as well.

Alcohols Alcohols are compounds containing a hydroxyl (—OH) group attached to a carbon. The remainder of the molecule has the structure of an alkane (contains only carbon and hydrogen with single bonds). Examples follow:

NAME	FORMULA	USE
Methyl alcohol (wood alcohol, methanol)	CH_3—OH	Solvent, disinfectant
Ethyl alcohol; see Figure 11-9 (grain alcohol, ethanol, denatured alcohol, proprietary solvent)	CH_3—CH_2—OH	Chemical solvent, alcoholic beverages, some rubbing alcohol formulations, solvent for medication, disinfectant
Isopropyl alcohol (isopropanol, 2-propanol)	$(CH_3)_2$CH—OH	Disinfectant, most rubbing alcohols are 70% isopropyl alcohol and 30% H_2O.

*Note that benzpyrene consists of 5 benzene rings fused together.
†*para* means that the two Cl atoms are attached at opposite sides of the aromatic ring.

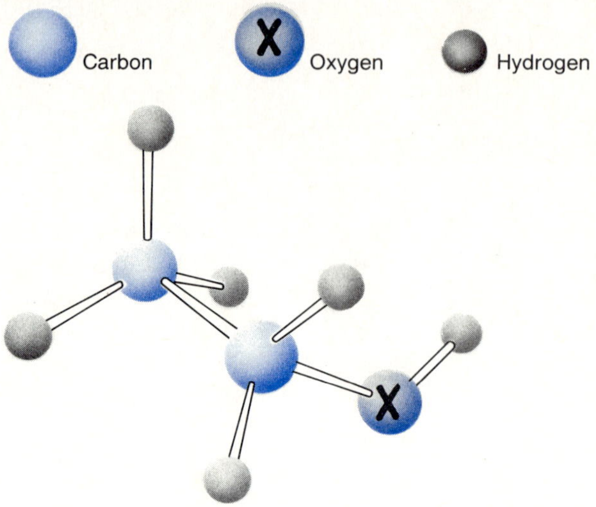

FIGURE 11-9
Molecular model of ethyl alcohol.

The $\begin{matrix} CH_3 \\ CH_3 \end{matrix} \!\!> CH-$ group is commonly called the *isopropyl group*. Some common products which contain alcohols are shown in Figure 11-10 below.

FIGURE 11-10
Commonly encountered products containing alcohols.

Alcohols were formerly named according to the alkyl group to which the —OH group is attached plus the word *alcohol:*

$$\underbrace{CH_3\text{—}CH_2}_{\text{Ethyl}}\text{—}\underbrace{OH}_{\text{Alcohol}}$$

These classical names are still frequently used.

The IUPAC system uses the prefix for the number of carbons plus the ending *-anol:*

$$\underset{\text{Ethanol}}{CH_3\text{—}CH_2\text{—}OH}$$

Carboxylic acids Carboxylic acids are compounds with the following group of atoms:

$$\text{—}\overset{\displaystyle O}{\overset{\|}{C}}\text{—}OH$$

Examples are:

NAME	FORMULA	COMMENTS
Acetic acid (ethanoic acid); see Figure 11-11	$CH_3\text{—}\overset{\displaystyle O}{\overset{\|}{C}}\text{—}OH$ or $CH_3\text{—}COOH$	Vinegar is 5% acetic acid.

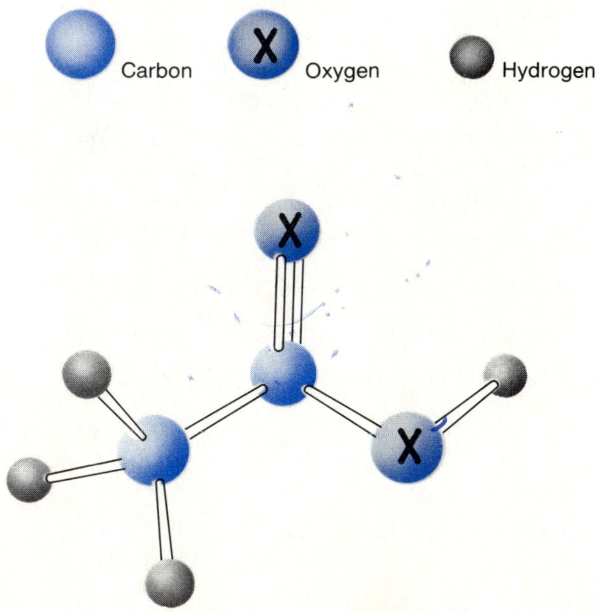

FIGURE 11-11
Molecular model of acetic acid.

FIGURE 11-12
Vinegar is a dilute solution of acetic acid.

NAME	FORMULA	COMMENTS
Butyric acid (butanoic acid)	$CH_3—CH_2—CH_2—\overset{\overset{O}{\|}}{C}—OH$	Responsible for the foul odor in perspiration and rancid butter
Formic acid (methanoic acid)	$H—\overset{\overset{O}{\|}}{C}—OH$	Causes the stinging in ant bites

Vinegar (Figure 11-12) is a dilute solution of acetic acid.

The more common carboxylic acids are referred to by their classical names (e.g., *acetic acid*), but the IUPAC system uses the prefix for the number of carbons plus the ending *-anoic acid*. For example:

$$CH_3—CH_2—CH_2—CH_2—\overset{\overset{O}{\|}}{C}—OH$$

Pentanoic acid

Esters Esters are compounds containing the $—\overset{\overset{O}{\|}}{C}—O—$ group. These compounds are prepared from carboxylic acids and alcohols. Examples are given in Table 11-4; and octyl acetate is shown in Figure 11-13.

Most fruit flavors, both natural and artificial, are esters. Thus "tasting" a banana is really "experiencing" the stimulation by pentyl acetate of the olfactory organs. Most fats and oils are triesters (three ester groups in the same molecule).

Esters are named from the acid and alcohol from which they are prepared. Thus, the ester prepared from ethyl alcohol and acetic acid is called *ethyl acetate*. The rule used to construct the names of esters is given in Figure 11-14, which shows the relationship between the various structural and name components. Note that a water molecule is eliminated or "split out" when the acid and alcohol join together to form the ester. Note also that it does not matter if the ester group is drawn "backward":

$$—O—\overset{\overset{O}{\|}}{C}—$$

as compared with the more common way:

$$—\overset{\overset{O}{\|}}{C}—O—$$

The bonding relationships are the same, and that is all that matters.

TABLE 11-4

Esters

NAME	ACID AND ALCOHOL USED TO PREPARE ESTER	STRUCTURE	USE
Ethyl acetate	$CH_3-\overset{O}{\overset{\|}{C}}-OH + HO-CH_2-CH_3 \longrightarrow$ Acetic acid Ethyl alcohol	$CH_3-\boxed{\overset{O}{\overset{\|}{C}}-O}-CH_2-CH_3$ ↑ Ester group	Solvent, fingernail polish remover
Methyl salicylate	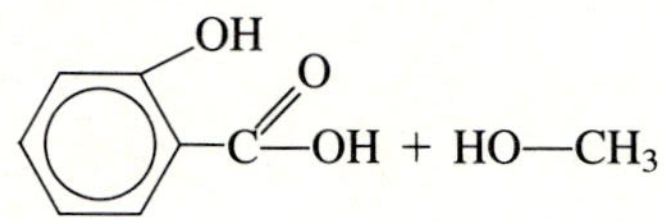	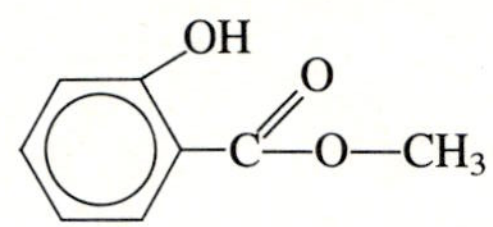	Oil of wintergreen
Pentyl acetate	$CH_3-\overset{O}{\overset{\|}{C}}-OH + HO-(CH_2)_4-CH_3$	$CH_3-\overset{O}{\overset{\|}{C}}-O-(CH_2)_4-CH_3$	Banana flavor
Octyl acetate (see Figure 11-13)	$CH_3-\overset{O}{\overset{\|}{C}}-OH + HO-(CH_2)_7-CH_3$	$CH_3-\overset{O}{\overset{\|}{C}}-O-(CH_2)_7-CH_3$	Orange flavor
Methyl anthranilate	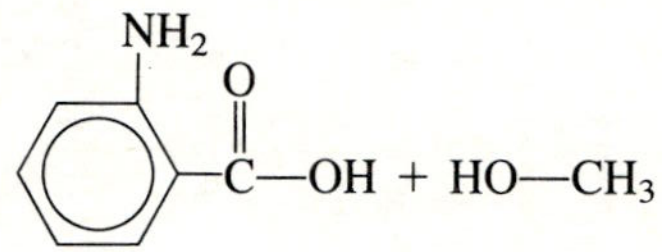	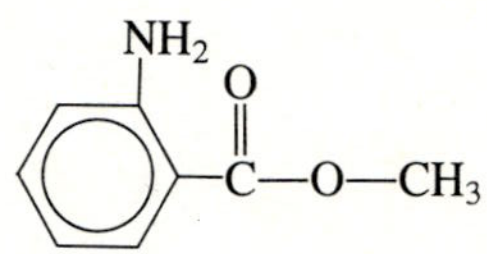	Grape flavor
		$CH_3(CH_2)_{14}-\overset{O}{\overset{\|}{C}}-O-(CH_2)_{30}CH_3$	Beeswax
		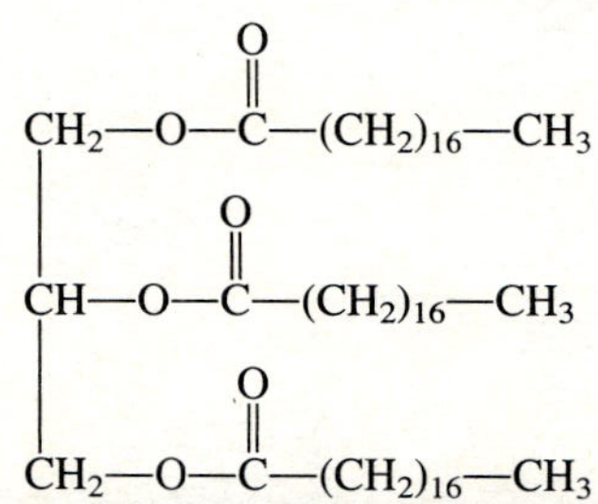	Lard, solid vegetable shortening

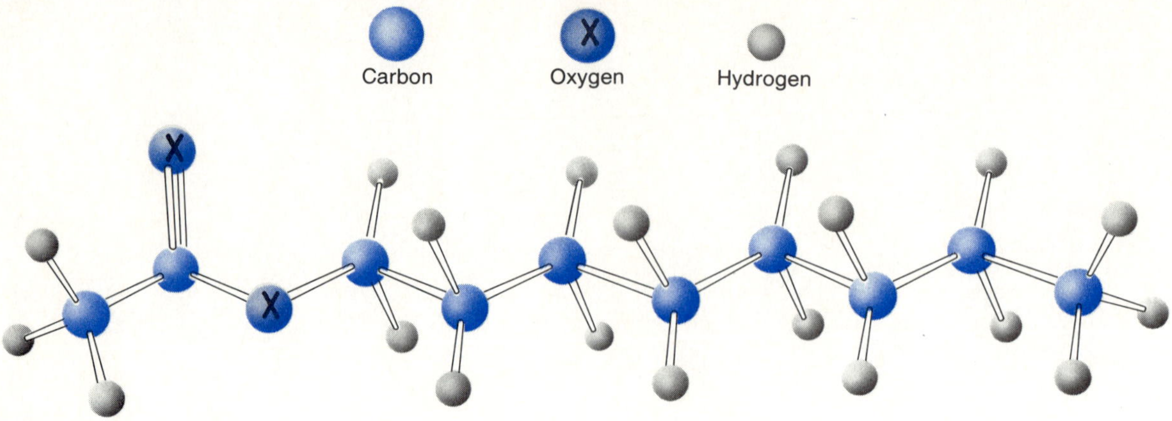

FIGURE 11-13
Above: Molecular model of octyl acetate (orange flavor).

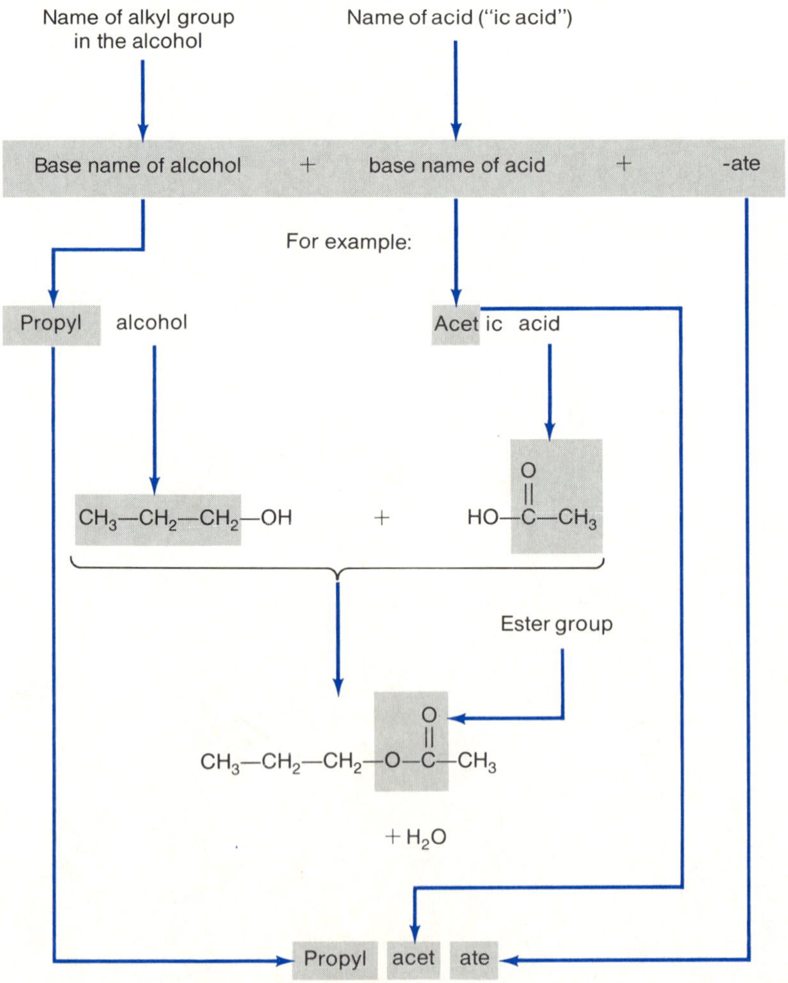

FIGURE 11-14
Right: The method used to name esters.

Ketones Ketones are compounds containing the $-\overset{\displaystyle O}{\overset{\|}{C}}-$ group bonded directly to carbon atoms. Examples follow:

NAME	FORMULA	USE
Acetone (propanone, dimethyl ketone); see Figure 11-15	$CH_3-\overset{\displaystyle O}{\overset{\|}{C}}-CH_3$	Solvent, fingernail polish remover, found in some glues
Methyl ethyl ketone (2-butanone)	$CH_3-\overset{\displaystyle O}{\overset{\|}{C}}-CH_2-CH_3$	Solvent, used in glues

Ketones are named in the *classical* system by the alkyl groups attached to the $-\overset{\displaystyle O}{\overset{\|}{C}}-$ group:

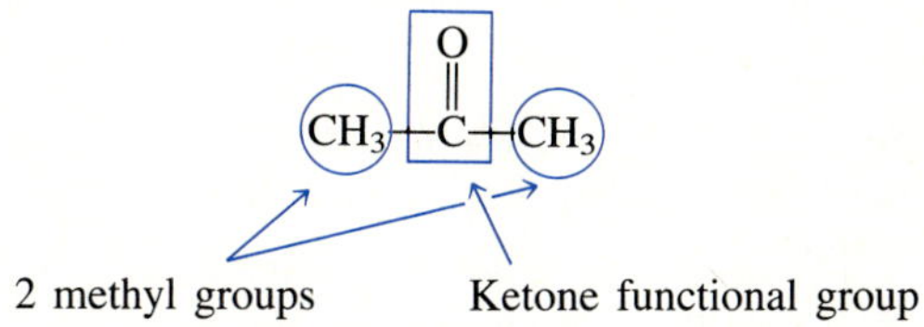

Thus: dimethyl ketone. Acetone is a *common* (nonsystematic) ''nickname'' for dimethyl ketone.

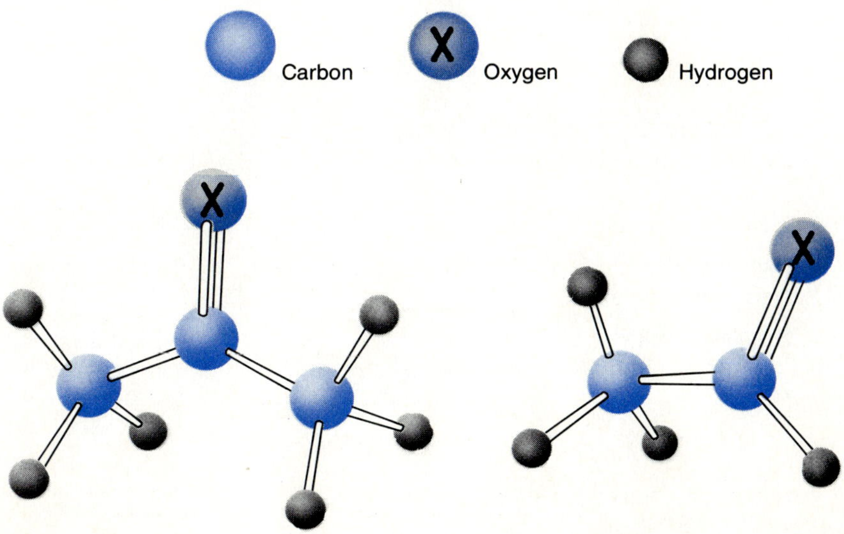

FIGURE 11-15
Molecular model of, *left,* acetone and, *right,* acetaldehyde.

The *IUPAC* name consists of a *prefix* designating the number of carbons in the longest chain, a *number* showing where the =O is attached to the carbon chain, and the ending *-anone*. Thus:

Location of C=O + prefix for number of carbons in longest chain + *-anone*

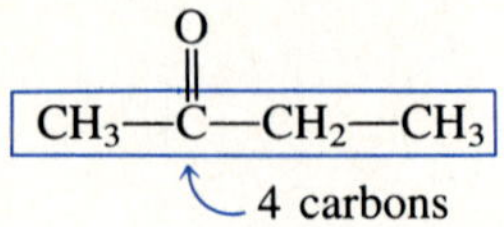

2-Butanone

Aldehydes Aldehydes are compounds containing the 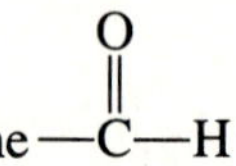 group. Examples follow:

NAME	FORMULA	USE
Formaldehyde (methanal)	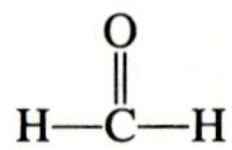	Preserving biological specimens, embalming fluid
Acetaldehyde (ethanal); see Figure 11-15	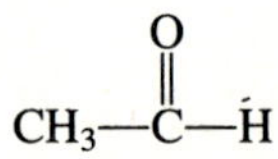	Chemical reagent

The common names must be memorized. The IUPAC names consist of a prefix indicating the number of carbons plus the ending *-al*. Thus:

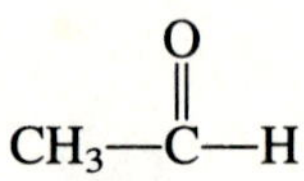

Ethanal

Ethers Ethers are compounds containing the —O— group. Two alkyl (alkane-like) groups are connected by an oxygen atom. Examples follow:

NAME	FORMULA	USE
Diethyl ether (see Figure 11-16)	$CH_3—CH_2—O—CH_2—CH_3$	Chemical solvent
Diisopropyl ether	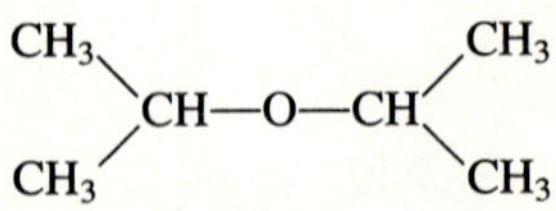	General anesthesia during surgery

Ethers are named according to the two alkyl group attached to the oxygen.

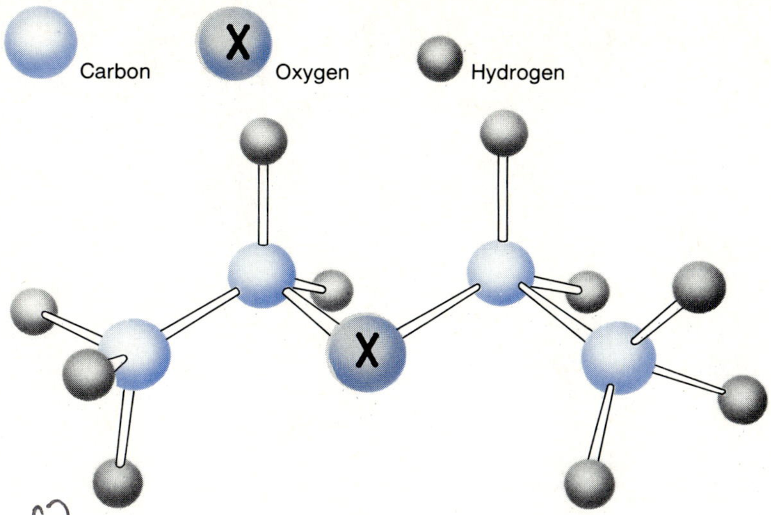

FIGURE 11-16
Molecular model of diethyl ether.

Amines Amines are compounds containing the —NH_2 group. Examples follow:

NAME	FORMULA	USE
Methyl amine	CH_3—NH_2	Flavor and odor of fish
Propyl amine (see Figure 11-17)	CH_3—CH_2—CH_2—NH_2	Chemical reagent

Amines are named according to the alkyl group to which the —NH_2 group is attached.

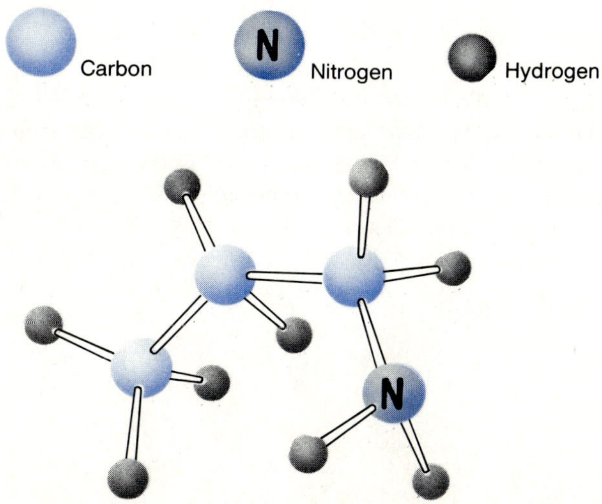

FIGURE 11-17
Molecular model of propyl amine.

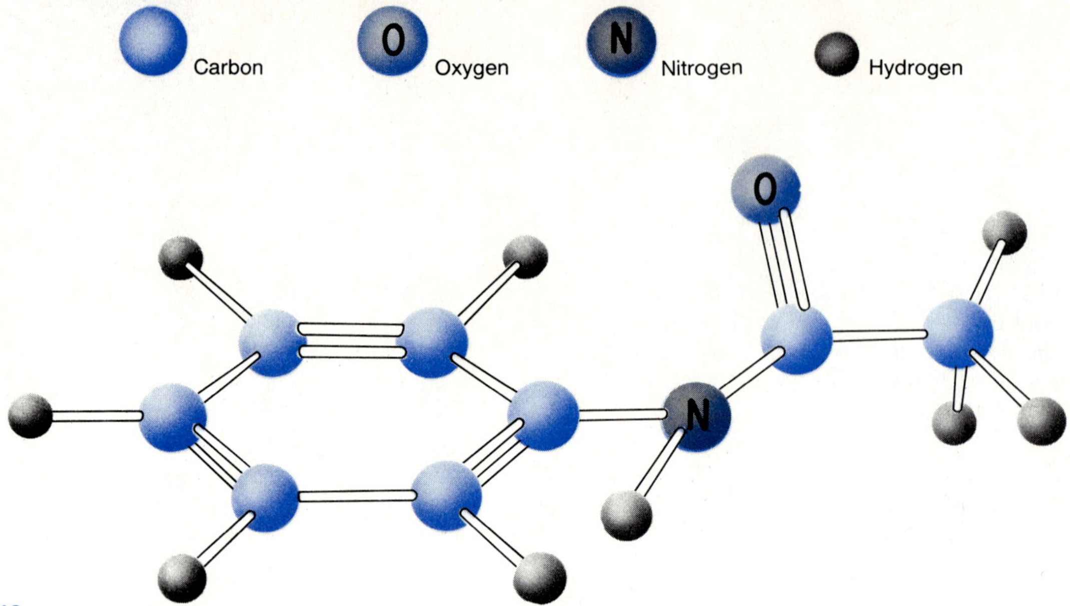

FIGURE 11-18
Molecular model of acetanilide.

Amides Amides are compounds containing the 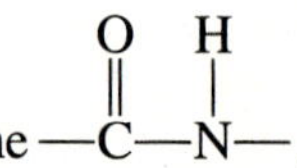 group. An example follows:

NAME	FORMULA	USE
Acetanilide (see Figure 11-18)	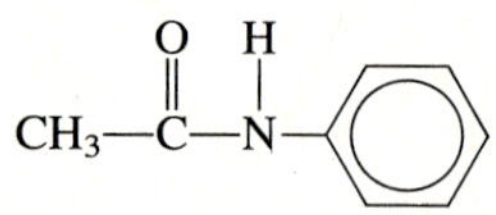	Analgesic (general painkiller)

Heterocyclics Heterocyclics are compounds containing a ring of atoms, *one of which* is *not a carbon*. When the structures are written as shown below, the points on the cyclic structures indicate a carbon atom and accompanying hydrogens. Examples follow:

NAME	FORMULA	USE
Pyrolidine		Chemical reagent used in synthesis of other compounds
Tetrahydrofuran		Solvent

NAME	FORMULA	USE
Chlorpromazine (Thorazine)	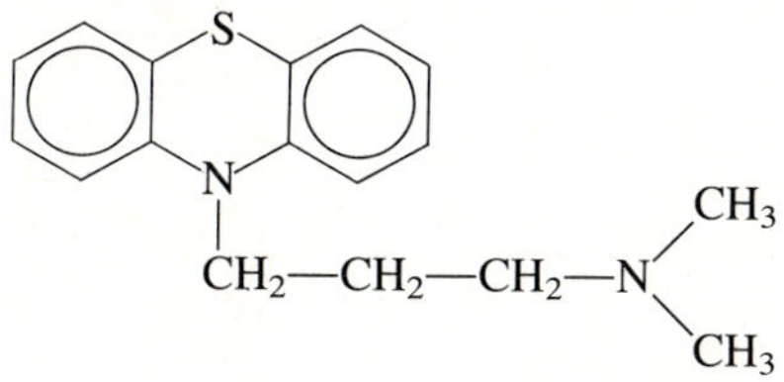	Tranquilizing drug

Phenols Phenols are compounds with an —OH group attached to an aromatic ring. Examples follow:

NAME	FORMULA	USE
Phenol*	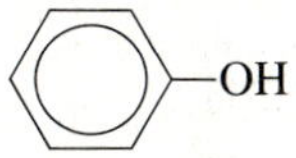	Antiseptic, disinfectant, mouthwash ingredient
Acetaminophen	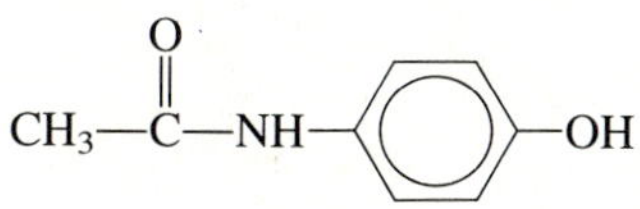	Pain killer and fever-lowering agent (active ingredient in Tylenol)

Learning the functional groups

Table 11-5 summarizes the functional groups. The student should do the following:

1 Memorize the name and structure of each example given in this chapter.

2 Learn the system of nomenclature for alkanes and substituted alkanes.

3 Learn to recognize the functional group to which any compound belongs by looking at its formula.

4 Learn the use of each compound given in this chapter.

It is suggested that a set of flash cards with the name of the compound on one side and either the structure or the use on the other would be a great help in learning items 1, 3, and 4. The program given earlier in this chapter will aid in developing skill in item 2. Some of the information presented in this chapter will be used later in this book in applying chemistry to various specialized areas, but for now, memory is probably the best approach.

---

*When the —OH group is attached to an aromatic ring the resulting compounds have different chemical properties from alcohols; therefore, this class of compounds is given a special name, phenols.

TABLE 11-5

Functional groups

NAME OF GROUP	CHARACTERISTIC FUNCTIONAL GROUP
Alkanes	—C—C—H
Alkenes	C=C
Alkynes	—C≡C—
Aromatics	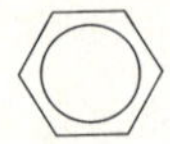
Substituted Alkanes	—C—X X=F, Cl, Br, I
Cycloalkanes	$(CH_2)_n$
Alcohols	—OH
Carboxylic acids	—C(=O)—OH
Esters	—C(=O)—O—
Ketones	—C(=O)—
Aldehydes	—C(=O)—H
Ethers	—O—
Amines	$—NH_2$
Amides	—C(=O)—N(H)—
Heterocyclics	$(CH_2)_nZ$
Phenols	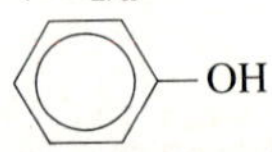

SYNTHETIC ORGANIC CHEMISTRY

The task of the synthetic organic chemist is to produce new organic materials of theoretical interest (that is, for testing a hypothesis) or practical value (for solving a problem or making life more convenient). It is sometimes desirable to produce synthetically a natural product which is rare or expensive to isolate from nature but which is of value as a vitamin, drug, flavoring, or research chemical. In carrying out this task, the chemist has available many different methods of *building carbon chains* and *changing functional groups*. By using a variety of different methods and reaction sequences, chemists can change available materials into new, heretofore unknown materials. Before beginning the actual synthesis, researchers must first familiarize themselves with the different methods available. This information is acquired from chemistry courses, from personal experience, and from a search of the chemical literature to find how another chemist effected a particular kind of change in the structure of a molecule. The researcher then has to decide if any of these methods can be applied to the problem at hand.

Envisioning a compound for a special purpose

The first step in any problem in synthetic organic chemistry is working out the general approach to the problem. This step can be illustrated by looking at a typical research project. Let us consider a search for new drugs to control seizures in people with epilepsy. The National Institutes of Health announced such a search because the approximately 15 drugs currently used either have adverse side effects or do not adequately control some patients.

A chemist working on such a problem would first examine the molecular structure of known anticonvulsants (compounds that stop seizures). Two of the most widely used are shown below:

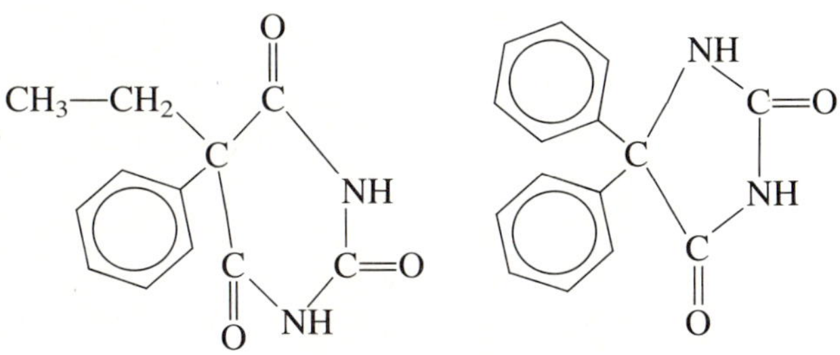

Phenobarbital

Diphenylhydantoin (Dilantin)

Notice that both these compounds have *benzene* portions and each has a *ring* of atoms containing *nitrogens* and *carbons doubly bonded to oxygen*.

Applying the structure-function principle (see page 234), a chemist seeking to synthesize previously unknown compounds that could reasonably be expected to have anticonvulsant properties would first envision molecular structures that had these same features and search for methods of preparing these compounds. This attempt may or may not be successful; and even if it is successful, the new compound may not prove to be an anticonvulsant.

Preparing the envisioned compound

The second aspect of a synthesis problem is the technique used in the laboratory to actually prepare the envisioned compound. Although you are not expected to learn the reactions commonly used in organic chemistry, it would be instructive to examine the procedures used in the synthesis of at least one compound.

Following is a standard presentation of the synthesis of sulfanilamide, one of the sulfa drugs used as an antibiotic. Each step, and how it is carried out physically, is explained.

Step 1:

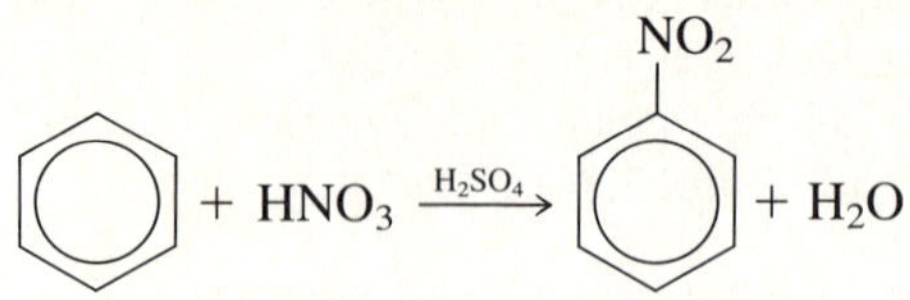

Step 2:

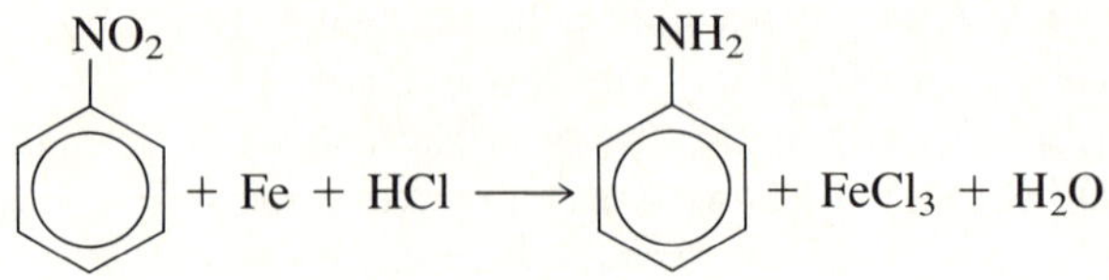

Note that we are concerned here not with balancing the equation but only with indicating the materials used and produced.

Step 3:

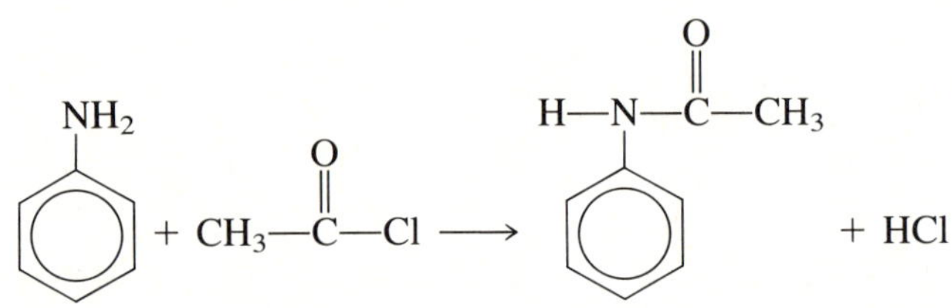

Step 4:

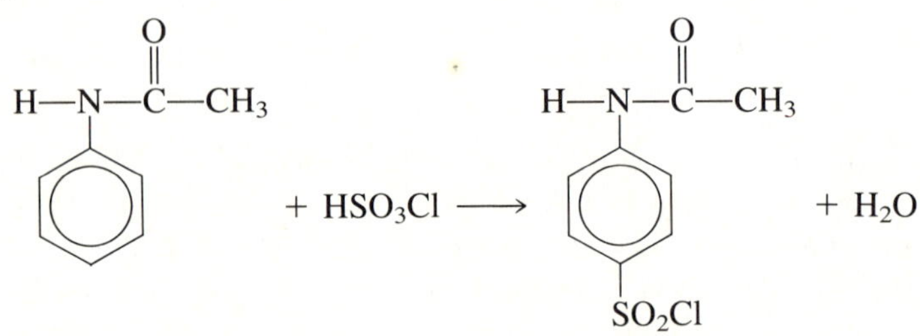

Step 5:

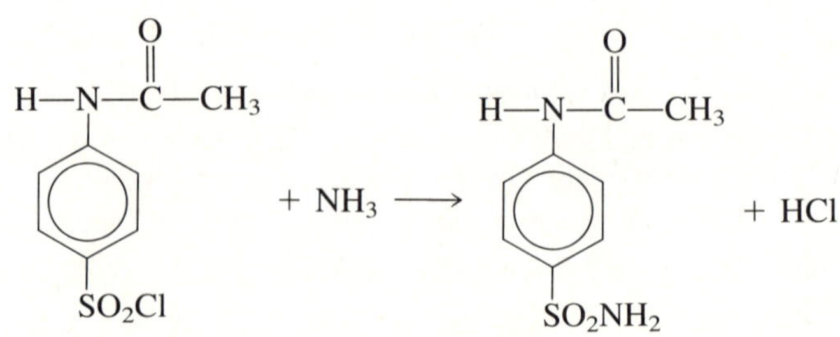

Step 6:

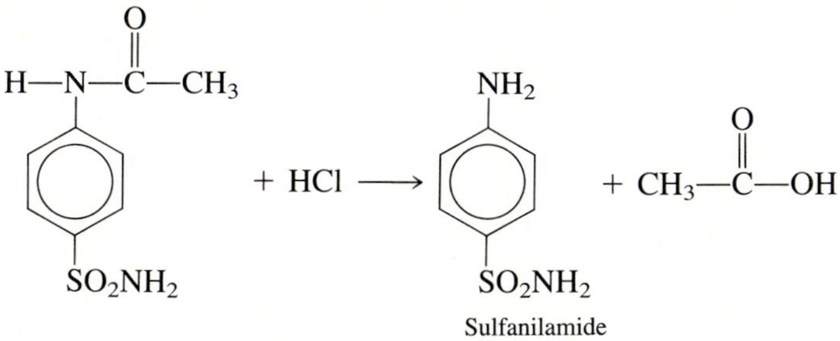

This series of steps probably looks quite complex to a beginning chemistry student. However, it will become clearer as we look at each step in turn.

Reaction 1 Benzene is mixed with nitric acid plus a small amount of sulfuric acid and heated in a flask for 1 h. This causes an $—NO_2$ group to add to benzene. Figure 11-19 shows the apparatus used for this procedure.

Reaction 2 The nitrated benzene is mixed with powdered iron and hydrochloric acid and stirred for 30 min. This causes the $—NO_2$ group to change to $—NH_2$.

Reaction 3 The 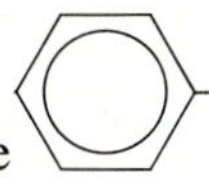 $—NH_2$ is mixed with the $CH_3—\overset{\displaystyle O}{\overset{\|}{C}}—Cl$, and almost immediately, the product shown is formed. This product is a solid and is isolated (removed from the other materials) by filtering. This material is called *acetanilide*. Reaction 3 is carried out to prevent the NH_2 group from reacting with the reagents used in steps 4 and 5.

Reactions 4 and 5 The acetanilide is then placed in another flask and mixed with HSO_3Cl. This adds the $—SO_2Cl$ group to the benzene ring. Without actually isolating this product, NH_3 (ammonia) is added and the —Cl is replaced with an $—NH_2$ group. The result is a solid material.

Reaction 6 This solid is then boiled in water to which a little acid has been added. This causes the $—\underset{\underset{\displaystyle O}{\|}}{C}—CH_3$ group to be removed. The result is solid sulfanilamide (sulfa drug), which after purification is placed in suspension or pressed into tablets for ingestion by those needing this medication.

Chemistry—indeed, all science—is a creative endeavor involving technical skill and creativity. Both art (creativity) and science (technical skills) are required of a good research chemist. For this reason, some synthetic organic chemists are better than others—they have an inborn or developed ability that

FIGURE 11-19
A typical apparatus used to synthesize organic compounds in a laboratory.

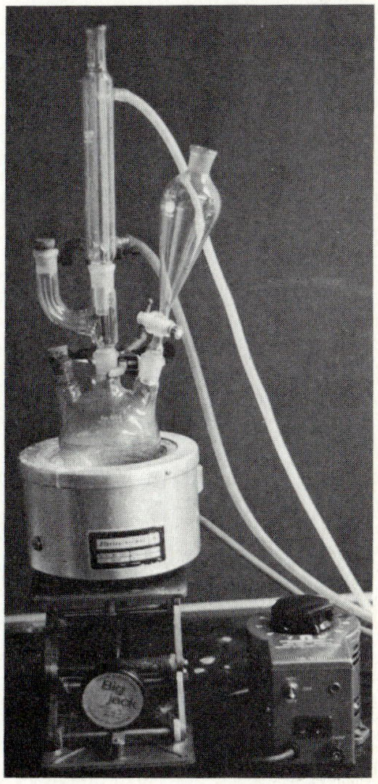

FIGURE 11-20
R. B. Woodward. (From a drawing by the author.)

their peers do not have (this is true in most areas of life). This ability to bring together bits of knowledge from various places and to think about new ways of effecting chemical changes, which no one else has conceived of before, is what makes great chemists. Such a man was the late R. B. Woodward of Harvard University (Figure 11-20), who, along with Albert Eschenmoser of the Swiss Federal Institute of Technology, over a period of 7 years developed a 90-step synthetic procedure for vitamin B_{12} and was able to prepare in the laboratory a vitamin B_{12} that was identical to the natural material. As one can see from Figure 11-21, this is an extremely complex molecule; a lesser chemist would not have even attempted to synthesize it.

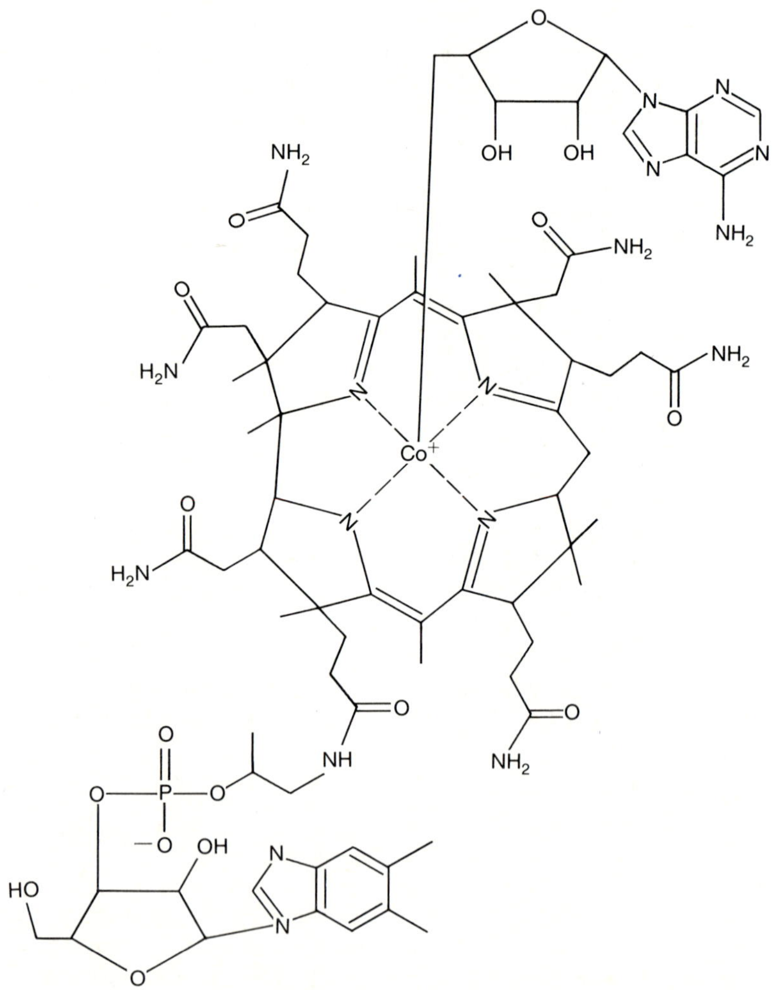

FIGURE 11-21
Molecular structure of vitamin B_{12}.

COMMON APPLICATIONS OF ORGANIC CHEMISTRY

Petroleum chemistry

Since most organic chemicals are prepared from petroleum products, the petroleum industry is one of the largest in industrial societies. Because we will refer to certain aspects of this area in later chapters, we should now take a brief but reasonably comprehensive look at petroleum refining processes.

Crude oil is found in pockets in sedimentary rock (rock laid down in layers by the action of water). It was formed, as far as we can tell, from large quantities of decaying plants. It is a mixture of alkanes and aromatic hydrocarbons. Crude oils from the eastern United States and the Texas-Oklahoma region are mostly alkanes; California oils contain a large amount of aromatics as well.

The process of refining crude oil begins with a separation into the various components. Actually, each compound is not purified; rather, groups of compounds with similar physical and chemical properties are isolated as a mixture for particular commercial uses. The method used in the separation is called *fractional distillation* (see Figure 11-22). This involves heating the crude mixture to successively higher temperatures until the components boiling at that temperature boil out of the mixture (i.e., change to the gaseous state). This vapor is cooled, and the cooling causes it to change back to the liquid state again (condensation). Table 11-6 shows the various boiling temperatures used, the name of the material collected at that temperature (fraction), and the commercial use of that material.

Figure 11-23 shows an oil refinery. Once the various fractions have been collected by the refinery, they are sent by pipeline, truck, or railroad tank car to distribution points.

Because more heavy motor-oil-type compounds are produced than are needed, and because so much gasoline is required, a process has been developed to convert heavy oils, composed mostly of molecules characterized by high molecular weight and long carbon chains, into gasoline composed of relatively short-chain (six- to ten-carbon) compounds. The oils are heated in the absence of oxygen (to prevent them from burning) in the presence of a *catalyst*. A catalyst is a material that does not actually participate in a change (it does not get changed itself) but causes a change in other materials that would be practically impossible without it (the reaction would be too slow to be useful). When the oil is heated catalytically, the long oil molecules break down into shorter molecules of the gasoline type. This is called *catalytic cracking:*

$$C_{18}H_{38} \xrightarrow{\text{catalyst}} 2C_9H_{18} + H_2$$

In addition to a mixture of low-molecular-weight alkanes and alkenes, hydrogen gas is also produced. This is the main commercial source of hydrogen. Figure 11-24 shows a catalytic cracker, and Figure 11-25 shows the catalyst used to bring about this change.

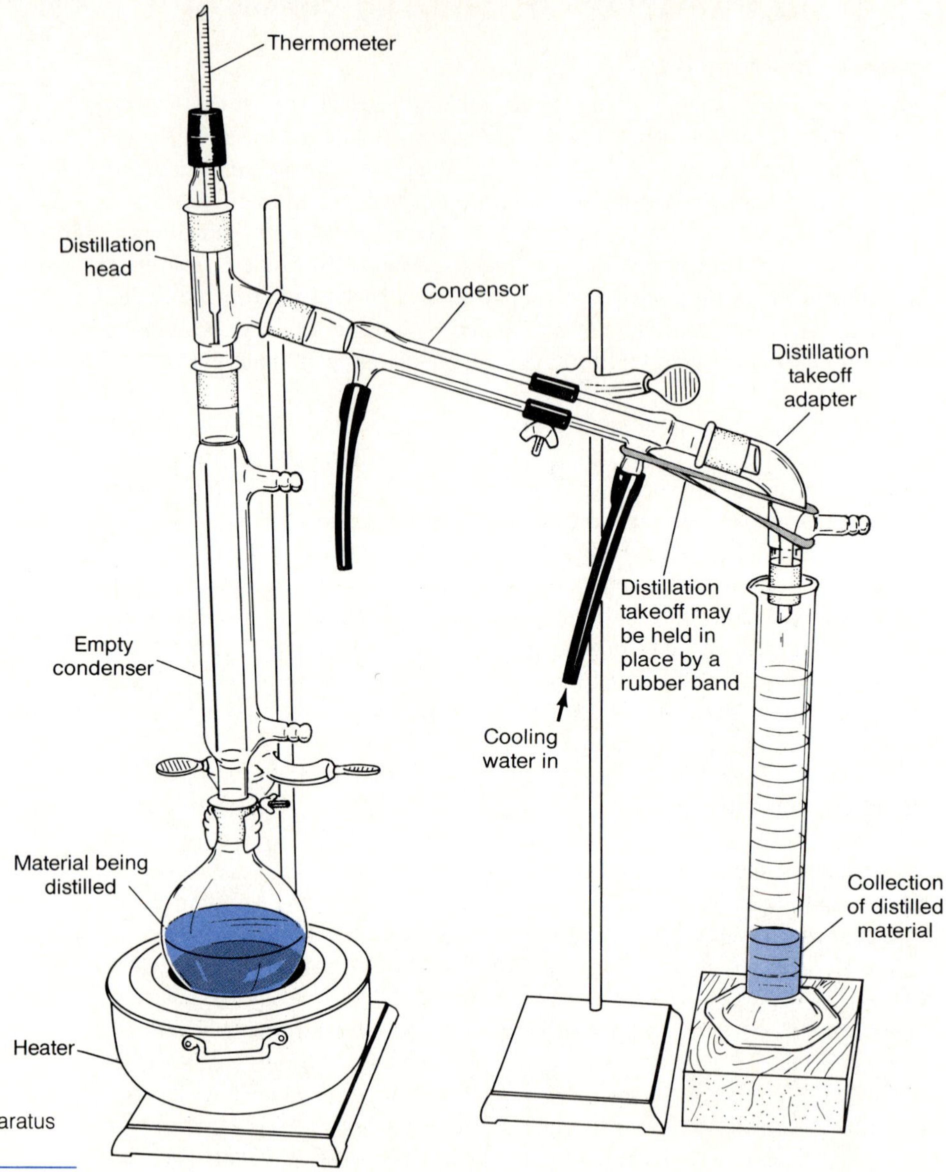

FIGURE 11-22
Fractional distillation apparatus used in a laboratory.

TABLE 11-6

Various fractions isolated in the refining of crude oil

TEMPERATURE, °C	COMPOSITION	USE
Below 20	Methane, ethane, propane, butane	Natural gas for fuel
30–200	Alkanes containing from 6 to 10 carbons per molecule	Gasoline
200–325	Alkanes containing from 12 to 15 carbons per molecule	Kerosene (coal oil)
325–400	Alkanes containing from 15 to 18 carbons per molecule	Diesel fuel and fuel oil for fuel and heating
Above 400	Alkanes with greater than 18 carbons	Paraffin, lubricating oils
Solids left after distillation	Complex mixtures	Asphalt

FIGURE 11-23

Petroleum refinery. (Photograph from Ashland Oil Company.)

FIGURE 11-24
Catalytic cracking unit. (Photograph from Ashland Oil Company.)

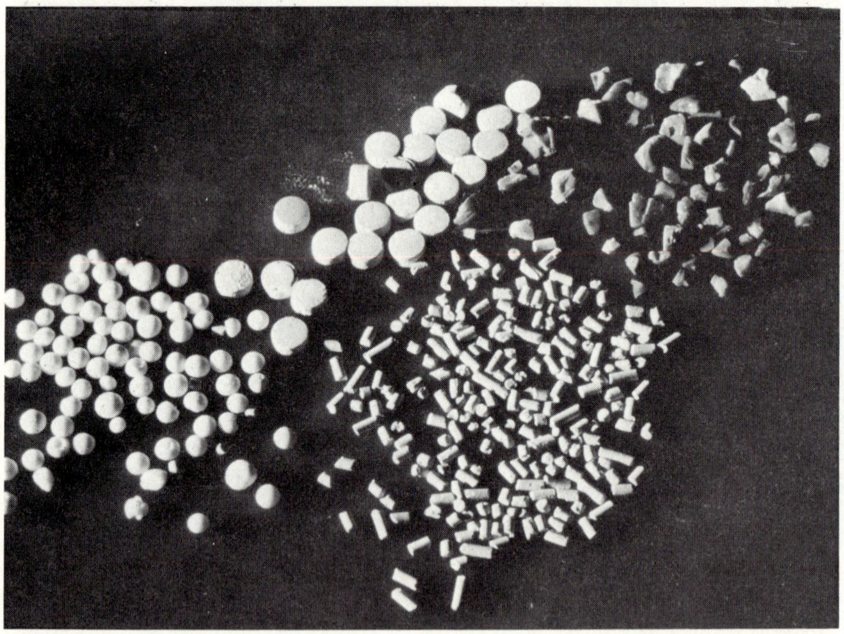

FIGURE 11-25
Solid catalyst used in cracking petroleum products. (Photograph from Ashland Oil Company.)

Plastics and coatings

Let us consider a family of commonly encountered materials called *polymers*. These include plastics, rubbers, varnishes, paints, and synthetic fibers. Polymeric materials are produced by chemically linking many individual small molecules (called *monomers*) together by covalent bonds. This produces very long, high-molecular-weight molecules that in many cases have unusual properties. By designing different molecular structures, chemists have been able to make many polymeric compounds with a wide range of properties and uses. In this section, we will look at several of the many such materials on the market today and examine the types of chemical techniques and processes used to produce them.

Polyethylene Polyethylene (see Figure 11-26 for polyethylene products) is used in most modern inexpensive household containers. The formula of all polyethylene is the same, but the manufacturing process determines the exact properties of the material:

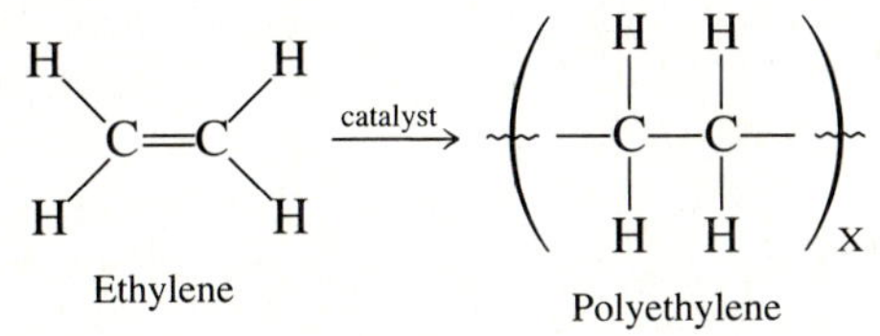

The softer translucent type is called *low-density* polyethylene; the colored, opaque harder type is *high-density* polyethylene.

Polyvinyl chlorides Vinyl floor tile and clear-plastic food wrapping material (Saran) are formed by polymerizing vinyl chloride and vinylidene chloride, respectively:

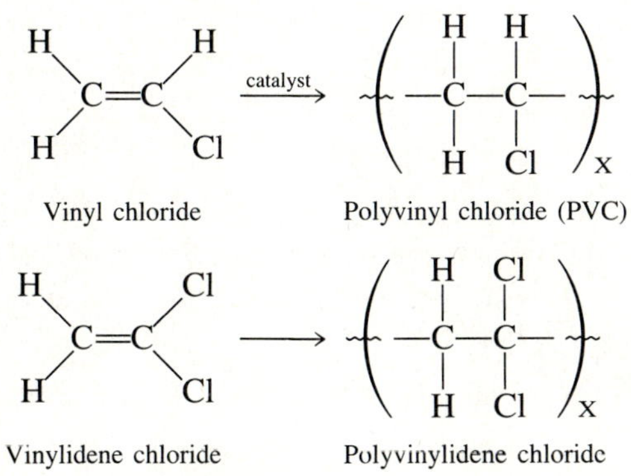

Polyvinyl chloride can be mixed with other materials to form a number of familiar products, such as water pipes (see Figure 11-27) and electrical insulation.

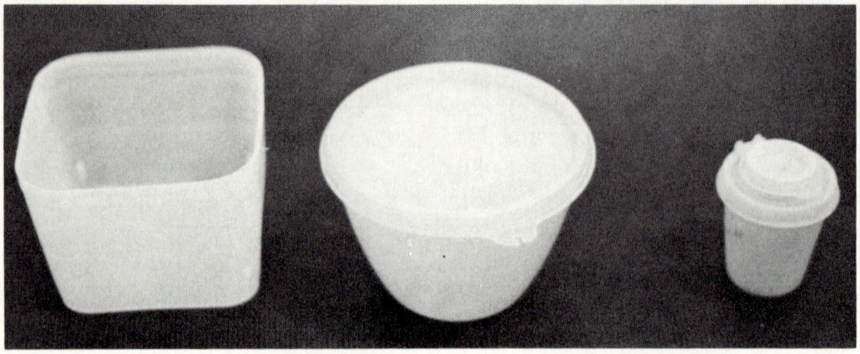

FIGURE 11-26
Right: Polyethylene products.

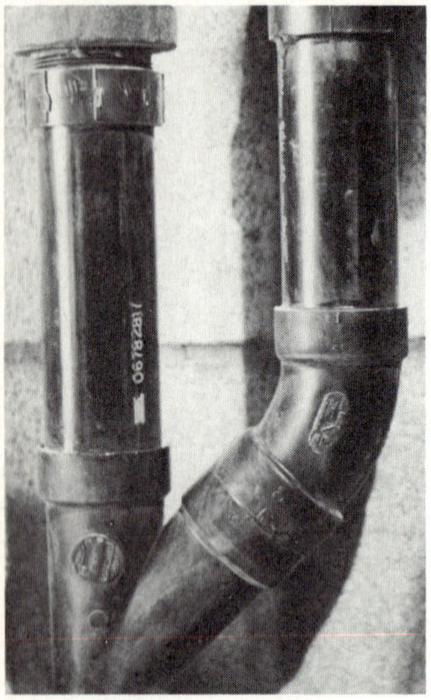

FIGURE 11-27
Above: Drainpipe made from polyvinyl chloride.

Lucite A lightweight, transparent material that is better than glass for some applications is the synthetic polymer whose trade name is Lucite (Plexiglas). It does not shatter on impact and is thus safer for functions such as windows in storm doors. A Lucite door will bend rather than break if something falls against it (as a person or animal might do). It can also be cut into pieces and glued to make transparent models of buildings or machines. However, it scratches easily and will dissolve if nonpolar organic solvents are used in contact with it.

The chemical name and formula for Lucite, along with the corresponding monomers, are shown below:

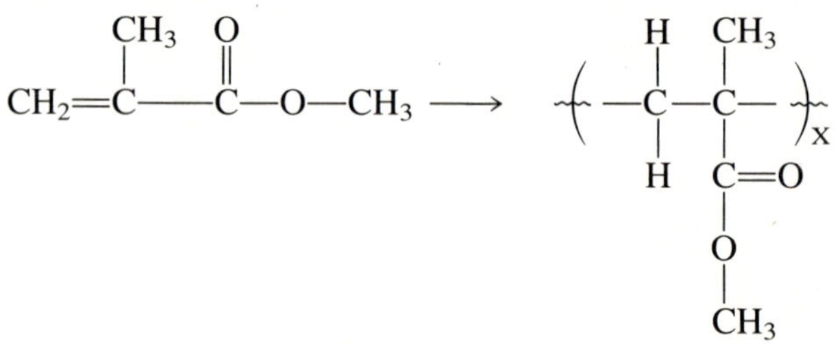

Methyl methacrylate — Poly(methyl methacrylate) or Lucite

Bakelite The easily breakable plastic used frequently for light switch covers (see Figure 11-28) and construction of electrical receptacles is called *Bakelite*. It is formed by reacting a phenol with formaldehyde, and is thus called a *phenolic resin* (resin = plastic):

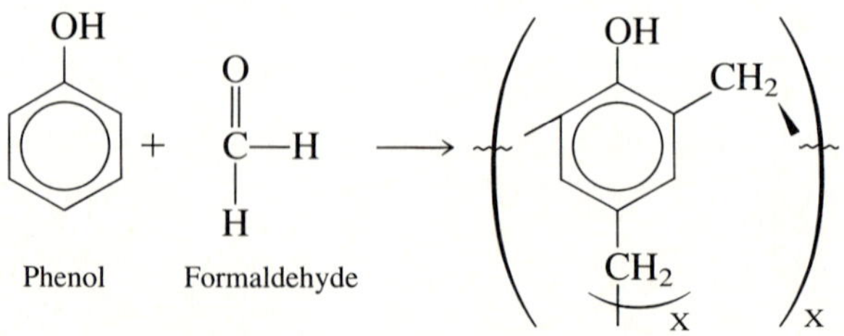

Phenol — Formaldehyde

By using different phenol compounds (with a methyl group attached to the benzene ring, for example), polymers with various properties can be made. Because these shrink when placed in a mold, sawdust or some other "filler" is added to reduce the shrinking and reduce the amount of polymer needed to prepare a given object. This, of course, reduces the cost, but it does not lower the quality.

FIGURE 11-28
Bakelite electric switch.

Laminated plastic Kitchen countertops and other work surfaces are usually made of Formica or other brands of the same material:

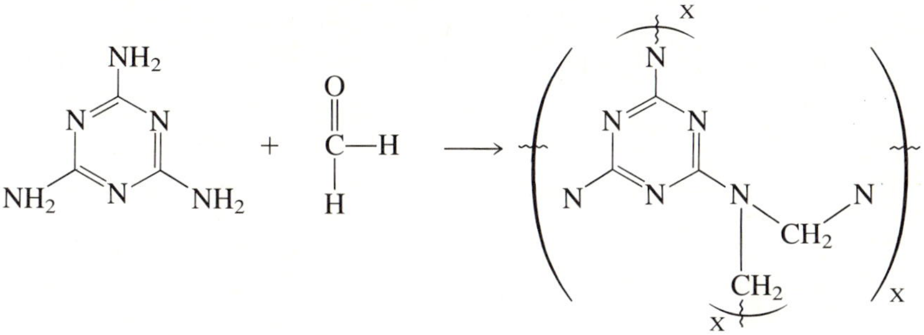

Melamine Formaldehyde Formica

This rather hard, tough plastic is prepared from melamine and formaldehyde. These laminates are usually decorative in appearance, and the total product is prepared by soaking a photograph of wood or a piece of paper with a colored design in partly polymerized Bakelite, coating this with several layers of the melamine-formaldehyde resin, and heating it in a press to bind the layers together and harden the polymeric materials. Thus when you look at a countertop (as in Figure 11-29), you are looking through a thin, clear layer of melamine-formaldehyde resin at a resin-saturated photograph or piece of paper.

FIGURE 11-29
Laminated polymeric kitchen countertop.

Coatings For purposes of protection or decoration (or both) most metals and wooden and plastic objects and surfaces are coated. These coatings include paint, varnish, shellac, lacquer, and fingernail polish. The last three of these can be thought of as temporary coatings, since they are applied as a solution of natural or synthetic polymers. They harden because the solvent evaporates, leaving a coating of the polymer. This polymer coating can be removed by redissolving it in the original solvent.

Fingernail polish is dissolved by rubbing with a cloth moistened in ethyl acetate or acetone. A little oil is usually added to these solvents to prevent excessive drying of the skin with which they come into contact. Shellac is readily soluble in 95% denatured ethyl alcohol, which is what shellac thinner is. One can remove shellac from furniture or other wooden objects by rubbing with coarse steel wool which is wet with alcohol. Lacquer is soluble in a mixture of ketones and can likewise be removed from furniture, when one is preparing to refinish a piece, by rubbing with coarse steel wool and lacquer thinner.

Most commercial furniture (furniture which has not been custom-made or manufactured by one of the fine-handcrafting organizations) is coated with lacquer. It can be sprayed on, and it drys quickly and smoothly. But (contrary to claims made for it), it is not waterproof or alcohol-proof: either water or alcohol will leave an unsightly mark. These marks can sometimes be removed by rubbing gently with lacquer thinner and, after drying, polishing with fine steel wool and waxing. Wood coated with spar (marine) varnish is essentially waterproof (no mark will be left by the condensation from a drinking glass, for example).

The more permanent coatings such as paint and varnish involve the application of a partly polymerized mixture in a solvent. During hardening, both the solvent evaporates and the O_2 in the air causes complete polymerization to occur. Such a coating cannot simply be redissolved in the same solvent because the new material formed (polymerization is a chemical process and a new substance is formed) is not soluble in the same solvent used for the original monomer.

Latex paints, which use alcohol and an acid as the monomers, are water-soluble, and a polyester (alkyd) is formed on the wall or other surfaces painted. Acrylic latex paints use methyl methacrylate monomers and Lucite for forming the polymer that coats the wall. Latex paints contain small particles of polymerized material. (That is why latex paints are never high-gloss; the coarse structure is built in.) They also contain enough monomer so that additional polymerization can occur as the paint cures. The polymer itself is essentially colorless, and a pigment such as TiO_2 is added to produce a white base, which is then tinted various colors. Oil-base paints and all varnishes are prepared with either natural or synthetic monomers dissolved in hydrocarbon solvents.

Paints often have linseed oil, and varnishes sometimes have tung oil. Both these oils are in a class called *drying oils,* which react with oxygen (from the air) to form a polymeric coating which holds the pigments. These oils, like others, are composed of triesters, which are molecules whose fatty acid portion contains double bonds. Natural (raw) oils are not *conjugated* (that is, they do not have alternating double and single bonds) and would, if untreated, produce a very slow-drying product. When these oils are boiled, their double bonds rearrange into a conjugated relationship, and these boiled oils form polymers much more rapidly than the untreated oils.

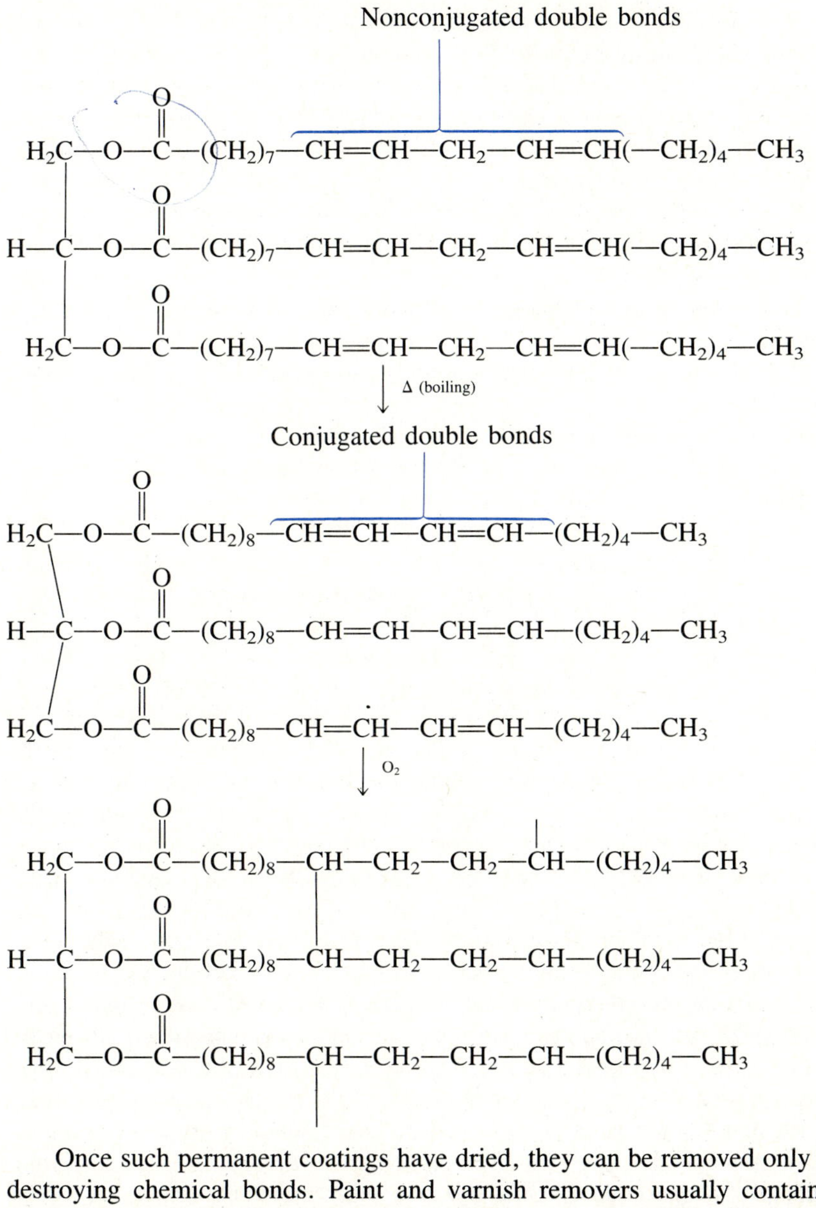

Linoleic ester (raw linseed oil contains 61 percent of this ester.)

Boiled linseed oil

Polymeric drying oil (in hardened or dried paint)

Once such permanent coatings have dried, they can be removed only by destroying chemical bonds. Paint and varnish removers usually contain a strong base and organic solvents such as dichloroethane or CCl_4. Some must be scraped off; others can be washed off with water. They should always be used with plenty of ventilation, and contact with the skin should be minimized.

Medical applications of polymers

Plastics (polymeric materials) have been used for a number of medical applications including artificial heart valves, contact lenses, and dental fillings. This field is just opening up, and many new products will be available in the near future.

As an example of such materials and of how a chemist develops them, consider dental filling material. The material used must be compatible with the mouth—that is, it should not irritate the mouth and must not decompose when wet. It should also be hard, look basically white, and come in various shades to match natural tooth colors. A dentist should be able to mix it easily, and after mixing it should become quite hard in a relatively short time. This hardening process is a chemical reaction in which the monomer molecules react under the influence of a catalyst to form the polymeric dental filling material.

This reaction not only must occur rapidly but also must do so at body temperature (that is, no heat greater than body temperature heat should be required) and must not give off excess heat, since that would be painful and could cause burns.

The monomer found to be the best is shown below; it has ends that have the same structure as methyl methacrylate, and the resulting polymer is similar to Lucite:

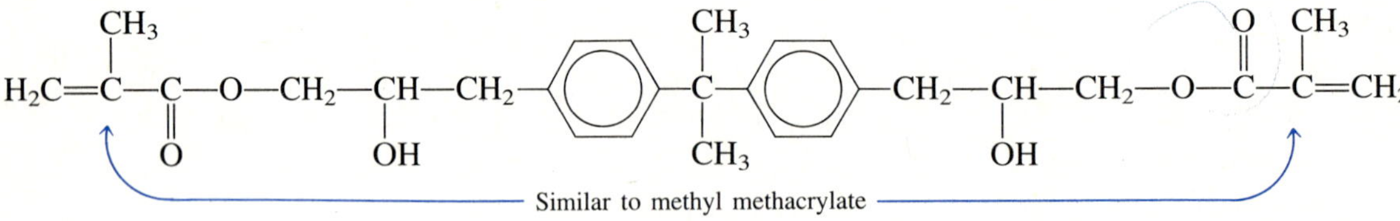

The polymer that forms from this monomer is relatively soft and gives off heat when polymerization occurs. These problems are solved by adding finely powdered sand to the polymer. The sand adds hardness and greatly reduces the amount of organic material present, thus reducing the amount of heat liberated.

When the usual initiator catalyst (a peroxide) is used, some heat is required from an external source, such as a flame. This, of course, will not work in the mouth. When an amine is present along with the peroxide, the reaction goes too fast and the material hardens in the container before it can be used. It is also true that the dentist is not usually a chemist and does not want to have bottles of the needed chemicals around. These problems are solved by dividing the monomer mixture into two parts and adding peroxide to one part and an amine to the other part. The resulting two mixtures—one containing monomer and peroxide and the other, monomer and an amine—are then stored in separate containers. The dentist has only to mix equal amounts of material from the two containers.

Once the peroxide and amine (along with monomer) are mixed, the mixture is placed in the cavity and hardens in about 10 min:

Monomers + peroxide + sand
Monomer + amine + sand $\xrightarrow{\text{mixing}}$ monomer + sand + amine + peroxide

$\xrightarrow{\text{10 min}}$ polymer reinforced with sand (hard filling)

Organic chemicals for farm and garden use

One of the reasons American farmers and gardeners produce more and more food on less and less land with less work is the availability of chemical substances to control pests. Most of these compounds are synthetic organic materials. The terms *pests* and *pesticides* should, of course, be used with care, since a given organism is not inherently bad. It is only that its presence in a particular place at a particular time is working in opposition to the goal of a particular person or group of people. For example, a rose is desirable in a rose garden but is a pest in a wheat field. Perhaps the terms *biologically active compound* or *biogenic material* would be more precise. (The term *xenobiotic* is also now being used.) At any rate, we have available many materials that will kill or repel most plants and insects without harming the plants or animals we want to grow. We will only deal with a few examples here, since a comprehensive treatise of this subject would fill an entire book.

A material used a great deal in the past was DDT (*d*ichloro*d*iphenyl*t*richloroethane), whose structure is given below; it kills many insects that damage plants and livestock:

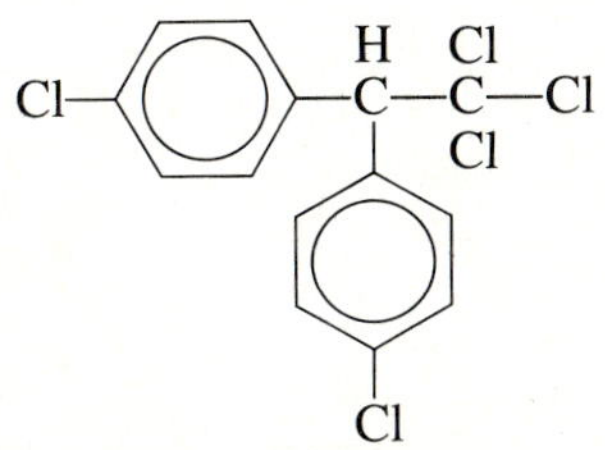

DDT

DDT is in the class of chlorinated hydrocarbons which are very stable. This means that these compounds do not decompose in nature but stay around and become concentrated in certain fish and birds (see Chapter 12 for a more complete treatment of the pollution problem). DDT served well for a time; but once its polluting effect was detected, its use was essentially stopped.

Another currently and commonly used *herbicide* (a substance that kills selected plants) is 2,4-D (*2,4-d*ichlorophenoxyacetic acid). It is used to spray shoestring plants, sumac plants, etc., and kills them by stimulating their growth to the point of destruction (they grow to death). The young plant cells divide before maturing completely; an unhealthy plant is produced, which eventually dies. The structure of 2,4-D is shown below:

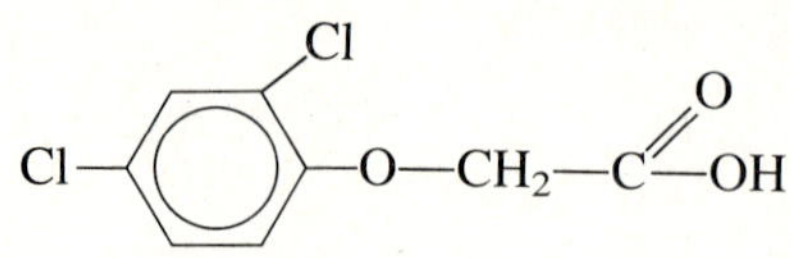

2,4-D

As can be observed, it is a chlorinated compound, an ether, and a carboxylic acid. It is slowly degraded by nature, so that it does not pose an environmental problem as far as is presently known. It kills broadleaf plants but does not harm grass or grain. Thus it can be sprayed on lawns or cornfields, and it will kill dandelions and plantain and many other weeds but will not harm the desired plant. This selectivity of action is what makes it so valuable. It has been used since the 1940s, and areas to be treated require only about 1 to 2 lb per acre for it to be effective.

Maleic hydrazide (MH) is representative of another kind of functional group and another biochemical effect. It is a *growth inhibitor* and is used for several widely divergent applications. It is sprayed on partially grown tobacco plants to prevent new leaves from starting. The new leaves are called *suckers;* their presence reduces the growth of the tobacco leaves already growing and reduces their quality. The structure of maleic hydrazide is shown below:

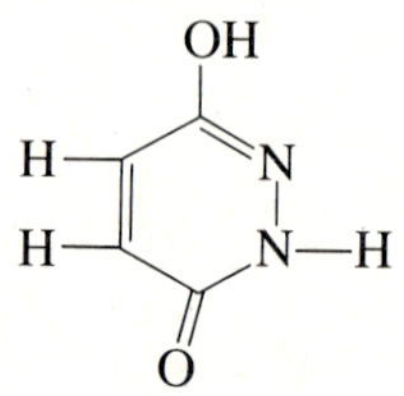

Maleic hydrazide

This material is also sprayed on potatoes to prevent sprouting and thus increases the length of time the potatoes can be stored. The result is that the potatoes arrive at the market in better condition and that shipping can be more leisurely and therefore less expensive. Maleic hydrazide is also sprayed on grass along highways to slow its growth so that mowing does not have to be done as frequently.

Another material, Atrazine, which was introduced for agricultural use in the late 1950s, can be sprayed on the ground before corn is planted. It prevents weeds and grass from growing but does not harm the corn (or any other crop). This means that the farmer does not have to cultivate as often (or does not have to cultivate at all); this reduces costs and increases the acreage one person can till. Atrazine is biodegradable and remains in the environment only a very short time. The structure of Atrazine is shown below:

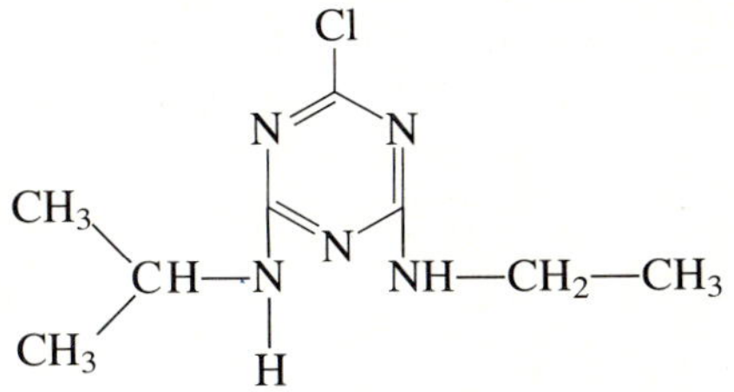

Atrazine

Another compound, commonly known as *butylate* and sold under the tradename Sutan, came out in 1969. It is a complex sulfur-containing material, as shown below:

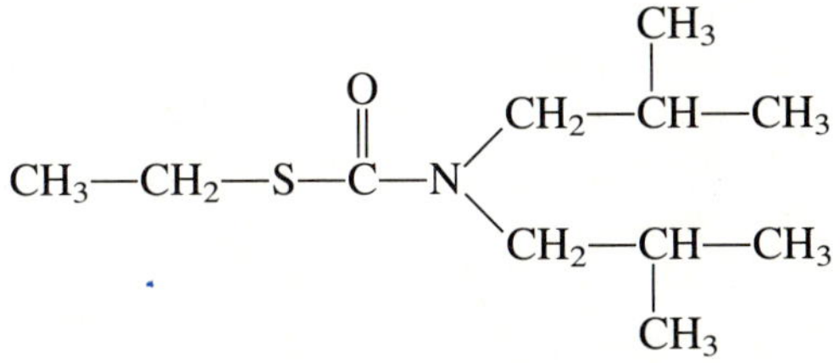

Butylate

Research continues in this area, and there are now over 50 herbicides on the market.

In addition to DDT and herbicides, there are many other materials used to control insects. A common one is Chlordane:

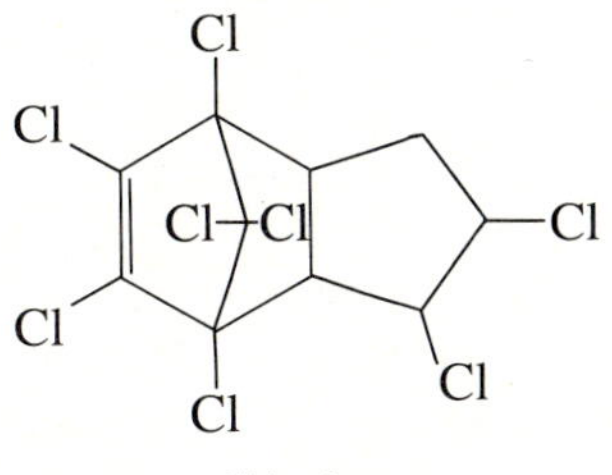

Chlordane

Chlordane controls ants, roaches, and spiders in the home and garden and insects on livestock. Its structure is complex.*

* In fact, most of the compounds we have been examining have complex structures. A beginning student will, in most cases, not be required to memorize these structural formulas. They are given to illustrate the complexity involved in the synthesis of these materials and thus one of the reasons for their relative expense. They are also given for general interest—to satisfy the curiosity of those who are interested in seeing (but not necessarily learning) the molecular structure of everyday products. A given material and its use are usually more familiar to one segment of society than to another.

It was noted earlier that DDT is not destroyed in the environment. Chlordane is also a chlorinated hydrocarbon, but it is degraded slowly in the environment and thus does not pose as much of a long-range environmental threat. It does persist for many years, however, and should be used carefully. The response to environmental factors is a direct result of the precise molecular structure (the structure-function principle).

Two other important insecticides are Sevin and Malathion, which also are complex organic materials:

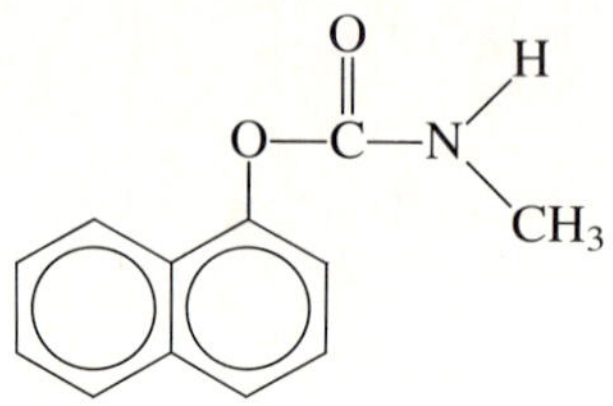

Sevin

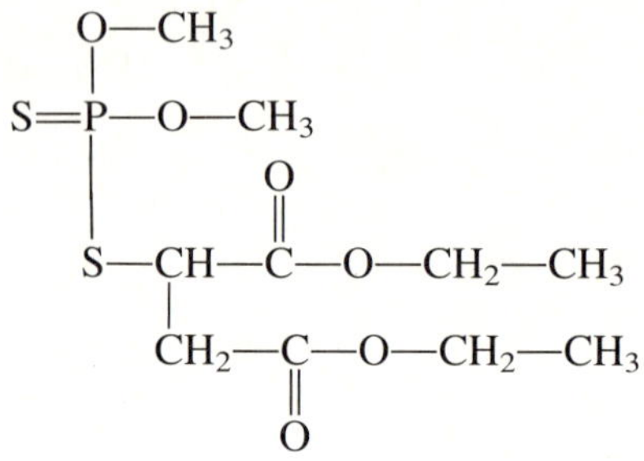

Malathion

Older, simpler organic materials, which are used to *repel* rather than kill moths which eat holes in valuable cloth, are naphthalene and *para*-dichlorobenzene:

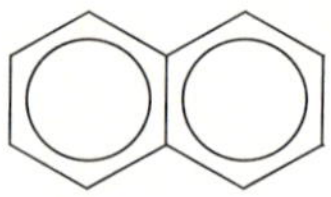

Naphthalene

para-Dichlorobenzene

These same materials can be used to repel dogs from flower beds and rabbits and deer from gardens. This is a better approach in some respects than killing because the materials are not toxic to other nearby animals such as pets.

Still another approach to insect control is the use of *pheromones*. These are materials emitted by the female of some species to attract the male during the mating season (or vice versa in some species). These materials can be placed in traps into which the males are drawn. The trapped insects cannot escape and can be destroyed in environmentally safe ways. With few males left, breeding essentially ceases and the insect population is greatly reduced. Since no toxic material is used, this approach is environmentally more desirable. This is also a very selective method. Further research is needed, however, to discover the pheromones for each species one wishes to control.

As an example, consider the control of bark beetles by the Scandinavian countries. A mixture of the three compounds shown below, used in trace amounts, caused the migrating beetles to enter special devices and to be trapped there:

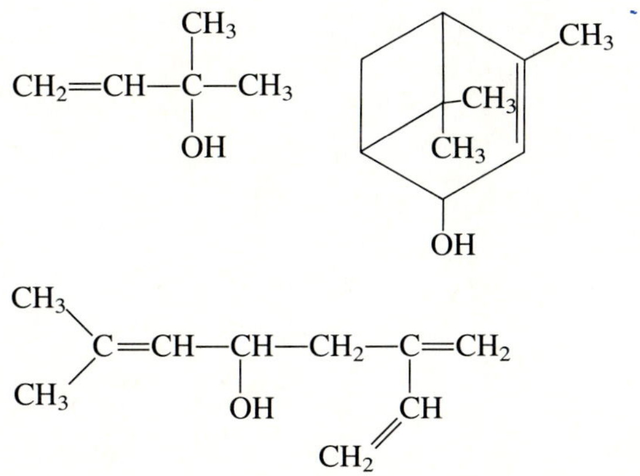

All three compounds had to be present. Note that all three are alcohols and alkenes, which are common functional groups in pheromones, and that they represent a range of complexity. These materials had to be produced synthetically, since they are found in trace amounts in nature. By their use, a great deal of economic loss is being prevented in the lumber industry without introducing toxins into the environment and without harming other insect life or interfering with the mating cycle of other species.

STUDY GUIDE

Terms to learn

- Alcohol
- Aldehyde
- Alkane
- Alkene
- Alkyne
- Amine
- Aromatic
- Bakelite
- Biogenic
- Carboxylic acid
- Catalyst
- Catalytic cracking
- Chemical synthesis
- Cycloalkane
- Ester
- Ether
- Fractional distillation
- Functional group
- Herbicide
- Heterocyclic
- Isomer
- Ketone
- Lucite
- Plastics
- Phenol
- Pheromone
- Polyethylene
- Polymers
- Polyvinyl chloride
- Polyvinylidene chloride
- Structure-function principle

Skill practice problems

1 Name the following compounds:

(a) 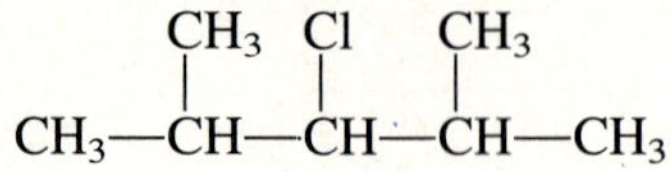

(b) $CH_2{=}CH{-}CH_2{-}CH_3$

(c) 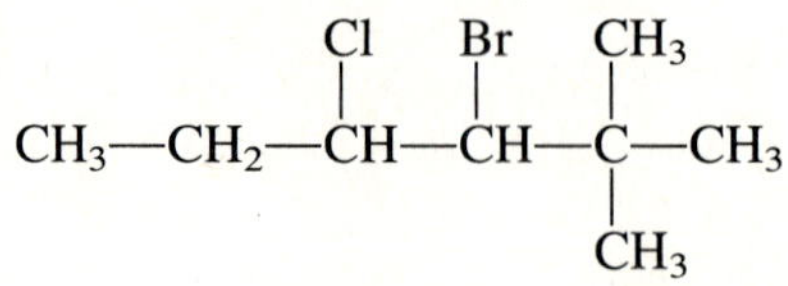

(d) 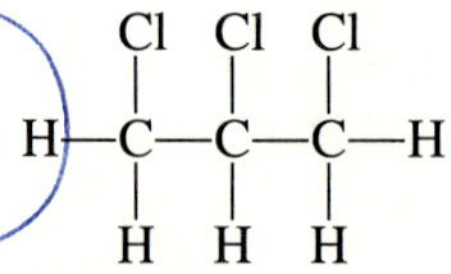

(e) CHI_3

2 Give structures for the following compounds:

(a) Hexabromoethane
(b) 5-Chloro-2,2,3-trimethylhexane
(c) 3-Ethyl-2-methylpentane
(d) 1,2-dichloro-2,3-dibromopropane
(e) 2,2,4-trimethylpentane

3 Name the functional group or groups to which each of the following compounds belongs. (Some may belong to more than one group.)

(a) $CH_3{-}CH_2{-}CH_2{-}O{-}CH_2{-}CH_2{-}CH_3$ __________

(b) 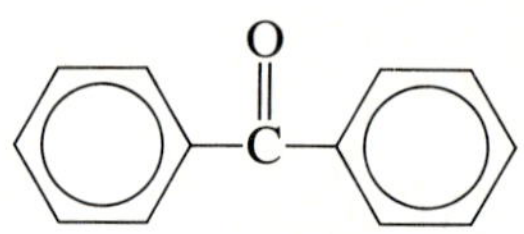 __________

(c) 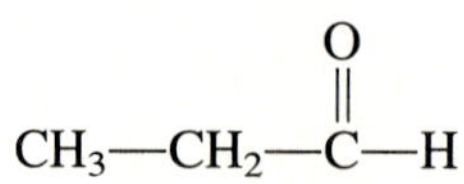 __________

(d) 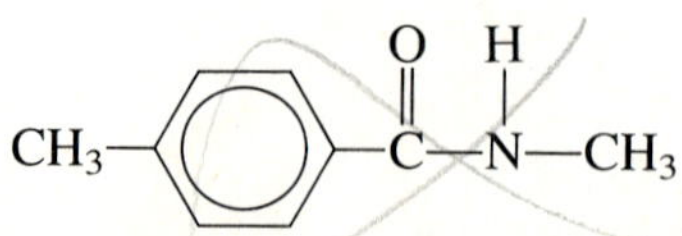 __________

(e) 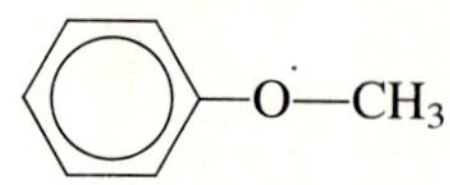 __________

(f) 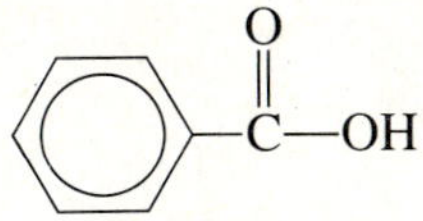 ________________

(g) $CH_3—CH_2—CH_2—CH_2NH_2$ ________________

(h) 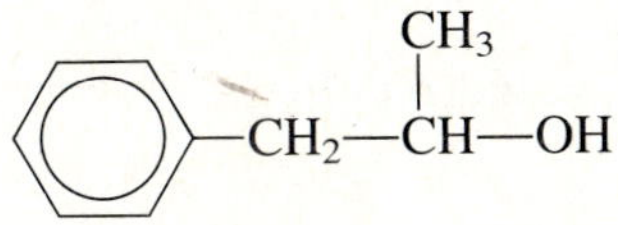 ________________

(i) 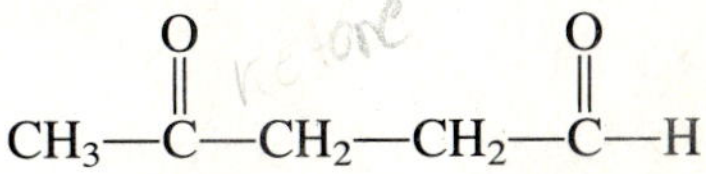 ________________

(j) 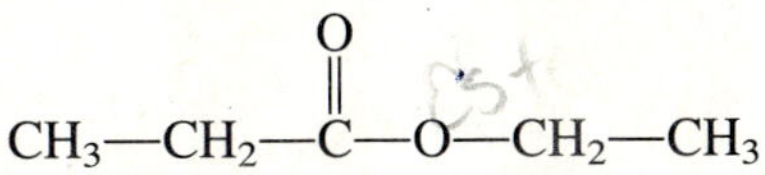 ________________

4 Identify the use or occurrence of each of the following and its common name.

(a) $H—C{\equiv}C—H$

(b) 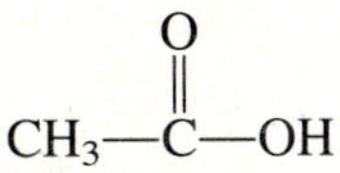

(c) $CH_2Cl—CH_2Cl$

(d) C_8H_{18}

(e) 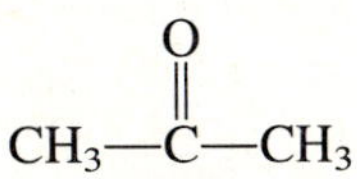

(f) 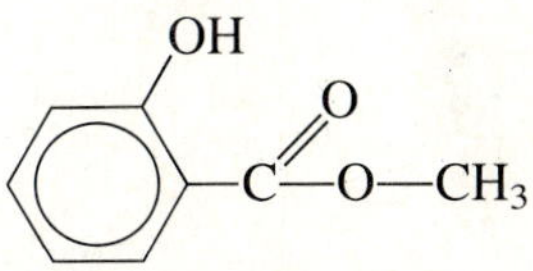

5 Define the following terms or describe the processes.

(a) Catalytic cracking
(b) Alkene
(c) Catalyst
(d) Organic synthesis
(e) Fractional distillation

6 What is the structural formula of PVC?

7 What is the purpose of using boiled linseed oil in paints, varnishes, and other permanent coatings?

8 Which of the following finishes does *not* easily allow a ''ring'' to form when a glass of cold drink is placed directly on it?

(a) Shellac
(b) Varnish
(c) Lacquer
(d) Paint

9 List some characteristics of a polymeric material needed for construction of an artificial heart valve.

10 Why is a pheromone trap better than DDT or Chlordane for insect control? Why is 2,4-D environmentally safer than DDT?

11 List three ways in which maleic hydrazide is currently used. Can you think of other ways it could be used?

Goals

To understand that pollution is a problem of modern culture, not the result of malicious industries

To gain insight into the relative damage of various pollutants

To learn how pollution problems can be solved

To understand the chemistry of the internal combustion engine—chemistry in action

To understand that pollution problems can be solved only by identifying and understanding the chemical substances involved

12

CHEMISTRY AND THE ENVIRONMENT

In this chapter, we will be examining environmental concerns and considering the chemical nature and source of various important environmentally active materials. Possible solutions, both personal and social, will also be covered. Most of the chapter will be devoted to air pollution, but we will also consider water pollution in some detail and take a brief look at land pollution.

THE PROBLEM OF POLLUTION

Let us begin by examining several instances where pollution is a major concern. In 1952 a severe air stagnation in London, England, caused the materials given off by factories to be trapped within the city. This condition lasted for several days, and the levels of poisonous materials became so high that several thousand people died. Most were elderly people, children, or those who were already ill, especially those with respiratory diseases. In 1962 another such incident in London killed about 700 people, and that same year in Donora, Pennsylvania, 20 people died and several thousand became sick when a similar incident occurred.

A number of automobile accidents can be attributed to drowsiness caused by a buildup of carbon monoxide (CO) from rush-hour traffic in cities. Attention is dulled and reflexes become slower under these conditions, even though no acute toxicity is noticed. Now that pollution control devices are used in automobiles, this problem is less serious than it once was.

In western Tennessee a copper refinery produces such large amounts of sulfur dioxide (SO_2) that no vegetation is seen for several miles. In Charleston, West Virginia, a piece of fresh, shiny iron wire becomes red with rust overnight if exposed to the outside air, and window screens deteriorate rapidly in that city. In many places, including Washington, D.C., one can literally smell and taste the SO_2 in the air on certain days. It is produced from the burning of sulfur-containing coal and fuel oils.

The Environmental Protection Agency (EPA) estimates that over half the cancer deaths in the United States each year are caused (at least partially) by air pollution. About 2 million Americans—1 percent of the population—die from all causes each year. Of that number, 15 to 20 percent (400,000) die of cancer. It is estimated that over half these deaths (more than 200,000) are caused by air pollution. This claim can be supported by considering the cancer rate at various locations. Invariably, the higher rates are in the heavily populated, industrialized areas with a high concentration of air pollutants.

Figure 12-1 is a bar graph showing rates of stomach cancer in several countries. Japan has the highest rate—about seven times that of the United States. The causes of this difference are environmental (they include lifestyle), as is shown by the fact that "pure" Japanese people living in the United States have the same rate of stomach cancer as other American citizens. Within the United States, some areas have a much higher cancer rate than others. For example, Maine has a rate of over 200 per 100,000 population, while Alaska has about 70 per 100,000. The difference is attributed mainly to environmental factors.

Many lakes and rivers are becoming choked with algae and other plant

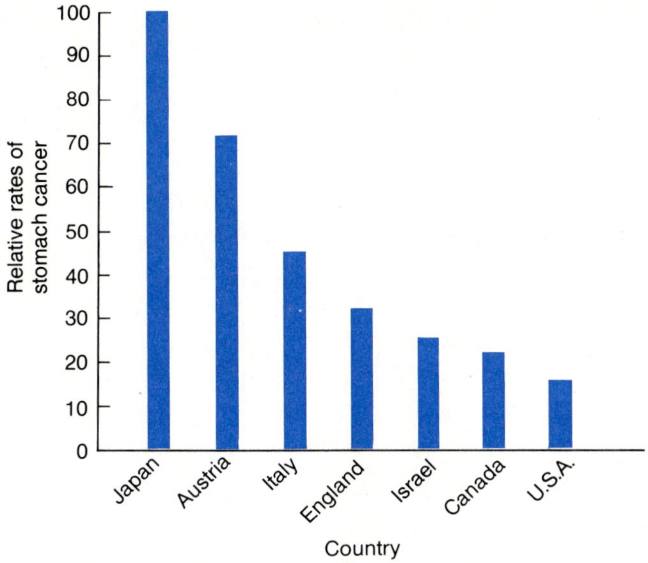

FIGURE 12-1
Rates of stomach cancer in several countries.

growth as a result of their pollution with high levels of fertilizers from farmland runoff. These fertilizers provide nutrients that stimulate the rate of growth of algae and other aquatic plants. Other bodies of water will not support any life because of pollution with industrial waste. In Japan, some years ago, during two separate periods, a number of people became sick, and some died, from mercury (Hg) that had been put into a river upstream from where they lived. In Minamata, 46 out of 121 poisoned people died from eating fish caught from the polluted water. In Niigata, 7 out of 46 died.

In the late 1970s and early 1980s it became apparent that at certain locations, the land was polluted. In western New York and western Tennessee, ill health in children and poisoned drinking water were the result of chemicals being buried in these areas years earlier. The metal containers holding these chemicals had rusted or had otherwise become broken, and their contents were released into the environment.

Children who live in inner cities where they are constantly exposed to motor vehicle exhausts are found to have higher levels of lead (Pb) in their bodies. High levels of lead are also found in children living near lead-refining sites. Lead in some of these children's bodies is said to be near the toxic level, and parents and teachers say that the children show learning disabilities.

Agricultural production is decreased by the presence of air pollutants. Grapes that produce about 20 lb per year for each plant in clear air yield only about 10 lb per year in the currently polluted California air.

The list could go on, but these examples illustrate the seriousness of the problem caused by the materials produced in our quest for a higher standard of living and the effect of these materials on human welfare and the natural ecosystems.

In 1816, huge volcanic explosions caused large quantities of dust to be released into the atmosphere. This dust spread nearly worldwide, and caused colder-than-normal weather in the northern and eastern United States because of the partial blocking by the dust cloud of heat from the sun. The effect was so severe that temperatures at times reached freezing throughout the summer as far south as Georgia. Several years later, however, the atmosphere was back to normal, with no permanent ill effects. Natural sources of pollution introduce a staggering 10^{12} tons of potentially harmful material into the air each year worldwide, while humans put only about 10^{8} tons in the atmosphere per year. This means that pollutants made by humans are only 0.01 percent of those produced in nature. If that is the case, why is there such concern over human-made pollutants? There are three reasons:

1 Many of the materials used by humans are synthetic, and nature cannot destroy, inactivate, or accommodate some of them. Nature's own pollution (if indeed it should even be called that) is recyclable—there is a self-cleaning provision built into nature. Human pollution, on the other hand, can build up in a system, since there may be no natural way of getting rid of it or it may be removed more slowly than it is produced.

2 Human-made pollutants are produced in the largest quantities where the largest number of people live. A small amount of pollution, if produced and kept in a small area, can build to a high *concentration* in the atmosphere or in soil and harm those people who happen to live in that area. Thus, the problem is not the total amount of material produced but the uneven distribution of that material.

3 Many of nature's pollutants are not extremely harmful to living systems, while many of the industrial and motor vehicle pollutants are highly toxic, even in relatively small amounts.

Before proceeding, let us consider the meaning of the word *pollution*. For example, does pollution necessarily cause harm? The working definition that we will use here is this: *Pollution is a condition where a substance is present at a location, or in such quantities, or both, that it can cause harm to living systems*. For example, soil causes no harm as long as it is on the ground, but in the atmosphere it is pollution. In the atmosphere, it can cause harm to humans, animals, or plants (although it would not necessarily be extremely dangerous). Lead metal is appropriate when used to seal joints in drainpipes, but as particles in the food of small children, it can cause harm and constitutes pollution. Salt in food (in small amounts) is desirable, but pollution results from large quantities in public water supplies (resulting from groundwater seepage or spills from oil well operations). Pollution, then, has more to do with the situation than with the nature of the material itself.

The other important aspect of the pollution problem is that the earth's atmosphere is not a vast, endless sea into which we can pour any amount of matter and expect it to be diluted to a harmlessly low concentration. Most of our pollution stays within 2 mi of the earth's surface, as shown in Figure 12-2. This means that most toxic materials remain in the area where we live. They do not escape and spread over the whole depth of the earth's atmosphere.

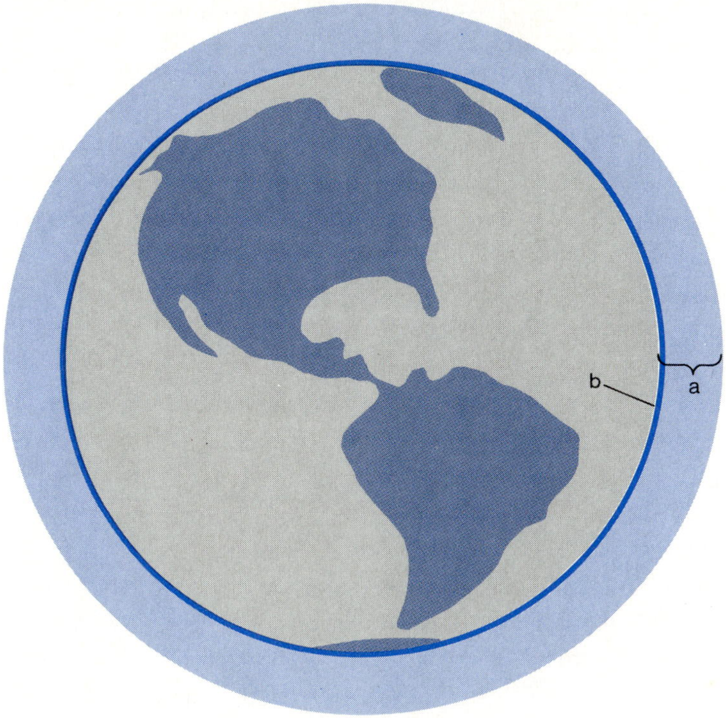

FIGURE 12-2
Cross section of the earth, showing the 2-mi depth of atmospheric dilution. The area indicated by *a* is about 1000 mi deep; *b* indicates the 2-mi layer where most pollutants remain. (Not to scale.)

Likewise, the seas and rivers were once thought to be so vast that we could use them as convenient waste disposal sites. This is not a correct view, as shown by the rise in the concentration of mercury (Hg) in large tuna to dangerous levels as a result of the dumping of industrial waste material containing Hg into waterways.

It is apparent that waste materials introduced into the environment accidentally or on purpose cost a great deal in terms of health, quality of life, money, and time spent to clean the environment or to prevent contamination. We will now take a more detailed look at the major sources of pollution, the damage caused, and possible solutions in these problems.

AIR POLLUTION

Industrial air pollution

There are two main sources of air pollution—industrial and automotive (see Table 12-1 for a summary). *Industrial pollution* is primarily a result of the burning of fuel to produce the energy needed in manufacturing processes. The generation of electricity by burning coal is one of the worst offenders and will be used in the initial discussion of the problem. However, all fuel-burning facilities (such as heating plants at universities and apartment complexes) and all industries which use large amounts of heat (such as glass and steel manufacturing) are also important, on a more local level.

TABLE 12-1

Major sources of air pollution

SOURCE	POLLUTANT
Industrial:	
Burning of fuel	SO_2, SO_3, H_2SO_4 CO_2 NO_x Particulates
Ore refining	Metal oxides SO_3
Miscellaneous	Small (or large) amounts of the materials being produced at each factory
Automotive:	
Burning of fuel	CO_2 CO C_8H_{18} (unburned hydrocarbons)
Antiknock compounds	PbO_2 Polyaromatic compounds (benzpyrene; $C_{20}H_{12}$)
By-products of fuel burning	NO, NO_2 (NO_x) HNO_3

Carbon dioxide The two main fuels used for energy generation are coal and fuel oil; coal is the least expensive and most widely used. When these are burned, the following chemical reactions occur:

$$\underset{\text{Coal*}}{C} + O_2 \longrightarrow CO_2 + \text{heat}$$

$$\underset{\text{Fuel oil}}{C_{15}H_{32}} + 23O_2 \longrightarrow 15CO_2 + 16H_2O + \text{heat}$$

The heat these fuels release when they are burned is the product of interest; however CO_2 is also produced at the same time. CO_2 will be called a *pollutant,* since it is put into the atmosphere in large amounts, but it is not a particularly damaging one. It is not toxic, per se, to animal life and is used by plants for photosynthesis. It is removed fairly rapidly from the environment by plants and by its reaction with water to form carbonic acid, which in turn is converted to inorganic carbonates. The production and fate of CO_2 in the atmosphere are shown in Figure 12-3. While some of the carbohydrates produced by plants are digested by animals and changed into CO_2, and some of these carbohydrates decay, these sources of CO_2 are negligibly small compared with the production of CO_2 by the burning of fuel. Some plant carbohydrates, such as wood, are burned directly; but most, such as coal and fuel oil, come from ancient deposits of decayed and compressed plant materials.

Our industrialized society requires a great deal of energy (see Chapter 14), and we are burning fuel materials at a very rapid pace. We are slowly using up our coal and oil resources—that is, we are converting these ancient carbohydrates back into the CO_2 and H_2O from which they were originally

*Coal is actually a complex material whose structure is just beginning to be understood. It contains large molecules of covalently bonded carbon along with small amounts of hydrogen, oxygen, and sulfur.

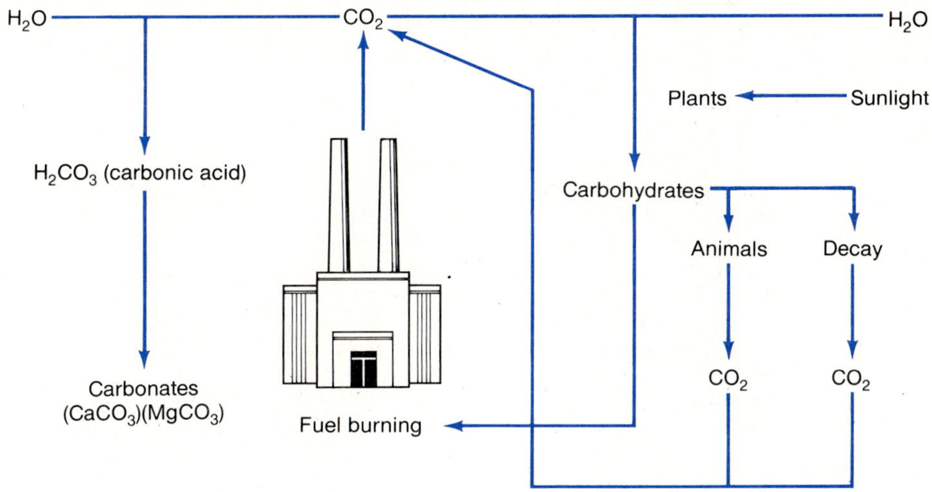

FIGURE 12-3
Production and fate of CO_2 in the atmosphere.

produced. Thus we are placing more CO_2 in the atmosphere than has so far been present during human life on earth. The normal concentration of CO_2 in the atmosphere is 0.035 percent. This value represents the difference between the production of CO_2 by animals and by a small amount of burning of fuel and the consumption of CO_2 by plants and oceans when production of CO_2 and its consumption are in dynamic equilibrium. Before the extensive use of fire by humans, when vast forests and other vegetation covered most of the earth, the CO_2 level was surely below the 0.035 percent observed in recent years. The more rapid production of CO_2 today is *probably* increasing the amount in the atmosphere. Plants are not able to remove it as fast as humans are producing it.

A problem arises because CO_2 molecules can absorb infrared rays more effectively than any other atmospheric component except water vapor. N_2, O_2, and He, for example, do not absorb infrared radiation.

Infrared radiation is best thought of as heat radiation because it causes molecules with polar bonds, such as CO_2 and H_2O, to vibrate more vigorously. Remember that heat is a measure of the *total* kinetic energy of atoms and molecules, and temperature is a measure of the *average* kinetic energy of atoms and molecules. When the CO_2 absorbs infrared radiation and vibrates more vigorously, the molecules have more energy and a higher temperature. Figure 12-4 illustrates the transfer of energy from infrared light to CO_2 molecules and their resultant increase in temperature. You can experience this phenomenon by holding your hand in bright sunlight. Your hand feels warmer because the infrared radiation is causing the molecules in your hand (mostly water) to vibrate faster. Your nerve endings experience this greater kinetic energy as warmth. Although the CO_2 molecules vibrate in several ways, bending is the vibration that absorbs the most radiation and is the one illustrated in Figure 12-4.

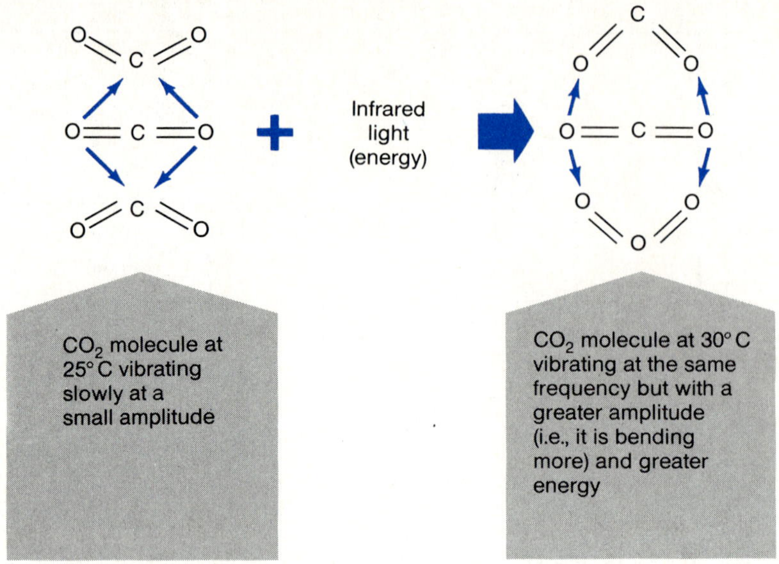

FIGURE 12-4
Absorption of infrared light by atmospheric CO_2.

The surface of the earth absorbs heat from the sun and normally radiates much of it back into space. Atmospheric CO_2 prevents this heat from escaping, however. The environmental result of more atmospheric CO_2 is that more of the sun's heat will be retained by the earth and the average temperature of the earth will tend to increase. This in turn could cause the polar ice caps to melt, raising the level of the oceans and causing the loss of fertile coastal farmland and the destruction of many of the world's largest cities. This increase in temperature due to more atmospheric CO_2 is called the *greenhouse effect*. The rapidity with which it is taking effect is not readily determinable, and whether other factors are counteracting it is a subject of current debate among scientists. It is a slow process, however, and humanity should have ample time to prepare for any future result. In addition to the slowness of the warming process, it is superimposed on natural temperature cycles which have always been present, but which are not well understood. Thus it is difficult to determine if a small change in temperature (several degrees over a period of 20 to 30 years) is part of a larger natural cycle or is caused by CO_2 and other air pollutants. CO_2, therefore, while its presence should be understood, is not considered to be of serious, immediate environmental concern. However, recent data released by the EPA suggest it may be a more serious and immediate problem than had been thought in the past.

Oxides of sulfur In addition to carbon or hydrocarbons, most fuels have small amounts of sulfur (S), either as elemental sulfur or as its covalent compounds. As the fuel burns, so does the sulfur it contains:

1 Burning of sulfur: $S + O_2 \longrightarrow SO_2$

2 Oxidation of sulfur dioxide: $2SO_2 + O_2 \longrightarrow 2SO_3$

3 Production of sulfuric acid: $SO_3 + H_2O \longrightarrow H_2SO_4$

The sulfur combines with oxygen to form sulfur dioxide, SO_2 (reaction 1). The dioxide slowly combines with atmospheric oxygen to form the trioxide, SO_3 (reaction 2). The trioxide reacts with water to produce sulfuric acid, H_2SO_4 (reaction 3). Coal is notorious for containing large amounts of sulfur, but various coals have greatly differing sulfur contents. Fuel oils usually have only small quantities of sulfur, and it is possible to remove this sulfur before burning the oil. The main source of the sulfur oxides, then, is coal, which is otherwise a desirable fuel because of its great abundance, wide distribution, and resultant low cost (see Figure 12-5).

When coal is burned, SO_2 is formed in the flame and emitted via the smokestack into the atmosphere. This material is offensive for several reasons. It has a pungent odor and taste and gives the sensation of a mild sore throat. On still, damp mornings near power plants one can experience this unpleasant odor. It also has the ability to bleach away color. It is the material used commercially to bleach the natural red color from maraschino cherries before the familiar bright-red color is added to these cherries with an artificial dye. It is also used to bleach wood pulp so that the paper made from it will be

FIGURE 12-5
Industrial plants which burn coal produce most of the SO_2 in the atmosphere.

white. When it is in the atmosphere, it will bleach clothing, especially wet clothes hanging outdoors to dry. The fibers of cloth are weakened by SO_2 also. Another undesirable effect is that of *weakening lung tissue* and *enhancing the uptake of carcinogenic* (cancer-causing) *materials* in tobacco smoke. This means that people who smoke have an increased chance of getting cancer if the air is polluted with SO_2 in the location in which they live. Plant growth is inhibited by the presence of SO_2. Grape production is cut approximately in half, and alfalfa and cotton production is reduced if SO_2 is present even in small amounts. It is estimated that the overall loss in American agricultural productivity is between \$0.5 and \$1.0 billion annually.

After the SO_2 is converted to SO_3 by combination with atmospheric oxygen, the latter easily dissolves in atmospheric water droplets and rain to produce a dilute solution of sulfuric acid (H_2SO_4). Sulfuric acid causes corrosion of marble (as in Figure 12-6) and mortar (both chiefly $CaCO_3$) and causes metals to corrode faster than they would in moist air alone. Following are some of the chemical reactions involved.

Corrosion of marble, mortar, or limestone:

$$CaCO_3 + 2H_2SO_4 \longrightarrow Ca(HSO_4)_2 + H_2O + CO_2$$

Corrosion of iron:

$$2Fe + 3H_2SO_4 \longrightarrow Fe_2(SO_4)_3 + 3H_2$$

The final problem is acid rain. Rain which falls through air containing SO_3 is made acidic by the H_2SO_4 which forms; it is then called *acid rain*. In addition to the undesirable corrosion mentioned above, this rain collects in

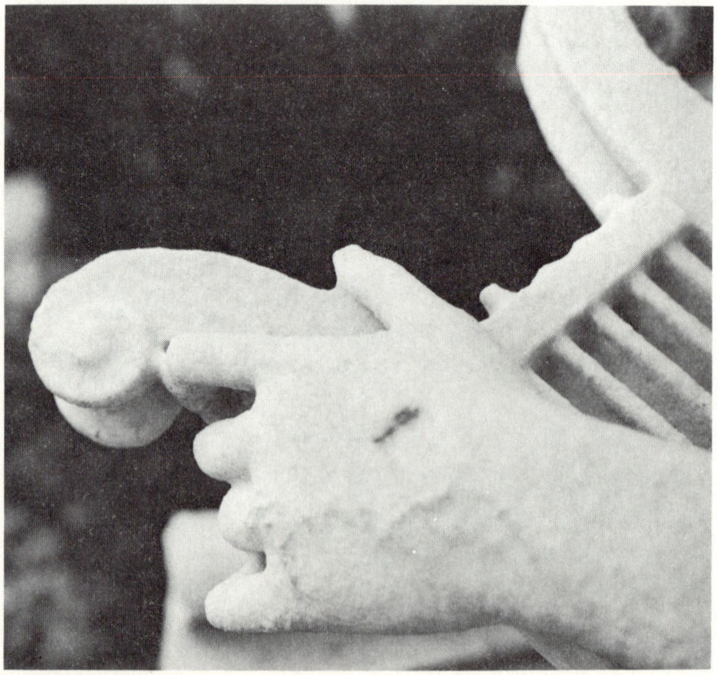

FIGURE 12-6
Marble statue corroded by acidity in the environment. Note the irregular contours and the ridges left from the removal of about 1⁄16 in of marble.

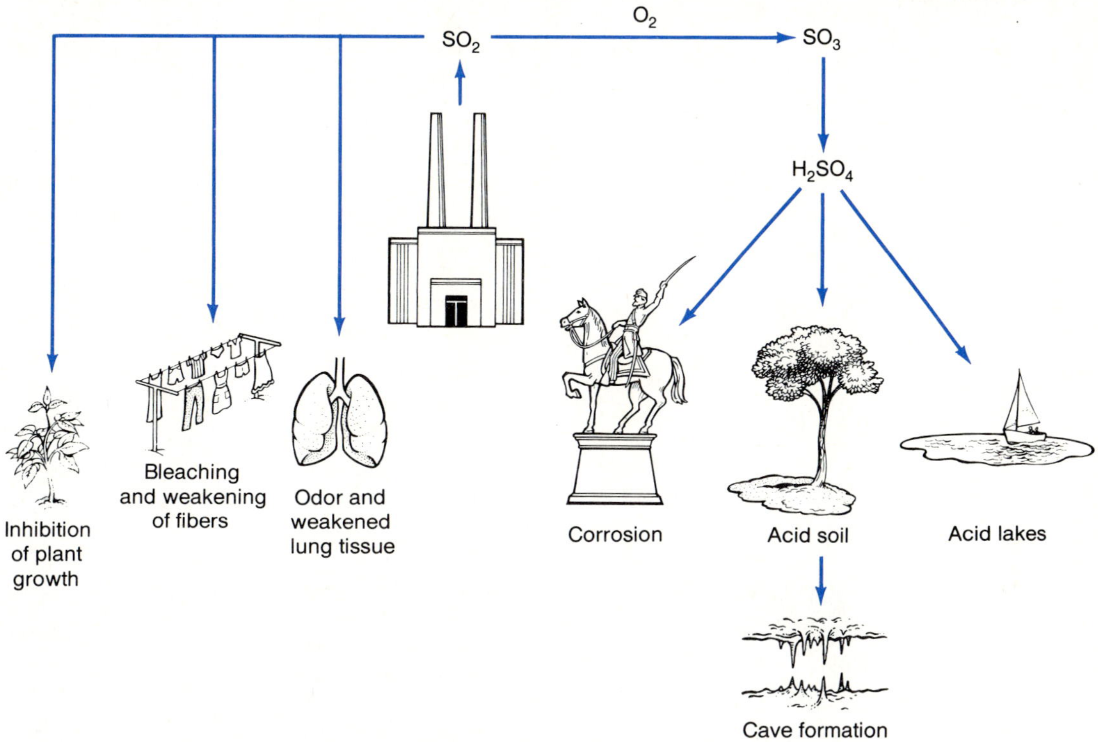

FIGURE 12-7
Sulfur oxides in the environment.

rivers and lakes, and in some cases the acidity is so high that plants and fish cannot live. This increase in acidity produces a body of water with less than optimum usefulness. Acid rain also makes the soil acid and prevents certain kinds of plants from growing. Figure 12-7 shows the overall fate of SO_2. The soil in some parts of the country contains limestone, which can react with the H_2SO_4 in the rain and neutralize it:

$$\underset{\text{From acid rain}}{2H_2SO_4} + \underset{\text{Limestone}}{CaCO_2} \longrightarrow Ca(HSO_4)_2 + H_2O + CO_2$$

The groundwater in these areas has more Ca^{2+}—from the $Ca(HSO_4)_2$—and is harder (see Chapter 10 for a detailed discussion of hard water). These areas are fortunate, however, since the acid is no longer present in the water to pollute lakes and streams and the soil acidity is not greatly affected by acid rain.

Table 12-2 shows the pH scale and the position of acid rain and normal rain on the pH scale. For a review of acidity, refer back to Chapter 10.

The pH of rain or other solutions is commonly measured by a device called a *pH meter,* shown in Figure 12-8. The solution to be measured is placed in a beaker, and the electrodes (sensing devices) are immersed in the solution. When the meter is turned on, the needle indicates the pH of the solution.

TABLE 12-2

The pH scale

pH	COMMON MATERIAL	CHARACTERISTICS
1	Dilute hydrochloric acid	Strongly acidic ↓
2		
3	Vinegar	Mildly acidic
4	Acid rain (pH 2–6)	↓
5	Normal rain (pH 5–6)	
6		
7		Neutral
8		Mildly basic
9		↓
10		
11	Household ammonia	
12		Strongly basic
13		↓
14	Lye	

FIGURE 12-8

A pH meter used to determine the acidity of water solutions.

Rain falling in air containing no SO_3 or other acidic pollutants has a pH between 5 and 6. It is slightly acidic because of dissolved CO_2, which produces H_2CO_3. In polluted air, the pH of the rain itself usually varies between 6 and 4 but sometimes is as low as 2. Lakes sometimes reach an acidity level as high as pH 4.5, which is below the range where most fish can exist (pH 6 to 5). Remember that *lower* numbers on the pH scale mean a *higher* acid concentration.

FIGURE 12-9
Smoke from burning leaves and home fireplaces contains particulate matter.

Particulate matter Another source of concern is the emission of small *particles* into the atmosphere. These may be soot, which is highly porous carbon, or dust-sized particles from the manufacture of various materials. These particles are commonly called *smoke*. These materials are small enough to stay suspended in the air indefinitely because of the motion of the gaseous air molecules (remember that molecules of gaseous materials are in constant, rapid motion) and wind currents. Such particles come not only from industrial sources but also from forest fires, domestic burning of leaves (Figure 12-9), and volcanic activity.

The harmful effects of this type of pollution are as follows. The atmosphere becomes hazy and visibility is reduced. This, in turn, reduces the enjoyment of natural scenery. In addition, some of these particles (the larger ones) eventually settle onto clean clothes, cars, vegetation, etc. Windows have to be washed more often, and the general presence of dirt becomes a nuisance. Some of these particles are themselves toxic, such as fine PbO_2 powder near a lead refinery; others, such as soot, carry other toxic materials. Because of the very porous nature of soot, it can absorb a large number of molecules of carcinogenic materials or SO_3; and when the soot particle comes into contact with a lung, the risk of cancer increases. When a piece of ash (a large piece of porous carbon) containing SO_3 falls on an automobile, the acid concentration can be high enough to damage the paint or cause other forms of corrosion. There is also the possibility that particulate matter will block sunlight and cause a cooling effect on the earth.

Miscellaneous problems of air pollution An important source of SO_2 and SO_3 pollution is ore refining. For example, silver, mercury, copper, and lead are most often found in ores containing compounds of these elements with sulfur. These metal sulfides must be converted to the oxides before the pure metal can be isolated.

Typical reactions are shown below:

$$PbS_2 + 3O_2 \longrightarrow PbO_2 + 2SO_2$$
Galena (lead ore)

$$2Ag_2S + 3O_2 \longrightarrow 2Ag_2O + 2SO_2$$
Argentite (silver ore)

$$HgS + O_2 \longrightarrow Hg + SO_2$$
Cinnabar (mercury ore)

Notice that each of these reactions produces SO_2, which is changed to SO_3 in the atmosphere. This produces an acidity so high that no vegetation can grow in an area surrounding these ore processing facilities (Ducktown and Copper Hill, Tennessee, are two such communities).

Near most factories that produce chemicals, peculiar odors are noticed by the local residents and by people traveling through the area. Each facility inadvertently releases small amounts of the starting materials or products into the air. Often this is not harmful to health, but it is quite unpleasant for those living downwind. However, in some cases these materials are highly toxic.

Another problem is that of local pockets of stagnant air. Normally, the sun warms the earth, which, in turn, warms the air near the earth. This warm air, being less dense, rises, and the cooler air moves toward the ground to replace it. This natural turbulence, plus the prevailing winds, usually prevents pollution from building to an acutely dangerous level. At certain times at some locations, an *air inversion* occurs in which cool air is near the ground with warmer air above it. This prevents the air from moving; and if there are sources of large amounts of pollution in that area, the buildup of pollutants can be great. Figure 12-10 shows a typical setting in which smog buildup occurred during an air inversion over a city located in a valley whose main industries are chemical manufacturing and the burning of coal for energy.

If the air inversion lasts for several days, the SO_2, SO_3, particulate matter, and chemical fumes become highly concentrated and a haze develops. This is called *industrial smog*. Industrial smog was the cause of the deaths referred to at the beginning of this chapter which occurred in 1952 and 1962 in London and in Donora, Pennsylvania. It causes most people to feel sick, but it most seriously affects those who are very young or very old or have chronic respiratory problems. Many cities now have smog alert warnings and announce an air quality index during such periods to warn people to take precautions, such as staying indoors, wearing a mask, or temporarily leaving the area.

FIGURE 12-10
Smog buildup during an air inversion.

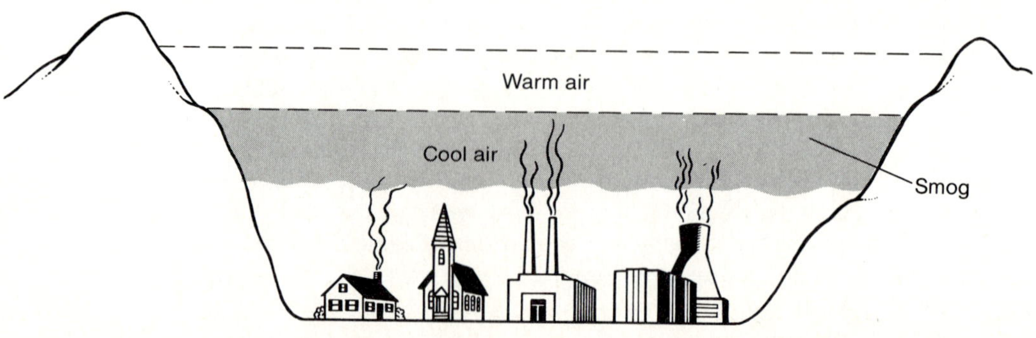

Solution to industrial air pollution Now that we have examined pollutants and the problems they cause, let's consider some possible remedies. The main concern is to make (and keep) the environment safe for humans, animals, and plants, in both the immediate *and* the distant future. The air quality act passed by Congress in 1972 has encouraged both the development and use of a number of pollution control methods. Air quality has improved since that time, but there is still more work to be done. Industry is understandably sometimes slow to cooperate because of the cost involved; taxpayers do not want to be overburdened; and each administration in Washington takes a slightly different view of the importance of the problem of pollution. As a result, long-term policies are inconsistent and improvement of air quality is retarded.

As was noted previously, carbon dioxide may or may not be a serious problem. It is unclear at this time if measures are needed to control it and if so what those measures are. Sulfur oxide emission is a serious problem and can be controlled in several ways. One is to reduce the use of metals and the use of electricity; this reduces the need for ore refining and coal burning at electrical generation plants. The second is to burn fuel oil or low-sulfur coal, but these are more expensive and raise the cost of electricity. A third approach is to remove sulfur from the fuel before burning it, but this either is expensive or is not feasible for other reasons.

A good way of solving the problem of SO_3 is the use of a *stack absorber,* as diagramed in Figure 12-11. This is a device that bubbles the gas coming out of the stack through a solution of sodium hydroxide, NaOH, or calcium hydroxide, $Ca(OH)_2$. Any SO_2 or SO_3 present reacts with the NaOH according to the following reactions:

$$SO_2 + 2NaOH \longrightarrow \underset{\text{Sodium sulfite}}{Na_2SO_3} + H_2O$$

$$SO_3 + 2NaOH \longrightarrow \underset{\text{Sodium sulfate}}{Na_2SO_4} + H_2O$$

After most of the NaOH has been changed to the sodium sulfite and sulfate, the solution is removed from the stack absorber and a new NaOH solution is added. The spent solution can be processed to recover and use, sell, or discard the salts produced. The installation of the stack absorber and the NaOH needed are added costs, but the production and use or sale of Na_2SO_3 and Na_2SO_4 can help to offset these costs.

Particulate matter can theoretically be removed by filters, but because the particles are very small, the filters quickly become clogged. An *electrostatic precipitator* is a much better device. It consists of closely spaced charged plates near the top of the stack, as shown in Figure 12-12. Most particles develop an electrostatic charge while moving through the air in the stack. Those with a positive charge are attracted to the negatively charged plates and vice versa. The plates can be hit periodically to remove the buildup of loose particles. This solid can then be buried, or if that causes a problem, treated and disposed of in another way. This is a fairly inexpensive process.

The technical information needed for controlling SO_2 and SO_3 pollution is available, although improved methods would be welcome. Implementation

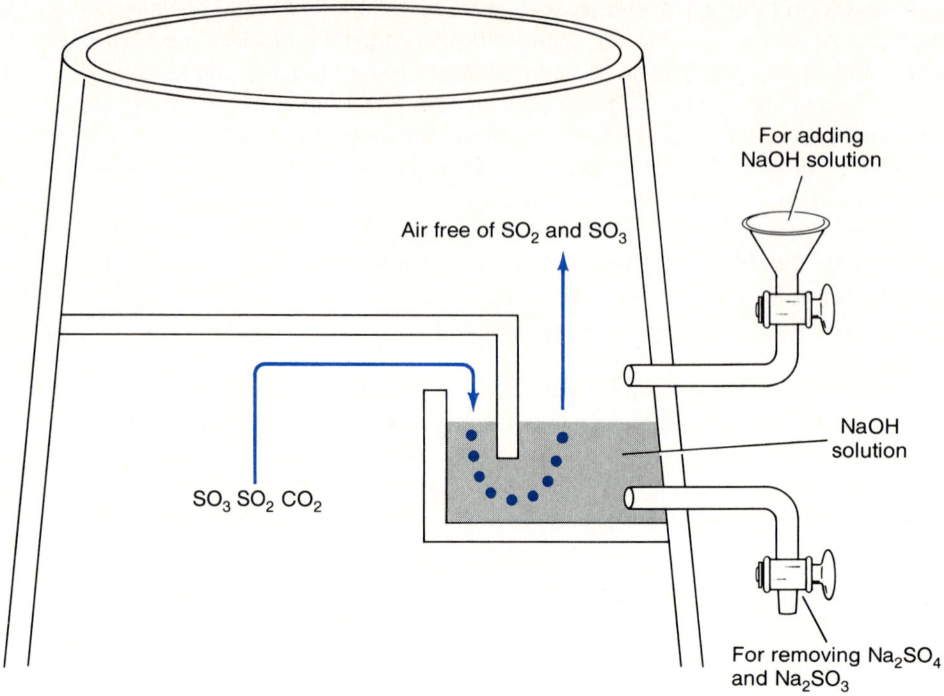

FIGURE 12-11
Using a stack absorber to remove sulfur oxides from combustion gases.

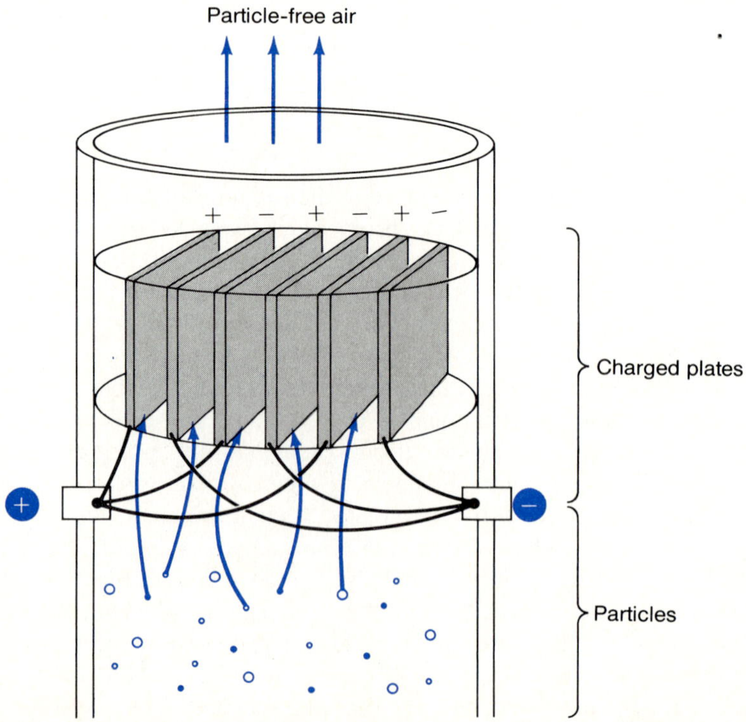

FIGURE 12-12
Electrostatic precipitator.

of these techniques requires the following. Industry must be willing to go to the effort and expense required to implement these techniques, and consumers must be willing to pay more for the product produced by that company. Finally, legislators, encouraged by constituents, must pass reasonable, fair laws to ensure a consistent trend toward clean air over a period of years.

Chemistry of the internal combustion engine: Automotive air pollution

The second major source of air pollution is motor vehicles that run on the internal combustion engine. Let us first consider how this engine works.

Design of the internal combustion engine A diagram of a cylinder of an internal combustion engine is shown in Figure 12-13*a*. Each engine consists of four, six, or eight of these cylinders, each of which is synchronized with the others. The power for the engine comes from the explosion of a mixture of gasoline and air in the chamber.

FIGURE 12-13
Operation of the internal combustion engine.

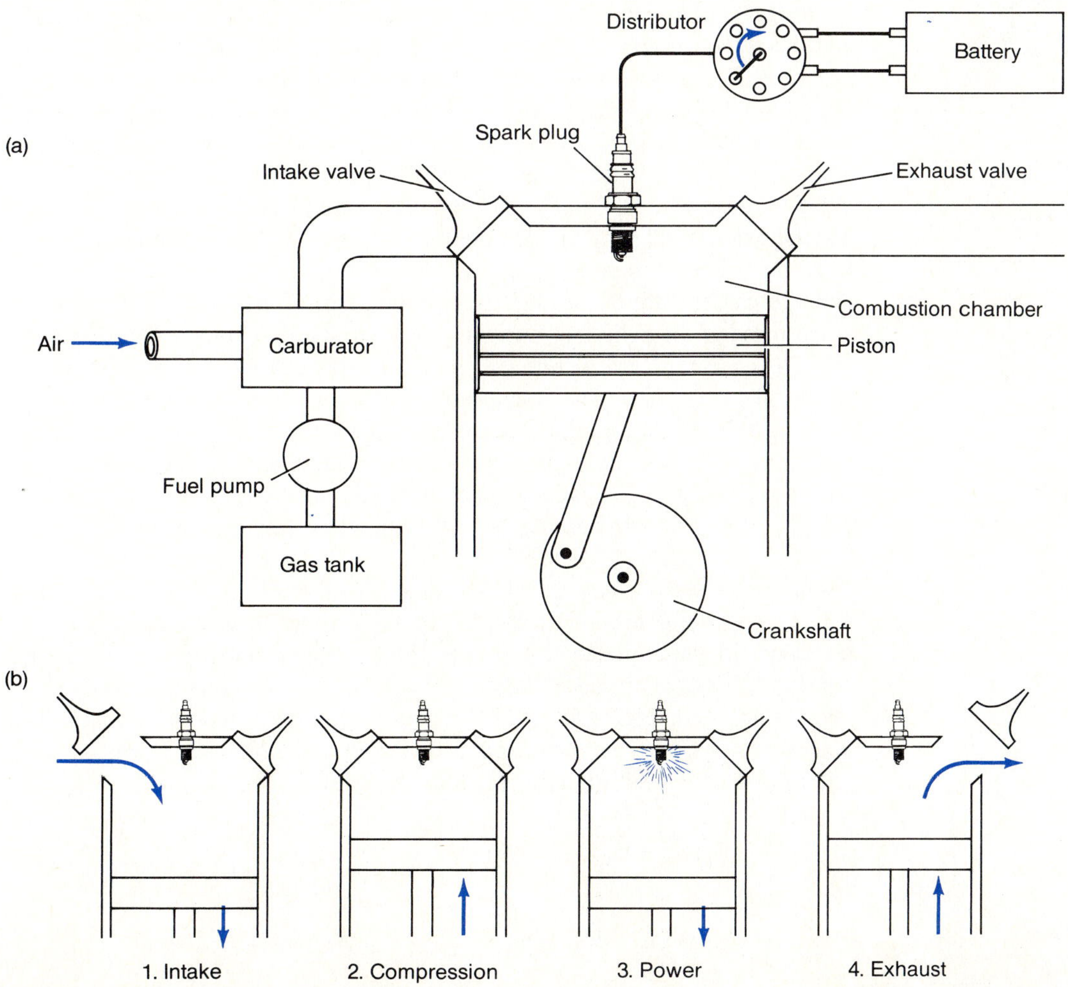

The reaction during this explosion is:

$$2C_8H_{18} + 25O_2 \longrightarrow 16CO_2 + 18H_2O + \text{heat}$$

Gasoline is actually a mixture of hydrocarbons, but we will assume that it is pure octane. The reactants, which are at room temperature, include 2 molecules of octane (C_8H_{18}) and 25 molecules of O_2, a total of 27 molecules. The products include 16 molecules of CO_2 and 18 molecules of H_2O in the vapor state, a total of 34 molecules. The product molecules are at a high temperature as a result of the heat energy released (see Chapter 14). There are more molecules after the reaction has occurred. This, along with the temperature increase, causes a large, rapid pressure buildup, which is what an explosion is. Figure 12-13*b* shows how this explosion is harnessed mechanically to do the work of moving an automobile. Four processes are involved, and an engine using them (most ordinary automobile engines, for example) is called a *four-stroke* or *four-cycle engine*.

During the *intake stroke,* the intake valve opens and the piston moves down. This draws gasoline vapor and air into the cylinder. (The gasoline is vaporized and mixed with air in the carburetor.) The intake valve then closes, and the piston moves up, compressing the mixture of gasoline and air (*compression stroke*). At this point a spark is sent to the spark plug, which ignites the mixture, causing the explosion which drives the piston down. This causes the crankshaft to turn (*power stroke*). The crankshaft is connected mechanically—by means of the clutch, transmission, drive shaft, differential, and axles—to the wheels. Finally, during the *exhaust stroke,* the piston moves up and the exhaust valve opens, letting the hot CO_2 and H_2O out through the exhaust pipe. Each engine has four, six, or eight cylinders, and each cylinder is undergoing the different strokes at different times. Thus, at any given instant at least one of the cylinders is in the power stroke, one in the exhaust stroke, etc. The camshaft controls the valves and assures that they open and close at exactly the right time. The crankshaft connections ensure that the pistons are moving in the right directions at the right time, and the distributor sends an electric spark to each cylinder when the piston has reached a maximum height during the compression cycle. It should be obvious from this description that correct timing is very important for the smooth running of an engine.

Motor vehicle pollution As already noted, neither CO_2 nor H_2O is a pollutant. A perfect engine run under ideal conditions would cause very little if any pollution. In actual practice, however, there are several sources of pollution.

First, because air rather than pure O_2 is used in the combustion (remember that air is 20 percent O_2 and 80 percent N_2), an insufficient amount of O_2 is present to support complete combustion. Part of the fuel is not burned at all, and part is converted to carbon monoxide (CO):

$$C_8H_{18} + O_2 \longrightarrow CO_2 + \boxed{CO} + H_2O + \boxed{C_8H_{18}}$$

Deficient amount (O_2) — Pollutants (CO, C_8H_{18})

The *unburned gasoline* and CO, along with the H_2O and CO_2, escape through the exhaust pipe into the atmosphere. The C_8H_{18} smells bad, and can react with other pollutants to produce smog (this will be examined in detail later). The CO is slowly converted by atmospheric O_2 to CO_2, but meanwhile the concentration of CO builds up in locations where motor vehicle traffic is high.

Carbon monoxide is toxic to animals. It reacts with the hemoglobin (an iron-containing protein found in red blood cells) in the blood in such a way that the hemoglobin cannot perform its normal function of transporting O_2 to the body's cells. When cells, especially brain cells, are deprived of O_2, they cannot function properly; if the deprivation is extensive, they die. Thus, when toxic levels of CO are present, no O_2 can be transported from the lungs to the brain and drowsiness, and finally death, occurs. This process is diagramed in Figure 12-14.

The O_2 becomes attached to the hemoglobin in the lungs. This complex (combination of O_2 and hemoglobin molecules) moves through the blood vessels to the cells, where the hemoglobin releases the O_2 for the cell's use. At the same time, hemoglobin picks up CO_2 from the cells—CO_2 is a normal end product of many cellular reactions. The deoxygenated hemoglobin (carrying CO_2 from cells to lungs for exhalation) goes back to the lungs to complex (combine) with more O_2.

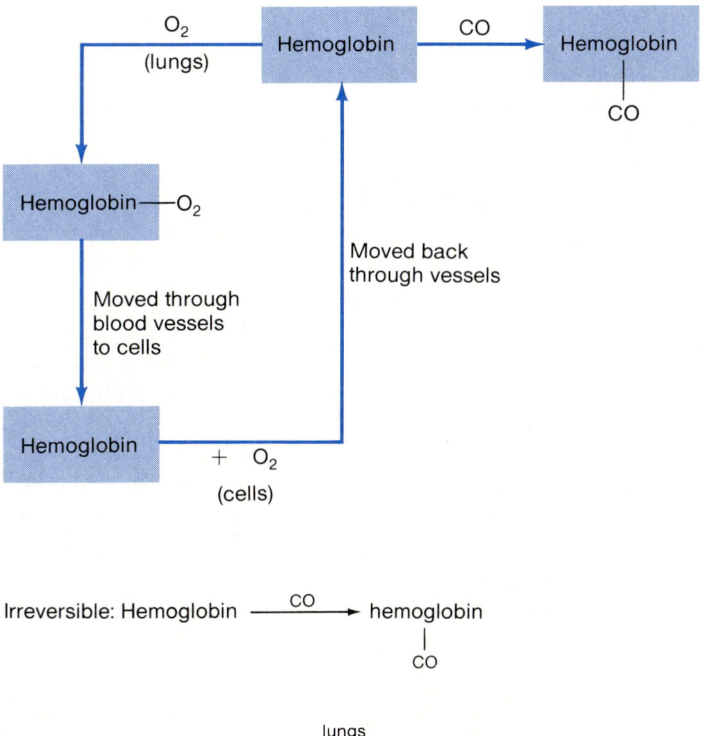

FIGURE 12-14
Action of CO in preventing O_2 transport in the human body.

The reaction of O_2 with hemoglobin is *reversible*—the O_2 can easily be released at the cell. When CO complexes with hemoglobin, the bond is a much tighter one which is not easily broken. This means that once a hemoglobin molecule combines with CO, the hemoglobin molecule is unavailable to be used to move O_2 to the cellular location where it is needed. When most of the body's hemoglobin molecules have been inactivated in this way, an internal suffocation occurs—the individual cells die and eventually the person dies. It is estimated that 9 ppm (parts per million) of CO in the air is safe. As the CO concentration rises above this level, the functioning of the brain is impaired. During the smoking of a cigarette, the level of CO in the lungs reaches as much as 100 ppm (11 times the safe level). Before encountering the catalytic converter, motor vehicle exhaust has several hundred parts per million of CO (as you may know, a number of people commit suicide every year by breathing motor vehicle exhaust fumes). The CO level builds up along city streets during rush-hour traffic, causes drowsiness among drivers, and no doubt contributes to accidents.

Second, since the temperature in the engine cylinder during combustion is several hundred degrees Celsius, the N_2 and O_2 combine directly to produce two *oxides of nitrogen,* NO_2 (nitrogen dioxide) and NO (nitric oxide). The mixture of these is commonly referred to as NO_x:

$$N_2 + O_2 \longrightarrow \underbrace{NO_2 + NO}_{NO_x}$$

The NO_x mixtures are unpleasant-smelling materials and contribute, along with the C_8H_{18}, to photochemical smog. They also, when dissolved in water (including that in lungs), produce nitric and nitrous acids and contribute to acid rain.

Third, gasoline usually contains a small amount of sulfur, and when burned in the presence of a normal air mixture, produces SO_2. This is not a serious problem, however, because of the small amount present.

A fourth problem encountered in the operation of the engine is that of *preignition*. When gases are compressed, they get hot; and during the compression stroke, if the engine is a high-compression one, the heat is high enough to cause the mixture to explode *before* the spark arrives. This means that the power stroke comes too soon and an unevenness is noticed in the engine's running. This is called *knocking*. To prevent this, and still keep the compression high enough to give the power to which we Americans have become accustomed, materials are added to prevent knocking. Such materials are called *antiknock compounds* and are either tetraethyl lead, $Pb(C_2H_5)_4$, in regular gasolines or benzene, C_6H_6, in nonleaded gasolines. The amount of antiknock compounds used determines how much compression the mixture can stand and still not knock (explode prematurely). Fuels are rated according to an arbitrary scale called the *octane number*. Heptane is the fuel which knocks the most and is given a 0 rating on the scale. Isooctane (2,2,4-trimethylpentane) is the compound that is the best fuel (as far as knocking is concerned) and is given a 100 on the scale. Thus a fuel with an octane rating of 90 is 90 percent as good as isooctane. This scale is shown in Table 12-3.

TABLE 12-3
Octane number scale

OCTANE RATING	MATERIAL
100	Isooctane
90	"High-test" gasoline
80	Regular gasoline
70	
60	
50	
40	
30	
20	
10	
0	Heptane

The presence of these antiknock compounds causes other problems, however. If a fuel containing lead is used, PbO_2 is produced during combustion:

$$Pb(C_2H_5)_4 + 14O_2 \longrightarrow PbO_2 + 8CO_2 + 10H_2O$$

This is the white deposit sometimes found on the inside of exhaust pipes. It is a toxic material, and shows up as high blood levels of Pb in inner-city children who are constantly exposed to motor vehicle fumes, which are more concentrated near the ground. PbO_2 is a solid and slowly settles out of the air. It is more prevalent near highways, and cattle grazing near a road give milk with a significant Pb content. When nonleaded gasoline, which contains benzene antiknock materials, is used, the benzene is not burned completely and some of it fuses into polynuclear aromatic compounds such as benzpyrene:

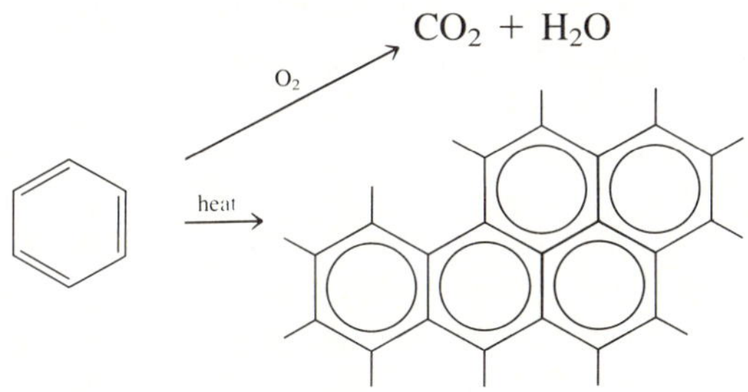

Benzene (C_6H_6) Benzpyrene ($C_{20}H_{12}$)

Benzpyrene is a highly carcinogenic material. If one is going to drive powerful high-compression cars that run smoothly, one has a choice between lead poisoning and increased risk of cancer.

Photochemical smog (sometimes called *Los Angeles smog*) has already been mentioned. This is a condition of hazy, pollution-laden air caused by sunlight shining on a stagnant air mass containing C_8H_{18} (along with other hydrocarbons from gasoline) and NO_x from motor vehicle exhaust (see Figure 12-15). The stagnation of air is the result of a temperature inversion, just as with industrial smog. The reactions that occur are approximately as follows:

$$NO_2 \xrightarrow{\text{light}} NO + O^{\bullet}$$
$$O_2 + O^{\bullet} \longrightarrow O_3$$
$$O^{\bullet} + C_8H_{18} \longrightarrow \text{complex organic materials (eye irritants)}$$

However, the situation is actually somewhat more complex than this. The sunlight causes an $O^{\bullet}$ to break off from an NO_2 molecule. This atom does not have its octet rule fulfilled; it is called a *free radical* (an atom or group of atoms that has one or more unpaired electrons) and is very reactive. It reacts with molecules of O_2 in the atmosphere to form O_3 (ozone), which causes rubber tires to break down and lung tissue to disintegrate, which aggravates or causes respiratory diseases. The $O^{\bullet}$ also reacts with the unburned gasoline to produce complex organic materials that burn one's eyes. These sources of pollution were summarized earlier, in Table 12-1.

FIGURE 12-15
Photochemical smog. (Photograph from EPA.)

Solution to automotive pollution There are several possible ways to reduce undesirable automotive emissions. One obvious one is to reduce the number of automobiles being driven. This would require car pooling and extensive use of public transportation. Americans now drive about 1.5×10^{12} (1.5 trillion) mi per year—about 7500 mi per person each year. While many cities have adequate public bus transportation, some do not; and in any case most Americans prefer the independence and convenience of driving. Still, the high price of gasoline and improved intercity mass transportation, along with public concern, have led to a reduction in total fuel consumption in recent years. Further efforts in this direction will not only reduce our dependence on foreign oil supplies but also aid in reducing automotive air pollution.

A return to low-compression engines is another possibility. This would eliminate the need for antiknock compounds, reduce the temperature of operation, and thus reduce the formation of NO_x. It would also reduce the fuel efficiency of the engine, however, and most American drivers would not want to drive cars that accelerate slowly. (For some people, fast cars are symbols of virility or sex appeal.) Perhaps, with time, this too will change.

Another possibility would be to use pure O_2 rather than air in the combustion process. This would allow better burning and more power, which would reduce the C_8H_{18} and CO emission and eliminate the NO_x, since no N_2 would be present to react with the O_2. The main disadvantages would be the expense of using tanks of compressed O_2 and the danger, in the event of a collision, of fire and explosion if the O_2 tank ruptured. If this approach were taken, the PbO_2 or polynuclear hydrocarbons would still be present, however, and the energy needed to produce the O_2 would cause pollution.

The best method currently available, and the one used on all automobiles since 1976, is a device called a *catalytic converter,* shown in Figure 12-16. It is composed of finely divided metal catalyst particles embedded in a ceramic material. The ceramic support has small channels through which the exhaust gases can pass. The actual composition of the catalyst is proprietary information (information that is protected by patent), but it is made from precious metals such as gold, silver, and platinum, which account for most of the cost of the converter.

FIGURE 12-16
Catalytic converter.

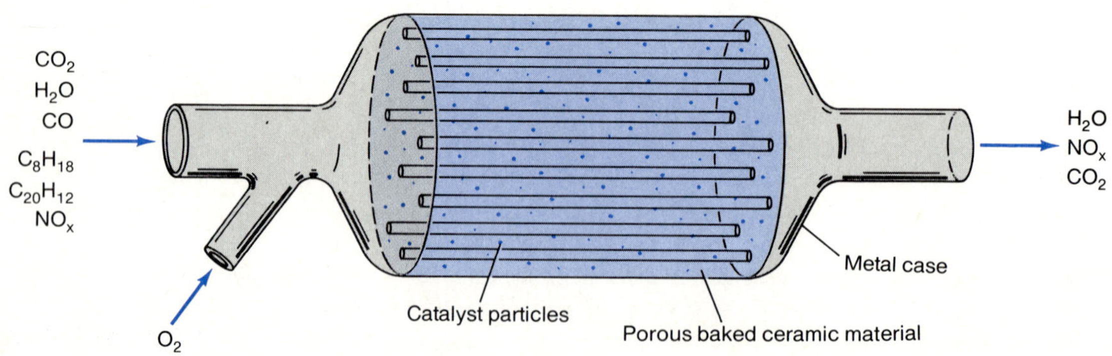

A *catalyst* is a material that causes a chemical reaction to go more rapidly without itself being consumed in the process. The exhaust gases containing CO, C_8H_{18}, benzpyrene ($C_{20}H_{12}$), and NO_x are first mixed with air. When this mixture travels through the converter, it comes into contact with the catalyst and the following reactions occur:

1 $$2CO + O_2 \xrightarrow{\text{catalyst}} 2CO_2$$
Carbon monoxide

2 $$2C_8H_{18} + 25O_2 \xrightarrow{\text{catalyst}} 16CO_2 + 18H_2O$$
Octane

3 $$C_{20}H_{12} + 23O_2 \xrightarrow{\text{catalyst}} 20CO_2 + 6H_2O$$
Benzpyrene

4 $$NO_x + O_2 \longrightarrow \text{no reaction}$$

One should not use lead-containing antiknock compounds in cars with catalytic converters, because the PbO_2 will form a coating on the catalyst particles and inactivate them. This is called *poisoning the catalyst.* If lead-containing antiknock compounds were not added to gasoline, PbO_2 could be eliminated from the list of pollutants. The carbon-containing pollutants all react with oxygen and are changed to CO_2 and H_2O, which are harmless. This one device can cause destruction of most of the CO, C_8H_{18}, and $C_{20}H_{12}$ (at least when it is relatively new), and its use prevents the use of Pb. Only the NO_x remains; it is toxic and contributes to acid rain, but with C_8H_{18} not present in the atmosphere, no photochemical smog can form. So far, no device has been discovered that will reduce NO_x. The catalytic converter is not a panacea because its efficiency is reduced with increased time of operation and it is virtually useless after about 50,000 mi, but it is a major advance toward clean air.

Another approach is to change the type of engine used in automobiles to diesel or turbine engines. This is being done on a small scale, but consumers are not willing, in most cases, to bear the added cost of manufacture and maintenance of such engines, even though their fuel efficiency is better. Manufacturers seem reticent about promoting these engines, although diesel automobiles are becoming more common.

Air-quality standards Table 12-4 and Table 12-5 give some pertinent information about air pollution: SO_2 levels and sources of pollution, respectively. Table 12-6 shows what are considered safe levels of common pollutants, along with the level of these pollutants in normal "clean" air and urban air before 1970. ("Clean" air is air away from any known source of pollution.)

TABLE 12-4
SO_2 pollution level as a function of population (pre-1970 levels)

POPULATION	SO_2, mg/m^3
Rural	10
<10,000	30
10,000–25,000	30
50,000–100,000	40
100,000–1,000,000	50
>1,000,000	100

TABLE 12-5

Relative amounts of pollution from various sources

SOURCE	PERCENT OF TOTAL (BY WEIGHT)	PERCENT OF DAMAGE (ESTIMATED)
Automobile	55–60	10
Electric power generation	20–25	50
Manufacturing	10–15	35
Domestic waste disposal	5	5

TABLE 12-6

Air quality standards for common pollutants

MATERIAL	"CLEAN" AIR*	MAXIMUM SAFE LEVEL, ppm	URBAN AIR†
NO_x	0.001 ppm	0.05	0.05–0.2 ppm
Hydrocarbons (C_8H_{18})	1.5 ppm	0.24	2–3 ppm
CO_2	318 ppm		
CO	0.1 ppm	9	0.004–17
O_3	0.02 ppm	0.08	0.03 ppm
PbO_2	0.001 mg/m^3		2 mg/m^3
SO_2	0.2 ppm		25–250 ppm

*Air away from any known source of pollution.
†Before 1970.

WATER POLLUTION

Sources and effects of water pollution

Water is the solvent for all living systems and is used in very large quantities by all advanced cultures. We tend to assume, because of our everyday use of water, that it is available, and will continuously be available, in adequate quantities. We use it in cooking and drinking, which are the uses requiring the greatest freedom from bacteria and toxic materials (pollutants). Relatively pure water is needed in even larger amounts for water-heating systems, bathing, flushing toilets, washing cars, watering lawns, and irrigating crops.

It had been assumed that rivers flowing into bays and oceans presented a convenient way to flush away our domestic wastewater (sewage and wash water) and industrial wastes. Even when it became obvious that this was not the case, there was an alarming lack of a sense of responsibility on the part of those wishing to get rid of waste materials. The city of Washington, D.C., used to (before 1973) dump all its untreated sewage into the Potomac River. The sewage was washed into Chesapeake Bay, where swimmers frequently encountered chunks of feces floating near them in the water. Not only does this cause unpleasant sights and smells; it is unsafe to swim in such water because of the bacteria which are present. In addition to health-related and aesthetic problems, toxic materials build up in lakes and accumulate in river beds primarily as a result of the dumping by industry of small concentrations of acids, toxic metals, and organic solvents into rivers flowing by their manufacturing facilities.

Rainwater running over fertilized farmland tends to dissolve the nutrients. When this water, called *farmland runoff,* gets into rivers and lakes, it cause algae and other water plants to grow in profusion. The materials used as fertilizers include $NaNO_3$ (sodium nitrate, or nitrate of soda), Na_3PO_4 (sodium phosphate), and K_2CO_3 (potassium carbonate, or potash), and these are the offending nutrients in lakes. Domestic sewage includes human wastes, which upon decaying also cause plants to grow; and $Na_5P_3O_{10}$ from laundry detergents, which also serves as a nutrient. When a lake becomes too rich in nutrients, and when the plants grow more rapidly than fish can eat them, the water becomes choked with vegetation—the water volume becomes mostly occupied by plant growth. The top layers of plants block the sunlight from reaching the lower layers, and the lower layers die. These dead plants rot and use the oxygen dissolved in the water. Fish cannot live in water that does not contain dissolved O_2, and the rotting plants make the lake very unpleasant and useless as a water supply, sporting area, or scenic area. This process is called *eutrophication*. In addition, the level of nitrate (NO_3^-) in drinking water taken from polluted lakes and rivers sometimes reaches 80 to 100 ppm, which is near the toxic level. This condition occurs in the spring after a period of rain, following fall and spring application of fertilizer. Table 12-7 summarizes types of water pollution, and Figure 12-17 illustrates sources of water pollution.

The purity of the water used for irrigation or flushing toilets—both uses require large volumes of water—does not have to be high, and most of the pollutants already mentioned would cause no damage. Water for drinking and food processing, on the other hand, must have a very low level of bacteria, toxic metal ions, and organic solvents, since these materials are poisonous to humans even at fairly low levels.

TABLE 12-7

Types of water pollutants and the damage caused

SOURCE	POLLUTANT	DAMAGE
Industry	Acids (H_2SO_4, HCl, HNO_3), bases (NaOH)	Change in pH of water
	Metal ions	Toxicity
	Organic solvents (CCl_4, $CHCl_3$, C_6H_6)	Toxicity
Farmland runoff	$NaNO_3$, Na_3PO_4	Eutrophication, toxicity
	Pesticides (chlorinated hydrocarbons, DDT, PCB)	Toxicity, interference with reproduction of birds of prey
Sewage	Solid waste, $Na_5P_3O_{10}$	Eutrophication
	Detergents	Suds on water
	Bacteria	Disease

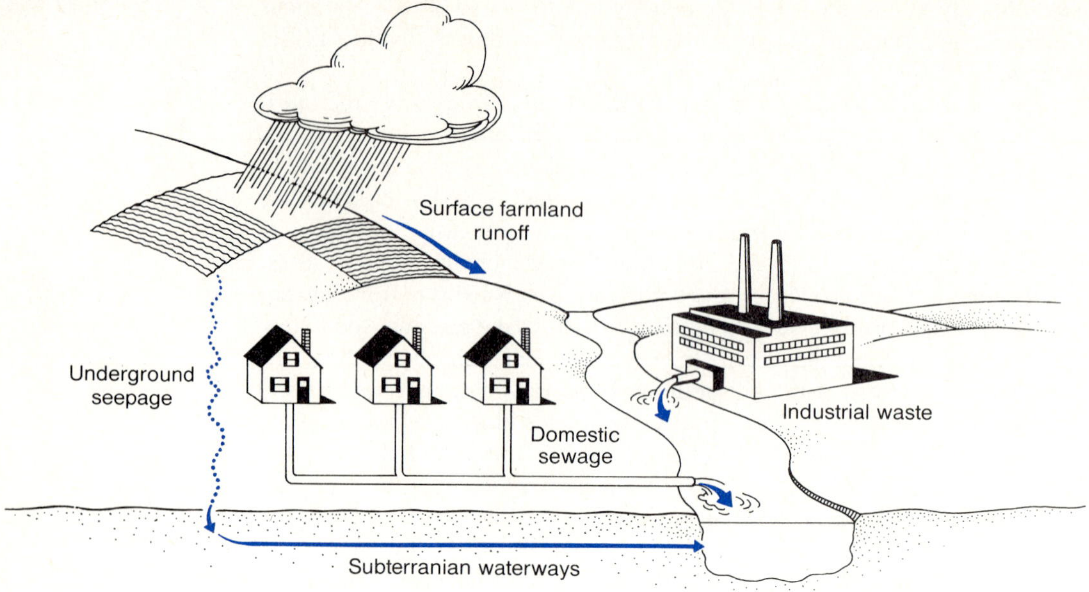

FIGURE 12-17
Sources of water pollution.

The food chain

Some of these materials just mentioned pose a special problem for higher animals and humans. Several metal ions are used for many industrial purposes. One that is particularly troublesome is mercury, which will be used in this explanation of the food chain. However, many other metals and some organic solvents and insecticides enter the food chain by the same process.

The mercury ions (or metal) that are discarded settle in the mud at the bottom of a river. Bacteria convert this mercury into dimethyl mercury:

$$Hg \xrightarrow{\text{bacteria}} Hg(CH_3)_2$$

The water itself develops a certain concentration of $Hg(CH_3)_2$. In the early twentieth century, ocean water had about 0.03 ppb (parts per billion) of Hg. Now the level is close to 0.1 ppb (a threefold increase), and many lakes have as high as 0.4 ppb Hg. The bacteria have a higher concentration than the water, however. Small water animals eat these bacteria, and the small animals are eaten by small fish, which are then eaten by larger carnivorous fish. These larger fish are eaten by humans or birds of prey. This is called the *food chain,* and this process tends to concentrate the undesirable material as the material goes up through the food chain. This concentrating of pollutants along the food chain is shown in Figure 12-18, along with the approximate concentration of dimethyl mercury in the bodies of organisms along the food chain.

Consider a small water animal, such as a crayfish, which may have 25 ppb of Hg in its body. A small bass, which weighs 0.5 lb, may eat 1 lb of crayfish during a year. Mercury accumulates in the body—it is not elimi-

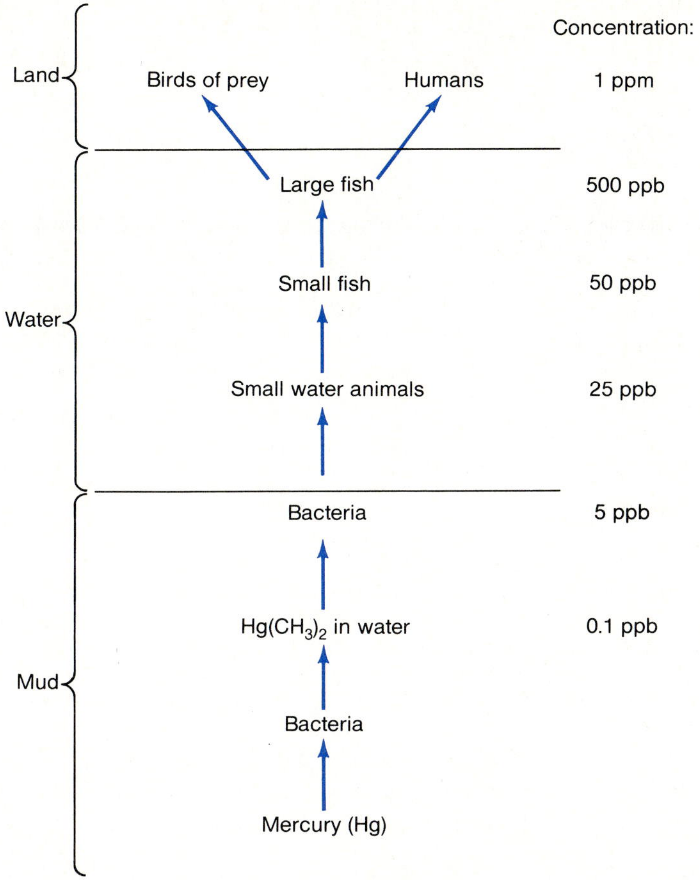

FIGURE 12-18
Concentrating of pollutants in the food chain.

nated—which means that all the Hg in the 1 lb of crayfish is now in the 0.5-lb bass. This means that the concentration of Hg in the bass's body is twice as high as it was in the crayfish (50 ppb compared with 25 ppb). A large fish, such as a 5-lb bass, eats about 50 lb of small fish during a year. Again, all the Hg in the 50 lb of small fish will be in the 5-lb body of the large fish; thus the Hg concentration goes up by a factor of 10, to 500 ppb (or 0.5 ppm). Thus the higher on the food chain a given organism is, the more concentrated a nonexcreted pollutant becomes. It is also true that the older and larger a person or fish becomes, the higher the concentration of the pollutant in the body becomes if the diet remains the same. Table 12-8 shows the amount of Hg in ocean fish at various levels in the food chain. Notice that swordfish, which consume tuna, are at the top of the chain and have the highest Hg levels, while flounder, which are bottom feeders, have the lowest Hg levels. Also notice that the larger tuna have more Hg for a given weight of flesh than the small ones do. This is because it takes more food to add a given weight to a larger fish.

TABLE 12-8

Methyl mercury concentration in ocean fish at various locations in the food chain

FISH	POSITION IN FOOD CHAIN	CONCENTRATION OF Hg,* ppm
Swordfish	High	1.0
Tuna:	Intermediate	
100-lb		0.5
50-lb		0.25
25-lb		0.10
Flounder, cod	Low	0.05

*These are experimental values. The legal limit of Hg in fish sold for human consumption is 1 ppm, according to federal regulations.

For example:

1 A 10-lb tuna eats 30 lb of cod (0.05 ppm Hg) and gains 15 lb of body weight. The Hg in the 30 lb of cod is now in the 25 lb of tuna. The tuna now weighs 25 lb, and its new flesh has a concentration of Hg of about 0.06 ppm:*

$$\frac{30 \text{ lb of cod} \times 0.05 \text{ ppm}}{25 \text{ lb of tuna}} = 0.06 \text{ ppm}$$

2 A 25-lb tuna eats 125 lb of cod (0.05 ppm Hg) and gains 25 lb of body weight. The Hg in the 125 lb of cod is now in 50 lb of tuna. The tuna now weighs 50 lb and has a level of Hg of approximately 0.125 ppm:

$$\frac{125 \text{ lb of cod} \times 0.05 \text{ ppm}}{50 \text{ lb of tuna}} = 0.125 \text{ ppm}$$

3 A 50-lb tuna eats 500 lb of cod (0.05 ppm Hg) and gains 50 lb of body weight. The Hg in the 500 lb of cod is now in 100 lb of tuna. The tuna weighs 100 lb and has a level of Hg of around 0.25 ppm:

$$\frac{500 \text{ lb of cod} \times 0.05 \text{ ppm}}{100 \text{ lb of tuna}} = 0.25 \text{ ppm}$$

This is a problem at the human level only for those who eat large amounts of fish, such as those on extended low-calorie diets that include tuna or swordfish. Such people should eat tuna of the smaller variety to avoid becoming victims of increased Hg levels. Those who eat several tuna sandwiches or swordfish steaks every day can develop symptoms of Hg poisoning, which include lethargy and the inability to think clearly. Up to one tuna sandwich each day should cause little if any problem for most people. Table 12-9 shows the effect of eating various amounts of tuna per day. Notice that most of the Hg shows up in the hair.

*In each of these three examples, the result is only approximate, since the Hg in the original 10-lb tuna is also still present but not included in the result.

TABLE 12-9

Mercury levels in humans who eat tuna

SANDWICHES PER DAY	Hg, μg/DAY	BLOOD LEVEL, ppm	HAIR LEVEL, ppm
0		0.001	3.1
<1	9–16	0.006	5.3
2	17–26	0.007	4.9
3	27–38	0.012	9.4
>4	40–70	0.018	14.1

Another problem related to the food chain is the presence of chlorinated hydrocarbons, which are used in agriculture as pesticides. Dichlorodiphenyltrichloroethane (DDT), shown below, was commonly used in the past:

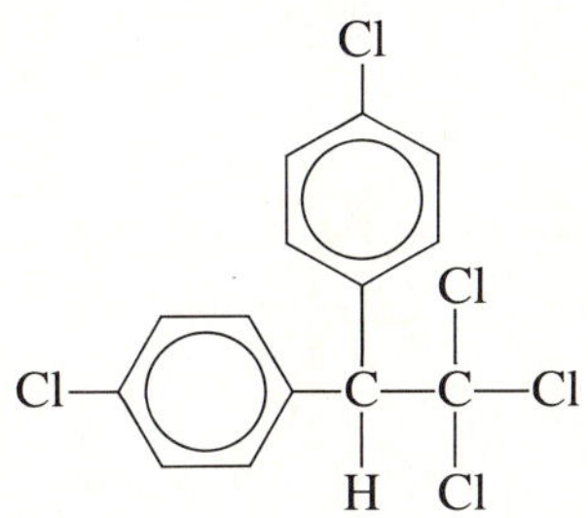

DDT

It dissolves in water only to a small extent, but it is present in trace amounts in farmland runoff. The food chain concentrates this material, however, and birds of prey such as eagles are at the top of the chain. If DDT is present, the eagle eggs break before hatching can occur, which endangers the future existence of the species.

Solution to water pollution

Now that we have seen some of the toxic materials that appear in our waters and the harm they do, let's consider what can be done and, in some cases, is now being done. Very little can be done to improve the quality of the already contaminated ocean water because of the tremendous volume involved. The quality of small lakes and rivers can be changed over a relatively few years, however, by the following practices:

1 Controlling domestic waste.

(a) Building sewage treatment plants to remove solids and kill bacteria.

(b) Using low-phosphate laundry detergent. Most manufacturers are now using less phosphate under pressure from state laws. An acceptable level is 0.5 percent phosphorous.

2 Reducing farmland runoff.

(a) Practicing contour plowing and using cover crops to hold back rainwater and prevent large volumes of runoff.

(b) Applying of fertilizers, including organic ones such as manure, to cultivated ground and working them in (plowing them under, etc.) to prevent them from being washed away rapidly by surface water. Such nutrient materials should never be applied on top of snow, because when the snow melts most of them will be washed into the waterways. This is not only a poor practice ecologically but also a poor practice economically, since the fertilizer is wasted.

(c) Developing and using pesticides that are not harmful. With the aid of research funds from the agricultural chemistry industry and the federal government, pesticides can be developed that are degraded by O_2 and H_2O over a several-month period after the pest has been destroyed. These pesticides would thus never be given a chance to build up in the environment, and the food chain problems would be virtually eliminated.

3 Improving industrial practices.

(a) Cleaning industrial wastewater of contaminating pollutants using available methodology. Most industrial pollution is a result of a constant releasing of plant wastewater which contains small amounts of acids, bases, metal ions, or organic solvents. If technically possible, these wastes should be cleaned of their contaminants. For some contaminants chemical methods are available, and for others (such as organic solvents) the wastewater can be placed in holding ponds where special bacteria can be used to degrade the organic solvents into CO_2, H_2O, and other environmentally safe materials.

(b) Developing alternative techniques for removing pollutants from industrial wastewater. If metal ions cannot be removed by available methods, other industrial processes should be developed which do not use these toxic compounds, or the contaminated water should be stored or recycled. *Under no circumstances* should hazardous wastes be released into waterways as a means of easy, cheap disposal.

LAND POLLUTION

Because the solid earth is much less mobile than water and air, less has been said in recent years about its pollution. Usually solid domestic garbage is buried in landfill areas near towns and cities. This garbage contains food, paper, wood, and plastic, which decompose to produce nontoxic products that are then absorbed by the earth with no ill effects.

Some manufacturing processes, however, produce solid or liquid materials that are too toxic to put in waterways, and sometimes these have been buried in steel drums in remote (or not-so-remote) areas. Now, it does not take a lot of intelligence or training to know that steel will rust if placed underground, since the ground is always wet (or at least damp) for several feet down. It is inevitable that in several years (if not sooner) the material in the drums will eventually leak from the disintegrated containers and find its way into wells and streams. Simply burying waste does not solve the problem but only complicates and postpones it.

In western New York State, a group of children suffered serious damage to their health when organic fumes escaped from containers which had been buried years before in the ground on which their homes were later built. This location is known as Love Canal. In western Tennessee, a chemical company buried millions of gallons of chlorinated hydrocarbons underground before being forced to stop. These materials caused sickness in local residents because the materials were leaking from the drums into water supplies. The Environmental Protection Agency (EPA), whose task it is to enforce environmental quality laws, has engaged the Army Corps of Engineers to design safe disposal sites for solid and liquid chemical waste. A suggested design is shown in Figure 12-19. The landfill area should not be used for other purposes during the foreseeable future. Incineration (contained burning) is an alternative, especially for very hazardous organic materials like dioxin. (See the discussion in Chapter 3.)

FIGURE 12-19
Design of disposal site for solid and liquid chemical waste.

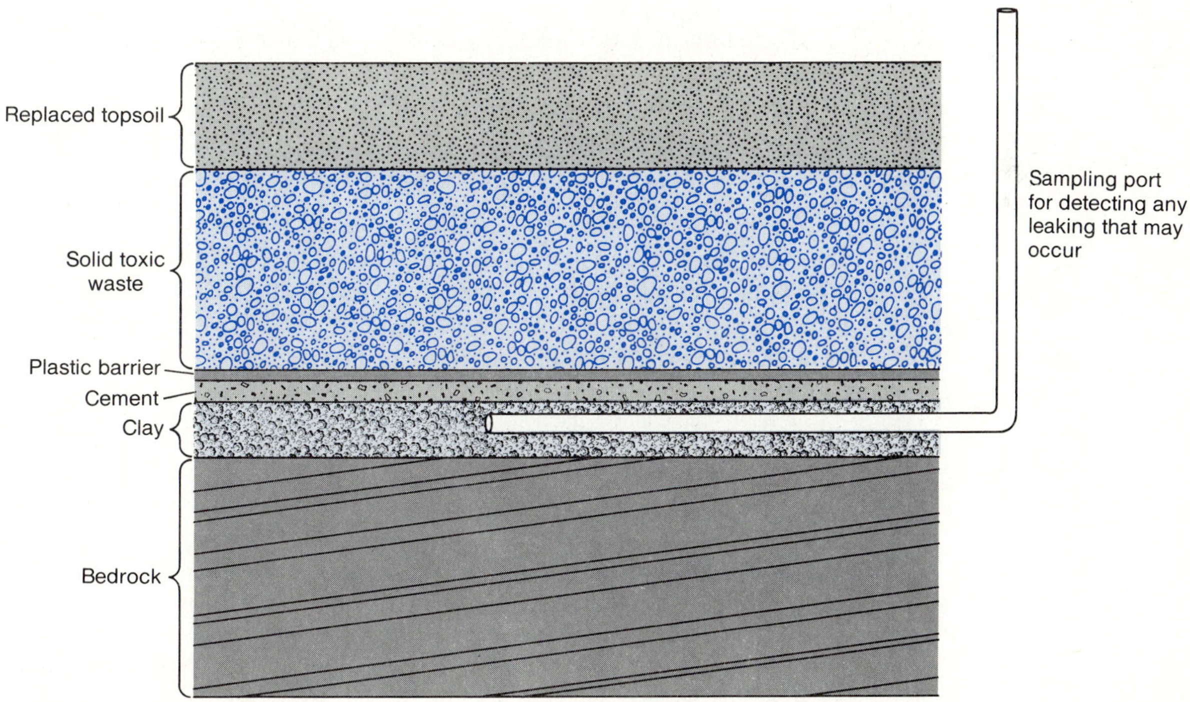

FIGURE 12-20
Industrial facilities like this one make both products we use and need and some undesired pollution.

THE POLLUTION GAME

Following is an interesting activity that can be used in a classroom setting or in a small group:

1 Two teams are selected of two to four people each. The teams are to compete with each other.

2 Each team is given a kind of activity such as golfing, swimming, an evening at the opera, a quiet drive in the country, or a cross-country motorcycle trip.

3 During round 1 of the game, each team is given 2 min to prepare a list on the board or on paper of the reasons why its own activity is *not* polluting.

4 During round 2, each team is given 2 min to list the ways that the opponent team's activity *is* polluting. This often brings out creativity and a sense of competition among the team members. The winning team is the one that performs its two tasks with the greatest skill (as determined by the instructor or a vote by class members).

This game can be used to compare the effects on the environment of different types of recreational, occupational, or (as in Figure 12-20) industrial activities.

STUDY GUIDE

Terms to learn

Acid rain	Catalytic converter
Air inversion	Electrostatic precipitator

Eutrophication	Photochemical smog
Food chain	Pollution
Greenhouse effect	Preignition
Octane number	Runoff
Particulate matter	Stack absorber

Skill practice problems

1 Why do small children in large cities have more Pb poisoning than the adults in the same area?

2 Give three reasons why human-made pollution, while *smaller* than natural pollution, is *more harmful*.

3 List the harmful effects of SO_2 and SO_3 in the atmosphere.

4 If a river has a concentration of H^+ of 10^{-5} mol/L, what is the pH of that water?

5 Evaluate the following statements.
(a) Industry is responsible for most of the pollution which we experience.
(b) Small fish have a higher concentration of Hg than large ones.
(c) Sulfur has a higher oxidation state in SO_3 than in SO_2.
(d) Tetraethyl lead is used as a gasoline additive to provide lubrication in the motor.
(e) Industrial smog contains C_8H_{18} and PbO_2.
(f) The currently accepted maximum safe level of CO is 9 ppm.

6 An intake of 30 μg per day is considered the maximum safe dose of Hg for a 150-lb person. How much (weight in ounces) tuna containing 0.50 ppm Hg can be safely eaten each day by such a person?

7 What pollution is caused by playing golf?

8 If an automobile uses one-fifth as much fuel per mile as a city bus and the bus carries 50 people, what percent savings in fuel usage and pollution would be realized if everyone rode a bus instead of driving two people per automobile? Assume that the automobile also produces one-fifth as much pollution per mile.

9 If a car gets 20 mi per gallon of gasoline and each gallon has 4 g of $Pb(C_2H_5)_4$, how many grams of PbO_2 are emitted into the atmosphere for each mile the car travels? The following reaction occurs during combustion. (Hint: See Appendix E.)

$$Pb(C_2H_5)_4 + 14O_2 \longrightarrow 8CO_2 + 10H_2O + PbO_2$$

Questions for discussion

1 Is it technically possible for most companies to operate their facilities in a nonpolluting manner? Is it economically feasible? Assuming that the company exists to make a profit, where would the cost of pollution control come from? If it means a higher-priced product, would *you* be willing to pay for a cleaner environment?

2 Do you think the EPA should strictly enforce all regulations, or should it "cooperate" with industry to prevent undue economic hardship on certain companies or whole segments of our society?

3 The author recently heard a statement by a government official to the effect that, "We need more scientific studies on the acid rain problem before forcing coal-burning industries to eliminate most of the sulfur dioxide emissions." Do you think more scientific study is needed, or is it simply a matter of making some difficult and possibly unpopular decisions and then enforcing them?

Goals

To understand the nature of nuclear processes

To understand the process by which the ages of ancient artifacts are determined

To understand the advantages and disadvantages of nuclear energy

To gain an understanding of the danger of large amounts of radiation and of possible precautions in case of nuclear incidents, and to be able to intelligently assess the risks of nuclear weapons

13

NUCLEAR CHEMISTRY

All the chemical changes encountered so far in this text have involved a transfer or sharing of electrons or the movement of entire atoms or ions from one bonding relationship to another. These can be illustrated by the following reactions; in these ordinary chemical reactions, no changes occur in the nuclei of the atoms, and the *symbols* stay the same on both sides of the equation:

$$4Na + O_2 \longrightarrow 2Na_2O$$
$$C_3H_8 + 5O_2 \longrightarrow 3CO_2 + 4H_2O$$
$$AgNO_3 + NaCl \longrightarrow AgCl + NaNO_3$$

NUCLEAR REACTIONS

There is a class of reactions in which the nuclei do change, however. These are much less common than ordinary chemical reactions, but they are the subject of considerable interest for both theoretical and practical reasons. They are called *nuclear reactions*. During nuclear reactions, the atomic nuclei change and new elements are formed. They are the only kind of reaction studied by chemists where elements themselves change, and the only exceptions to Dalton's original statement that atoms are indestructible.

A brief review of atomic structure is in order here, since we will be expressing changes in the composition of atoms. The nucleus contains protons and neutrons, which compose most of the mass of the atom (the electrons have negligible mass). Isotopic notation is used to represent nuclear changes. For example:

Isotopic notation $^{20}_{10}Ne$

- Mass number (total atomic mass to nearest whole number; total number of protons plus neutrons)
- Atomic symbol
- Atomic number (number of protons or positive charges in the nucleus)

The three subatomic particles can be expressed using this same notation:

Neutron	$^{1}_{0}n$	1 amu No charge
Proton	$^{1}_{1}p$	1 amu Charge of +1
Electron	$^{0}_{-1}e$	Negligible mass Charge of −1

These concepts will be of use as we examine the four basic types of nuclear reactions: (1) natural radioactivity, (2) transmutation, (3) nuclear fission, and (4) nuclear fusion.

NATURAL RADIOACTIVITY

In 1896, a French scientist, Henri Becquerel, made an interesting discovery. He placed a piece of uranium ore in the same drawer with a container of unexposed photographic film. When the film was removed for use, it was found to have already been exposed. A good deal of thought and experimentation revealed that invisible rays from the uranium ore had penetrated the container wall and exposed the film. This was a new idea at the time, but shortly after that Pierre and Marie Curie (Figure 13-1) isolated the first pure element to exhibit this same ability to penetrate matter and expose film. They patiently worked through 5 tons (4,540,000 g) of an ore called *pitchblende* (Figure 13-2) to isolate about 1 g of the element radium (Ra). The materials giving off these penetrating rays were called *radioactive* materials, and the rays themselves *radiation*.

An example of a nuclear chemical process which produces radioactivity is shown below:

$$\underset{\text{Radium-226}}{^{226}_{88}\text{Ra}} \longrightarrow \underset{\text{Radon-222}}{^{222}_{86}\text{Rn}} + \underset{\text{Alpha particles}}{^{4}_{2}\text{He}} + \underset{\text{Gamma rays}}{\gamma}$$

The atoms of Ra-226 have nuclei which are inherently unstable. They change spontaneously into Rn-222 atoms at a rate characteristic of this material; in so doing, they give off (or *emit*) alpha particles and gamma rays. The Ra-226 atom ''spits'' out an alpha particle, which consists of two neutrons and two protons, and gives off energy (gamma rays). It is the alpha particles and gamma rays that constitute the radioactivity (penetrating rays) of the radium. This is shown in Figure 13-3.

FIGURE 13-1
Pierre and Marie Curie—pioneers in nuclear chemistry. (From drawings by the author.)

FIGURE 13-2
Pitchblende—the ore from which radium was first isolated.

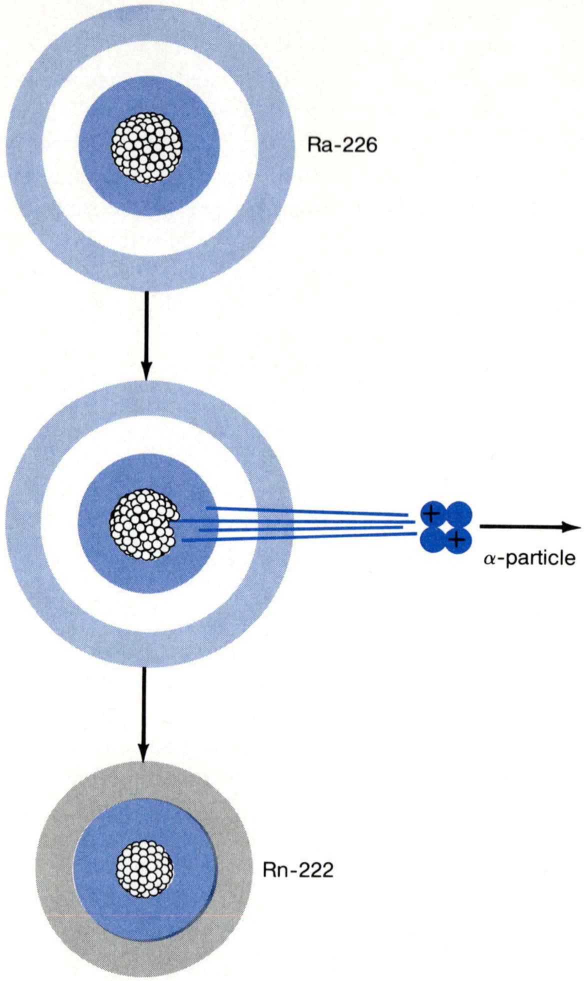

FIGURE 13-3
Emission of alpha particles by an atom of Ra-226.

Another isotope of radium, Ra-228, undergoes another type of nuclear change:

$$\underset{\text{Radium-228}}{^{228}_{88}\text{Ra}} \longrightarrow \underset{\text{Actinium-228}}{^{228}_{89}\text{Ac}} + \underset{\text{Beta rays}}{^{0}_{-1}\text{e}}$$

Here, the radium atoms change one of their neutrons into a proton by giving up an electron. A neutron behaves as though it were composed of a proton and electron combined. Thus when it gives up its electron—emits it at high energy—the neutron is changed to a proton, and that causes the atomic number of the atom to increase by 1. These high-energy electrons are called *beta* (β) *rays*. This natural changing of the nucleus is sometimes referred to as *nuclear decay*.

TABLE 13-1

Natural radioactivity

NAME	SYMBOL	NATURE	CHARACTERISTICS
Alpha	α	$^{4}_{2}He$	Most dangerous if ingested
Beta	β	$^{0}_{-1}e$	Least dangerous
Gamma	γ	Short-wavelength electromagnetic radiation	Most dangerous from an external source

Table 13-1 shows the three kinds of natural radioactivity and their characteristics.

All three types of radiation can penetrate matter and cause chemical bonds to break when the energy they possess is transferred to molecules. Nuclear radiation is, therefore, dangerous to living systems. Beta rays are the least dangerous because they are stopped by the skin and the particles (electrons) travel such small distances that little damage is done. Alpha particles are large, can release their energy at the point of impact, and can do considerable damage. They are also usually stopped by the skin; however, if they are ingested, they can cause internal damage. Gamma rays are the most penetrating, and if they have a high energy such as those given off by cobalt-60, they are lethal to some kinds of living cells.

Radiation can cause biological damage in three basic ways. If a large enough dose is taken in, so many biomolecules are destroyed that the cells cannot function and the organism dies. If the biomolecules of genetic material are damaged but the cell continues to live, that cell has a good chance of becoming cancerous (perhaps the cancer will not actually develop for several years). It is supposed that if the damaged genetic material is in the reproductive system (male or female), the offspring stand an increased chance of being deformed. This latter effect has not been proved however. It is known that high doses of radiation during pregnancy greatly increase the incidence of deformity in humans. This is known from observations of the effects of radiation resulting from the atomic bombs dropped on Japan at the end of World War II.

Measuring radioactivity

Photographic film is exposed by radiation and can thus be used to detect radioactive materials in certain situations. For example, badges containing unexposed film can be worn by people exposed to possible sources of radiation (see Figure 13-4). These badges are developed periodically to check the extent of exposure and thus the level of radiation to which the worker was subjected. Any condition causing excessive radiation exposure can then be corrected, but damage due to previous exposure cannot be reversed. This method is cumbersome, slow, and only semiquantitative. It is not the method used routinely for experiments.

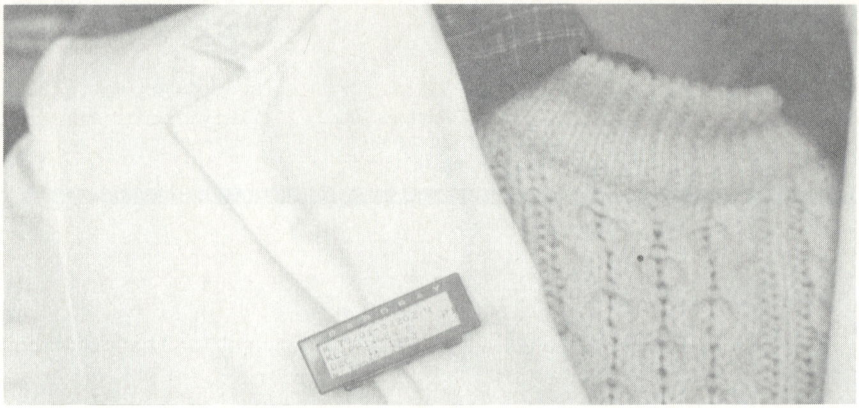

FIGURE 13-4
Film badge worn by a chemist who is working with radioactive material. This badge allows determination of the radiation dose to which the wearer has been exposed.

The Geiger-Müller tube, shown in Figures 13-5 and 13-6*a*, is the most commonly used radiation detecting device. It is a cylindrical metal device into which a wire is placed at one end. A thin window covers the opening at the other end. The space in the tube is filled with an inert gas to which a small amount of a more reactive material such as oxygen (O_2) or chlorine (Cl_2) has been added. Radiation causes the inert gas atoms to become ionized. During operation a high voltage—1000 to 2000 volts (V)—is placed across the outside of the tube and the wire in the center. When a radioactive material is placed near the thin window, the alpha, beta, or gamma rays penetrate into the tube chamber and cause ions to form in their path. These migrate and cause other ions to form, which allows a current to travel between the wire and side of the tube. Each particle entering the tube thus creates a current impulse, which can be transformed electronically to an audio signal (bleeps) or be read on a dial or digital counter, or both.

Radiation intensity is usually measured in counts per minute (cpm), which is proportional to the number of rays that enter the Geiger-Müller tube chamber. There is a small amount of radioactive material in most common substances, and the radiation from these sources along with cosmic rays (from space) constantly gives a low level of radiation. This is called *background radiation*. It is in the range of 30 to 50 cpm for each square inch of surface. When the radiation from a radioactive material is to be measured, the background radiation in that laboratory must be measured with the counter to be used in the experiment and this background amount subtracted from the amount obtained from the material. For example, suppose a sample of radium is to have its radioactivity determined. Before the measurement is made, the Geiger counter is run for 10 min and registers 350 counts. This means that there are 350 counts per 10 min, or 35 cpm. This is the background. Next the radium sample is placed near the thin window of the Geiger tube, and during 1 min, 456 counts are registered. Remember that 35 of these are due to background radiation. The radiation from the sample is $456 - 35 = 421$ cpm.

This points out the necessity of doing control experiments in connection with the collection of data and also shows that radiation is part of our normal environment. We are exposed to a low level of it constantly.

FIGURE 13-5

Left: Geiger-Müller tube, used to measure nuclear radiation.

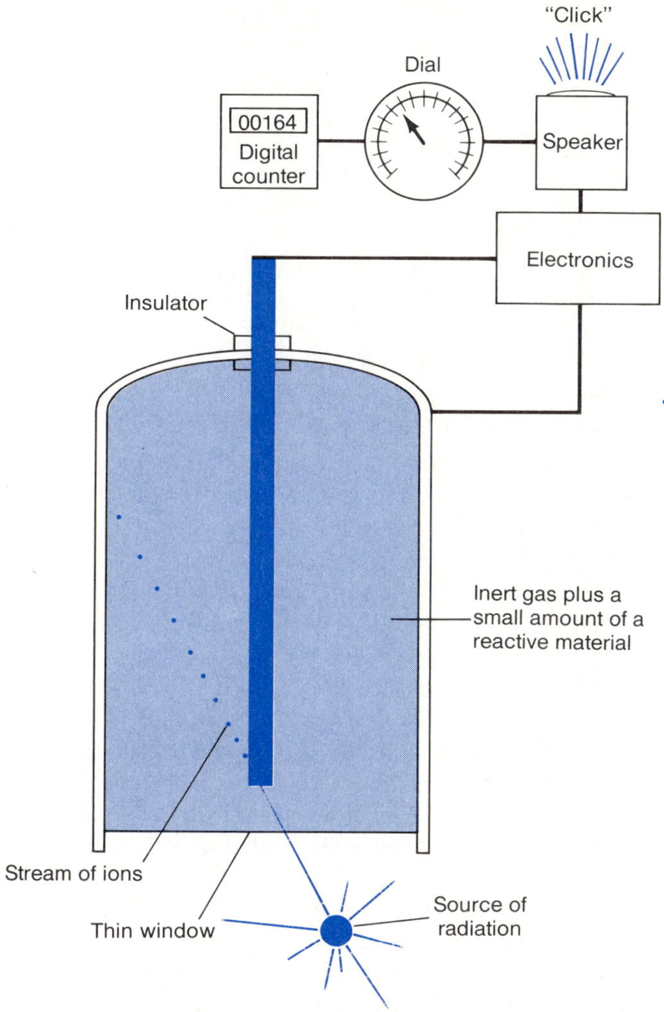

FIGURE 13-6

Below: Geiger counters. (a) Basic model. The Geiger tube is at the left of the console. (b) More sensitive research model.

(a)

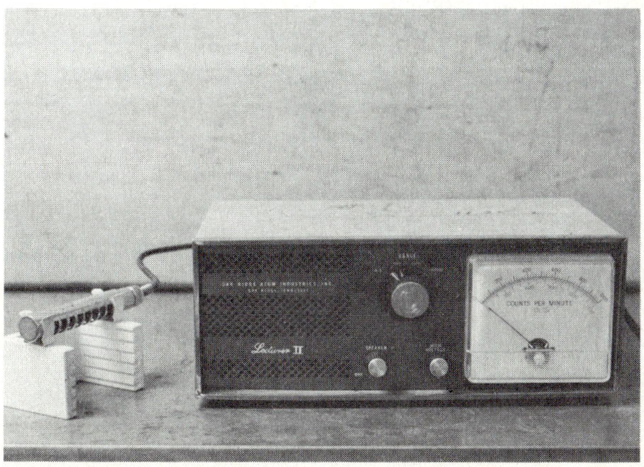

(b)

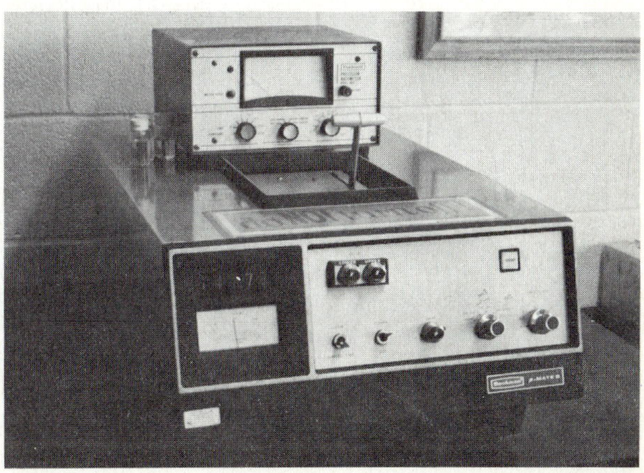

TABLE 13-2

Half-lives of some radioactive elements

ELEMENT	SYMBOL	HALF-LIFE
Polonium-204	Po-204	3.8 h
Polonium-211	Po-211	0.5 s
Radium-226	Ra-226	1620 years
Radium-228	Ra-228	6.7 years
Carbon-14	C-14	5700 years
Radon-222	Rn-222	3.82 days
Uranium-238	U-238	4.5 billion years

Half-life and radioisotope dating

It has been found that different radioactive materials decay and give off radiation at different rates. For example, it takes 1590 years for half of Ra-226 to change to Rn-222. The length of time needed for half the material to change is called its *half-life* (symbolized by $t_{1/2}$). The half-lives of several materials are shown in Table 13-2 and illustrate the wide range one finds in nature. The materials with the longer half-lives are more stable and therefore change at a slower rate.

The concept of half-life can be used to determine the age of some materials. One of the most common methods uses carbon-14 to date carbon-containing artifacts from ancient cultures. Figure 13-7 diagrams the route of carbon-14 in the earth's system.

Neutrons produced by cosmic rays from the sun and other extraterrestrial sources constantly bombard the earth's upper atmosphere, which is largely (80 percent) nitrogen (N_2). A small percentage of the nitrogen atoms ($^{14}_{7}N$) are changed to carbon-14 (^{14}C) by these neutrons. This isotope of carbon is radioactive and has a half-life of 5700 years. It reacts with the oxygen of the atmosphere to form carbon dioxide that contains carbon-14 atoms ($^{14}CO_2$). This radioactive carbon dioxide mixes with ordinary CO_2 and is incorporated into the tissues of plants and animals. Thus all *living* material has a very small amount (1 atom out of 10^{12}) of its carbon as carbon-14.

As long as a plant or animal is living, it is constantly taking in and giving off carbon-containing material. Plants consume CO_2 from the atmosphere and give off fruit, leaves, etc., while animals take in food and breathe out carbon dioxide. Thus living material is in equilibrium with the carbon cycle and contains the same ratio of $^{14}_{6}C$ as the atmosphere in general. Carbon-14 is slowly decaying (half of it in 5700 years) and constantly being re-formed in the upper atmosphere, so that a fairly constant amount of it is found in the biosphere. (See Figure 13-8.) Once the plant or animal dies, however, the carbon-14 ceases to be replenished and its level slowly decreases. Its level in a dead plant or animal is cut in half every 5700 years.

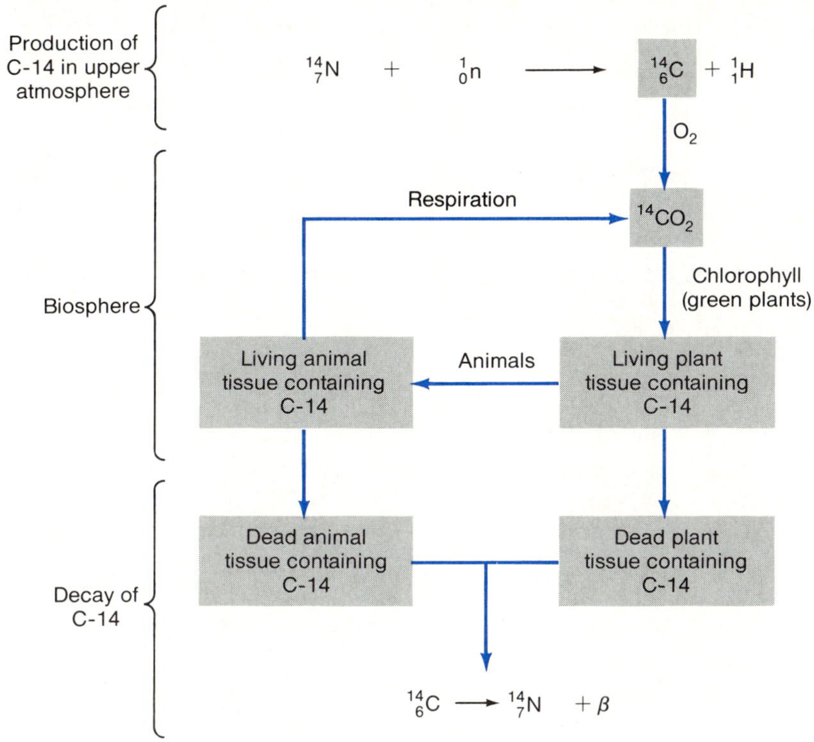

FIGURE 13-7
Formation and decay of carbon-14 in the environment.

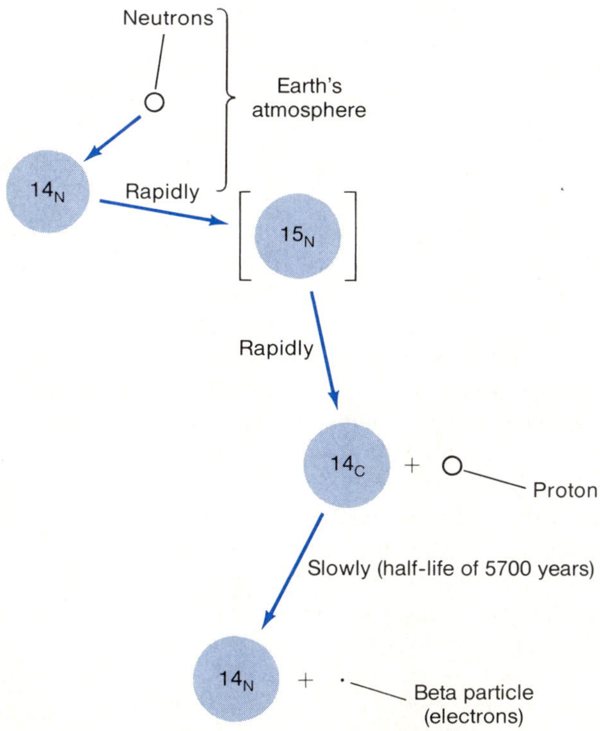

FIGURE 13-8
Atoms of carbon-14 being formed and decaying.

Total time since death			
	Living organism	1.0×10^{-10} % ^{14}C	(14 cpm for each gram of carbon)
	↓ Death		
0	Newly dead tissue (bone, wood, etc.)	1.0×10^{-10} % ^{14}C	(14 cpm/g)
	↓ 5700 years		
5700 years	Carbonaceous artifact	0.50×10^{-10} % ^{14}C	(7 cpm/g)
	↓ 5700 years		
11,400 years	Carbonaceous artifact	0.25×10^{-10} % ^{14}C	(3.5 cpm/g)
	↓ 5700 years		
17,100 years	Carbonaceous artifact	0.125×10^{-10} % ^{14}C	(1.7 cpm/g)
	↓ 5700 years		
22,800 years	Carbonaceous artifact	0.062×10^{-10} % ^{14}C	(0.8 cpm/g)

FIGURE 13-9
Variation in the carbon-14 content of an object as time passes.

By measuring the amount of carbon-14 in an object, one can tell how long it has been since the material died. Figure 13-9 shows how age is related to amount of carbon-14. Once the organism dies, much of it (such as fruit, leaves, and flesh) decays; but some parts (such as bone and hair in animals and wood in plants) can, under proper conditions, survive for a long time. Each *successive* 5700-year period changes the amount of carbon-14 to half of its previous value. Figure 13-10 shows a plot of these numbers. With this plot in hand, a scientist need only experimentally measure the percent of carbon-14 in a sample whose age is to be determined. This percent is matched up with an age on the plot.

Consider a piece of charcoal found in a campfire site during the excavation of an ancient American Indian village. It was found to contain 0.30×10^{-10} percent of carbon-14 (i.e., it gave about 5.1 cpm per gram of carbon). From the plot in Figure 13-10, this corresponds to an age of about 8000 years. This tells the archeologist approximately when that civilization existed. Several assumptions are made here, of course. One is that the wood used in the fire was burned within a few years of the time it died (remember that we are really dating from the time the tree died). We have to also assume that the rate of radioactive decay has been the same in the years since the tree lived and that the biosystem had the same percent of carbon-14 while the tree was living (8000 years ago) as it does today. This last assumption would be hard to prove except for one rather easily performed experiment. The age of different parts of old living trees can be determined by counting the growth rings (one for each year). Wood at the various points can also be dated by its carbon-14 level. Both methods give about the same ages, which increases our confidence in the carbon dating method.

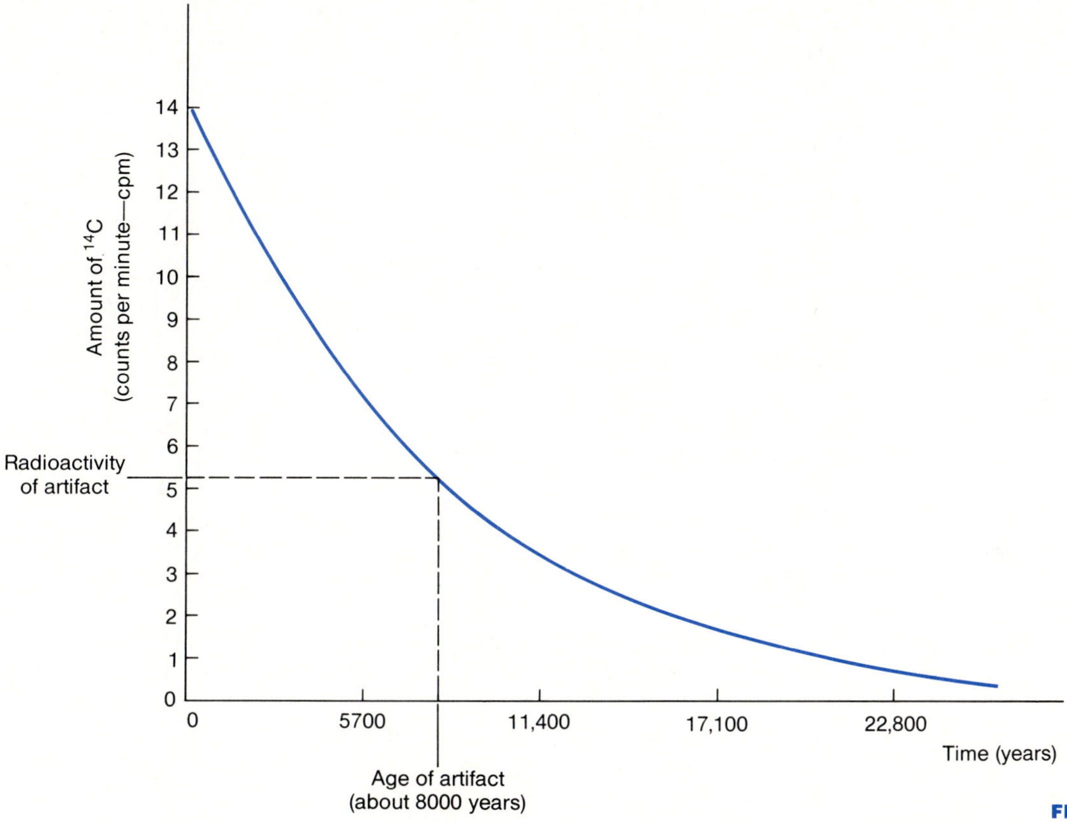

FIGURE 13-10
Plot of carbon-14 against time. (Counts per minute of beta radiation are shown for four different half-life periods.)

There are limitations to this method. First, because of rather large errors in measuring the small amount of carbon-14, these values are good only to ±50 years and can be used only for samples which are no older than about 20,000 years. After 30,000 or 40,000 years, essentially all the carbon-14 is gone; therefore the method is of little if any value for carbon-containing samples older than that, except to say that the sample is older than 40,000 years. Samples only 100 to 200 years old also show large errors, since large amounts of CO_2 from ancient fossil fuels are part of the biosphere. This CO_2 is lower in carbon-14 than CO_2 from living systems and makes the samples appear older than they actually are.

The most successful applications of this method are found in archeology, art, and literature. An archeologist can determine the absolute age of artifacts from ancient or prehistoric cultures: Egyptian, American Indian (as in Figure 13-11), or others. The validity of claims about supposedly old art objects can often be determined. For example, an original painting by Leonardo da Vinci would be about 600 years old. Perhaps a scrap of canvas from the back of a painting that might be by Leonardo is expendable and can be analyzed for carbon-14 to prove or disprove that the age of the canvas is consistent with the claim. (Of course this would not prove who had painted it.)

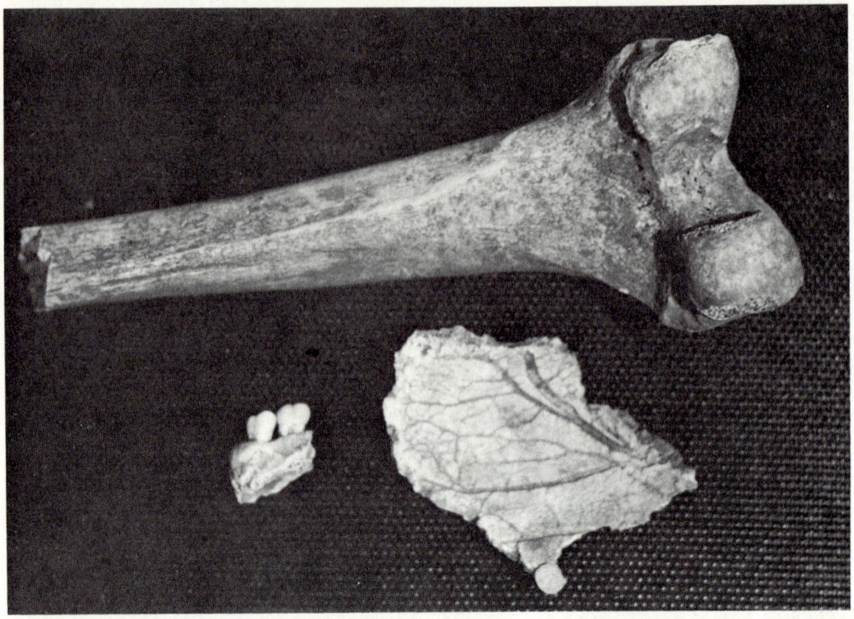

FIGURE 13-11
Ancient remains like these bones from an early American burial site can be dated by the carbon-14 method.

FIGURE 13-12
One of the Dead Sea scrolls. Its approximate age was determined by carbon-14 dating. (© John C. Trevor, 1970.)

A third area where isotope dating has proved to be of value is in ancient literature. About 1950, the Dead Sea scrolls (see Figure 13-12) were discovered in Palestine, as was discussed in Chapter 3. They contained many parts of the biblical Old Testament, including the entire book of Isaiah. At that time, the oldest manuscripts from which modern translations of the Bible had been made were written about A.D. 300. These Dead Sea scrolls were written on parchment (sheepskin), and the discovery site led scholars to suspect that they were quite old. Carbon-14 dating placed their origin at about 100 B.C., which was 400 years older than the oldest extant manuscripts. A translation of these manuscripts *could* have proved to be a more accurate representation of the original than those based on later ones. However, they turned out to be virtually identical to those written 400 years later, and no essential changes were required. They did increase confidence in modern biblical texts and had considerable impact on thinking in the field of bibliology.

Other isotope methods that can be used in dating include the potassium-argon method. Potassium-40 ($^{40}_{19}K$) is a minor isotope of this element (only about 0.01 percent), but it is radioactive and changes to $^{40}_{20}Ar$ with a half-life of 1.3 billion years. Since many minerals, especially feldspars, contain potassium, the ages of rocks in various strata can be estimated with this method. Likewise, uranium-238 ($^{238}_{92}U$) changes to lead-206 ($^{206}_{82}Pb$), with a half-life of 4.5 billion years. The *highest* ratio of lead-206 to uranium-238 is about 50 to 50, which might suggest that the earth is about one half-life, or 4.5 billion years, old. Since these rocks could have been partly Pb when the earth was formed, this age for the earth is dubious and represents only a *maximum* value.

Medical applications of natural radioactivity

Since the radiation given off by radioactive materials can be easily detected by film or Geiger counters, it has proved to be a good diagnostic tool in the field of medicine. A radioactive compound can be placed into the body and its fate determined. When radioactive iodine (iodine-125) is injected into the body, it concentrates in the thyroid gland. Photographic film placed next to the body shows a bright spot at the base of the throat, or a Geiger counter held at that spot responds vigorously. These observations indicate that all iodine entering the body concentrates in the thyroid gland, since *radioactive materials behave the same chemically and biochemically as ordinary isotopes of the same element*.

It is also possible to prepare glucose or other common biochemical substances in which nonradioactive atoms are replaced with their radioactive isotopes. (See Figure 13-13.) The radioactively labeled compound can be placed into the body and then detected by several instruments located at different positions next to the body. A computer can then reconstruct a three-dimensional map of the distribution of the radioactive material. Various mental disorders can be diagnosed by measuring the distribution and rate of metabolism of *radiolabeled* glucose. An important advantage of these techniques is that they do not require surgery. The radioactive material is given at a low concentration; and since the half-lives of the radioactive isotopes used are short, the patient is in little danger from the radiation itself. It is not completely safe, of course, but the consequences of not properly diagnosing the malady would be much worse.

Another all too often necessary application of radioactivity is the use of cobalt-60 in the treatment of cancer. The gamma rays of this powerful source of radiation kill living cells. When it is not possible to surgically remove all the cancerous tissue, or if surgery is not feasible, cobalt can be used to destroy malignant cells. It also kills normal cells in the vicinity, and the high dose of radiation is likely to induce more cancer. This new cancer will not appear for a long time, however, and the risk of developing a new malignancy is not high compared with the danger of death from the present one. The decision to use

FIGURE 13-13
A vial of a radioactive isotope used in medical diagnosis.

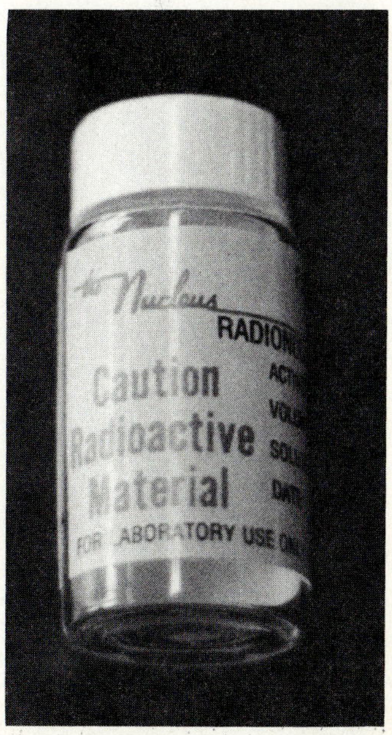

cobalt-60 to treat cancer is an example of risk-benefit analysis; that is, the possible risk is low compared with the likely benefits.

Finally, one can use radiation, especially gamma rays, to kill microorganisms in situations where heat sterilization would adversely affect a product. Raw fruit and vegetables can be passed through a strong radiation source. All bacteria are killed, but the fruit is not made radioactive, nor is its taste or color changed. The nutrient value may be affected somewhat, however. One can also seal an item, such as a surgical blade, in a container and irradiate it with gamma radiation. All bacteria inside the sealed container are killed, and the item can be shipped and used without fear of bacterial contamination.

TRANSMUTATION

It is possible to bombard atoms of one element with high-energy particles and change them into atoms of another element. In one sense, that is the alchemist's dream come true—to change one element into another. The process, however, is so expensive that no one could get rich doing it; but some isotopes that are used for military purposes are produced in large quantities by this process (plutonium is an example).

If ${}^{25}_{12}Mg$, for example, is bombarded with high-energy alpha rays, the following process occurs:

$$ {}^{25}_{12}\text{Mg} + \underset{\text{Alpha rays}}{{}^{4}_{2}\text{He}} \longrightarrow [{}^{29}_{14}\text{Si}] \longrightarrow {}^{28}_{13}\text{Al} + {}^{1}_{1}\text{H} $$

The two protons and two neutrons of the alpha particles can be thought of as momentarily combining with the nucleus of the magnesium atoms. This produces a highly unstable nucleus containing 14 protons and 15 neutrons (mass of 29). This is a silicon atom, since it has 14 protons; the brackets indicate its instability. This silicon atom almost immediately changes spontaneously to atoms of aluminum by giving off protons. Thus, magnesium has been changed to aluminum. This type of change is called a *transmutation*—the changing of one element into another. Two more examples are shown below:

$$ {}^{14}_{7}\text{N} + \underset{\text{Alpha rays}}{{}^{4}_{2}\text{He}} \longrightarrow [{}^{18}_{9}\text{F}] \longrightarrow {}^{17}_{8}\text{O} + {}^{1}_{1}\text{H} $$

$$ {}^{35}_{17}\text{Cl} + {}^{1}_{0}\text{n} \longrightarrow [{}^{36}_{17}\text{Cl}] \longrightarrow {}^{35}_{16}\text{S} + {}^{1}_{1}\text{H} $$

Here nitrogen is changed into a rare isotope of oxygen (oxygen-17), and chlorine upon bombardment with neutrons becomes sulfur.

The high-energy particles needed for such processes are produced by naturally radioactive materials or are produced artificially. Often these particles are accelerated in one of three devices—a cyclotron, betatron, or linear accelerator. These instruments, which are large and quite expensive, speed up the alpha or beta particles and aim them at the material to be changed during the transmutation. Figure 13-14 is a diagram of alpha particles being accelerated in a cyclotron, hitting a piece of magnesium (the "target"), and effecting its change into aluminum; Figure 13-15 shows an actual cyclotron.

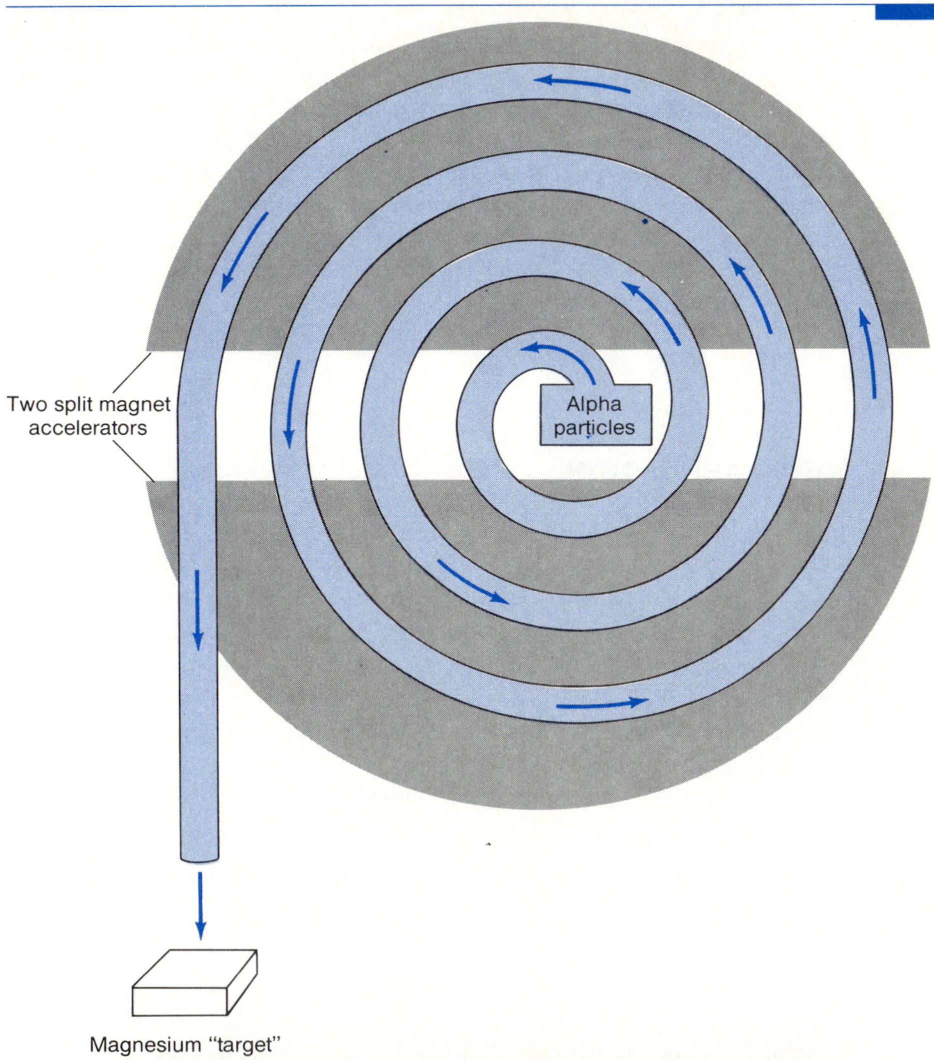

FIGURE 13-14
Operation of a cyclotron.

FIGURE 13-15
A cyclotron. Shown here are E.O. Lawrence (*right*) and his coworker M.S. Livingston with their 30-in cyclotron in 1930. (Lawrence Berkeley Laboratory, University of California.)

The purpose of such transmutation procedures, in addition to providing basic scientific information, is to produce isotopes for medical use or other research purposes. These materials, such as the oxygen-17 produced from nitrogen, are not found in nature, are very rare, or are difficult to isolate from natural sources. Isotopes that are found in nature can also be produced artificially from other elements by transmutation. All the artificial elements, those with atomic numbers greater than 92, have been produced by similar processes. For example, californium (Cf) is prepared by bombarding uranium-238 with carbon nuclei as follows:

$$^{238}_{92}U + ^{12}_{6}C \longrightarrow ^{245}_{98}Cf + 5^{1}_{0}n$$

NUCLEAR FISSION

So far we have considered two types of nuclear processes—natural radioactivity, where the nuclei of certain isotopes change spontaneously (without aid from humans); and transmutation, where a natural isotope is bombarded with particles and changed to a new element by humans. A third very important nuclear process involves the *splitting* of nuclei into two fragments to produce two different elements. This is called *fission*. It takes place naturally (at a very slow rate) only with the isotope $^{235}_{92}U$. The rate of this process can be greatly increased and the energy produced can be used by building a device called a *nuclear fission reactor*.

Although many processes occur during the fission of uranium, the following is one of the more common reactions:

$$^{235}_{92}U + ^{1}_{0}n \longrightarrow ^{139}_{56}Ba + ^{94}_{36}Kr + 3^{1}_{0}n + \text{gamma rays} + \text{energy}$$

During the process a neutron hits the nucleus of a uranium atom, causing it to split into atoms of barium (Ba) and krypton (Kr). In addition to these two new elements, *three* neutrons are released along with gamma radiation (see Figure 13-16). One of the most significant aspects of this reaction is that huge amounts of energy are also released. Careful measurement of the exact masses of all products shows that they weigh a little less than those of the reactants.

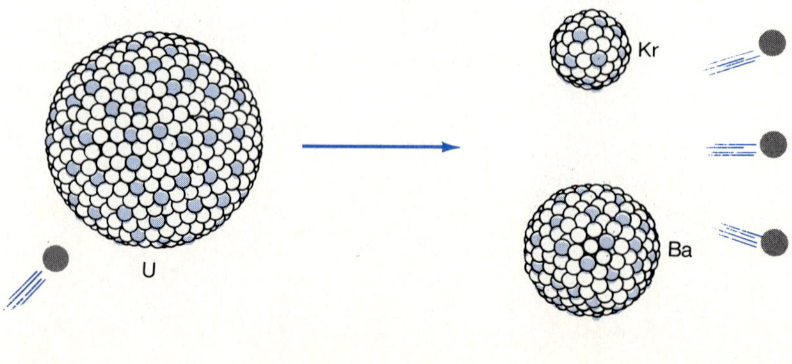

FIGURE 13-16
Uranium atom being split.

This appears at first to be a violation of the law of conservation of matter. In actual fact, however, *matter* is simply converted to *energy*. Albert Einstein dealt with this phenomenon and calculated it theoretically before it was discovered experimentally. His famous equation allows one to calculate the ratio of mass to energy. The equation is:

$$E = M C^2$$

Energy (J) | Mass (kg) | Speed of light (3×10^8 m/s)

Let's consider 1 g of matter and determine how much energy it would produce if totally changed to energy. Our 1 g has to be expressed as 0.001 kg to be used in this equation.

$$E = (0.001 \text{ kg})(3 \times 10^8 \text{ m/s})^2$$
$$E = 9 \times 10^{13} \text{ kg} \cdot \text{m}^2/\text{s}^2$$
$$E = 9 \times 10^{13} \text{ J}$$

A joule (J) is equivalent to a kilogram-meter squared per second squared. Converting to calories (cal):

$$9 \times 10^{13} \text{ J} \left(\frac{1 \text{ cal}}{4.18 \text{ J}}\right) = 2.14 \times 10^{13} \text{ cal}$$

A gram is not very large, but 2.14×10^{13} cal is a tremendous amount of energy—enough to heat 2.14×10^8 L (214 million liters or about 53 million gallons) of water from 0°C to 100°C. It is the liberation of this tremendous amount of energy that makes the fission process so intriguing.

Since one neutron is required to split an atom of uranium-235 but three neutrons are produced, a chain reaction is possible. The neutrons given off by the splitting of one atom can cause three more atoms to split, which can cause nine to undergo fission, etc. (see Figure 13-17). Since much of the uranium-containing material (either uranium metal or one of its compounds) is empty space (all atoms are mostly empty space), many of the neutrons escape the mass without encountering a uranium-235 nucleus and thus do not cause fission. If the piece of material is large enough, the chance of *most* of the neutrons hitting a nucleus before escaping is increased and a multiplying chain sequence can be maintained. Because of the tremendous energy liberated by fission, a sustained chain reaction of the type just discussed causes a large blast, which is the basis for atomic bombs. The mass that will just sustain a chain reaction is called the *critical mass*. It is diagramed in Figure 13-18. A chain reaction in which two of the neutrons escape and only one is left to continue the chain will not cause a blast.

The example that we have been using is that of $^{235}_{92}U$, but $^{233}_{92}U$ and $^{239}_{94}Pu$ are also capable of fission, and are both produced artificially in nuclear reactions from $^{232}_{90}Th$ and $^{238}_{92}U$, respectively, in transmutation processes. (See Table 13-3, which appears later in this chapter, for the production of $^{239}_{94}Pu$.)

FIGURE 13-17
Chain reaction.

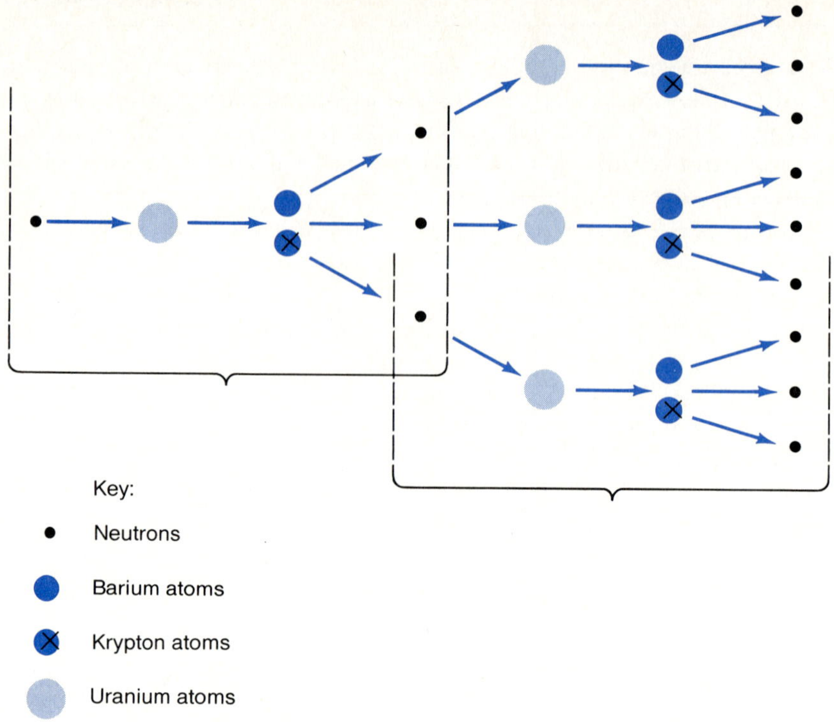

FIGURE 13-18
The critical mass concept in nuclear fission reactions.

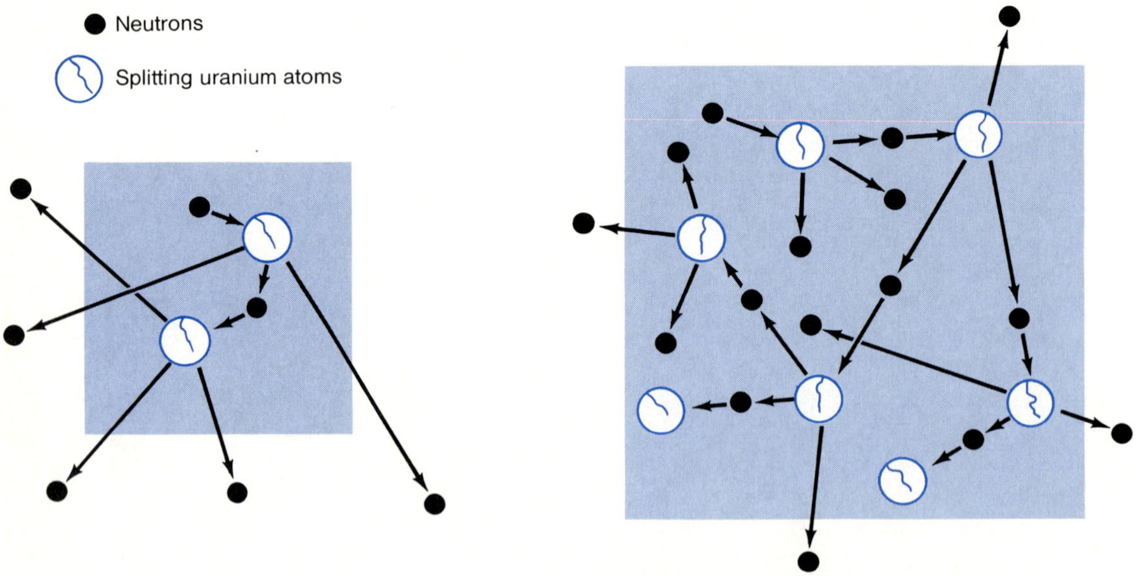

Subcritical mass—most of the neutrons are escaping the block and do not cause fission.

Critical mass—enough of the neutrons are causing fission to allow a multiplying chain reaction to occur.

Fission bombs

The two main applications of fission are *atomic bombs* (which will be discussed in this section) and *nuclear reactors* (discussed in the next section).

The bomb is based on the principle of *critical mass*. Natural uranium contains only 0.7 percent uranium-235 and must be enriched if a significant amount of fission is to occur. The percent of uranium-235 required varies from 25 to 40 percent depending on the bomb design. Subcritical pieces (pieces that weigh less than the critical mass) of the uranium (usually as its oxide) are formed for use in constructing bombs. While fission is occurring, no self-sustaining chain reaction is possible. Several of these subcritical pieces are placed separately within the bomb structure, and conventional explosives (gunpowder is one example) are put behind each one. When these explosives are detonated, the subcritical pieces are thrown together, producing a mass greater than critical; the result is a self-sustaining chain reaction that produces a nuclear explosion. This is diagramed in Figure 13-19. There are other methods of joining the pieces together, but the principle is the same. The critical mass is about 2 lb. As an example, assume that each of the three pieces shown in Figure 13-19 weighs about 1.5 lb. This is below the critical mass, and thus no chain reaction is possible. When all three are joined, however, the mass becomes about 4.5 lb, which will sustain a chain reaction and cause an explosion.

Bombs of this type have been manufactured and stored for use since the mid-1940s; thousands are now owned by the United States and many more thousands by other countries. Only two have ever been used in wartime, however. President Truman ordered that atomic bombs be dropped on two industrial cities in Japan (Nagasaki and Hiroshima) to bring a rapid end to World War II. The tremendous heat, radiation, and energy produced by a nuclear blast (such as that shown in Figure 13-20) are devastating to human

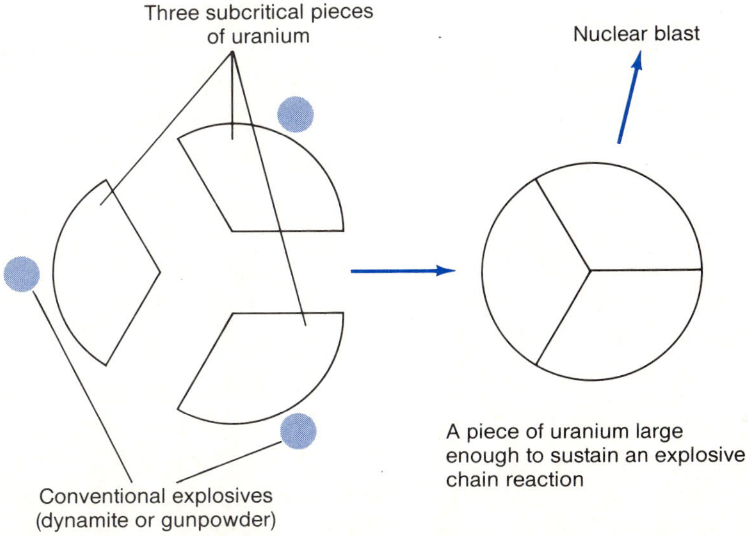

FIGURE 13-19
Mechanism used to detonate a fission ("atomic") bomb.

FIGURE 13-20
A nuclear blast. (Photograph from United States Air Force; used with permission.)

life as well as buildings. Temperatures reach 10,000,000°C at the blast site (steel actually evaporates), and there is enough energy in one of these bombs to kill 70,000 people. Since the possibility of an intentional or accidental nuclear bombing is very real (even a nuclear war is possible), we will consider (briefly) the possibility of precautions in such an event.

There are several areas, or zones, at or in the vicinity of the blast (see Figure 13-21). The first is the immediate blast area, in which everything is destroyed. For a 5-megaton bomb (one equal in power to 5×10^6 tons of TNT), this area is 5 to 10 mi in radius. Nothing can live in this zone.

Next is an area 10 to 20 mi in radius where the shock wave from the blast knocks down most buildings and where a tremendous amount of heat is produced several seconds after the blast. A great deal of gamma radiation is also produced by the fission reaction and is of high intensity in this zone. Any protection from the *direct* blast can be of great advantage in this area. Heat and radiation travel in a straight line, so that a curb or small embankment can be a protection from the intense radiation, the heat, and the flying objects. The main dangers in this zone are from falling buildings, heat (which can kill a person 5 to 10 mi from the blast), and gamma radiation. The gamma rays cause radiation sickness, which is usually fatal within several days, though it is not contagious. The intense radiation destroys so many molecules that the body cannot repair the damage before life ceases.

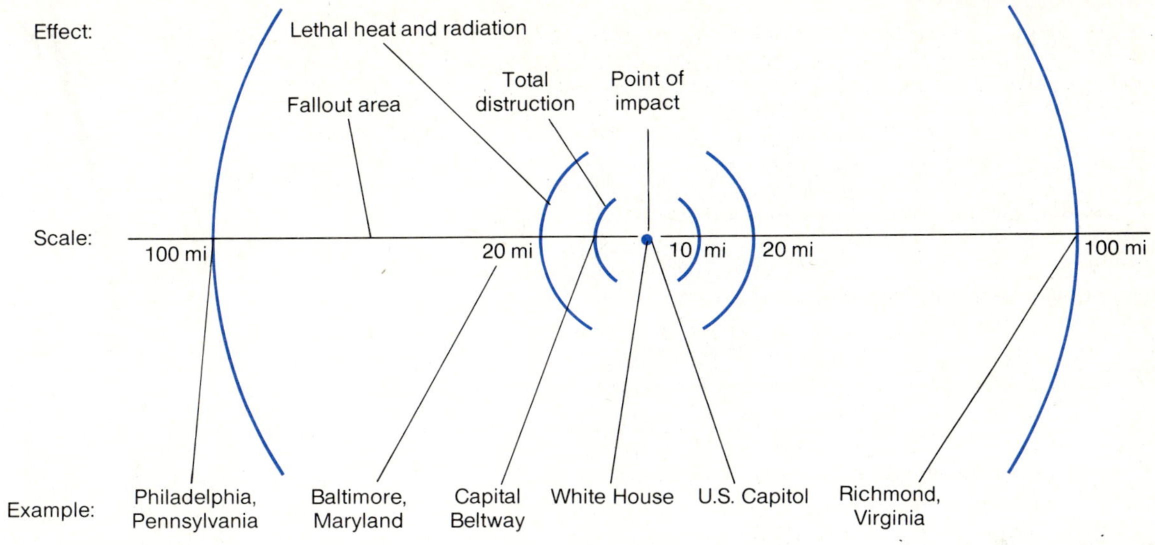

FIGURE 13-21
Danger zones after a nuclear blast (for a 5-megaton bomb).

The next area is 20 to 100 mi in radius and is called the *fallout zone*. When the blast occurs, it pulverizes large amounts of rock and soil into particles the size of fine soil and puts this material into the atmosphere. Wind carries these particles for 20 to 100 mi until they gradually fall to the ground downwind from the blast (in the United States, downwind is usually eastward). This fallout comes 15 to 24 hours (h) after the explosion and is radioactive; most of the fallout material has a half-life of several days. It is dangerous to be exposed to this fallout, but fallout does not make other materials radioactive. One should stay in a basement or other underground shelter for about 2 weeks until most of the radiation has been dissipated. Foods, such as apples, which have fallout on them can be carefully washed and then eaten; water can be filtered. In this zone the main dangers are radiation sickness from constant exposure to the fallout and cancer or genetic damage from intermittent exposure. One should, short of starving or dying of thirst, stay in a protected area for 10 to 14 days.

Fission reactors

The most important peacetime use of fission is in a nuclear reactor. In this device, diagramed in Figure13-22, the heat given off by the fission reaction is used to heat water to steam which can then be used to run electric generators (as in Figure 13-23), turn ship propellers, heat buildings, etc. The most important current application is generating electricity, and the second is powering nuclear ships.

The heart of a reactor (see Figure 13-22) is a quantity of a fissionable material such as uranium-235 or plutonium-239. Some reactors can use natural uranium oxide (0.7 percent uranium-235), but others require enrichment of the uranium from 5 to 25 percent uranium-235 to ensure that a significant amount of fission will take place. The amount or concentration is kept below

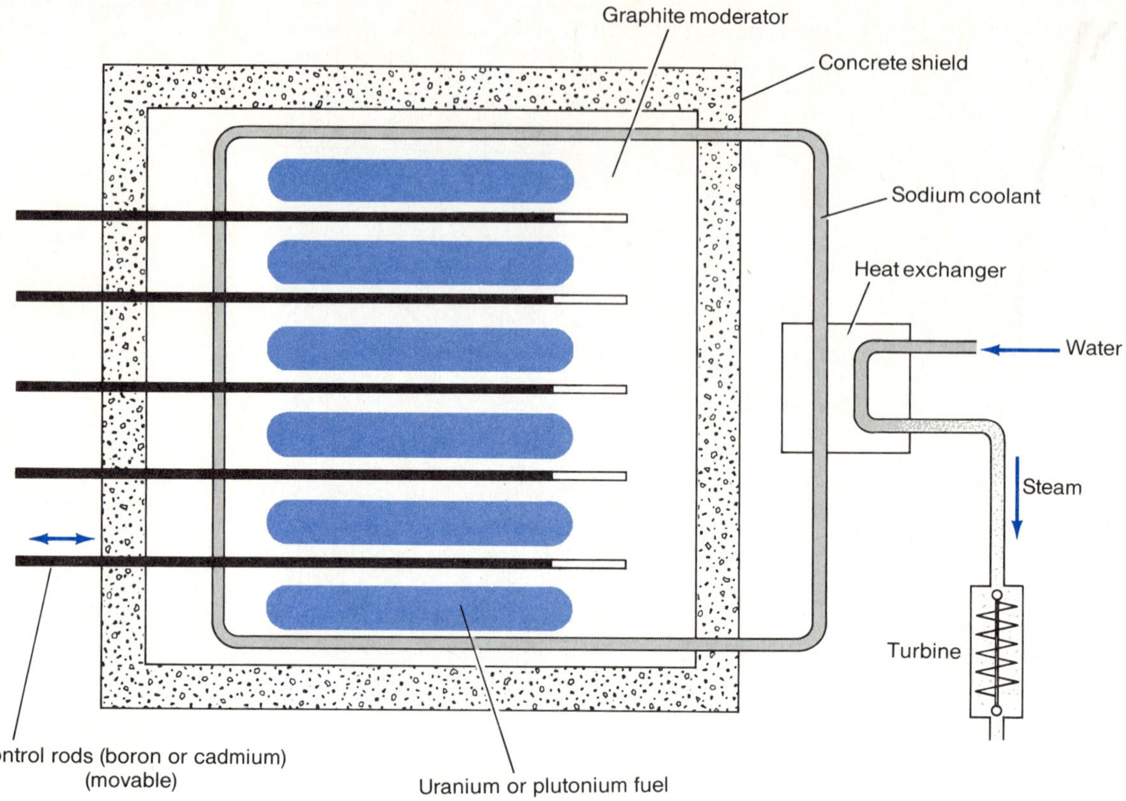

FIGURE 13-22
Above: Nuclear reactor.

FIGURE 13-23
Right: Electric generating plant powered by nuclear energy. (By permission of Electric Power Research Institute.)

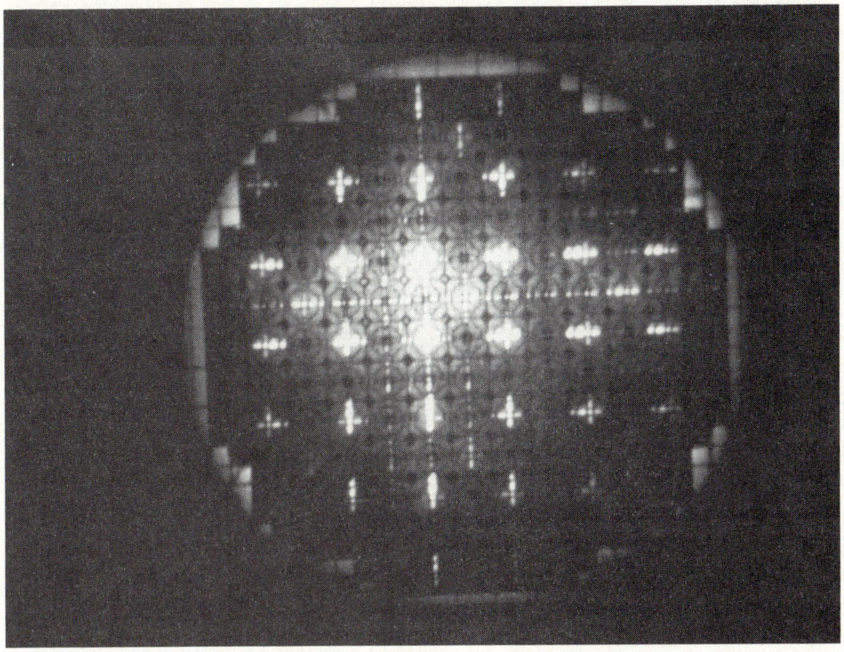

FIGURE 13-24
The glow produced by a nuclear reactor, called the *Crenkov glow.* (By permission of Electric Power Research Institute.)

the critical mass so that a nuclear explosion is impossible. This is called the *core* of the reactor. (The glow produced in a core is shown in Figure 13-24.)

Neutrons which are moving relatively slowly cause atoms to split more efficiently than rapidly moving ones. Since those given off by a splitting atom are going fast, they need to be slowed down (moderated) if they are going to cause more atoms to split. This is done by placing a material called a *moderator* between the several parts of the core. The common moderators are graphite (a form of carbon), "heavy" water, and ordinary water. "Heavy" water is D_2O, in which the hydrogen atoms are the isotope deuterium (D), ${}^2_1H$; it is obtained by fractional distillation of natural water.

Several elements, including cadmium (Cd) and boron (B), have the ability to absorb neutrons without splitting. Rods made of these materials can be inserted into the core to virtually stop energy production. The neutrons given off by splitting atoms are absorbed by the Cd or B rather than being absorbed by other uranium atoms. These rods are called *control rods,* and their position controls the rate of fission and therefore the energy output of a reactor.

The heat generated by fission is transferred by running cooling material through pipes in the core. This hot material (which may be liquid sodium or ordinary water) is in turn cooled by water, which is converted to steam. The steam pressure is used to run turbines. These cooling coils are needed not only to get energy from the reactor but also to keep the reactor from getting so hot that it melts.

Finally, the entire system must be shielded with concrete or some other dense material to protect workers from exposure to radiation.

Although there are many nuclear reactors in use, there are several problems with them. The construction cost of a nuclear electric generating plant, for example, is unusually high compared with that of a coal plant. The used cores are highly radioactive and must either be stored for a *long* period of time (the half-life is several hundred thousand years) away from populated areas or be reprocessed. Reprocessing allows part of the fissionable material to be reused; but transporting and handling it create the danger that it might be stolen by terrorists, who could then build a bomb. This is possible (although some people say that it is not); in fact, India developed nuclear weapons from fuel originally destined for reactors.

There are also some misconceptions about nuclear reactors. A reactor *cannot* produce a nuclear explosion, since there is less than a critical mass of fissionable material present. Relatively minor accidents do not release a dangerous amount of radioactivity into the environment. Remember that there is constant background radiation from natural and medical sources, and a small amount more is not significant. The health hazards associated with the release of small amounts of radioactive material by nuclear power plants are far less than those resulting from the pollution of the air by coal-powered electric generating plants. In fact, most coal-burning power plants probably discharge more radioactive material into the air than any nuclear power plant. This is a result of trace amounts of radioactive materials in the coal. Far fewer workers are harmed in the nuclear industry than in conventional power plants. There are a large number of safety devices built into reactors; these partly account for the expense involved.

In sum, reactors—like most other things—have advantages and disadvantages. Whether their use will become widespread in the immediate future will probably depend on safety and economics. (More will be said about the pros and cons of nuclear energy in Chapter 14.)

FUSION

The last kind of nuclear change we will examine is called *fusion*. It involves the joining (fusing) together of two nuclei to form a larger one. A common example is the reaction of deuterium ($^{2}_{1}H$) and tritium ($^{3}_{1}H$) to form helium:

$$^{2}_{1}H + ^{3}_{1}H \xrightarrow{\text{high temperature}} ^{4}_{2}He + ^{1}_{0}n + \text{heat} + \text{energy}$$

Here even more mass is converted to energy than in fission reactions (on the basis of the amount, or mass, of starting reactants and mass of final products). This kind of reaction is what occurs in the sun and other stars and is a potential source of energy if it can be harnessed.

The hydrogen bomb is one application of the fusion reaction. A small fission bomb is detonated in the presence of a quantity of deuterium and tritium. The heat generated by the fission causes fusion to occur, which gives off even more energy.

Fusion can be envisioned as taking place in four stages. These are shown

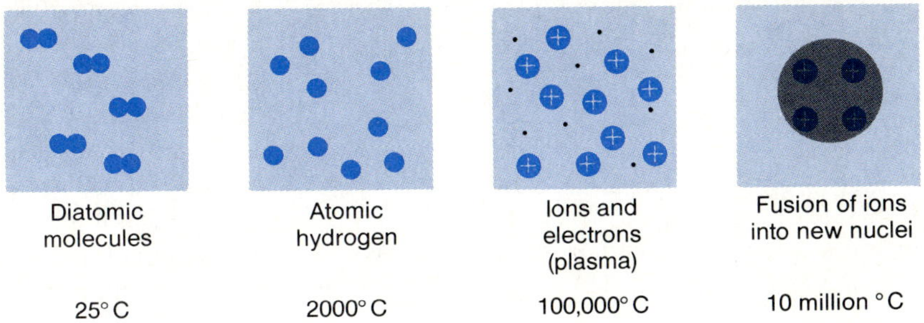

FIGURE 13-25
The stages leading to fusion.

in Figure 13-25. At ordinary temperatures (say, 25°C), hydrogen exists as diatomic molecules. As the temperature is raised to several thousand degrees Celsius, these molecules break apart to form individual atoms, since their kinetic energy is so high. At still higher temperatures, around 100,000°C, the nuclei and electrons separate to form positive ions in a cloud of electrons—which is actually a form of matter different from any we are used to, since it does not consist of atoms. This is called a *plasma*. As the temperature reaches 10^7 to 10^9°C, the nuclei are moving so fast that when they collide, they fuse to form larger nuclei and give off energy.

Thus, a fission device is needed to raise the temperature of hydrogen to the fusion level. Since most of the explosive force of a hydrogen bomb is due to fusion, which produces no radioactive isotopes, it is a so-called clean bomb. It does give off a small amount of radiation from the fission detonator, but fallout is not as serious a problem as with fission bombs.

Deuterium can be isolated from ordinary water. Natural hydrogen contains 0.016 percent deuterium, and tritium can be made from deuterium in nuclear reactors where D_2O is the moderator. Compared with uranium reactions, a fusion reactor would have several advantages. The fuel (hydrogen) is relatively abundant, and no radioactive wastes are produced. Although much research is being done, no fusion reactor has yet been built. One difficulty is raising the temperature to the required 10^7°C. Another is the containment of the plasma, since it would melt any "container." There is also the problem of getting the energy out of the plasma. Finally, there has to be a way of controlling the reaction so that an explosion does not occur. The present theoretical approach is to use magnetic fields to compress the fusion material until it is heated and then suspend the resulting plasma by magnetic fields. If this can be done successfully, it could go a long way toward solving our energy problems.

Table 13-3 lists the four basic types of nuclear reactions which we have studied and gives a way to distinguish between them. The column head *Characteristics* states the principle involved and (in parentheses) what one should look for in the equation in order to identify a reaction.

TABLE 13-3 Types of nuclear processes

TYPE	EXAMPLE	CHARACTERISTICS	USE
Natural radioactivity	$^{238}_{92}U \longrightarrow ^{234}_{90}Th + ^{4}_{2}He$ α (alpha) particle	One element spontaneously changes to another. (Only one symbol on the left side of arrow.)	Treatment of disease, sterilization, and various industrial uses
Transmutation	$^{238}_{92}U + ^{1}_{0}n \longrightarrow ^{239}_{94}Pu + 2\,^{0}_{-1}e$ β (beta) particle	One element is bombarded with subatomic particles and changes to another element. (An element symbol and a particle on both sides of the arrow.)	Production of rare or artificial isotopes for chemical and medical uses
Fission	$^{235}_{92}U + ^{1}_{0}n \longrightarrow ^{94}_{36}Kr + ^{139}_{56}Ba + 3\,^{1}_{0}n$ + heat + force	Neutrons cause atoms to split and to form two new elements with *lower* atomic numbers. (One element symbol on the left and two on the right.)	Energy production (reactors) and bombs
Fusion	$^{2}_{1}H + ^{3}_{1}H \longrightarrow ^{4}_{2}He + ^{1}_{0}n$ + heat + force	*Two elements* fuse together producing a *new element* with a *higher* atomic number. (Two element symbols on the left and one on the right.)	Bombs and a possible source of future energy

DEVELOPMENT OF NUCLEAR POWER

In Hiroshima, Japan, on August 6, 1945, the atomic bomb was first used for destructive purposes. In this explosion, equal to 20,000 tons of TNT, most of the city was destroyed and 66,000 people were killed. On August 9, 1945, 39,000 people were killed at Nagasaki in a second explosion, which destroyed 40 percent of that city. This was the climax of research that had been going on for many years. In this section we will examine some of the history and problems of developing a weapon as destructive as the atomic bomb, which was probably the most important discovery since gunpowder.

The history of radioactive elements goes back to the time when Becquerel discovered that there was such a thing as radioactivity by exposing some photographic plates to a sample of radioactive material. The first people to work with these new substances were Marie Curie and her husband Pierre. The Curies discovered radium in 1890, and by 1910 Marie Curie had succeeded in isolating a little of the precious metal by processing pitchblende, an ore containing uranium and radium.

In 1919, Ernest Rutherford produced the first artificial nuclear reaction by bombarding nitrogen with helium. In so doing he fulfilled the alchemist's dream of changing one element into another. He produced oxygen and hydrogen by this reaction:

$$^{14}_{7}N + ^{4}_{2}He \longrightarrow ^{17}_{8}O + ^{1}_{1}H$$

In 1934, E. O. Lawrence and M. S. Livingston (see Figure 13-15) invented the cyclotron, a device in which nuclear reactions could be produced by bombarding a substance with positive particles.

In 1938, it was discovered, by different laboratories at about the same time, that barium is produced when uranium is bombarded by neutrons. This confirmed the theory, suggested by O. R. Frisch and Lise Meitner, that uranium would split, or "fission," if bombarded by neutrons. In 1939, Niels Bohr suggested that since the combined weight of the products was less than the weight of the starting material, neutrons were given off. In 1940, it was discovered that only uranium with an atomic weight of 235 (^{235}U) would fission when slow neutrons were used (they were slowed by passing them through heavy water—D_2O—or paraffin). However, natural uranium was only 0.5 to 0.7 percent uranium-235, and no way was known to separate it at the time.

It was believed from theoretical calculation that an explosive could be made by using fission, and scientists in different countries were working on developing such an explosive. The United States felt that it must get ahead on such a project as part of its war efforts. In 1939, some prominent scientists, including Albert Einstein, approached President Roosevelt, who authorized research for this project. L. J. Briggs was put in charge of starting the project, and $6000 was made available. In 1941, a committee consisting of Briggs, J. B. Conant, G. B. Pegram, A. H. Compton, E. O. Lawrence, H. C. Urey, E. V. Murphie, H. T. Wensel, S. K. Allison, J. W. Beams, G. Breit, E. U. Condon, and H. D. Smyth was organized under the Department of Scientific Research and Development. Through negotiations, the United States established a cooperative program with Canada and Great Britain, and Brigadier General L. R. Groves was put in charge of the army activities in this field.

Since only a small part of normal uranium is uranium-235, a process had to be developed to isolate it. An electromagnetic process and a gaseous diffusion process, along with a third process called *thermal diffusion,* were put into operation at Oak Ridge, Tennessee. The gaseous diffusion method, which utilizes the different rates of diffusion of uranium-235 and uranium-238, was used extensively. UF_6 in the gaseous state was used as the starting material; and since the molecules containing uranium-235 were 3 amu lighter, these molecules diffused very slightly faster than the molecules of uranium-238 at a given temperature.

When enough uranium was produced, an atomic reactor was built at Stagg Field on the campus of the University of Chicago. This reactor used 6 tons of uranium and uranium oxide (UO_2), and was started in December 1942. In this reactor some plutonium (Pu) was produced. This was the first synthesized element, and it was found that it had fissioning properties similar to those of uranium-235. Although it was not produced in abundance, it was cheaper to produce it in a reactor and separate it chemically from uranium than to produce uranium-235 by gaseous diffusion.

On July 6, 1945, the United States exploded its first atomic bomb at Alamogordo, New Mexico. The explosion vaporized the metal tower on which the bomb was placed and fused the sand into glass for a radius of 800 yards (yd). That same year the United States exploded 40 bombs, most of them at test sites. President Truman ordered two of them to be used by the Air Force at Hiroshima and Nagasaki, Japan. In the face of such awesome weapons the Japanese had no choice but to surrender.

Uranium is found on the earth's crust in various forms. It is the heaviest natural element and was discovered in 1789 by Martin Kloproth. Its primary ores are UO_2, which is 85 percent uranium; pitchblende, which is chiefly a combination of UO_2 and UO_3; and Davidite. Besides being used for the production of atomic energy, uranium has been used in alloying steels and as a component of light filaments to prolong their lives. Its compounds were also used as dyes in the leather industry and as dyes for pottery glaze (a bright-orange one was common) but are no longer used because of the danger due to radioactivity.

In reactors, either cakes of UO_2 powder or pure uranium can be used. The UO_2 is refined from nature or produced from UO_3. Uranium metal can be produced from UO_3 by reducing with hydrogen.

$$UO_3 + H_2 \xrightarrow{>760^\circ} U + H_2O$$

It can also be produced from UO_2 by the following reactions:

$$UO_2 + 4HF \longrightarrow UF_4 + 2H_2O$$
$$UF_4 + 2Mg \longrightarrow U + 2MgF_2 + \text{heat}$$

Plutonium (atomic weight of 239) fissions like uranium-235 and has a half-life of 24,000 years. It is produced in nuclear reactions as follows:

$$^{238}_{98}U + ^{1}_{0}n \longrightarrow ^{239}_{98}U \longrightarrow ^{239}_{93}Pu + \underset{\text{Beta rays}}{^{0}_{-1}e}$$

The reactors in which this reaction and others take place are very expensive. For example, heavy water, which is used as a shield, costs about $50 per pound, and normal uranium costs $40 per kilogram. Uranium that is enriched with uranium-235 is leased at a rate of $25 for each gram of uranium-235 it contains. Since the plutonium is still combined with the uranium, the plutonium is refined to purity chemically by remote control (i.e., with machines so that human exposure is minimized). The old refining method was to dissolve the mixture of uranium-238 and plutonium-239 in nitric acid and sulfuric acid. The Pu, with an oxidation number of 4, was then precipitated with $Bi(PO_4)$. This precipitate was dissolved and precipitated several more times, and the unwanted material was left behind in solution. Now a better method using methyl-*n*-butyl ketone (an organic solvent) is used.

STUDY GUIDE

Terms to learn

Background radiation	Moderator
Carbon-14 dating	Natural radioactivity
Control rods	Nuclear decay
Critical mass	Nuclear reaction
Marie Curie	Nuclear reactor
Fission	Plasma
Fusion	Radioactive
Geiger-Müller tube	Radioactivity
Half-life	Radiolabeled drug
Isotope	Transmutation

Skill practice problems

1 List the four types of nuclear reactions.

2 List the three types of radioactivity given off by naturally radioactive materials.

3 What are the three ways that radiation can damage biosystems?

4 What is the purpose of a moderator in a nuclear reactor?

5 If a material has a half-life of 1000 years, and 6.25 percent of the original is present, what is the age of the sample?

6 A piece of wood from an Egyptian coffin gave off 2700 counts of beta radiation per hour. The piece of new wood of the same size gave 5400 cpm. How old is the wood in the coffin? Is there necessarily any relation between this age and the time the person died?

7 List several specific items for which carbon-14 dating is desirable.

8 What precautions should be taken by those living about 100 mi from a nuclear blast?

9 Why can a nuclear reactor not cause a nuclear explosion?

10 What are the main advantages of nuclear reactors as a source of energy? What are the disadvantages?

11 If a radioactive sample gave 270 counts in 3 min and background radiation was 40 cpm, what was the actual radiation from the sample?

12 How much energy (in joules) is produced from 10 g of matter if it is all changed to energy?

13 Classify each of the following nuclear changes as natural radioactivity, transmutation, fission, or fusion.

(a) $^{63}_{29}Cu + ^{1}_{1}H \longrightarrow ^{63}_{30}Zn + ^{1}_{0}n$

(b) $^{210}_{83}Bi \longrightarrow ^{210}_{84}Po + \underset{\beta \text{ (beta) rays}}{^{0}_{-1}e}$

(c) $^{3}_{1}H + ^{2}_{1}H \longrightarrow ^{4}_{2}He + ^{1}_{0}n$

(d) $^{244}_{96}Cm + \underset{\alpha \text{ (alpha) particles}}{^{4}_{2}He} \longrightarrow ^{245}_{97}Bk + ^{1}_{1}H + 2^{1}_{0}n$

(e) $^{235}_{92}U + ^{1}_{0}n \longrightarrow ^{139}_{57}La + ^{95}_{42}Mo + 2^{1}_{0}n$

(f) $^{144}_{60}Nd \longrightarrow ^{140}_{58}Ce + ^{4}_{2}He$

14 List several ways radioactive isotopes can be used in medicine.

Questions for discussion

1 Considering all factors, such as cost, pollution, safety, and production of nuclear wastes, what do you think about the advisability of extensive nuclear-powered electrical generation projects for nations with the technology to accomplish them?

2 What do you think about the wisdom of building large supplies of nuclear weapons? What situations would warrant the use of such devices? Do you think President Truman made the correct decision to use nuclear weapons to end World War II?

3 If a global nuclear war occurred, would you want to survive? What are some of the problems that would face those who lived through such an event?

14

THE CHEMISTRY OF ENERGY PRODUCTION

Goals

- **To understand the vast need for energy in modern society**
- **To understand that there are many sources of energy**
- **To comprehend the need for more scientific research to meet our future energy needs**
- **To understand the principles of heating and cooling processes and the use of electrochemical cells**
- **To examine some alternative energy sources**

Chemistry has already been defined as the study of matter, changes in matter, and *energies associated with these changes*. We have already examined the energies associated with certain physical changes such as changes of state. Now let's consider the energy changes that occur when materials undergo *chemical* changes.

ENERGY OF CHEMICAL CHANGES

When methane gas (CH_4) is reacted with oxygen to form carbon dioxide and water (see Figure 14-1), heat is given off; this chemical process is called *combustion* (or *burning*):

$$CH_4 + 2O_2 \longrightarrow CO_2 + 2H_2O + \text{heat}$$

When 16 g of methane (CH_4) is burned (as in Figure 14-2), 211,000 cal of heat (211 kcal) is given off. This is enough heat to change the temperature of 211,000 g (211 kg) of water by 1°C or 3 kg (about 1 gal) of soup from room temperature to boiling. Where does the heat come from? The law of conservation of energy says that energy cannot be created or destroyed—the total energy of the reactants before a chemical change must be equal to the total energy of the products after the reaction. Therefore the energy that is given off as heat must have been within the CH_4 and O_2 before the chemical change. It also means that the rearranged forms of these same atoms (CO_2 and H_2O) must have less energy within themselves than when they were CH_4 and O_2. The energy given off as heat (*kinetic* energy) was formerly stored as *potential* energy in the chemical bonds of the CH_4 and O_2 molecules. The potential energy contained in the bonds of molecules is called *chemical energy*.

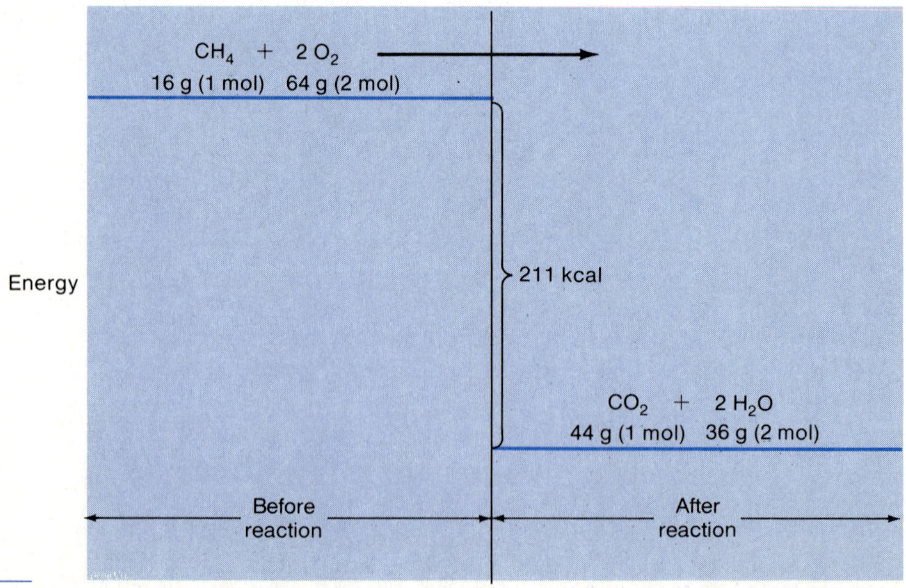

FIGURE 14-1
Energy diagram for the combustion of methane.

In the burning of methane:

$$\underset{16\ g}{CH_4} + \underset{64\ g}{2O_2} \longrightarrow \underset{44\ g}{CO_2} + \underset{36\ g}{2H_2O} + 211\ \text{kcal}$$

Total energy before reaction = total energy after the reaction

211 kcal of potential energy ⟶ 211 kcal of kinetic energy

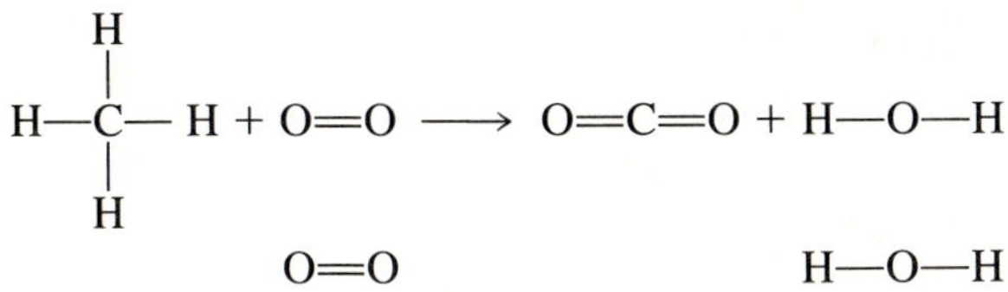

Four single C—H bonds and two O=O bonds have a certain amount of internal energy.

Two double C=O bonds and four single O—H bonds have 211 kcal less energy. They are more stable than the former arrangement.

Those types of chemical reactions in which energy is given off in large amounts can provide energy for various human needs. When fuel oil is burned, for example, it gives off 10.5 kcal/g according to the following reaction:

$$\underset{\text{Fuel oil}}{2C_{16}H_{34}} + 49O_2 \longrightarrow 32CO_2 + 34H_2O + 10.5\ \text{kcal/g}$$

When alcohol is burned, it releases only 6.5 kcal/g; thus on a weight basis, it is not as good a fuel for general heating purposes:

$$\underset{\text{Alcohol}}{C_2H_6O} + 3O_2 \longrightarrow 2CO_2 + 3H_2O + 6.5\ \text{kcal/g}$$

Although the primary product of combustion (the reaction of a compound with oxygen to form $CO_2 + H_2O$) is the heat generated, most chemical changes, in fact, give off heat. All those that do are said to be *exothermic*. The surroundings get *warmer* as a result of exothermic reactions. The change in heat during a chemical reaction is called the *enthalpy change;* it is symbolized by ΔH (the same symbol used to show energy needed to cause change in state). When a reaction gives off heat, the ΔH value is negative; $\Delta H = -211$ kcal/g for methane and -6.5 kcal/g for alcohol.

A few chemical changes actually *consume heat;* these are said to be *endothermic*. In endothermic changes, ΔH is positive and the surroundings get *colder*. An example is the dissolving of ammonium nitrate (NH_4NO_3) in water:

$$\underset{\text{Solid}}{NH_4NO_3} + H_2O + \text{heat} \longrightarrow \underbrace{NH_4^+ + NO_3^-}_{\text{Dissolved}} + H_2O$$

FIGURE 14-2
Gas flames serve as an important source of energy.

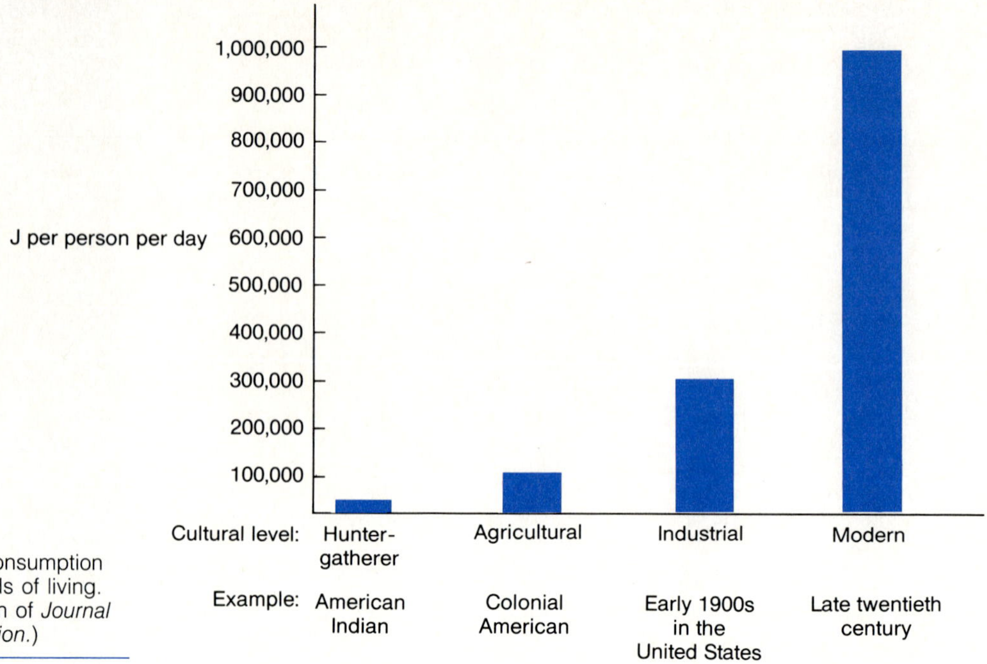

FIGURE 14-3

Individual energy consumption at different standards of living. (Used by permission of *Journal of Chemical Education.*)

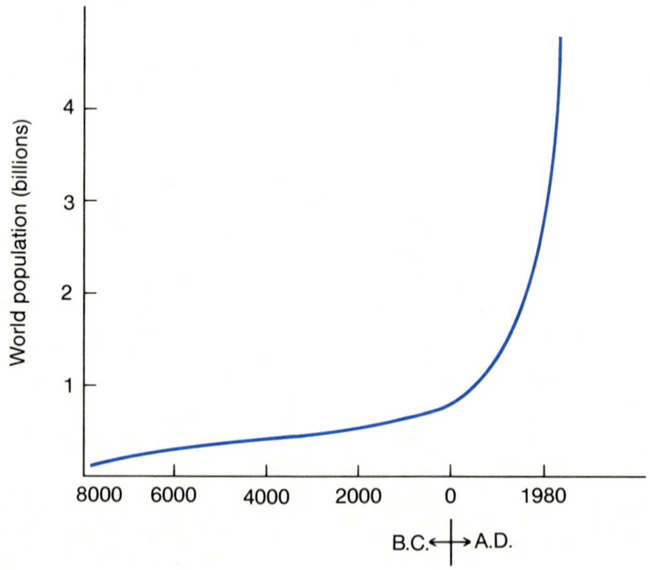

FIGURE 14-4

World population growth. (Used by permission of *Journal of Chemical Education.*)

The solution actually gets cold because the enthalpy of the dissolved ions is higher than that of the undissolved solid (the dissolved ions have greater potential energy than the undissolved solid).

In earlier chapters, we examined the air pollution problems associated with the production of energy. We also looked in detail at a major source of energy—nuclear power. In this chapter we will consider the chemistry involved in the production of energy from the various sources that are presently available.

ENERGY NEEDS

There is a direct relationship between the standard of living of a civilization or society and the energy it consumes. Think for a moment about the ways you have used some form of energy recently. This list would include various forms of heating, such as cooking and space heating of homes, offices, schools, and factories. It would also include forms of cooling, such as air conditioning and refrigeration, and items such as artificial lighting, communication devices (telephones and televisions), and vehicles for transportation. In addition, small appliances and power tools use energy, and a tremendous quantity of energy is used in the manufacturing of products. Americans use $400 billion worth of energy each year; at present growth rates, that figure could double in the next 50 years. About 40 percent of this energy is used for transportation, 35 percent is used for space heating, and 25 percent is used for manufacturing.

Figure 14-3 shows the amount of energy consumed by a person at different standards of living. Notice that each successive step takes three to four times as much energy as the previous one. Figure 14-4 shows the increase in world population. The combined influence of a high standard of living and increase in world population explains our current high use of energy. While not all the world's people have a high standard of living, most are living at a higher level than in previous years. Table 14-1 lists the main uses of energy in advanced cultures and shows why energy use at each successive level in Figure 14-3 is a good deal higher than at the previous level. Figure 14-5 (page 358) shows the relationships between various forms of energy.

TABLE 14-1
Outline of energy needs

Heating:
Space heating
Water heating
Industrial heating
Cooling:
Air conditioning
Refrigeration
Mechanical energy:
Cars (transportation)
Manufacturing
Power tools
Light:
Homes
Businesses
Streets

Heating

As our standard of living rises, comfort is thought of as more of a necessity than a luxury, and being warm in the winter is a major source of comfort. Heating the space in which we live and work represents a major portion of our energy needs. We also need to heat water for bathing and cooking and (in large amounts) for industrial processes such as refining metals.

Temperature has already been defined as a measure of the average kinetic energy of the molecules of a material. When we heat air in a room, we cause the molecules to move faster and therefore have more kinetic energy. Energy used for heating usually comes from the chemical bonds in a fossil fuel or from electricity.

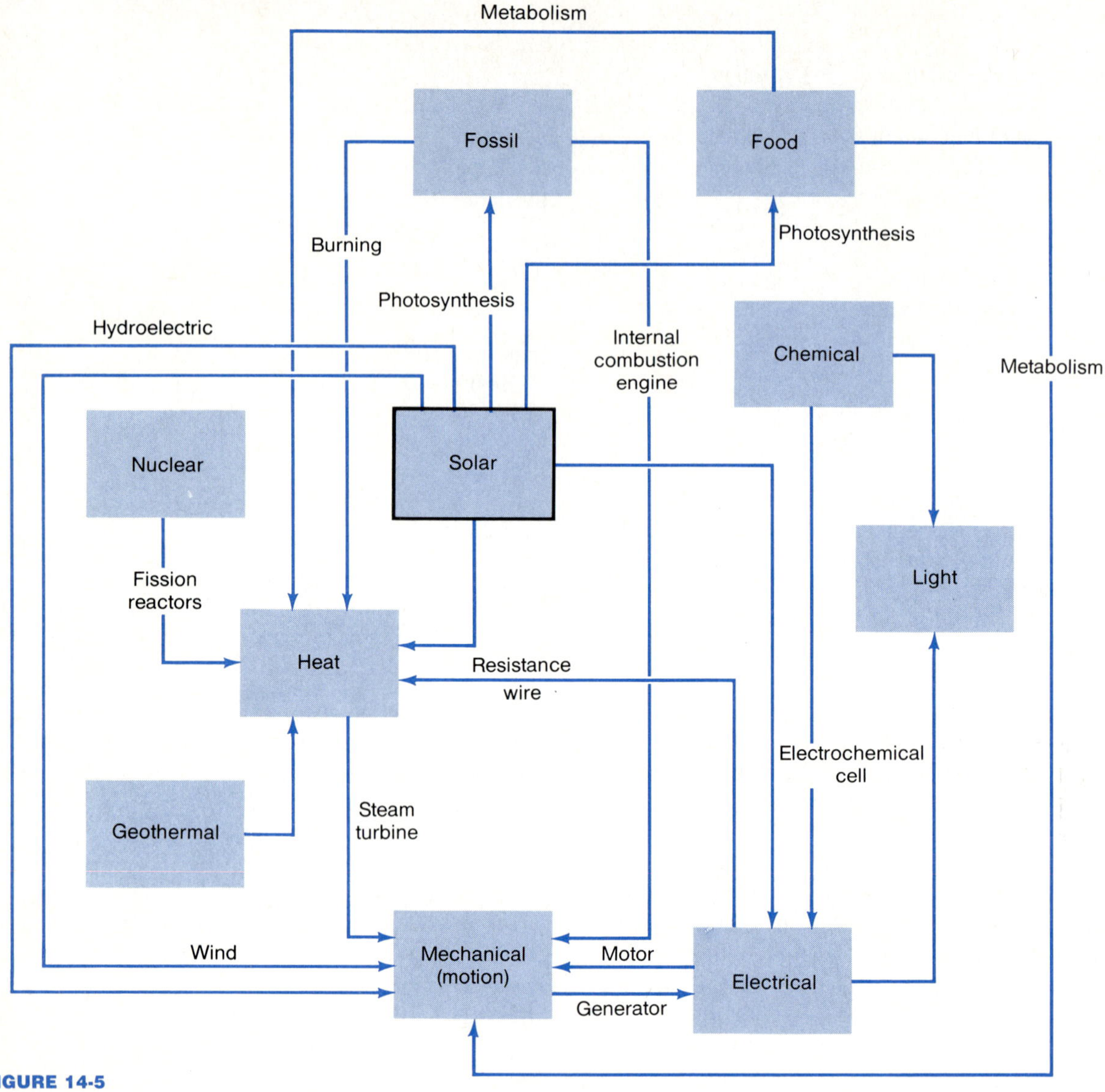

FIGURE 14-5

Relationships and interconversions between various energy sources. Note that although fossil and food fuels are both forms of chemical energy, they are shown separately here to reflect their use in our culture.

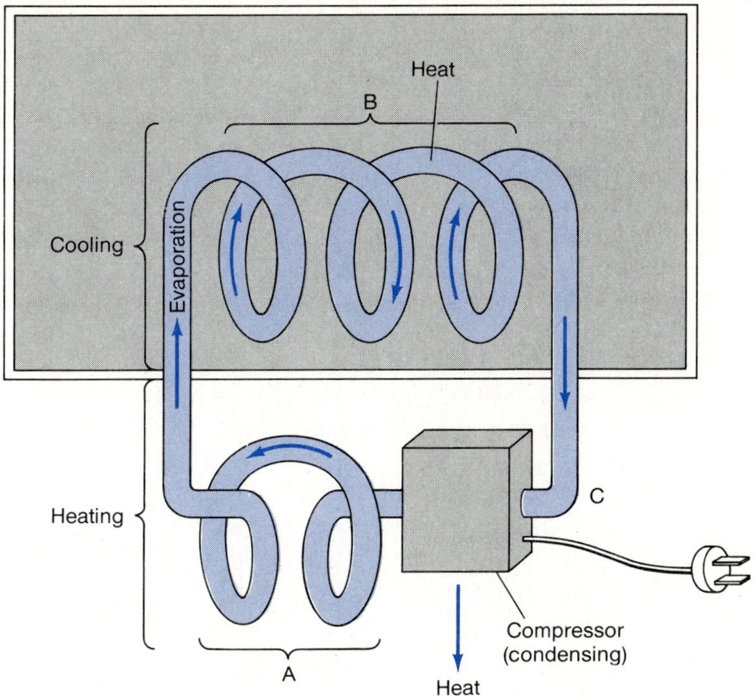

FIGURE 14-6
A cooling system.

Cooling

Since about 1960, cooling of living and work space in the hot summer months has been increasingly considered a necessity also. We are more productive on the job and less irritable at home if air conditioning is used. People in most locations in the central and southern United States air-condition their homes and offices for several months during the year. It is also important to cool foods during shipment and storage to prevent spoilage. Ice for drinks is considered highly desirable by many, as is cooling of milk and meat to prevent spoilage; all this freezing and refrigeration consumes a great deal of energy.

Both air conditioning and refrigeration are accomplished by the evaporation of a liquid. This process is shown in Figure 14-6. In cooling, we reduce the energy of molecules, and so heat energy must be moved from the area to be cooled to another area. Energy cannot be created or destroyed, but its form or location can be changed. If one area loses a given amount of heat energy, another area will gain that energy.

In a cooling system, a material that can be changed fairly easily from a liquid to a gas, and back again, is placed in a sealed tube system. The materials used are called *Freons;* they are substituted hydrocarbons such as trichlorofluoromethane (Freon-11) and dichlorodifluoromethane (Freon-12):

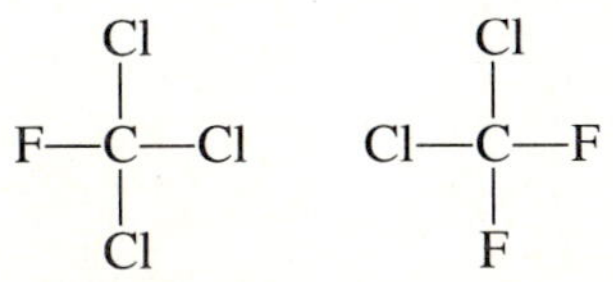

Freon-11 Freon-12

In our system (Figure 14-6) let's begin at point A. The Freon is in the liquid state and is pumped into the coils in area B, where it boils (evaporates), since the temperature in this area is higher than the boiling point of Freon. Remember from Chapter 9 that evaporation consumes energy; thus the gaseous Freon has more energy than the liquid Freon did (the molecules have more energy). This extra energy is taken from the air molecules in the closed compartment. These molecules are now moving more slowly and are cooler. The Freon gas is then pumped to point C, where a compressor is used to force the gas back into the liquid state. When a gas is condensed, its molecules have less energy and thus heat is given off. Electric energy is used to power the compressor, and the heat generated is transferred to the outside air. The liquid is then recycled and allowed to evaporate. This alternate condensing and evaporating is done continuously until the area to be cooled is at the required temperature. Since heat *normally* moves toward a cool area, removing heat from an area that is already cool takes energy. In this example, electric energy is being used to accomplish this.

It should be noted that the enclosed area is the inside of a refrigerator or the inside of an air-conditioned room (see Figure 14-7). If one stands barefoot near an operating cooling unit, one's feet can feel the heat. Likewise, the "outside" of an air-conditioning unit gives off heat. Thus, electric energy is being used to *move* heat from one place to another. Space heating can be accomplished using a similar unit called a *heat pump* (see Figure 14-8). With a heat pump the compressor operates inside the building and gives off heat, and the outside (usually the earth) is cooled. Thus heat is pumped from the

FIGURE 14-7
Refrigeration and air conditioning.

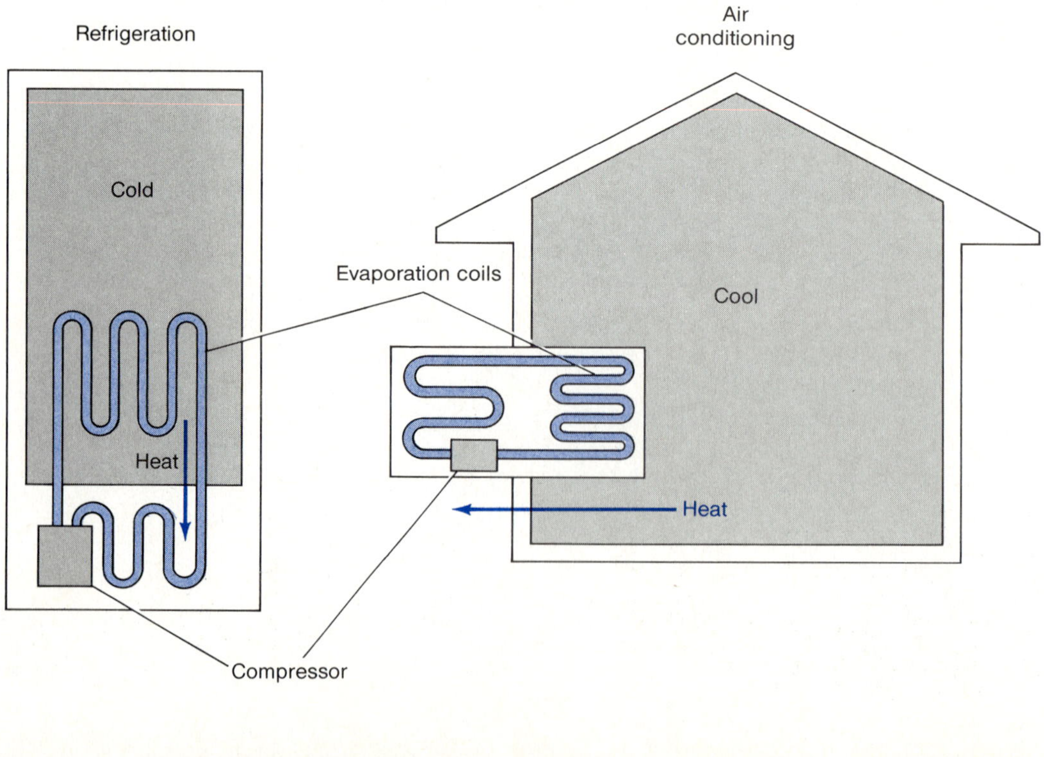

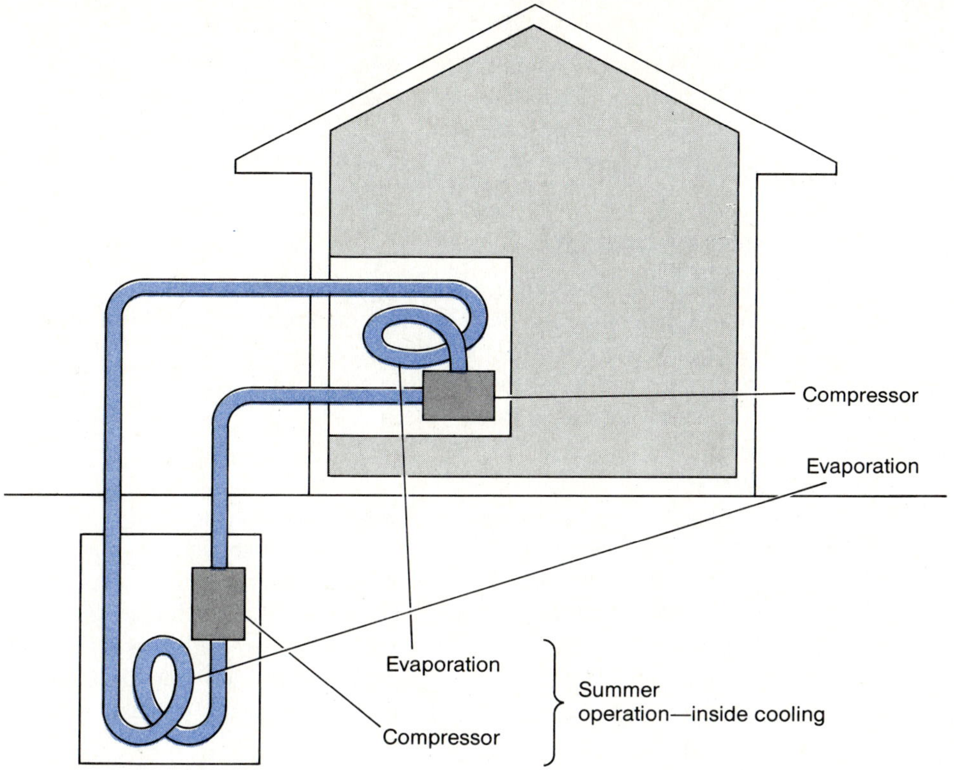

FIGURE 14-8
Heat pump system of heating and air conditioning.

outside (from the cool earth) into a building. Actually the same unit can be used for both heating in winter and cooling in summer by reversing the areas where compression and evaporation take place.

Mechanical energy

Our modern standard of living is to a considerable extent a result of the fact that we have learned to use machines to do the work that humans and animals did in earlier generations. Common physical work that is done mechanically now but used to be done by human muscles or animals includes washing clothes, transporting goods, polishing a floor, lifting heavy objects, mowing the lawn, cultivating farmland and small garden plots, transporting people, sawing and sanding wood, mashing potatoes, and cleaning rugs. With most physical work made easier, we have more free time and more energy to enjoy it.

Most sources of energy available to us are not in the physical form in which we need energy. It is therefore necessary to change electric or heat energy into forms that can be used to accomplish the many mechanical jobs that need doing. Figure 14-5 (refer back to page 358) shows that *steam turbines* and *electric motors* can be used for this purpose and that fossil fuel (stored chemical energy) can be converted directly to mechanical energy using an *internal combustion engine*.

Light

The hours of pleasant living during a 24-h period can be extended by providing lighted space in which to live, play, and work. Light is a form of electromagnetic radiation, which is one form of energy. Chemical and electric energy can be transformed to light energy, as shown in Figure 14-5.

ENERGY SOURCES

There are, of course, many sources of energy available in the world. Some are more efficient than others, some cheaper, some less polluting. Some we have not yet learned to use or are just learning to use. We will examine those that are currently used and those that hold some realistic potential for the future. Table 14-2 outlines these sources.

Fossil fuels

Fossil fuels include materials such as natural gas (methane, CH_4), oil (mixtures of hydrocarbons), and coal, as well as those synthetic fuels that can be produced from them. It could also include plants (such as trees) that can be grown as a continuous source of fuel (such as wood). (Such recently living material is sometimes called *biomass*.) The term *fossil fuel* is used because these materials were formed from once-living plants and animals (fossils).

In ancient plants, as in modern ones, the following process occurred during photosynthesis:

$$CO_2 + H_2O + \text{solar energy} \longrightarrow \text{carbohydrate} \xrightarrow[\text{pressure}]{\text{heat}} \text{fossil fuels}$$

Carbohydrates contain C—H and C—C bonds that have a higher potential energy than that of CO_2 and H_2O. This energy remains in the metamorphosed fossil fuels and is released when the fuel is burned.

TABLE 14-2

Outline of energy sources

SOURCE	FORM OF ENERGY
Fossil fuel (gas, oil, coal, wood)	Heat energy and mechanical energy are produced from chemical bonds (chemical energy).
Solar	Heat energy and light energy are produced from nuclear fusion taking place in the sun (mass converted to energy). The heat energy can be used directly, and the light energy can be used to produce electricity.
Nuclear	Heat energy is produced by fission (mass converted to energy).
Hydroelectric	Electric energy is produced from the mechanical energy of falling water.
Wind	Mechanical work is done or electricity is generated from the mechanical energy of the moving wind.
Geothermal	The heat energy of the earth's inner core is used directly.
Electrochemical	Electric energy is produced directly from chemical energy.

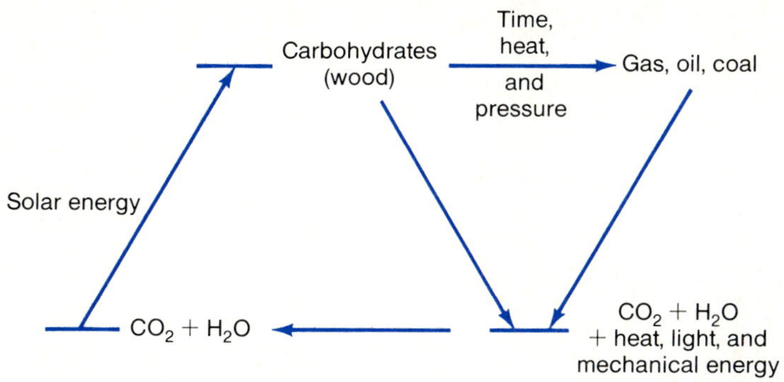

FIGURE 14-9
Capturing solar energy using the carbon cycle.

Figure 14-9 shows that the energy from ancient sunlight can be used now to meet our current needs. As we have seen, heating space, water, and various industrial mixtures (including steam for electrical generation) and running internal combustion engines (mechanical energy) are the main uses of energy derived from fossil fuel.

In addition to the usual oil wells, there are a number of alternative petroleum sources. These are sometimes called *bottom-of-the-barrel* sources; they include oil trapped in some shale deposits and oils that are very thick or otherwise of poor quality. They are more expensive to refine, but their cost could become competitive as fuel prices rise.

Solar energy

All energy sources, except possibly nuclear and geothermal energy, are ultimately derived from the sun. In this section, however, we will be considering the *direct* use of the sun's energy.

Devices called *solar collectors* allow us to capture the heat of the sun to heat water and living space. Most people have had the opportunity of being in a home attic or haymow (hayloft) during a hot summer day or have opened a closed automobile which has been sitting in the sun. The temperatures in these spaces can reach 140°F (60°C). It is also well known that a fire can be started if the energy from a 2-in-round area is focused using a magnifying glass onto a few square millimeters. Figure 14-10 shows how the heat from the sun can be used to heat air or water in a home system, and Figure 14-11 shows how the even higher temperatures needed for industrial purposes can be obtained from the sun by the use of reflecting mirrors. Solar heating is being used in the southern and central United States by innovative builders and homeowners as a supplement to conventional systems (see Figure 14-12). The industrial applications are still in the experimental stage.

Solar energy can be converted directly into electric energy by a *photovoltaic cell;* this device is diagramed in Figure 14-13. It is composed of two layers of semiconductors. The bottom layer is made by adding a small amount of a group IIIA metal such as aluminum (Al) to a group IVA material such as silicon (Si), which is the *same period* on the periodic chart. This produces a situation where aluminum atoms, with only three electrons in their outer shells, are trapped in a regular arrangement of silicon atoms, with four elec-

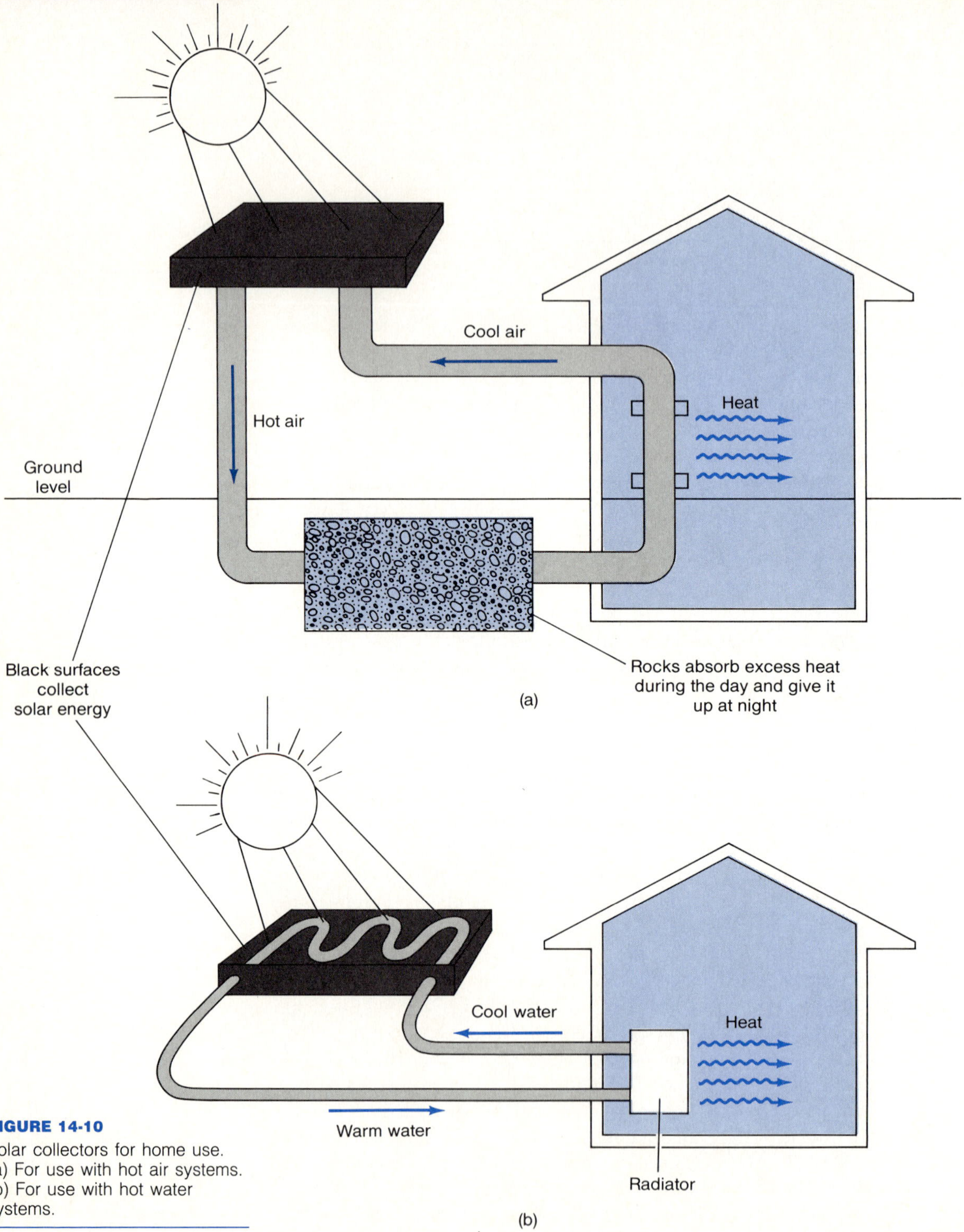

FIGURE 14-10

Solar collectors for home use. (a) For use with hot air systems. (b) For use with hot water systems.

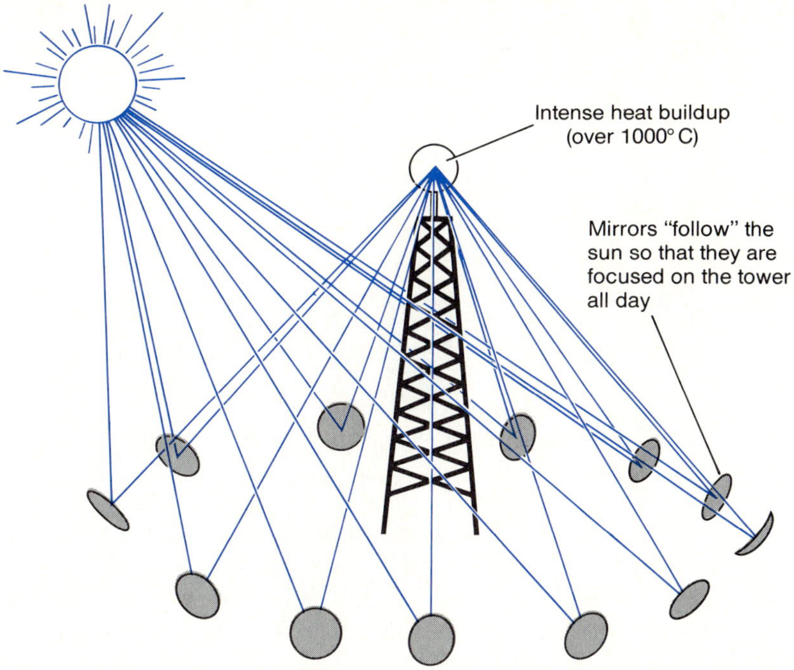

FIGURE 14-11
A solar collector for industrial use.

FIGURE 14-12
A domestic solar heat collector can be used to heat water for direct use or for space heating.

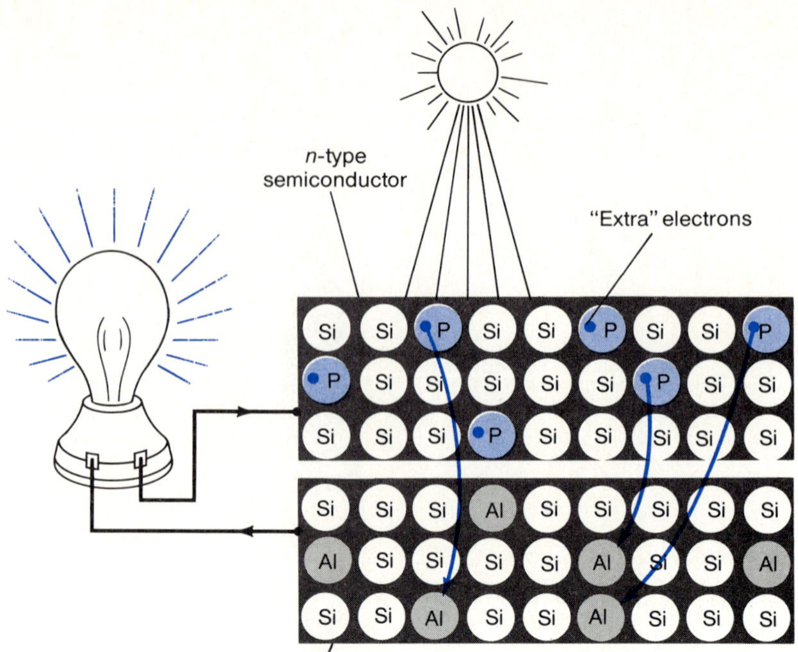

FIGURE 14-13
Principle of operation of a photovoltaic cell.

FIGURE 14-14
A photovoltaic cell (the area above the digital display) used to power a pocket calculator.

trons in their valence shell. Thus, in terms of the regularity of electron density, some areas of the solid are deficient in the number of electrons present, although no actual charge is developed. This is called a *p-type semiconductor*. The other layer is made by mixing silicon or another group IVA element with a group VA material such as phosphorus (P). Since the phosphorus atoms have five electrons in their outer shell, this produces areas where there are "extra" electrons over the four outer-shell electrons found in the silicon atom. This is called an *n-type semiconductor*.

When the sun shines on the top layer, the solar energy causes the electron in the n-type layer to move to the "holes" in the p-type layer. This causes the bottom layer to develop a negative charge and the top layer a positive charge. Once the charge buildup is sufficient, the electrons begin to flow through the external circuit back to the top layer. This is an electric current and can be used to run lights, electronic devices, or small motors. By using these two semiconductor layers, the photovoltaic cell is able to use the sun's energy to "pump" electrons and produce an electric current.

Present technology is at a point where electricity generated with a solar cell is more expensive than that generated by hydroelectric or nuclear plants or from coal. If the silicon layers could be made more cheaply (as no doubt they will be in the future), this could become an important source of electric energy. The main applications to date are space vehicles, novelty demonstration equipment, and pocket calculators (see Figure 14-14) that do not need batteries but can be operated in a well-lighted room.

Nuclear energy

In Chapter 13, the theory and design of nuclear fission reactors were discussed. These reactors are used as a source of *heat energy* and can probably be used for most applications where heat is needed. Current applications involve conversion to electric energy (see Figure 14-5) and to mechanical energy needed to run ships. It would be quite possible to heat whole areas of cities (or even whole cities) from the energy of one reactor. Heat for industrial processes could also be obtained this way.

It may also be possible to use electric energy, heat, or neutron radiation to split water into hydrogen and oxygen. The hydrogen and oxygen could then be transported in liquefied or gaseous form by pipelines or tanks (on trucks or trains) to the site where energy is needed. They could be burned cleanly (no pollution) to produce heat. This would save the electric energy that is presently lost when it is transported through wires. This process is shown below:

$$2H_2O \xrightarrow[\text{neutrons from nuclear reactor}]{\text{electricity, heat, or}} 2H_2 + O_2 \xrightarrow[\text{to home or factory}]{\text{transported}} 2H_2 + O_2 \xrightarrow{\text{burning}} 2H_2O + \text{heat}$$

Hydroelectric power

In certain parts of the United States, such as the Appalachian area and the Pacific northwest, water is a major source of electric energy. Water running over a dam is used to turn electric generators, which is an example of converting mechanical energy into electric energy (see Figure 14-15). Smaller dams could be used in many locations to provide part of the electric energy needs of a given community. As a result of inertia and probably also the negating efforts of the large utility companies, as well as the negative environmental effect of such action, this is not widely done even though it is the least expensive means of producing electricity. Another problem is the availability

FIGURE 14-15
A hydroelectric plant (Shasta Dam in California); the power of running water is used to generate electricity. (By permission of Electric Power Research Institute.)

of locations for building dams (many stream banks are inhabited). The building of dams to trap tidal water is feasible and is actually being used by Canada on the Atlantic coast. The water is released at low tide to run electrical generators. It should be noted that hydroelectric power, like other sources, comes ultimately from the sun. Solar energy causes the evaporation of water that then falls as rain at higher elevations and flows down through the dams. The sun is in effect ''pumping'' the water back uphill so that it can run down again, just as it pumps the electrons in the photovoltaic cell.

Wind

In many cultures wind energy has been used to do work. The Chinese and Dutch are the best-known examples, but before 1940 most American farms had a windmill which was used to pump water from wells and springs into cisterns for human use or into troughs for livestock. There is some interest now in using windmills to run electric generators (as in Figure 14-16) either as a primary source of electricity in areas where the wind blows continuously or as a supplemental source. Wind is the result of uneven heating of the earth by the sun, so that the actual energy comes from the sun. Commercial windmills with generators are available, but their use is limited at this time.

These windmills, along with their generators, can be used near private homes, but because of intermittent wind conditions there is the need to charge batteries in the home, much as an automobile generator charges a car battery. This current can then be used to supplement other electrical sources or used when there is no wind blowing. Large industrial wind-powered electric generators have been built, and more are being planned. These take advantage of local, constant, strong wind currents and are built so that the wind is forced through a vertical tower containing turbines. The turbine motion runs large generators.

FIGURE 14-16
A modern electric generator powered by a windmill: the DOE-NASA 200-megawatt (mW) machine at Block Island, Rhode Island. (By permission of Electric Power Research Institute.)

Geothermal heat

The inside of the earth is composed of hot molten rock and iron (Figure 14-17). At some places, notably those near earthquake and volcanic activity, the hot material is near the surface. (See Figure 14-18.) Pipes can be used to force water into the hot areas; the water returns to the top as steam, which can then be used as a source of heat or mechanical energy or as a source of energy to run generators. Small-scale use is practical; but there is some concern that if too much heat is taken from a given location, the unnatural, rapid cooling of that area may cause contraction of the earth, and earthquakes or other catastrophes could result.

Electrochemical energy

Electricity can be produced directly from chemical reactions, and while only a small percent of our electrical need is produced that way, it is the most economical and convenient method for certain mobile applications. Storage batteries, electric dry cells, and fuel cells are examples and are used to power such devices as automobile starters, flashlights, portable calculators, radios and tape recorders and to provide power in space vehicles. We will examine the basic ideas underlying electrochemical energy and then see how they are applied in several commonly used devices.

Figure 14-19 shows what is usually called a *Daniell cell* (after John Daniell, who invented it in 1836), and will serve to show how all electric cells work. First, if zinc metal (Zn) is reacted with a compound containing copper II ions such as $CuSO_4$, the following reaction takes place:

$$Zn + Cu^{2+} + SO_4^{2-} \longrightarrow Cu + Zn^{2+} + SO_4^{2-}$$

The Zn gives up its two outer-shell electrons to Cu^{2+} so that the Zn is changed to Zn^{2+} and dissolves in the solution. The Cu^{2+}, upon receiving the electrons, becomes Cu metal and forms a solid precipitate. This is shown in Figure 14-20.

The driving force behind this reaction is that the Zn wants to give up its electrons more than Cu does. The Zn is able to "force" its electrons onto Cu^{2+}. Zn is thus said to have a greater *oxidation potential* than Cu—it has a greater ability to give up its electrons and go to a higher oxidation state. *Oxidation* is defined as the process of giving up electrons and going to a higher oxidation number (refer back to Table 10-4 in Chapter 10). The Cu^{2+} ion accepts the electrons and goes to a lower oxidation state. This is called *reduction*.

In the chemical reaction just discussed, the Zn and Cu^{2+} ions are actually mixed in the same container and the transfer of electrons is a direct one: the two come into contact, the electrons simply move from one to the other, and the energy is released as heat. The *tendency* for electrons to be given up or accepted is a form of potential energy, and the moving electrons constitute a form of kinetic energy. An electric current is a flow of electrons through a circuit; if we could physically separate the Zn and Cu^{2+} ions (see Figure

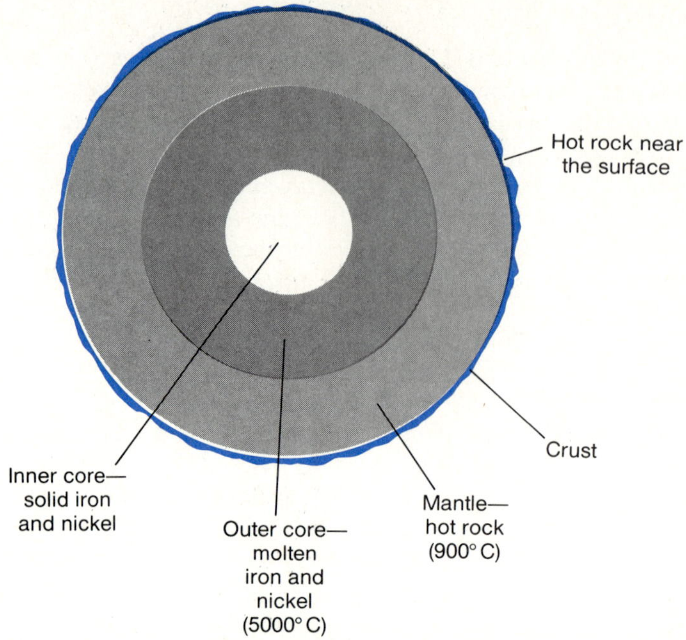

FIGURE 14-17
Cross section of the earth, showing the molten core.

(a)

(b)

FIGURE 14-18
(a) Hot water being emitted from the earth's crust—a potential source of energy in some locations. This is Old Faithful in Yellowstone National Park. (Photograph from Art Resource.) (b) A geothermal wellhead—pipes used to conduct hot water from the natural source. (By permission of Electric Power Research Institute.)

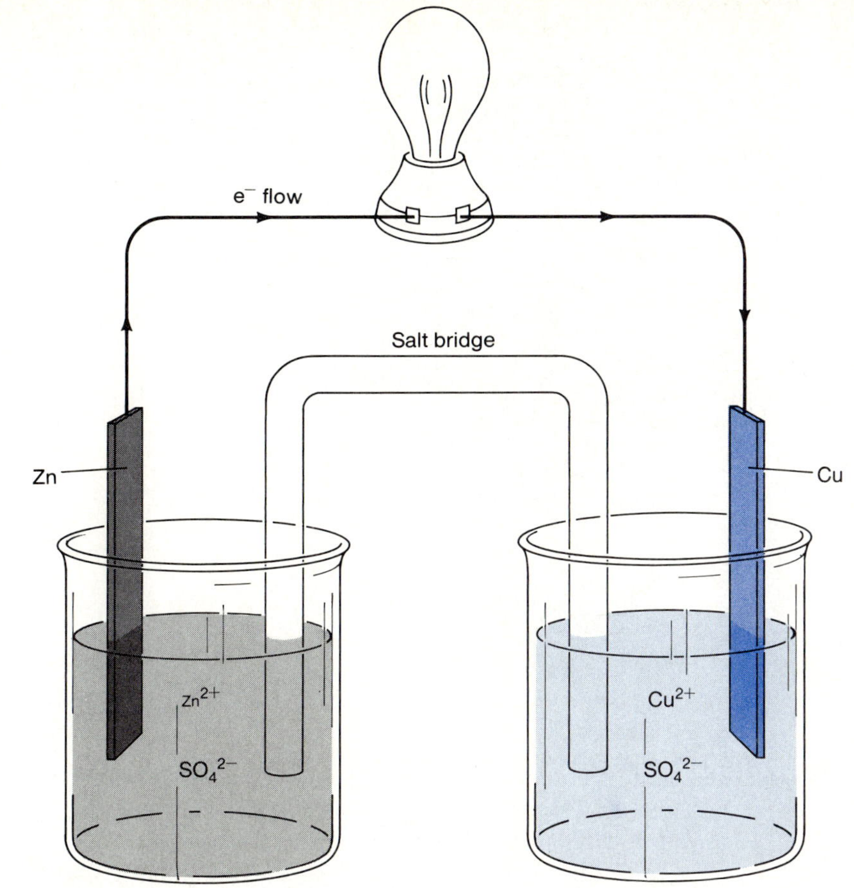

FIGURE 14-19

Left: Operation of a Daniell cell.

FIGURE 14-20

Below: Reaction of zinc with Cu^{2+} in solution by direct contact. (a) Before reaction: *left,* Zn; *right,* Cu^{2+}. (b) After reaction: resultant Cu metal is at the bottom of the beaker.

(a)

(b)

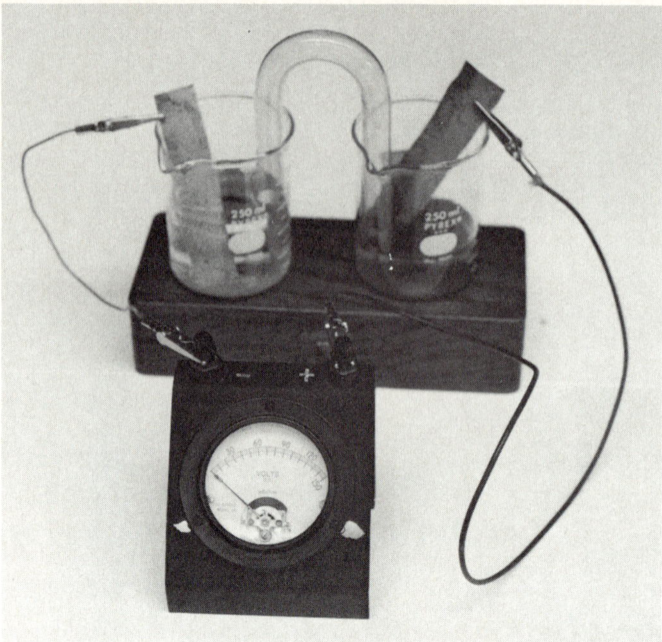

FIGURE 14-21

A Daniell cell generates a flow of electrons by keeping the reactants physically separated.

14-21) and allow the electrons to flow through a wire instead of being transferred directly, we could harness this energy. In Figure 14-21 one observes the Zn (along with a solution of $ZnSO_4$) in one container. The Cu^{2+} ions (as $CuSO_4$) along with Cu metal are in a *separate container*. The metal pieces (Cu and Zn) are called *electrodes,* and they serve a dual function: they conduct electrons in and out of the solutions as well as participate in the chemical process. The two electrodes are connected by means of a wire and an electrical device such as a light bulb. The Zn still has the same *ability* to give up two electrons as when it was in direct contact with the Cu^{2+}, and the same chemical reaction occurs. The Zn electrode slowly dissolves, and Cu metal is plated out onto the copper electrode, but the electrons now travel through the external circuit and cause the bulb to glow. The salt bridge is necessary to complete the electric circuit and to prevent the Cu^{2+} and Zn from coming into direct contact. It is usually made of a gelatinlike material.

The oxidation process taking place at the left-hand electrode in Figures 14-19 and 14-21 is shown by the following equation:

$$Zn \longrightarrow Zn^{2+} + 2e^-$$

This reaction depicted by this equation is called a *half-reaction* and shows only the electrons given up. The electrode where oxidation occurs (electrons are given up) is arbitrarily called the *anode* (it attracts anions). The right electrode is where reduction occurs and is called the *cathode* (it attracts cations). The half-reaction taking place at the cathode is:

$$Cu^{2+} + 2e^- \longrightarrow Cu$$

The voltage (the force with which the electrons tend to flow) of a standard

Daniell cell (where the concentration of the Cu^{2+} and Zn^{2+} ions is the same) is 1.1 volts (V).

The Daniell cell uses solutions and would therefore be inconvenient as a portable source of electricity. For portable electricity, other chemical systems which involve moist chemicals are used. Several of these are shown in Figures 14-22 to 14-26. The carbon-zinc dry cell, shown in Figure 14-22, is used for inexpensive flashlight batteries and comes in a variety of sizes (AAA, AA, A, C, D). Alkaline cells are used for the same purposes (see Figure 14-23) but have a longer life and are more expensive. The chemical reactions are the same as in the zinc-carbon cell, but the porous zinc and the KOH allow the voltage to stay at a high level longer. The voltage, like that of the carbon-zinc cell, is 1.5 V. The nickel-cadmium cells shown in Figure 14-24 come in the same sizes as the carbon-zinc cells, but once the reaction has occurred, they can be placed in a recharger which uses household current to reverse the chemical process forming Cd and NiO_2 again. Manufacturers claim that these can be recharged hundreds or even a thousand times.

The *lead storage battery* used in automobiles has a somewhat complex chemistry into which we will not go in detail. Basically, there are six individual cells, each one generating 2 V. The battery delivers 12 V total. During electricity generation, the Pb anode is oxidized, giving its electron to PbO_2, which is part of the cathode. These two half-reactions are shown in Figure 14-25. During recharging, which happens when the alternator or generator in the motor begins to run, electrons are fed back into the anode, causing the $PbSO_4$ to change to Pb metal once more:

$$\text{Recharging anode reaction:}\quad PbSO_4 + 2e^- \longrightarrow Pb + SO_4^{2-}$$

Likewise, the alternator forces electrons to move from the cathode and thus regenerates the PbO_2:

$$\text{Recharging cathode reaction:}\quad PbSO_4 \longrightarrow PbO_2 + 2e^-$$

After a time, the anode is only Pb and the cathode only PbO_2. The battery is now fully charged and ready to produce current again on demand.

Fuel cells are also devices in which chemical reactions are used to produce electricity directly, but they differ from the other commonly used devices in two ways. They are designed so that fuel or fuels can be fed into them continuously, and the electrodes (which are platinum) are not part of the chemical change. The platinum electrodes do not react; they serve only to conduct the electron to and from the solution. In the fuel cell diagramed in Figure 14-27, hydrogen is the fuel, and it reacts with oxygen. The reaction is similar to combustion, but there is no flame. Current is produced continuously as long as the fuel supply lasts. In this cell, the product of the reaction is water, which is one reason why fuel cells are used to supply part of the current on spacecraft—the water can be used for drinking. Thus, a given weight of H_2 and O_2 can be used for two purposes—to produce electricity *and* to provide water. If water has to be taken aboard and storage batteries are used, the overall weight (and thus the expense) is greater. In contrast to batteries, fuel cells produce an electric current only when fuel is being fed into them; no storage of electric energy is possible.

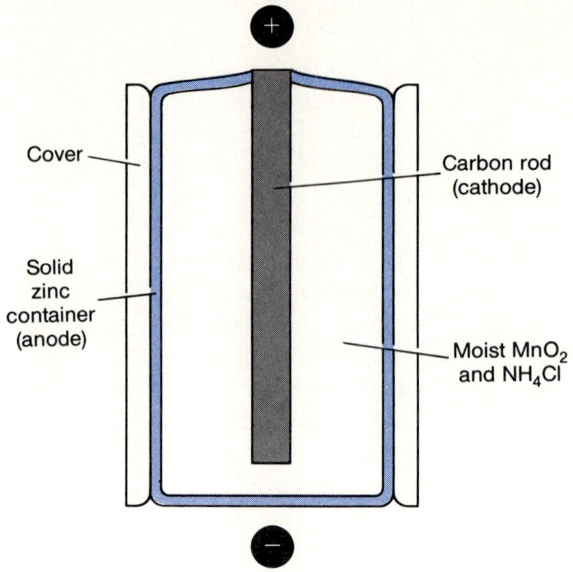

Half reactions: $Zn \rightarrow Zn^{2+} + 2e^-$

$$2MnO_2 + H_2O + 2e^- \rightarrow Mn_2O_3 + 2OH^-$$

FIGURE 14-22
Carbon-zinc cell.

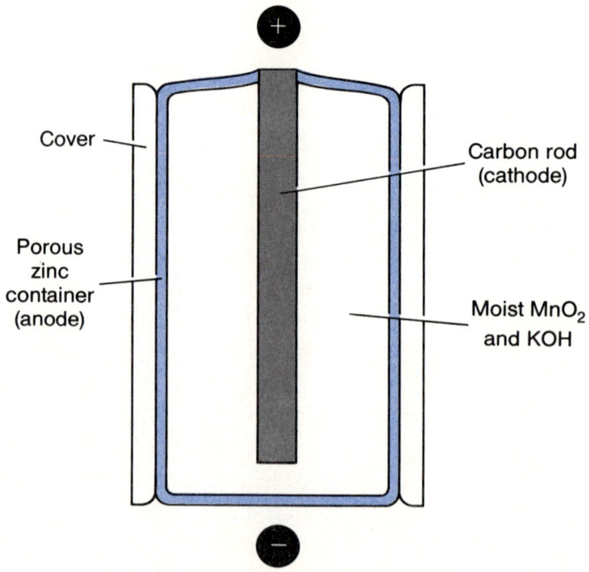

Half reactions: $Zn \rightarrow Zn^{2+} + 2e^-$

$$2MnO_2 + H_2O + 2e^- \rightarrow Mn_2O_3 + 2OH^-$$

FIGURE 14-23
Alkaline cell.

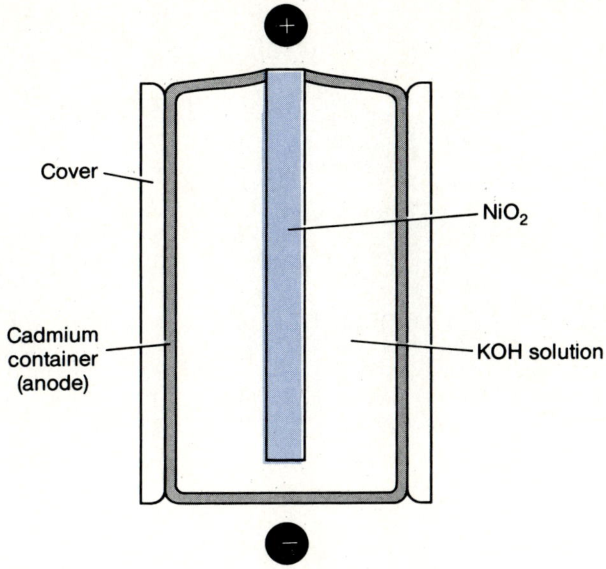

Half reactions: $Cd \rightarrow Cd^{2+} + 2e^-$

$NiO_2 + 2e^- \rightarrow NiO$

FIGURE 14-24
Nickel-cadmium rechargeable cell.

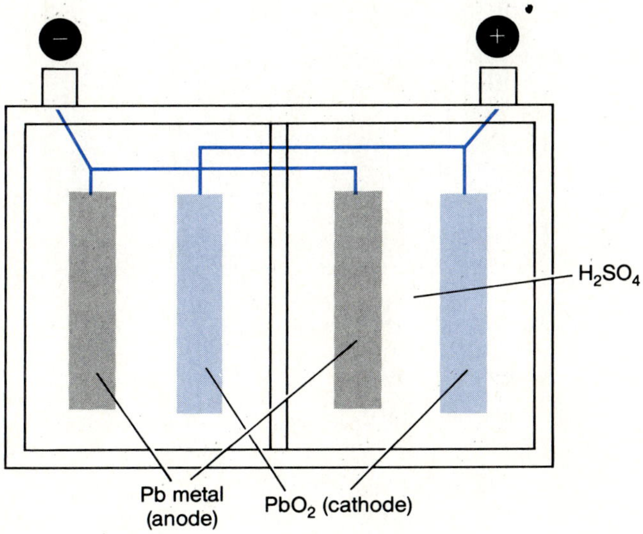

Half reactions: $Pb + SO_4^{2-} \rightarrow PbSO_4 + 2e^-$

$PbO_2 + 4H^+ + SO_4^{2-} + 2e^- \rightarrow PbSO_4 + 2H_2O$

FIGURE 14-25
Lead storage battery.

FIGURE 14-26
Commonly encountered storage cells and batteries.

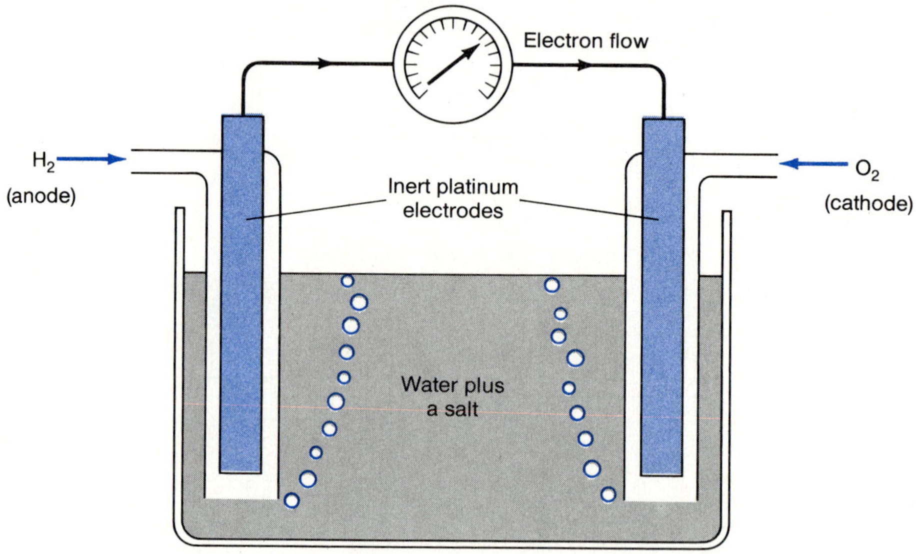

Half reactions: $2\ H_2 \rightarrow 4H^+ + 4e^-$ $\quad$ $4e^- + O_2 \rightarrow 2\ O^{2-}$

Supplimentary reaction: $4H^+ + 2\ O^{2-} \rightarrow 2\ H_2O$

FIGURE 14-27
Fuel cell.

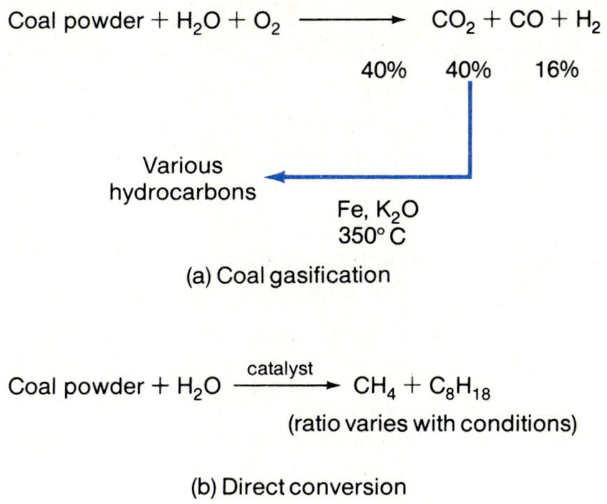

FIGURE 14-28
Production of synthetic fuel.

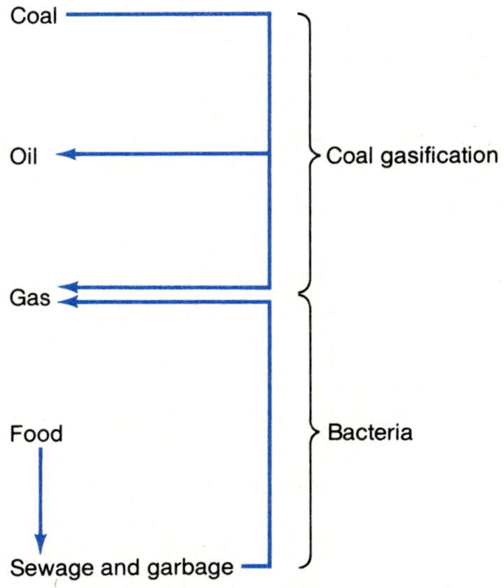

FIGURE 14-29
Interconversion of fossil fuels.

Synthetic fuels

Because coal is so abundant, there has been considerable interest since the late 1960s in developing methods of converting it into methane or gasoline and fuel-oil-type hydrocarbons, which not only are less polluting but are usable in internal combustion engines and furnaces. It is possible to expose powdered coal and water to heat and a catalyst to obtain coal gas, which can then be changed to CH_4 or C_8H_{18} depending on the exact conditions and catalysts used. Two of these schemes are shown in Figure 14-28, and an overall view is shown in Figure 14-29. Although the technology has been developed and pilot plants have been built, there has regrettably not been a large effort on the part of government or private industry to produce large amounts of these "synfuels." Such an effort, if successful, could release the United States from dependence on other oil-producing countries. In South Africa all petroleum needs have been met for years with synfuel from coal. It produces a form of economic independence, but it keeps the price of petroleum high and requires large government subsidies. One has to ask which is more important: the free-market concept or economic independence.

Renewable energy sources

Another interesting possibility is to develop a synthetic catalyst that will act as chlorophyll does in green plants and cause CO_2 and water to be converted to carbohydrates in reaction vessels, using sunlight as the energy source. These carbohydrates could be used as food or burned for fuel. At the present time, this approach is still at the theoretical stage, and the necessary catalyst has not yet been prepared.

Synthetic methane (CH_4) can be produced from sewage by certain bacteria. A plant in India uses human sewage to produce CH_4 from fermentation of waste materials. The CH_4 is then burned to produce steam, which produces electricity. This approach solves the problems of pollution of waterways (since the sewage is not dumped into rivers; see Chapter 12) and air pollution (since methane burns cleanly). It also provides a constant source of fuel that is roughly proportional to the population in a given area and thus to the energy needs of that area. The author knows of no such major public facility in the United States, although some individuals operate private ones. Garbage can also be treated in a similar fashion.

Grains such as corn can be fermented to produce ethyl alcohol (see Chapter 16 for a fuller discussion of this process), and methyl alcohol can be (and is) produced by chemical means in large quantities relatively inexpensively. These make good fuels for internal combustion engines, especially when they are mixed with gasoline. The octane rating is improved without addition of tetraethyl lead or benzene.

There are other renewable plant-derived energy sources. The jojoba plant grows well in desert areas and produces an oil which can be isolated by pressing the plant or heating it to distill the oil off. This oil can be burned directly or further refined. A South American tree which is in the same family as the rubber tree produces a hydrocarbon material which can be obtained by cutting the tree and allowing the material to drip out in much the same way as a sugar maple or rubber tree is tapped. The sap can be used directly (i.e., poured into a fuel tank right from the collecting bucket) in diesel vehicles. The widespread cultivation of such plants could produce a significant portion of our fuel.

Another approach to renewable fuel is the production of H_2 gas from water. This can be done only at very high temperatures:

$$2H_2O \xrightarrow{\Delta\Delta\Delta} 2H_2 + O_2$$

At these high temperatures, the H_2 and O_2 tend to recombine to form H_2O. Methods must be used which allow separation of the H_2 from the O_2 so that the H_2 can be transported and used as a fuel elsewhere. One kind of energy such as solar energy is used to split the water, and then this energy is gotten back again when the H_2 is burned.

One possible way of accomplishing the splitting of H_2O and the separation of H_2 and O_2 is to use a series of chemical reactions called *thermochemical cycles*. One such cycle is:

1 $4H_2O + 2SO_2 + 2I_2 \longrightarrow 2H_2SO_4 + 4HI$

2 $2H_2SO_4 \longrightarrow 2H_2O + 2SO_2 + O_2$

3 $4HI \longrightarrow 2H_2 + 2I_2$

The O_2 and H_2 are produced during different chemical reactions and so can be separated. You will notice that the SO_2 and I_2 used in step 1 are gotten back in step 2 and step 3, respectively. The H_2SO_4 produced in step 1 is used in step 2, and the HI produced in step 1 is consumed in step 3. Of the four H_2O used in step 1, two are produced again in step 2, so that only two H_2O molecules are actually split. The only real change that occurs during this cycle is:

$$2H_2O \longrightarrow 2H_2 + O_2$$

The energy required for this cycle could be produced from fossil fuel or solar or nuclear sources.

IS THERE AN ENERGY CRISIS?

To answer this important question, we must first look at the extent of our resources in each of the areas now available. Table 14-3 gives an overall picture. It can be seen from Table 14-3 that there are many natural sources of energy available. Some are more abundant and less expensive, some are less polluting to use, and some have been developed to a higher efficiency. We have, because of past and present economic advantages, tended to base our current fuel economy on fossil fuels: coal (20 percent), oil and gasoline (45 percent), and natural gas (30 percent). The remaining 5 percent is mostly from nuclear and hydroelectric sources. Because of the relative scarcity of these fuels and the fact that much of the oil is in a few nonindustrialized countries, the cost of using these fuels has risen in recent years.

We had become used to being fuel wasteful (''fuelish,'' if you will). The higher costs and temporary low supplies of gasoline and electric power have encouraged a tendency to *conserve* on our energy consumption. Americans now consume less gasoline than 10 years ago. Better insulation in homes and more efficient electrical appliances are lowering use of electric power. It would probably be possible to reduce our present energy needs over the next 30 or 40 years by 25 to 30 percent by using conservation alone.

The answer to the question ''Is there an energy crisis?'' is *no*. By conserving, by developing higher-efficiency devices, and by going ahead at full speed to develop solar, nuclear, and other alternative energy sources, we can supply our needs and wants indefinitely.

TABLE 14-3

Estimated energy resources

SOURCE	QUANTITY AVAILABLE*	NOTES
Fossil		
Natural gas	20 years	The least-polluting fuel. We have enough for 50 years or more at current use levels. There may be very large quantities of CH_4 in deep deposits, but this has not been proved.
Petroleum	50 years	No one knows for sure how much there is.
Coal	500 years	Most polluting if burned directly. The United States has more coal than any other country.
Biomass (living plants)	Indefinite	Plants grow slowly and take a large amount of land.
Solar	Unlimited	Not equally distributed geographically or seasonally. Technology is not yet fully developed. This source is not polluting, but the production of collecting devices could be.
Nuclear		
Fission	25 years for uranium-235. 1000 years for breeder reactors using plutonium-239	Expensive. This source produces radioactive, toxic fissionable products.
Fusion	Unlimited. (T_2 and D_2 from the oceans are available in large quantities.)	Clean. Technology is still undeveloped.
Hydroelectric	Unlimited duration. This source, however, is limited to certain areas and a low volume of consumption.	Dams cause some ecological disruption, and valuable farmland and historical sites as well as whole towns are frequently destroyed.
Wind	Unlimited duration. Limited locations and unreliability are serious disadvantages.	
Geothermal	Availability quite limited.	If a large amount of heat is removed from an area, the geological implications are unknown.
Electrochemical	Indefinite. Fuel cells could produce electricity indefinitely.	This source is limited to small-volume applications.

*Years of energy available if this were the source of all energy needed.

The use of *coal* (remember that we have about a 500-year supply of it) and the *synthetic* methane and gasoline that can be made from it will be important during the years while alternative energy sources are being developed. There is cause for concern over energy supply and use, and there is a clear need for more research. However, no crisis is at hand, nor is there any need to panic or be held hostage by other oil-producing countries or by the lobbies for the large oil-producing companies in this country.

A word now about energy conservation. There are a number of ways in which individuals and companies can reduce energy consumption. They include moderating temperatures in living and working space. In the winter, the air temperature needed for comfort is about 70°F, and this temperature can be lowered by wearing heavier clothing and raising the humidity. The tendency is for the relative moisture content in the air to decrease as the air is heated. Heating air in a cold month can cause as low as 30 percent relative humidity (since warm air can hold more water). This not only causes furniture and skin to dry out but causes the air to *feel* colder at a given temperature. If the relative humidity is raised to about 50 to 60 percent, the same comfort can be experienced at a temperature at least 5°F lower and possibly as much as 10°F lower. Growing plants are good humidifiers, and even a mechanical humidifier or one placed in the heating system (if it is a forced-air type of heating system) uses less energy than keeping the temperature higher.

Another way of conserving energy is to use public transportation and car-pooling arrangements. Buying more energy-efficient cars also makes a tremendous difference. As an example, a person driving to work 20 mi each day five times a week can save about $60 (almost 50 gal of gasoline) each month with a small car getting 35 mi/gal compared with a large, older car which gets only 12 to 15 mi/gal.

Home insulation and storm windows are cost-effective. In mid-America the cost of insulating can be recovered in 3 to 5 years and absolute savings can be realized after that. One should be careful not to fall prey to salespeople who claim that their very expensive products such as double-pane windows are a great deal better as insulators than ordinary storm windows. They may be a little better, but they cost a lot more. Salespeople may use charts and tables in an attempt to convince you that the $5,000 to $10,000 needed to outfit your home with their product will be recovered quickly. One such claim is that even after insulating walls and ceilings, 70 percent of the heat loss is through windows and doors (which is no doubt true), with the implication that you will save 70 percent on your energy bill if you install the new product. You should always think carefully for at least several days about any data presented by a salesperson. This should be done after the salesperson has left, so that the manipulative techniques salespeople are trained to use will not affect your clarity of mind and your emotions will not control your decisions. Resist pressure, such as "I can give you a special offer today but only if you decide now. Sorry, I can't wait for a phone call." Any legitimate company interested in establishing sound customer relations by selling and servicing a good product will not place undue pressure on you. It is your money, and you would be foolish to make a decision about a product (or anything else) at the spur of the moment or under pressure.

Another way to save energy is to use energy-efficient appliances. Some air conditioners, refrigerators, water heaters, furnaces, etc., do use less energy than others to produce the same amount of usable energy. Water heaters, for example, should be very well insulated. But here again one should be careful of companies (even some reputable ones) that attempt to sell products with the claim that they will reduce energy costs. Not all products will do that. For instance, there is an insulating jacket for a water heater. It costs $30 to $50, but the heat loss in a fairly new high-quality heater is so small that it would take a long time to recover the cost of this device. Eventually, one would save a little money (a few dollars each year). On a national scale this device is not very helpful after one considers the energy needed to manufacture it and the fuel needed to transport it to the warehouse and to the customer's home. There are ways of significantly conserving energy, and we should all do our part; but not all claims about energy savings are valid.

In conclusion, Figure 14-30 shows the relative importance of the various energy sources over our nation's history. Note that use of coal dipped after 1930 but is up again now, and that hydroelectric power production has been constant for almost a century. Note also that nuclear power has been in use only since about 1965 and contributes only a small amount. Gas and oil are the most-used energy sources. It should be pointed out that production of electricity from fossil fuel is only about 30 percent efficient; most alternative sources are much more efficient than this. Thus for each unit of electricity produced, two or three units of fossil fuel could be saved.

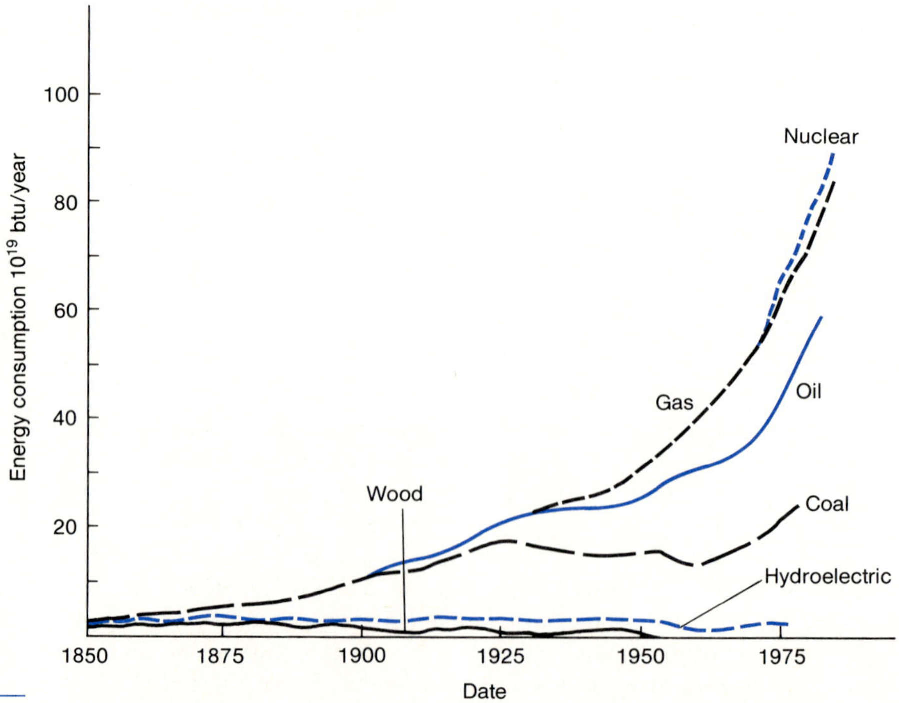

FIGURE 14-30

Energy sources in the United States since 1850. (Used by permission of *Journal of Chemical Education.*)

STUDY GUIDE

Terms to learn

Chemical energy	Kinetic energy
Daniell cell	Nuclear energy
Electrochemical cell	Oxidation
Electrodes	Photovoltaic cell
Endothermic	Potential energy
Energy	Reduction
Exothermic	Renewable energy sources
Fossil fuel	Solar collector
Freon	Solar energy
Fuel cell	Synfuel
Geothermal energy	Voltage
Heat pump	Wind energy
Hydroelectric energy	

Skill practice problems

1 Why is air conditioning considered a necessity? Is it really a necessity?

2 Is there an energy crisis? Is there an energy problem?

3 From memory, diagram a cooling system and explain how it works.

4 What is the formula for Freon 12?

5 Which is the best description of molecules of cool air compared with molecules of warm air?

(a) They are moving faster.
(b) They are smaller.
(c) They are moving slower.
(d) They have more energy, since energy was needed to cool them.

6 What are the two main reasons for the rise in energy consumption in recent years?

7 List three activities you have been involved in this week that used mechanical energy. Do the same thing for electric, heat, and chemical energy.

8 Can the following energy conversions be made *directly?* What device is used in each case?

(a) Nuclear to electric
(b) Heat to mechanical

(c) Electric to heat
(d) Solar to electric
(e) Heat to electric
(f) Chemical (fossil) to mechanical

9 Which of the following are natural (spontaneous) processes? Which require energy from another source to happen?
(a) A rock rolling *down* a hill
(b) A rock rolling *up* a hill
(c) An ice cube melting when placed in a pan of water at 50°C
(d) An ice cube getting colder when placed in a room at 15°C
(e) A turbine *starting* to spin

10 Which of the following is an *n-type* semiconductor?
(a) Si plus Al
(b) Ge plus As
(c) Ge plus Ga
(d) Ge plus P

11 What is the *least* expensive method of producing electricity?

12 Most of our energy today comes from which major terrestrial source?

13 What are the advantages and disadvantages of the use of solar energy?

14 What are the advantages and disadvantages of extensive use of coal burning for energy production?

15 What are the advantages and disadvantages of fusion reactors?

16 Is the following statement *true* or *false?* Solar energy is a totally nonpolluting energy source.

17 Write and balance the equation for the burning of butane.

18 What reaction takes place at the *anode* of a fuel cell that uses H_2 and O_2?

19 What is the half-reaction at the *cathode* of a carbon-zinc electrochemical cell?

20 Which of the following half-reactions takes place at the *anode?*
(a) $Fe^{2+} \longrightarrow Fe^{3+} + 1e^-$
(b) $Cu^{2+} + 1e^- \longrightarrow Cu^+$

21 Which of the materials are *reduced* during the following chemical changes?
(a) $2Na + Cl_2 \longrightarrow 2NaCl$
(b) $PbSO_4 + 2e^- \longrightarrow Pb + SO_4^=$
(c) $Mg + F_2 \longrightarrow MgF_2$

15

BIOCHEMISTRY AND NUTRITION

Goals

- **To understand that all living systems are controlled by the laws of chemistry**
- **To learn practical information about personal nutrition**
- **To understand that our bodies are extremely complex yet work amazingly smoothly**
- **To learn the principles of nutrition and weight control**
- **To understand that scientists can gather data on controversial issues but cannot always give final answers**

One of the most important areas in which chemistry has made significant contributions to our well-being is nutrition. Most (though not all) Americans, even if they do not eat enough food or the best kinds of food, are relatively well nourished. Our understanding of nutritional needs and the implementation of food enrichment programs can be thanked for this. As with many other chapters in this book, the discussion here will by no means be an exhaustive treatment. It will cover basic biochemical materials and processes, especially those that relate to food, and selected aspects of nutritional science which are of special interest or practical value to the consumer.

BIOCHEMISTRY—AN OVERVIEW

The human body can, for many purposes, be thought of as an extremely complex chemical system. Many different kinds of chemical changes are occurring simultaneously on many biochemical substances. Consider the simple act of putting a piece of food into your mouth. You must *see* the location of the food, which involves light hitting the back of the eye and causing chemical changes in the eye. These chemical changes in turn initiate chemical changes at each nerve cell, which result in the transmission of signals to the brain. The brain, through chemical changes, recognizes the combination of signals as "food." Then more chemical changes have to occur in order for you to decide to eat the food. Other signals that originate from chemical changes go to the muscles, which cause the molecules in the muscles to contract, each one contracting just the right amount so that you can grasp the utensil and move it to your mouth. The simple act of putting a piece of food into your mouth involves changes in literally thousands of different substances, all taking place in a few seconds, and it all happens so smoothly that you are usually not aware of its complexity. All life processes, even relatively inactive ones, such as your reading (and understanding!) this chapter, are quite complex. Biochemistry is the most incompletely understood of all areas of chemistry, and while the structure and function of many biochemical substances have been unraveled in recent years, there are many more found each year. We have only just begun to fathom the complexity of the chemical compounds of living systems, and we hardly understand most of the processes except on a superficial level.

Basic categories of biomolecules

There are four basic categories of biomolecules: lipids (fats), carbohydrates, proteins, and nucleic acids. Figure 15-1 shows how these fit into the overall scheme of chemical knowledge. Each of these will be discussed in some detail later in this chapter. In the scheme shown in Figure 15-1, the complexity increases as one goes down. Elements combine to produce inorganic compounds, which can in turn combine to produce relatively simple organic materials. This does not mean, however, that organic materials are necessarily produced from inorganic compounds. It does mean that organic materials are more complex than inorganic materials. As the organic molecules polymerize (form large chains), biomolecules are formed. These combine to form the

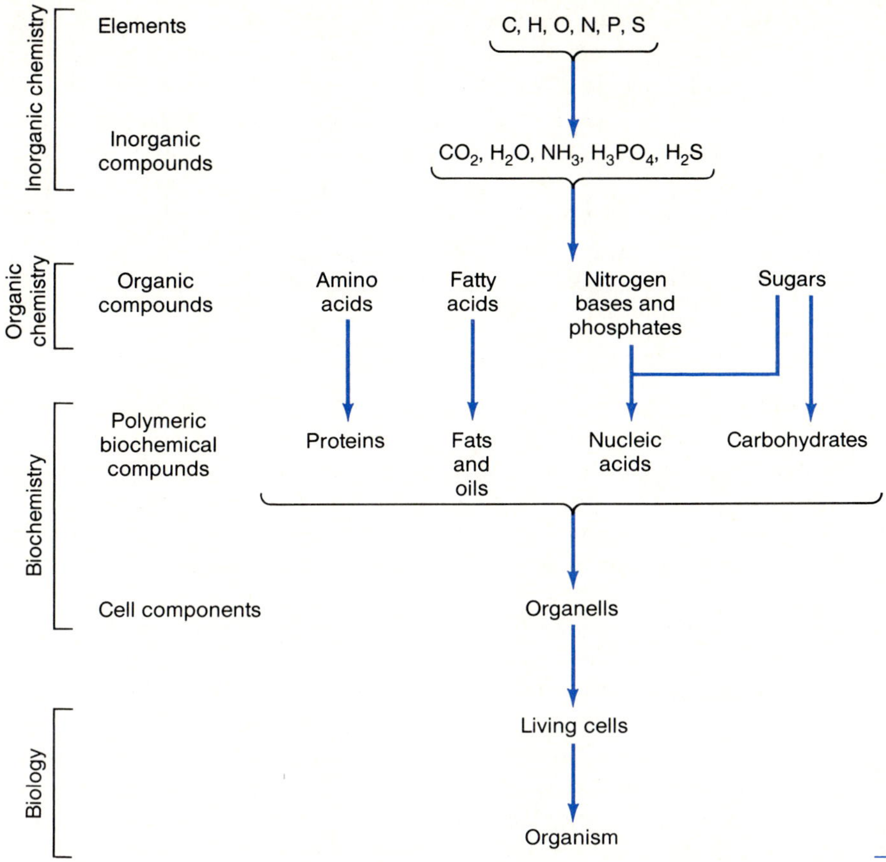

FIGURE 15-1
Overview of biochemical materials and their relationship to other areas of chemistry.

components of cells. The cells form multicellular organisms. The classical areas of science that have traditionally studied each level are shown on the right.

Human nutritional needs

Human nutritional needs fall into six categories: fats, carbohydrates, proteins, vitamins, minerals, and water. Figure 15-2 shows the fate of these nutrient materials in the human body. Dietary fats are used to produce hormones, body fat, and energy, while digestible carbohydrates are almost all changed to glucose. The glucose is converted to energy or changed to body fat or glycogen if the energy produced is not needed by the person to function. Nondigestible materials (fibers—mostly carbohydrate fibers) aid in the digestive process. Proteins in food are broken down into amino acids, which are built back up again into body protein materials including muscles and structural parts, such as bones and tendons, or changed to enzymes (catalysts needed for each biochemical process). Vitamins are organic materials needed in trace amounts, and minerals are inorganic ions also needed in small amounts compared with fats, carbohydrates, and proteins. Vitamins and minerals serve

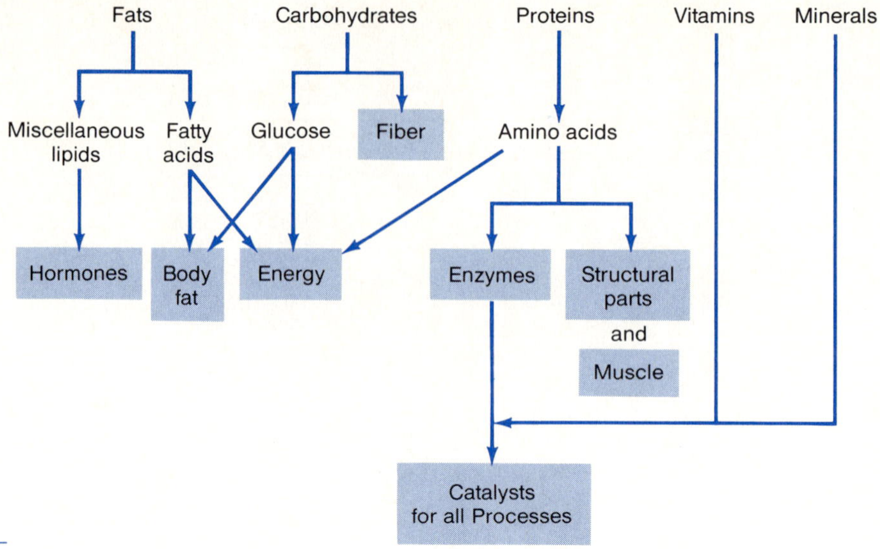

FIGURE 15-2
Fate of nutrient materials in the human body.

many functions, but frequently they are needed by enzymes in order for the enzyme to function. Their role is to activate the enzymes so that the various chemical processes that these enzymes catalyze can be maintained at the proper rate. Water is needed as a solvent or medium for the entire system and as a reactant in many chemical changes.

Basic categories of biochemical processes

There are four basic kinds of biochemical processes. The first is *breaking food down* into simple compounds (see Figure 15-3). The second involves *building* (synthesizing) these simple materials into the complex ones needed in the human biosystem. The third is the *production of energy*. Energy is needed by most body processes; it is needed to move (for muscle action) and to provide heat to maintain proper body temperature. The fourth is *regulation* of all the above processes and involves *hormones* and *enzymes*. All biochemical processes are collectively referred to as *metabolism*.

The living cell

Table 15-1 shows the approximate composition of a living cell. Different organs have cells that vary somewhat from these, but the composition of each type of cell is usually in the range given. The important point is that there are several thousand different materials in each cell. Proteins, for example, are all synthesized in the cell as they are needed in just the correct amounts for the functioning of that cell at a given moment. As we will see later in this chapter, each protein is quite complex itself, and its synthesis requires many steps (50 to 100 in most cases). All proteins are produced synthetically in a minute or less. There are no unwanted by-products, and each reaction has a nearly 100 percent yield. This is nothing short of amazing.

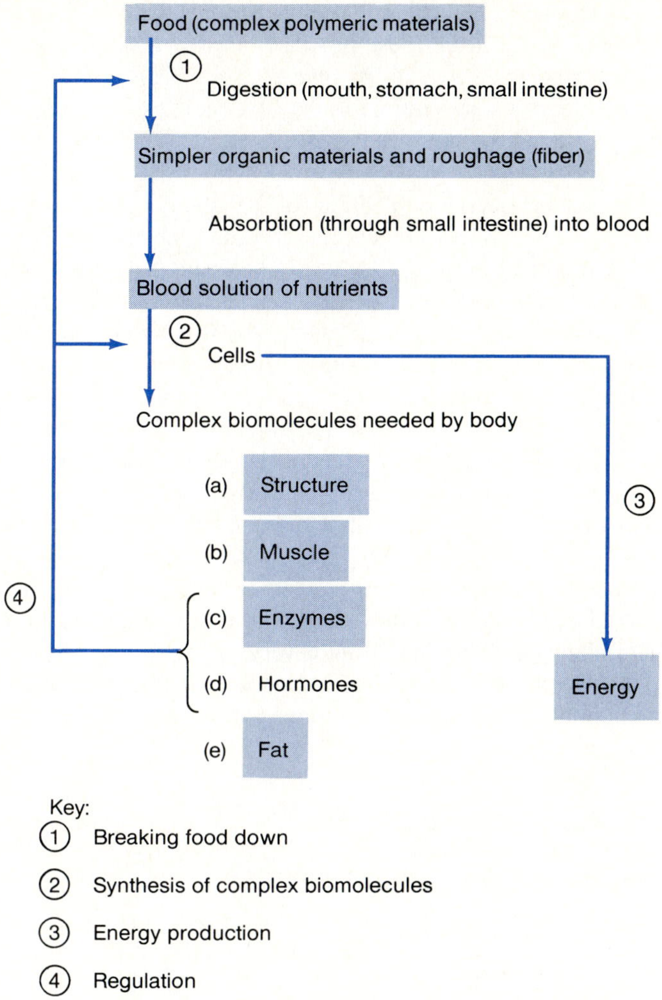

FIGURE 15-3
Overall metabolic scheme.

TABLE 15-1
Biochemical composition of living cells

TYPE OF MATERIAL	PERCENT BY WEIGHT	NUMBER OF DIFFERENT MATERIALS OF THIS TYPE
H_2O	70–80	1
Proteins	13–17	3000
Nucleic acids	6–9	1000
Carbohydrates	3–5	50
Lipids	1–10	40
Inorganic ions	1	10
Miscellaneous	2	500

Enzymes are proteins that serve as catalysts for *all* biochemical processes and greatly speed up these processes. Each of the steps in the synthesis of each biomolecule requires its own enzyme. Thus a large number of enzymes are required at a given time in the functioning of a living cell. Each enzyme has to be synthesized itself before it can cause other syntheses to occur.

This general overview of biochemistry should give a sense of the extreme complexity involved. When one considers all the processes producing and using biomolecules in all cells of all organs in the human body, one must surely stand in awe. An appreciation for the wonder of life from a chemical point of view is certainly one goal of the first part of this chapter. The goal of the biochemist is to isolate all cell components and determine the molecular nature of each, as well as understand all processes in which they are involved. There is much yet to be learned before this goal can be realized. We do not yet completely understand *normal* biochemistry; the various disease states are usually understood even less well. Eventually, of course, one would hope not only to have a thorough understanding of the biochemistry of each system but also to know the chemical basis for malfunctions. This could lead to a rational approach to designing drugs or other remedies for each medical problem encountered. We are not even close to reaching this goal, nor is the prospect of realizing it in the near future very great.

With this overview in mind, we will look more closely at each category of biomolecule and then at various aspects of nutritional interest.

BIOMOLECULES

Amino acids and proteins

Proteins are composed of many smaller units called *amino acids* covalently bonded to each other. Large molecules formed by linking together smaller ones are called *polymers,* or *macromolecules,* and the units of which they are composed are called *monomers*. Amino acid molecules contain both the carboxylic acid (COOH) group and the amino (NH_2) group (see Chapter 11). The general formula is:

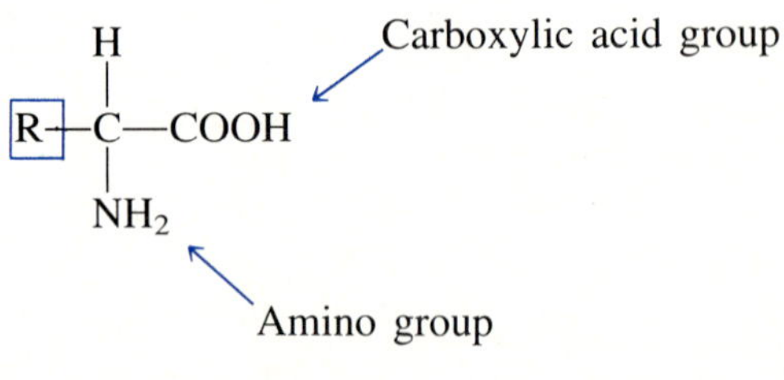

where R stands for any one of 20 different groups. All amino acids (except proline) have the portion *outside* the box in common and differ only in the nature of the R group. Table 15-2 gives information about the 20 amino acids that occur in proteins. There are several others found in biosystems, but they

TABLE 15-2

Amino acids

NAME	STANDARD SYMBOL	FORMULA*
Alanine	Ala	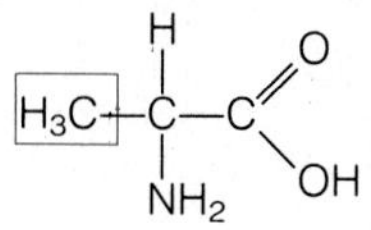
Arginine	Arg	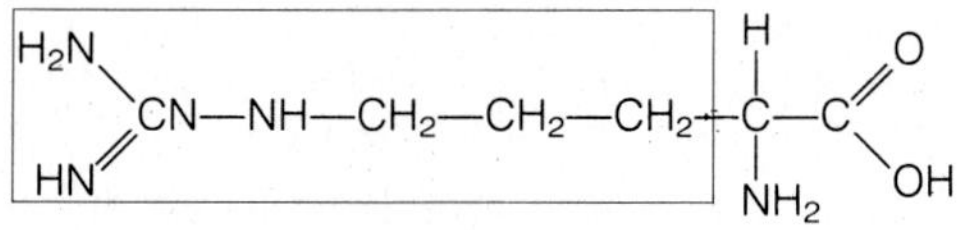
Asparagine	Asn	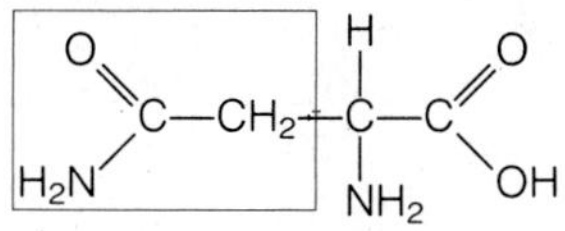
Aspartic acid	Asp	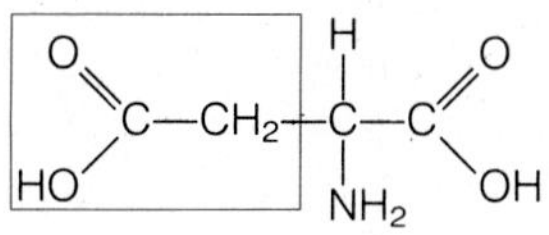
Cysteine	Cys	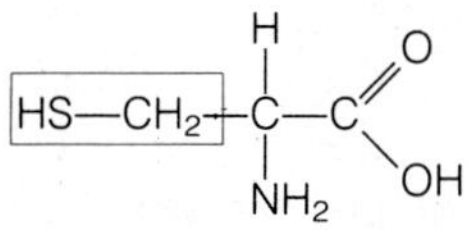
Glutamic acid	Glu	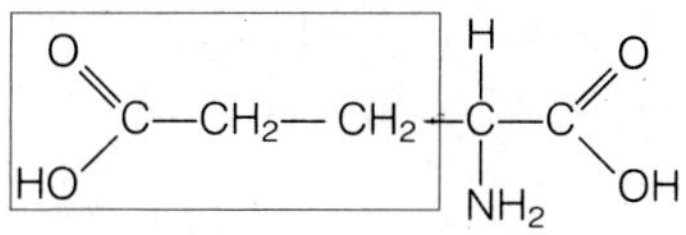
Glycine	Gly	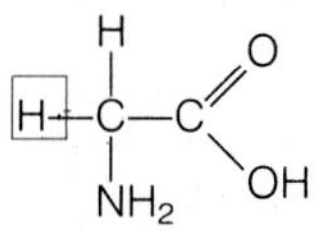
Glutamine	Gln	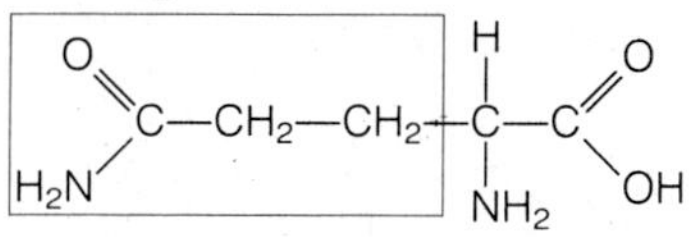
Histidine	His	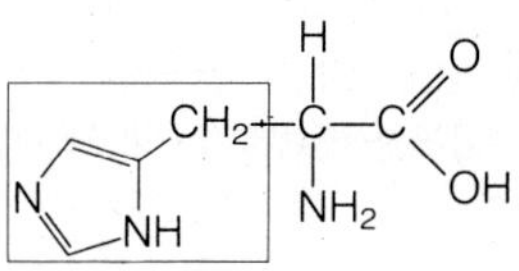
Isoleucine	Ile	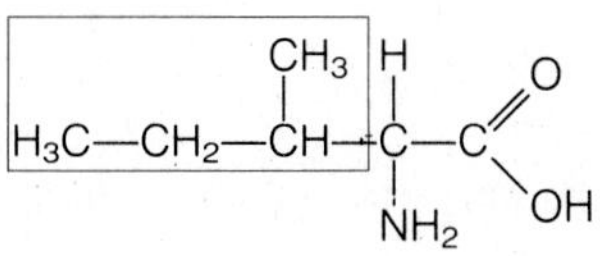
Leucine	Leu	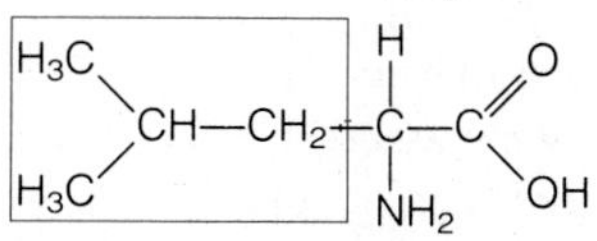
Lysine	Lys	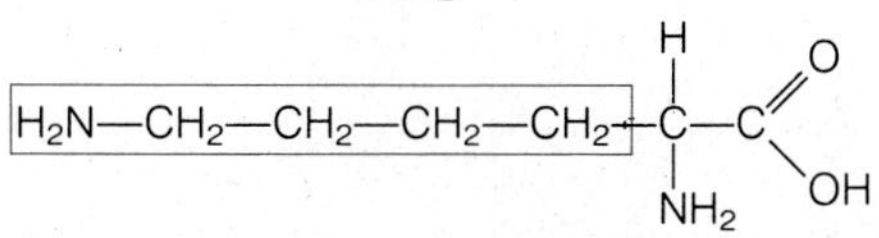

(Continued)

TABLE 15-2

Continued

NAME	STANDARD SYMBOL	FORMULA*
Methionine	Met	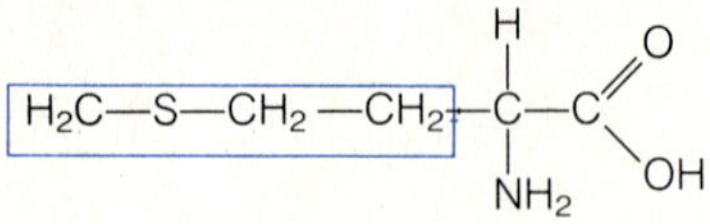
Phenylalanine	Phe	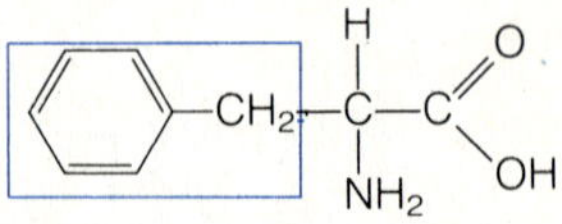
Proline	Pro	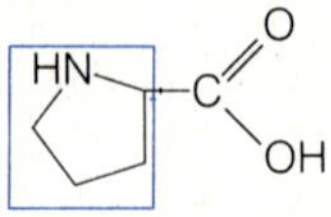
Serine	Ser	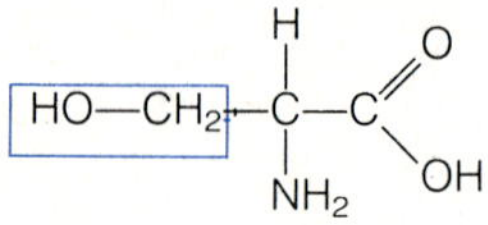
Threonine	Thr	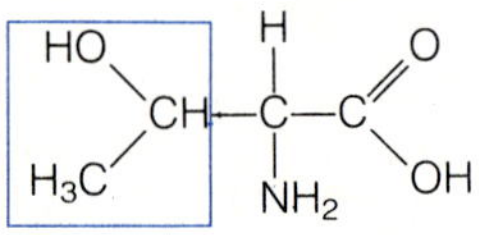
Tryptophan	Trp	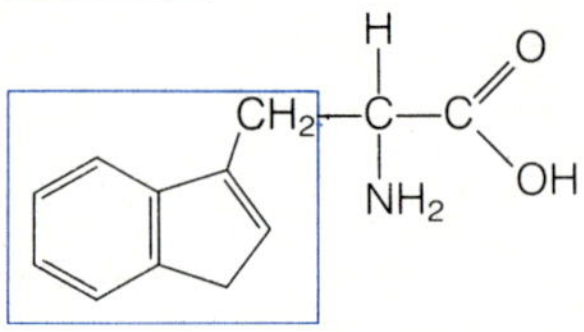
Tyrosine	Tyr	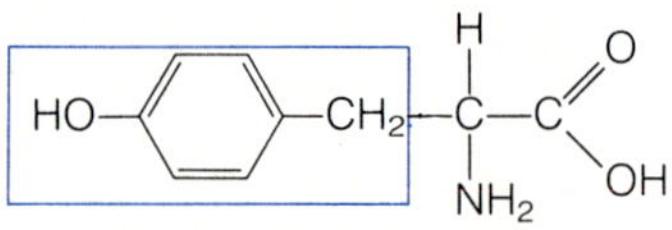
Valine	Val	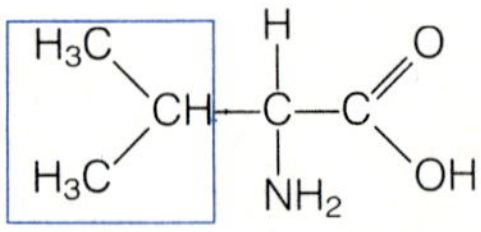

*R group is enclosed in box.

are not used to produce proteins.* Notice that each one has a standard abbreviation.

You will notice that in glycine, which is the simplest amino acid, the R group is an atom of H, while alanine has a methyl group at that location. Each amino acid has an R group with a different *size, shape, polarity* (some are

*Several other amino acids such as cystine and hydroxyproline are formed from these 20 once they have been incorporated into the protein molecule. Only 20 amino acids are used by the body to produce proteins however.

polar and some are not), and *chemistry* (some have an alcohol group, some have an acid group, some have an aromatic portion). It is these factors (size, shape, polarity, and chemistry) which allow different amino acids to function in various ways and which determine the shape and function of the proteins containing them. The human body can synthesize all but eight of the twenty amino acids from other amino acids. These eight, listed in Table 15-3, cannot be synthesized by humans and must be obtained in the diet. Not all proteins contain all the essential amino acids. The combination of bread, which is 5 to 15 percent protein (contrary to popular opinion), and peanuts, which are a source of vegetable proteins, provides a nearly complete complement of the essential amino acids. A good source of this combination is a peanut butter and jelly sandwich. Eggs and milk have the best balance of the essential amino acids, and in general, animal proteins have an essential amino acid balance that is better than that of vegetable proteins. Brown rice is the best of the vegetable protein sources.

TABLE 15-3
The essential amino acids

Adults:
- Isoleucine (Ile)
- Leucine (Leu)
- Lysine (Lys)
- Methionine (Met)
- Phenylalanine (Phe)
- Threonine (The)
- Tryptophan (Trp)
- Valine (Val)

Children (less than 6 years old):
- All of the above 8 amino acids plus:
- Arginine (Arg)
- Histidine (His)

Proteins are large polymeric molecules containing 50 to 5000 amino acid units. These units are linked together by an *amide bond* (see Chapter 11), which is usually called a *peptide bond* by biochemists:

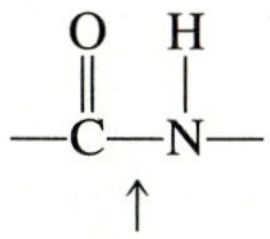

Amide bond

For example, glycine and alanine can react using the carboxylic acid group of one and the amino group of the other to form an amide bond which links the two molecules together:

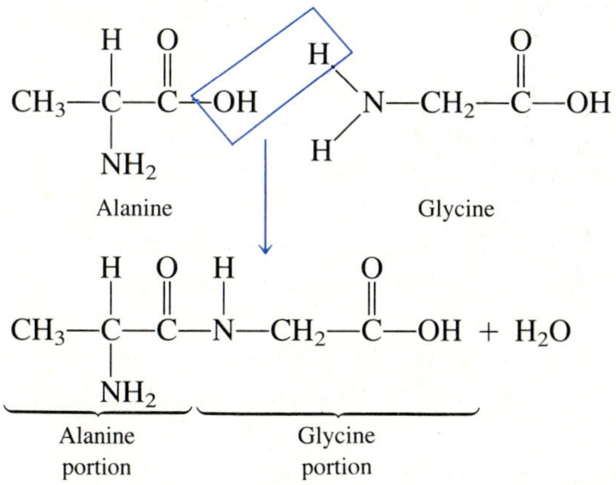

The carboxylic acid end of the glycine can then react with the amino group of a third amino acid to form another amide bond. This can continue until there are 100 or more amino acids in a sequence. Since there are 20 amino acids, the possible combinations and therefore the possible protein molecules that can be produced are virtually limitless.

To illustrate this point, let's consider just three amino acids: glycine, alanine, and serine. The possible combinations of these three amino acids in a small molecule containing only three amino acid units linked together are shown below.* (By convention, the amino acid with the free —COOH group is written to the right; the one with the free NH_2— group is written to the left.)

1	Gly-Gly-Gly	**10**	Gly-Gly-Ser	**19**	Ala-Ala-Ser
2	Ala-Ala-Ala	**11**	Gly-Ser-Gly	**20**	Ala-Ser-Ala
3	Ser-Ser-Ser	**12**	Ser-Gly-Gly	**21**	Ser-Ala-Ala
4	Gly-Ala-Ser	**13**	Gly-Gly-Ala	**22**	Ser-Ser-Ala
5	Gly-Ser-Ala	**14**	Gly-Ala-Gly	**23**	Ser-Ala-Ser
6	Ser-Ala-Gly	**15**	Ala-Gly-Gly	**24**	Ala-Ser-Ser
7	Ser-Gly-Ala	**16**	Ala-Ala-Gly	**25**	Ser-Ser-Gly
8	Ala-Gly-Ser	**17**	Ala-Gly-Ala	**26**	Ser-Gly-Ser
9	Ala-Ser-Gly	**18**	Gly-Ala-Ala	**27**	Gly-Ser-Ser

Each of these 27 small molecules would have different physical, chemical, and physiological properties. The possibilities increase dramatically as the number of amino acids used is increased and as the number of them in the sequence becomes greater. When one calculates the possible proteins that have 50 amino acids in a sequence using all 20 possible amino acids, one arrives at the staggering value of 10^{65}, which is relatively meaningless as a number, since it cannot readily be conceptualized. It does serve to show, however, that the number of possible proteins is virtually beyond comprehension. Yet organisms synthesize only those proteins tailored to their own specific needs!

Proteins are divided into two broad groups. The first is called the *soluble, or globular, proteins*. These are materials in which the chain formed by the amino acids is bent and folded much like a clump of string, as shown in Figure 15-4. The exact nature of the bending is dictated by the nature of the R groups on each amino acid in the sequence. The more polar R groups are on the outside folds so that they can dissolve in the polar water medium. Each amino acid sequence produces a different folding pattern. Enzymes are globular proteins, and their ability to catalyze a given reaction is the result of the overall three-dimensional structural pattern and the spatial relationship of R groups to each other.

The second class of proteins is called the *structural proteins* and is found in structural materials of the body including bone, muscle, hair, ligaments, cartilage, fingernails, and skin. In these materials the protein chains are found

*Note that the sequences shown in numbers 4 and 6 are *not* the same, since in number 4 the Ser has the free (uncombined) COOH and in number 6 the Gly has the free COOH. Likewise, 10 and 12, 13 and 15, and 16 and 18 (as well as other similar combinations) are not identical.

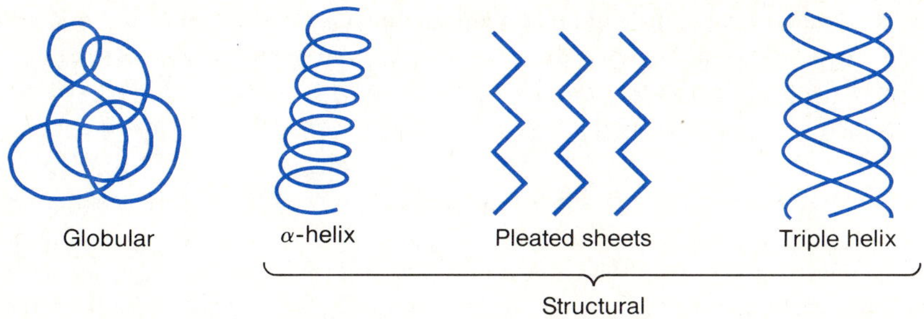

FIGURE 15-4 General arrangement of globular and structural proteins.

in a helix (much like a coiled wire) or flat sheets. These special arrangements and the nature of the R groups (mostly nonpolar groups) prevent the molecules from dissolving in water, and the interactions between adjacent coils or between adjacent parts of one chain make these materials strong.

Hair, for example, is composed of three closely linked parallel helices called a *fibril*. A hair strand (a strand visible to the eye) is composed of many of these parallel fibrils. Collagen, a protein found in skin, tendons, and bones, is composed of three mutually interlacing protein strands called a *triple helix*. This triple helix gives great strength to collagen-containing structures. The ability of different parts of the body to function in their own unique ways is a result of the molecular structure of the materials of which they are composed. In the case of proteins, the amino acids present and their sequence determine the structure of the protein chain and consequently the protein's function.

Carbohydrates

The second major class of biomolecules is the carbohydrate group. These include simple and complex sugars as well as starch, cellulose, and glycogen. All carbohydrates contain carbon atoms associated with the elements of water (hydrated carbon or "carbo-hydrates"). The most important member of this class, and the one most others are made from, is glucose, $C_6H_{12}O_6$. This formula could be written $C_6(H_2O)_6$, which illustrates the fact that for each carbon there are two hydrogen and one oxygen atom. The structure of glucose is shown below:

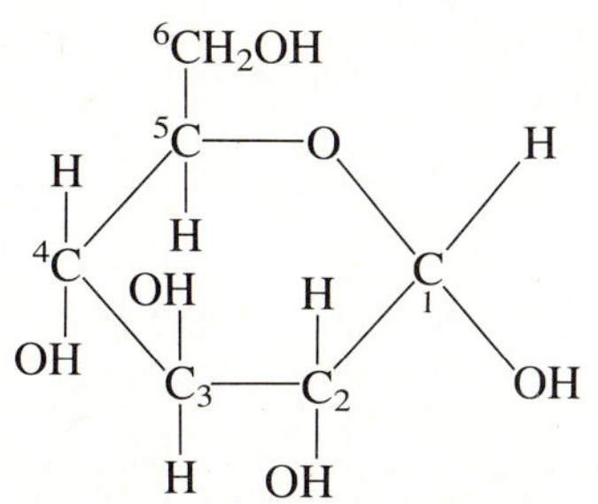

α-Glucose

or

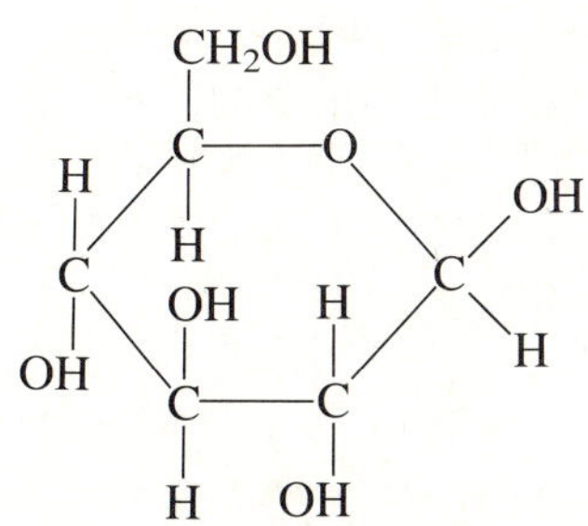

β-Glucose

Glucose is a cyclic material which has oxygens and hydrogens bonded to each carbon. It is an alcohol and an ether. The steriochemistry (orientation in space) is important for most biomolecules, and glucose is no exception. If any one of the OH groups except that on carbon 1 is switched up or down from the way it is drawn here, it would be a different molecule. The OH group attached to carbon 1 may be either up or down, however, and the designations β (Greek letter beta) and α (Greek letter alpha) are used to distinguish between them (β indicates up, α down). Glucose is called a *simple sugar* (*or monosaccharide*) because it is the building block for other carbohydrates. Two glucose units may link together as shown below:

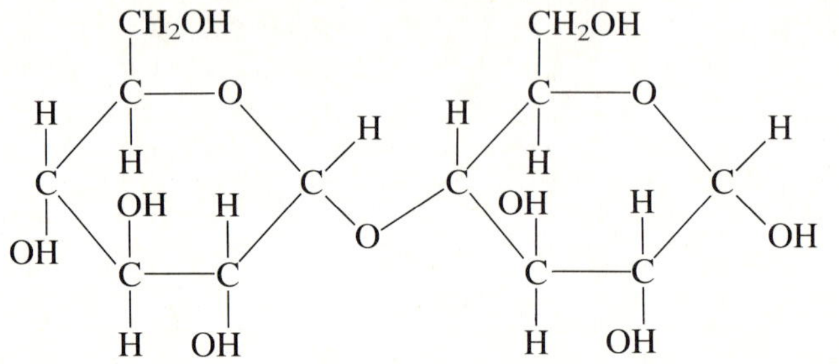

This can be written in an abbreviated form in which the carbon atoms are represented by points in the geometric figures and the hydrogen atoms attached directly to carbon atoms are not written. The oxygen link (actually an ether) is called a *glycosidic link* and can have two forms. The form shown below is called an *α-glycosidic link* because the OH group of carbon 1 in the left glucose was pointing *down* before it formed the glycosidic link:

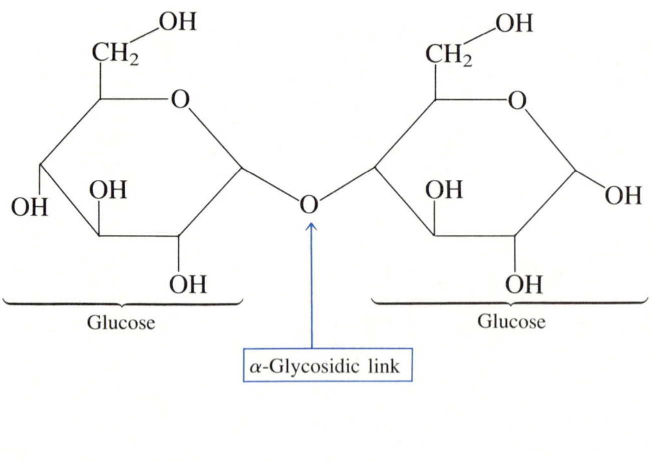

The other way these can be linked is with a β-glycosidic bond, which is shown below:

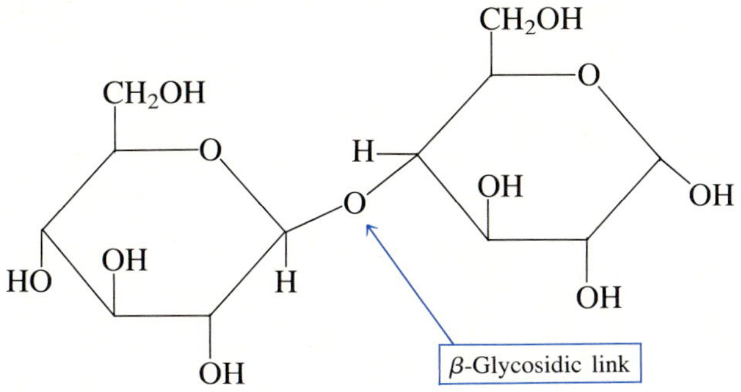

This results because the OH group of carbon 1 (left glucose) is in the β (up) position.

The compound formed when two glucose molecules are linked by an α link is called *maltose* and is one of the several common disaccharides. Other important disaccharides are lactose (milk sugar), which constitutes 5 to 8 percent of whole milk, and sucrose, which is common table sugar or cane sugar.

The more common complex carbohydrates (polysaccharides) are polymers of glucose. *Starch* is composed of two materials, amylose and amylopectin. *Amylose* is a series of very long molecules made from straight chains of glucose linked by α links. Amylopectin, the other component of starch, is branched polyglucose (see Figure 15-5). *Glycogen* is a highly branched polyglucose connected by α links and is used to store a small amount of energy in the liver and muscle for quick release when needed. *Cellulose,* the structural component of plants, is linear polyglucose connected by β links. An enzyme in human saliva (salivary amylase) is able to break α-glucosidic links. Thus starch can be broken down by human saliva into maltose, which is changed into glucose in the small intestine. β links of glucose cannot be broken by human enzymes, however, and thus cellulose passes through the body undigested and constitutes what is commonly called *fiber* or *roughage*. Enzymes found in bacteria in the digestive system of termites and ruminants (such as cattle) are able to break β links. This is why termites can use wood (sawdust) and ruminants can use hay as food—the wood and hay are converted to glucose just as starch is in humans. This illustrates the extreme specificity of most enzymes. The very small difference in the cellulose molecule (a β versus an α linkage) *completely* changes the ability of a given enzyme to cleave it. It should be pointed out that human enzymes can cleave the β links in lactose (milk sugar), however. These processes are summarized in Figure 15-5. Some complex carbohydrate materials such as pectin have units other than glucose present.

One thinks of sugars as sweet, but actually only sucrose and fructose are very sweet. Several others have a mild sweetness, and some are not sweet to most people (however, individual taste buds vary in their sensitivity to sweetness). Table 15-4 lists several carbohydrates and their relative sweetness.

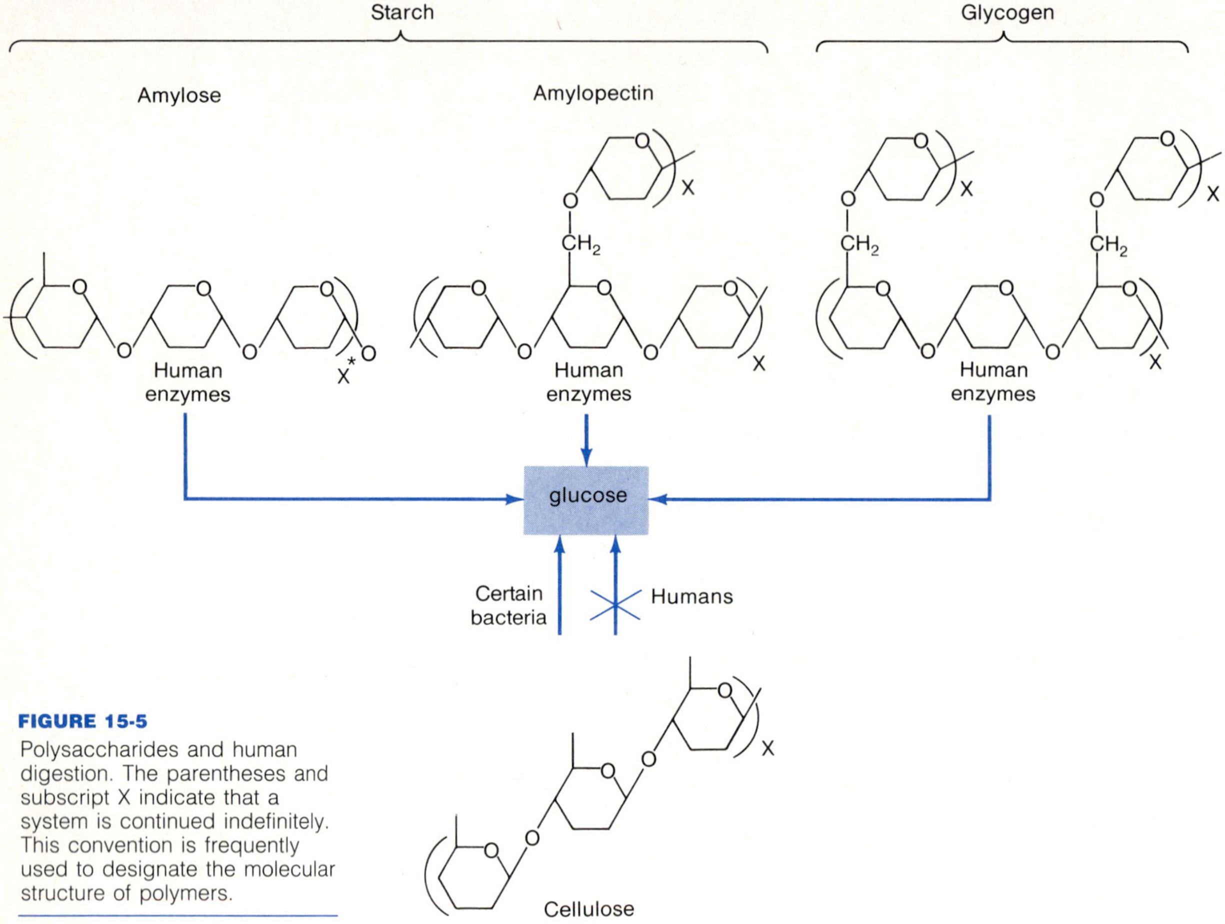

FIGURE 15-5

Polysaccharides and human digestion. The parentheses and subscript X indicate that a system is continued indefinitely. This convention is frequently used to designate the molecular structure of polymers.

TABLE 15-4

Sweetness of carbohydrates and synthetic sweeteners

SUGAR	RELATIVE NUMBER	DESCRIPTION
Monosaccharides:		
Glucose	69	Sweet
Galactose	63	Sweet
Fructose	120	Very sweet
Disaccharides:		
Sucrose	100	Very sweet
Maltose	46	Slightly sweet
Lactose	39	Slightly sweet
Polysaccharide:		
Starch	0	Not sweet
Artificial sweeteners:		
Saccharin	300	Extremely sweet
Aspartame	200	Extremely sweet
Acesulfame	10,000	Extremely sweet
Talin	200,000	Extremely sweet

Fats

Those materials in the biosystem that dissolve in nonpolar substances such as benzene are called *lipids*. There are many subcategories of lipids, but most of them are *fats* (solid) and *oils* (liquid), which have the same general formula. Most fats have structures similar to that of lard or solid vegetable shortening, shown below:

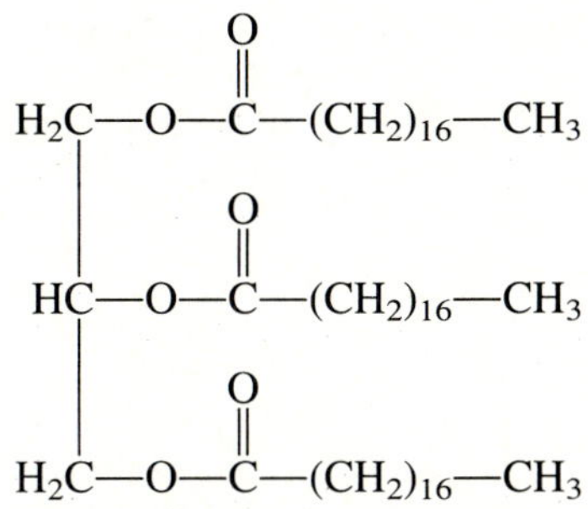

This is a triester (it has three ester groups) and contains three —$(CH_2)_{16}$—CH_3 side groups or 18 carbon atoms in a sequence. Various fats contain slightly different numbers of carbon atoms in the side groups.

Oils (liquids at room temperature) have the same basic structure except that the side groups contain carbon-carbon double bonds or very short chains of carbon atoms. Fats and oils which do not contain double bonds are said to be *saturated,* while those that do are called *unsaturated*.

Vegetable oils, such as cottonseed oil, have double bonds in each side chain, as shown below:

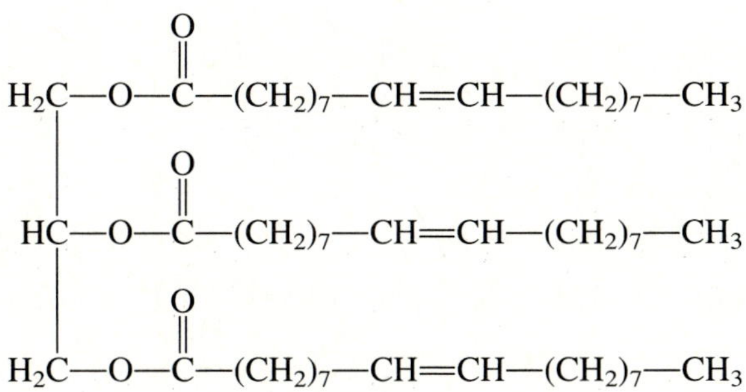

During digestion the three ester links are broken in the intestine by pancreatic enzymes, and the molecules of fat are changed into glycerin and three molecules of carboxylic acid; these long acids are called *fatty acids* because they are derived from fats:

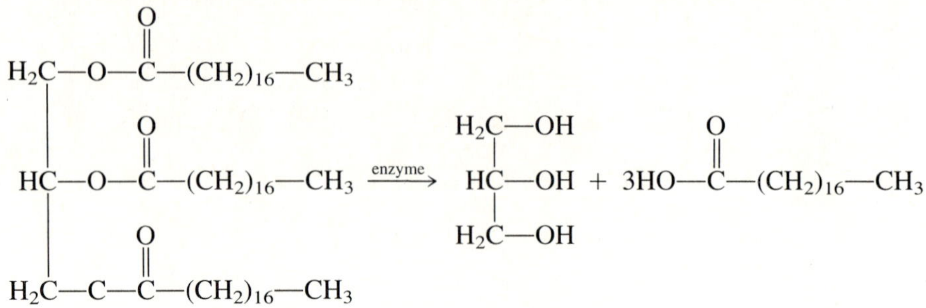

Glycerin

Stearic acid (a fatty acid)

Body fats are then synthesized from glycerin and various fatty acids. Most of the several common acids can be synthesized by the body from other materials, but one of them, linoleic acid, cannot be made by the human body and must be part of the fat content of the diet. (Although the human body cannot synthesize it, other organisms can make it.) Linoleic acid is called an *essential fatty acid.*

Another major lipid important in the biosystem, and in dietary considerations, is cholesterol. Cholesterol is needed in cell membranes and can be readily synthesized in the body if no dietary cholesterol is present. Its structure is shown below:

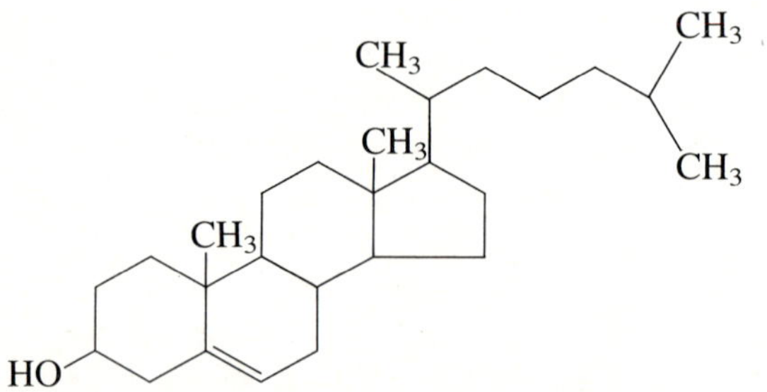

The role of this material in cardiovascular disease will be discussed later in the chapter.

Vitamins and minerals

Small amounts of several organic materials called *vitamins* and inorganic materials called *minerals* (mostly metal cations) are needed for normal body processes to occur. In the absence of these vitamins and minerals, many enzymes are unable to function and catalyze necessary biochemical reactions (see Figure 15-2). The structures of all vitamins are known, but only a few of them will be presented here.

One need not memorize these structures; they are given for general interest.

Vitamin C Vitamin C is known chemically as *ascorbic acid*. It is highly polar, because of the —OH groups, and is therefore soluble in water. It prevents scurvy, a disease of the teeth and gums, and is needed by the body to regenerate connective tissue during healing processes. The best source of this vitamin is citrus fruit, but it is found in many fruits and vegetables and is produced by most organisms except primates (including humans) and guinea pigs. The structure of vitamin C is shown below:

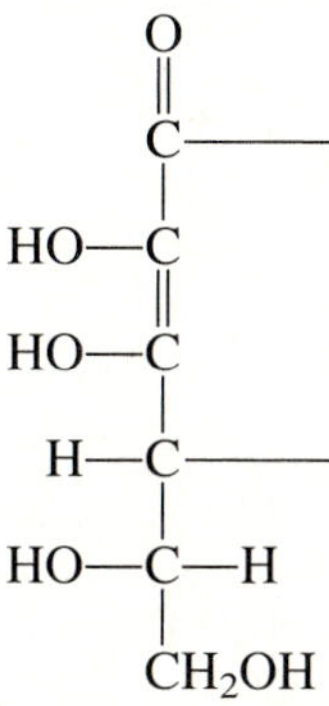

Vitamin A Vitamin A is needed for good vision, since the materials used by the eye during the visual process are made in vivo from vitamin A. Carotene, found in many yellow and green vegetables, is changed into vitamin A in the body, and the vitamin itself is found in the fat portions of meat, eggs, and milk. Vitamin A is a fat-soluble vitamin, since it has only one polar group and a large nonpolar part. The concentration of vitamin A in liver is quite high, and some sources of liver (e.g., polar bear liver) have so much vitamin A that ingestion of the liver can actually cause poisoning. Vitamin A is one of the few vitamins that are toxic in large doses. The structure of vitamin A is shown below:

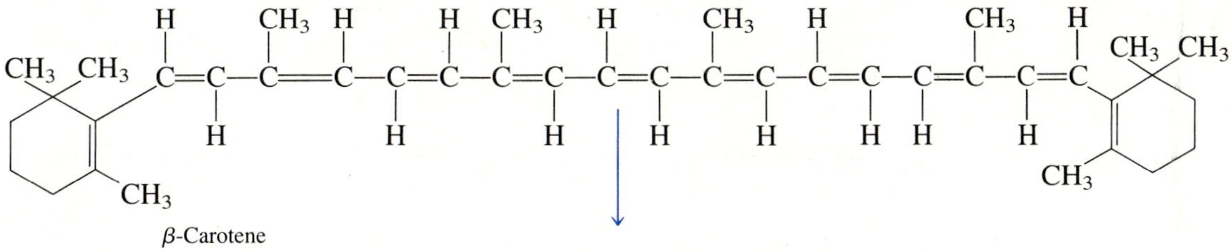

β-Carotene

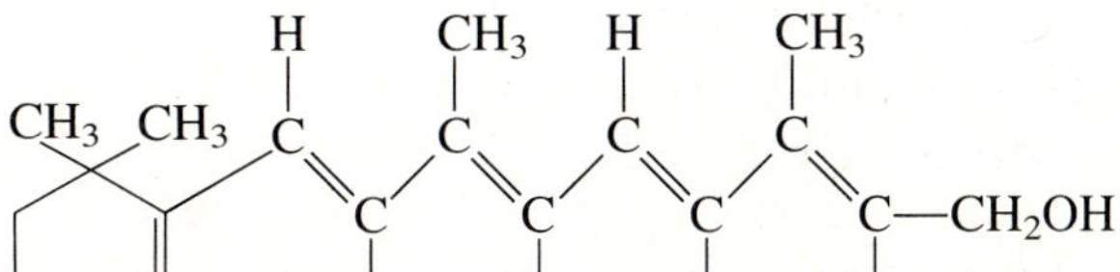

Vitamin A

Vitamin D There are several compounds, called the *D vitamins,* which prevent rickets (a weakening of the bones due to poor calcium metabolism). One of these, vitamin D_3, is shown below. Several closely related materials can be changed into one of the D vitamins in the body and are called *provitamin D*. One such material is 7-dehydrocholesterol; it is found in human skin and is converted, as shown below, into vitamin D_3 by the ultraviolet (UV) rays of the sun:

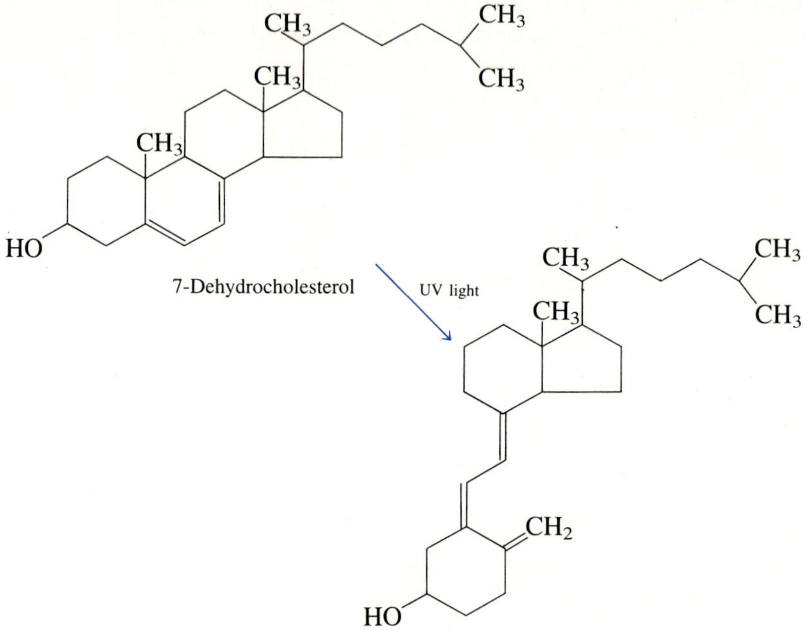

7-Dehydrocholesterol

Vitamin D_3

Thus, exposure to sunlight is an important source of this vitamin. Other sources are cod-liver oil and enriched whole milk. Since vitamin D_3 is mostly nonpolar, it is a fat-soluble vitamin.

Vitamin B_{12} The structure and synthesis of vitamin B_{12} are discussed in Chapter 11. It is found in most animal tissue and is useful in preventing pernicious anemia. It is water-soluble.

Vitamin B_1 Vitamin B_1, called *thiamine,* is the water-soluble material needed to prevent beriberi. Beriberi is a condition of general weakness and debilitation. Thiamine is found in grain hulls and meat. Its structure is shown below:

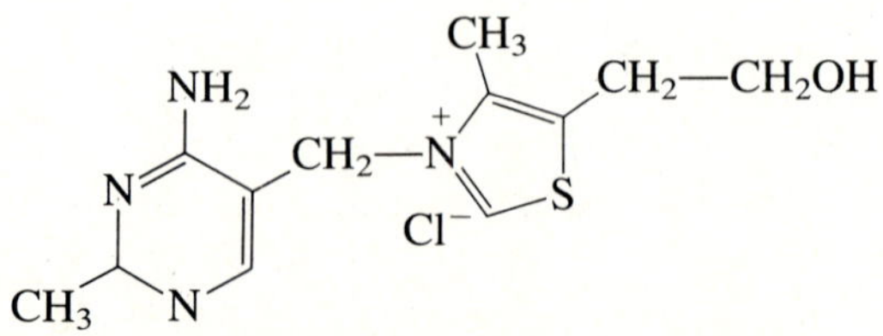

Vitamin B_2 The alternative name for vitamin B_2 is *riboflavin*. It is found in eggs and meat. It is water-soluble and is needed as a coenzyme for many biochemical functions in the body. A *coenzyme* is a material needed along with the enzyme to cause the enzyme to be active (the enzyme cannot function in the absence of the coenzyme). The structure of vitamin B_2 is shown below:

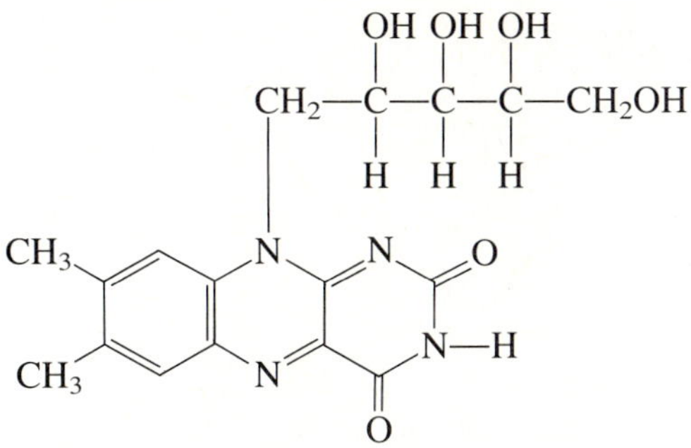

Vitamin B_3 There are two compounds, known chemically as *nicotinic acid* and *nicotinamide,* either of which can be used by the body as vitamin B_3. They are found principally in meat and serve as coenzymes. Their structures are shown below:

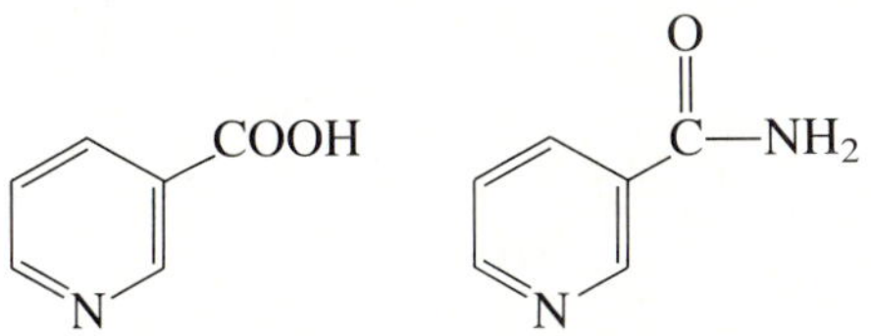

Nicotinic acid

Nicotinamide

Nucleic acids

There are two types of nucleic acids: ribonucleic acid (RNA) and deoxyribonucleic acid (DNA). DNA is found as a single large molecule in each cell and contains all the body's genetic information. Each cell in a given individual has an identical DNA molecule; the DNA of no two people is alike. One of the important roles of RNA is the control of protein synthesis.

The general structure of DNA is shown below:

$$\sim\sim(\text{phosphate}-\underset{\text{Base 1}}{\text{sugar}}-\text{phosphate}-\underset{\text{Base 2}}{\text{sugar}}-\text{phosphate}-\underset{\text{Base 3}}{\text{sugar}})_x\sim\sim$$

There are four possible bases that occur in different sequences in the DNA of each individual, and it is the different arrangement of these bases alone that controls the characteristics of that individual. Your skin color, your hair color, the shape of your nose, etc., are all controlled by this conceptually simple coding system.

This is shown more completely below:*

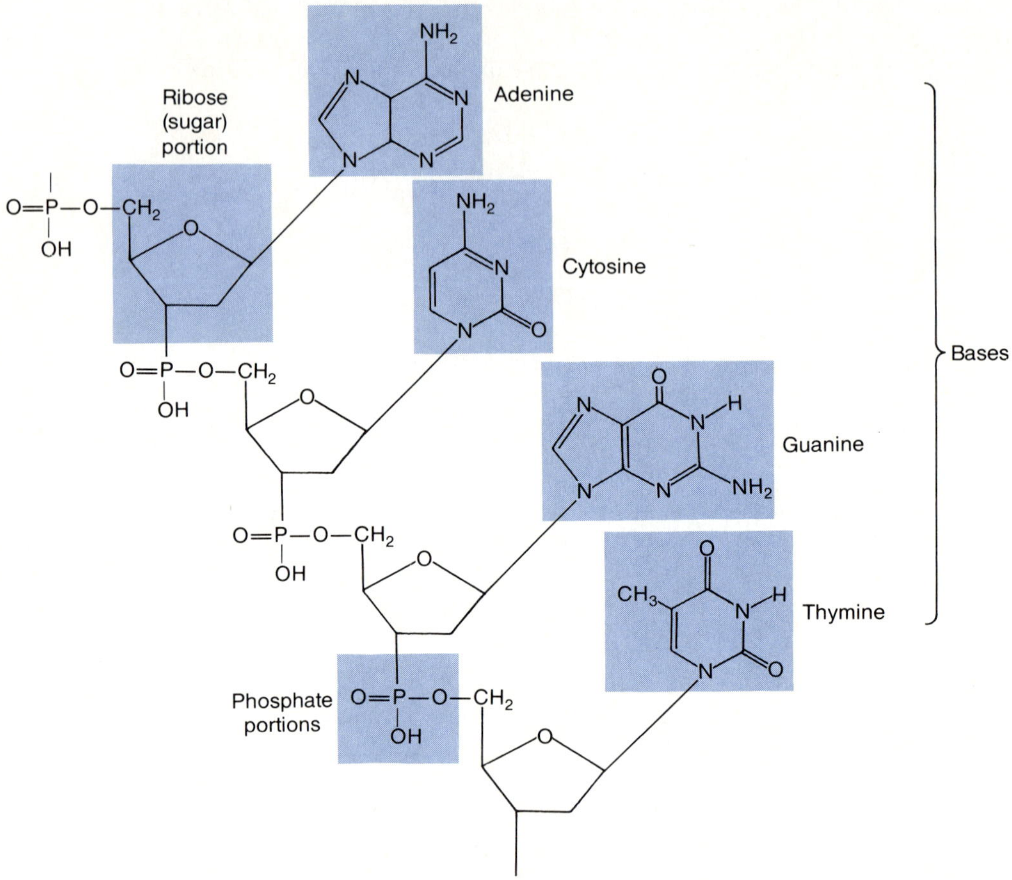

The four bases (nitrogen-containing heterocyclic compounds) are shown attached to the sugar-phosphate chain. Each base has a slightly different shape, polarity, and ability to form hydrogen bonds. These differences are used to store (code) information that directs the synthesis of different RNA molecules. The various RNA molecules synthesized, in turn, direct the synthesis of different cellular proteins. This sequence of events determines, directly or indirectly, all the genetic characteristics of species and individuals. Since the DNA is such a large molecule, there are many thousands of base units present, and coding can be stored for *all* characteristics and functions. It is truly an amazing molecule. These four bases (adenine, cytosine, guanine, and thymine) are the only ones found in DNA, and it is simply various combinations of them (i.e., different sequential orders) that determine ultimately all heredity and all biochemical processes in each organism.

*The phosphate-sugar chain, with the attached base units, continues for thousands of units—each with its base side group.

HUMAN NUTRITION: SELECTED TOPICS

Nutritional needs—that is, currently accepted values—are summarized in Table 15-5, and the caloric and general nutrient values of a number of foods are given in Tables 15-6 and 15-7. Figure 15-6 shows bar graphs of selected nutritional values of several common foods whose nutrient values are frequently misunderstood or dramatically high or low.

Vitamin and mineral requirements are stated in two ways. Formerly the *minimum daily requirement* (MDR) was used, which was the amount needed to prevent obvious symptoms of deficiency. Now the *recommended daily allowance* (RDA) is used; it is usually about twice the MDR. Some vitamins such as vitamin C can be taken in large doses (even several grams per day) without apparent harm, while others such as vitamins A and D are toxic, even fatal, in high quantities.

A diet consisting of a variety of foods from each of the classic food groups—meat, vegetables and fruit, milk and eggs, and bread—usually contains about the correct amounts of each nutrient. Multivitamins and mineral tablets or liquids are a good safeguard in times of stress or illness but are not necessary for most healthy adults on a balanced diet.

TABLE 15-5

Adult human nutritional needs—recommended daily allowances (RDAs)

Energy:	
Calories	2000–3000 kcal (Calories)
Mass nutrients:	
Carbohydrates	At least 50 g
Proteins	50–60 g
Fats	At least 50 g
Vitamins:	
A (carotene)*	5000 IU†
C (ascorbic acid)	60–70 mg
D (ergosterol)‡	400 IU
E (α-tocopherol)	15–30 IU
B_1 (thiamine)	1.5 mg
B_2 (riboflavin)	1.7 mg
B_3 (niacin)	20 mg
B_6 (pyridoxine)	2 mg
B_{12} (cobalamin)	0.006 mg
M (folic acid)	0.4 mg
Pantothenic acid	10 mg
H (biotin)	0.3 mg
K	Value unknown
Minerals§:	
Calcium (Ca)	600–1000 mg
Iron (Fe)	10 mg (men); 18 mg (women)
Zinc (Zn)	15 mg
Copper (Cu)	2 mg
Iodine (I)	0.15 mg
Phosphorus (P)	1000 mg

*Carotene is changed to vitamin A in the body.
†International units.
‡There are several materials that upon irradiation with ultraviolet light are changed to vitamin D.
§Many other metals are needed in small amounts, but the RDA has not been established.

TABLE 15-6
Selected nutrient values of some pure foods

FOOD	STANDARD SERVING WEIGHT	STANDARD SERVING SIZE	DIGESTIBLE CARBO-HYDRATE,* g	FAT,* g	PROTEIN,* g	MINERALS,* mg			VITAMINS†				
						Ca	Fe	P	C, mg	A, IU	B_1, mg	B_2, mg	B_3, mg
Milk	250 g	1 cup	10	8	6	300	0.4	250	2.5	350	0.1	0.4	2.5
Cheese	30 g	1 slice	1	12	9	300	0.4	200	—	450	0.01	0.15	—
Eggs	50 g	1 egg	0.4	7	9	30	2	120	—	570	0.05	0.15	0.05
Fish (cod)	100 g	¼ lb	0	0.5	20	10	0.5	200	2	—	0.06	0.09	2.2
Liver (beef)	100 g	¼ lb	5	3	20	7	8	300	30	44,000	0.26	3.3	14
Steak (chuck)	100 g	¼ lb	0	20	20	14	4	170	—	—	0.05	0.2	4
Hot dogs	100 g	2 hot dogs	1	20	20	12	3	50	—	—	0.16	0.18	2.5
Chicken	100 g	¼ lb	0	2	20	6	3	200	—	—	0.1	0.24	5.6
Bread	20 g	1 slice	10	0.5	1.5	15	0‡	15	—	—	0.05	0.03	0.4
Oatmeal	250 g	1 bowl	30	3	6	2	12	170	—	—	0.25	0.05	0.5
Potatoes (cooked)	200 g	1 potato	50	1	4	60	1	100	40	15,000	0.2	0.1	1.2
Carrots (cooked)	100 g	⅔ cup	8	0.4	1	30	1	30	0.3	17,000	0.02	0.02	0.3
Peas (green)	100 g	1½ cups	15	0.5	7	25	2	130	15	720	0.25	0.14	2.3
Beets (cooked)	100 g	½ cup	10	0.1	2	30	1	40	7	20	0.02	0.04	0.3
Spinach (cooked)	100 g	1¼ cups	3	0.5	2	120	4	30	30	12,000	0.08	0.2	0.6
Broccoli	100 g	1 cup	5	0.3	3	120	1.5	70	120	3,500	0.1	0.2	1.
Lettuce	50 g	½ cup	2	0.2	1	20	1	20	4	270	0.02	0.04	0.1
Apple	150 g	1 apple	20	0.5	0.5	10	0.5	15	8	150	0.06	0.04	0.3
Orange	100 g	½ orange	10	0.2	1	45	0.5	20	50	200	0.08	0.03	0.2
Fudge	50 g	⅛ lb (2 pieces)	40	5	1	20	0.2	30	—	120	0.005	0.035	0.0
Cake (yellow, plain)	50 g	1 slice	20	4	5	60	0.3	100	—	60	0.001	0.04	0.1
Butter	10 g	1 pat	0	9	0.1	2	—	2	—	330	—	—	—
Popcorn	10 g	1 cup	8	0.5	1	1	0.3	30	—	—	0.04	0.01	0.2

*Composite or average value from many sources.
†Information calculated from data found in B. T. Burton, *The Heinz Handbook of Nutrition*, 2d ed., McGraw-Hill, New York, 1965.
‡Some bread is enriched with iron to about 0.4 mg per slice.
Note: Dash indicates insignificant amount.

TABLE 15-7

Caloric values of pure foods

FOOD	CALORIES, kcal/100 g*
Milk products	
Milk (whole, about ½ cup)	75
Cheese	400
Cottage cheese	100
Meats	
Fish	100–200
Liver	125
Steak	200
Pork (loin)	250–300
Bacon	600–700
Chicken	100
Turkey	150
Hot dogs (about 2–3)	250
Breads†	
White (5–7 slices)	250
Pancakes (about 5)	200
Biscuits (about 5)	400
Crackers (soda, about 35)	425
Cereals‡	
Cornflakes	400
Wheat flakes	350
Cream of wheat	350
Oatmeal	400
Starches	
Potatoes (boiled, 1 small)	100
Rice (boiled)	100
Spaghetti, macaroni, noodles	350
Popcorn	400
Vegetables	
Corn	100
Carrots	50
Green beans	40
Lima beans	130
Navy beans (dried)	350
Peas	60
Beets	40
Spinach	60
Lettuce (raw)	20
Tomatoes (raw)	20
Fruit	
Apple (raw, ½ of a fruit)	75
Banana (raw, ½ banana)	100
Pear and peach (raw, ½–¾ of a fruit)	50–60
Orange (½ orange)	50
Grapefruit (⅓–½ of a fruit)	50
Strawberries (raw)	40
Raisins	300
Desserts	
Chocolate fudge	400
Peanut brittle	550
Lemon meringue pie	250
Apple pie	300
Tapioca pudding	100
Sugar cookies	425
Angel food cake	300
Devil's food cake	400
Doughnuts	500
Gelatin dessert	375

(*Continued*)

TABLE 15-7

Continued

FOOD	CALORIES, kcal/100 g*
Miscellaneous	
Nuts (most kinds)	700
Peanuts, chestnuts	400–500
Jelly	300
Butter, margarine (8 tblsp)	800
Oils	900
Mayonnaise	750
Granulated sugar (sucrose)	400
Honey	325

*All values are for 100-g (about ¼-lb) serving and are a composite value taken from many sources including *The Heinz Handbook of Nutrition* (see footnote to Table 15-6) and the *CRC Handbook of Chemistry and Physics,* CRC Press, Inc., Cleveland, Ohio. More familiar units are given in some cases.

†No spread.

‡Dry—no milk or sugar.

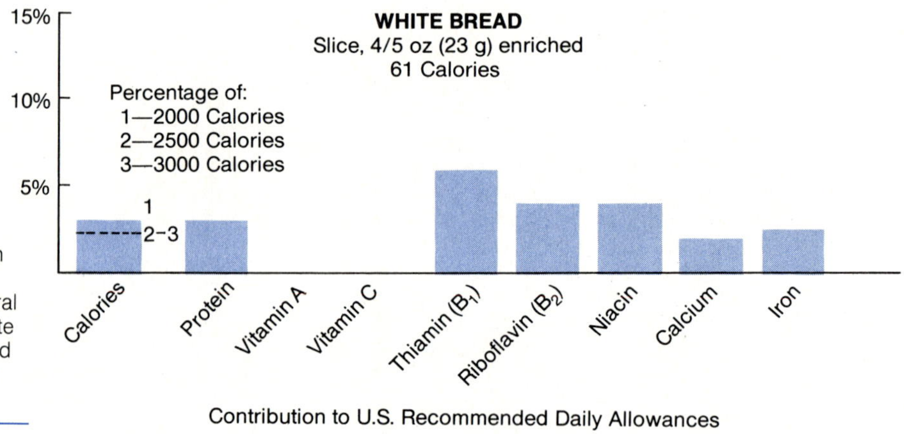

FIGURE 15-6
This graph and the graphs on pages 409–414 show percentages of RDA for several nutrients in common composite foods. (These graphs are used by permission of the National Dairy Council.)

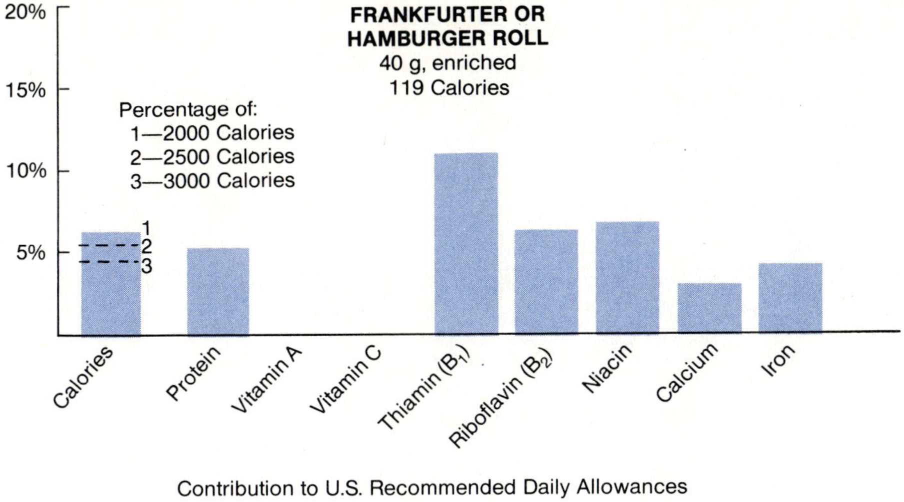
FRANKFURTER OR
HAMBURGER ROLL
40 g, enriched
119 Calories
Percentage of:
1—2000 Calories
2—2500 Calories
3—3000 Calories
20%
15%
10%
5%
1
2
3
Calories
Protein
Vitamin A
Vitamin C
Thiamin (B_1)
Riboflavin (B_2)
Niacin
Calcium
Iron
Contribution to U.S. Recommended Daily Allowances

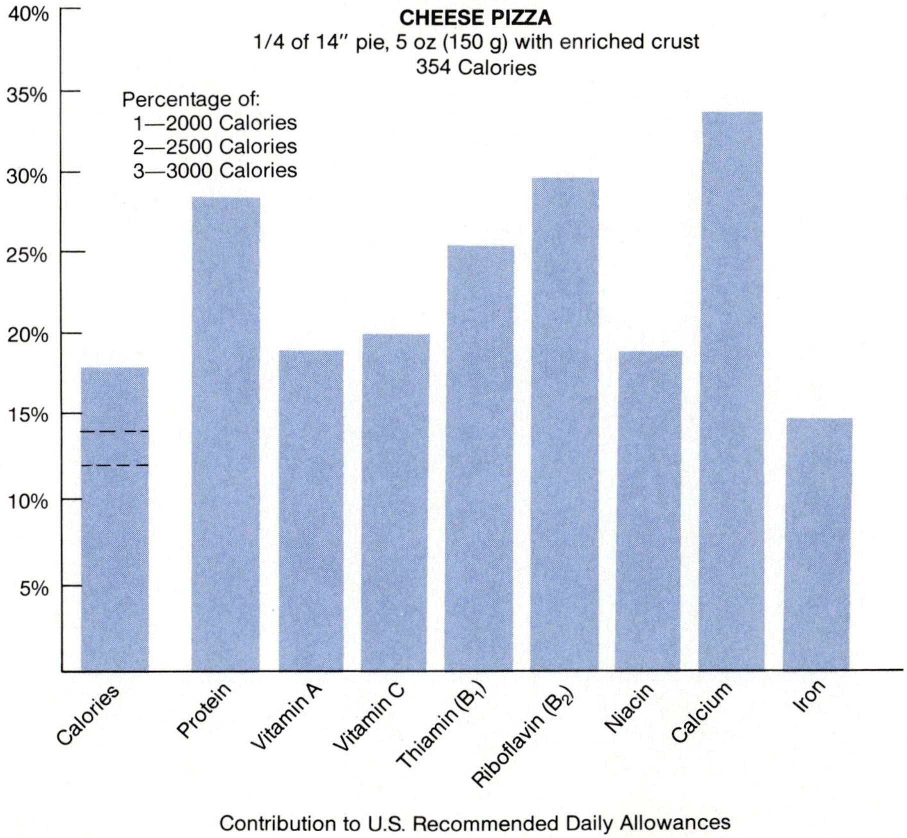
CHEESE PIZZA
1/4 of 14" pie, 5 oz (150 g) with enriched crust
354 Calories
Percentage of:
1—2000 Calories
2—2500 Calories
3—3000 Calories
40%
35%
30%
25%
20%
15%
10%
5%
Calories
Protein
Vitamin A
Vitamin C
Thiamin (B_1)
Riboflavin (B_2)
Niacin
Calcium
Iron
Contribution to U.S. Recommended Daily Allowances

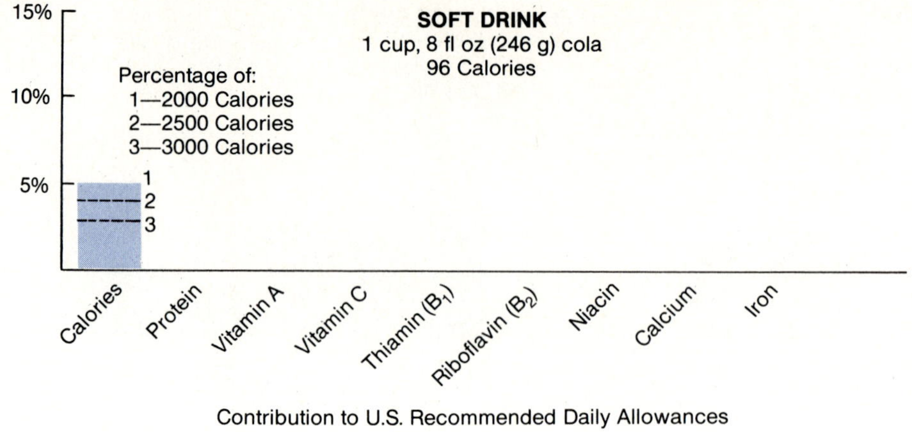
SOFT DRINK
1 cup, 8 fl oz (246 g) cola
96 Calories
Percentage of:
1—2000 Calories
2—2500 Calories
3—3000 Calories
15%
10%
5%
1
2
3
Calories
Protein
Vitamin A
Vitamin C
Thiamin (B1)
Riboflavin (B2)
Niacin
Calcium
Iron
Contribution to U.S. Recommended Daily Allowances

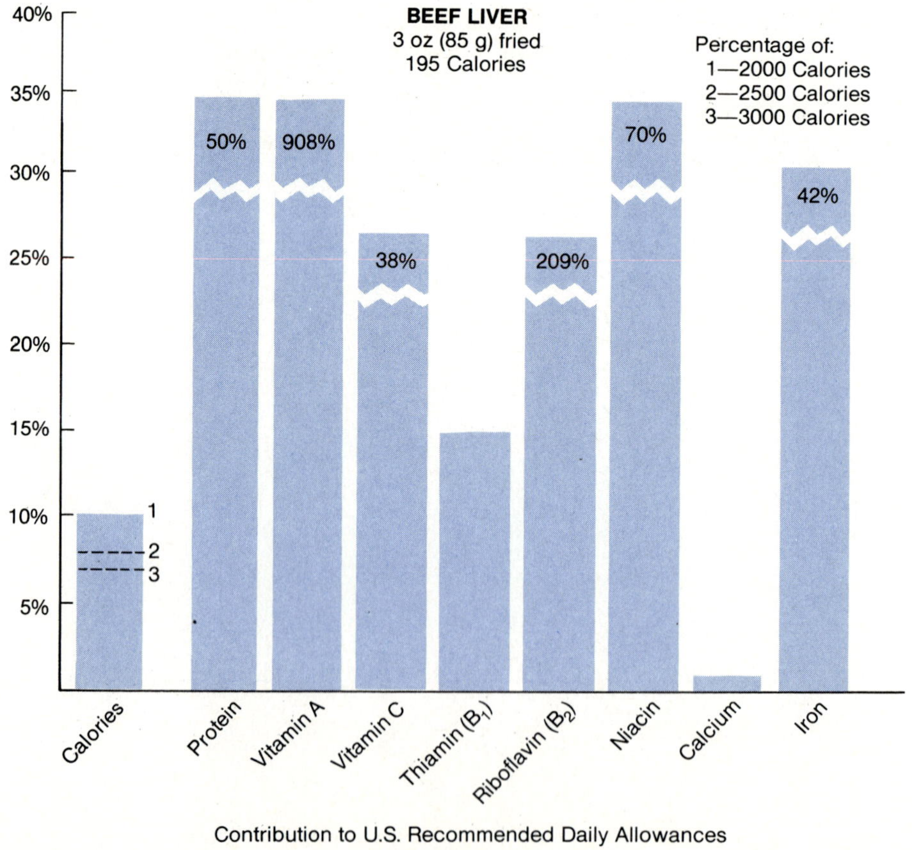
BEEF LIVER
3 oz (85 g) fried
195 Calories
Percentage of:
1—2000 Calories
2—2500 Calories
3—3000 Calories
40%
35%
30%
25%
20%
15%
10%
5%
50%
908%
38%
209%
70%
42%
1
2
3
Calories
Protein
Vitamin A
Vitamin C
Thiamin (B1)
Riboflavin (B2)
Niacin
Calcium
Iron
Contribution to U.S. Recommended Daily Allowances

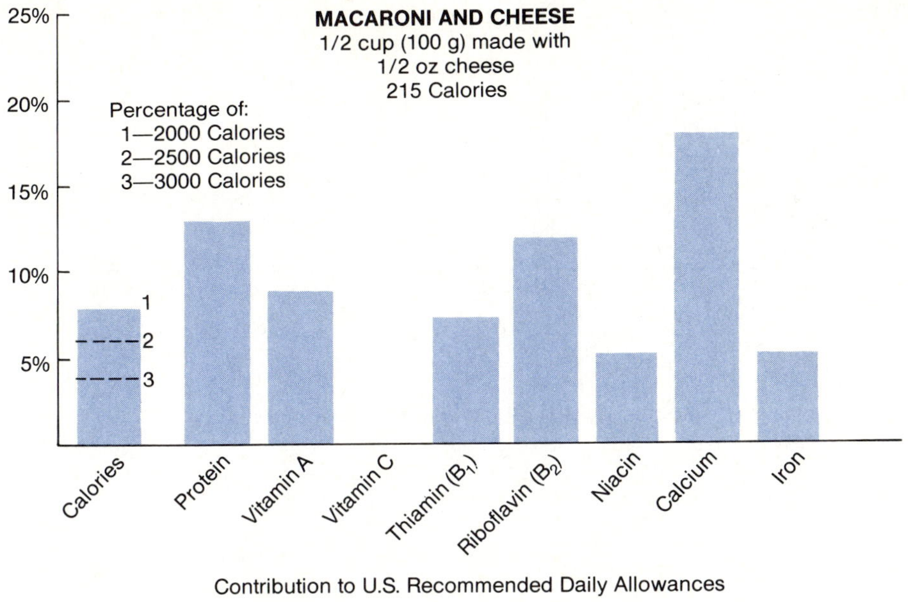
MACARONI AND CHEESE
1/2 cup (100 g) made with
1/2 oz cheese
215 Calories
Percentage of:
1—2000 Calories
2—2500 Calories
3—3000 Calories
25%
20%
15%
10%
5%
1
2
3
Calories
Protein
Vitamin A
Vitamin C
Thiamin (B_1)
Riboflavin (B_2)
Niacin
Calcium
Iron
Contribution to U.S. Recommended Daily Allowances

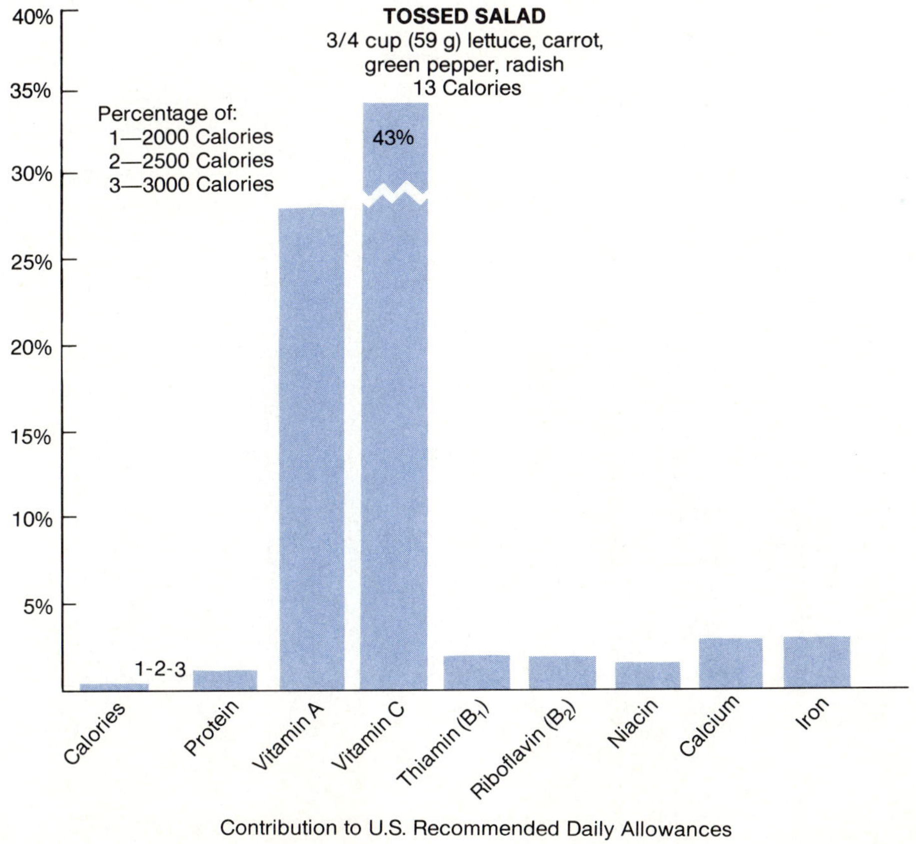
TOSSED SALAD
3/4 cup (59 g) lettuce, carrot,
green pepper, radish
13 Calories
Percentage of:
1—2000 Calories
2—2500 Calories
3—3000 Calories
40%
35%
30%
25%
20%
15%
10%
5%
43%
1-2-3
Calories
Protein
Vitamin A
Vitamin C
Thiamin (B_1)
Riboflavin (B_2)
Niacin
Calcium
Iron
Contribution to U.S. Recommended Daily Allowances

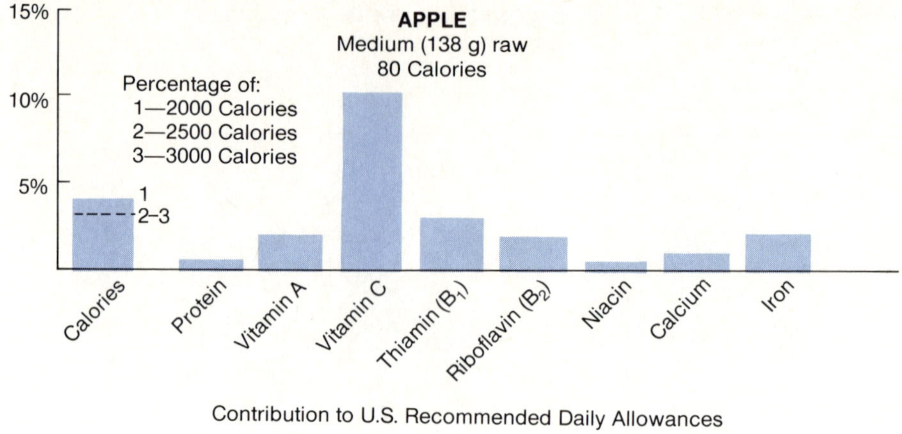
APPLE
Medium (138 g) raw
80 Calories
Percentage of:
1—2000 Calories
2—2500 Calories
3—3000 Calories
15%
10%
5%
1
2-3
Calories
Protein
Vitamin A
Vitamin C
Thiamin (B1)
Riboflavin (B2)
Niacin
Calcium
Iron
Contribution to U.S. Recommended Daily Allowances

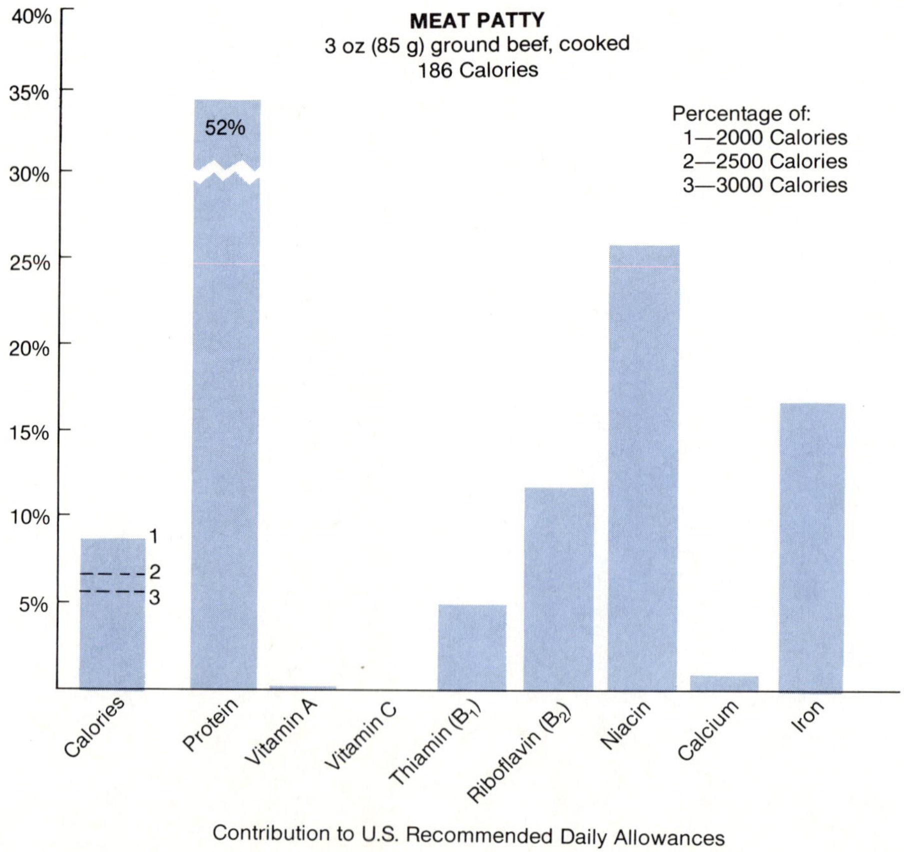
MEAT PATTY
3 oz (85 g) ground beef, cooked
186 Calories
Percentage of:
1—2000 Calories
2—2500 Calories
3—3000 Calories
40%
35%
30%
25%
20%
15%
10%
5%
52%
1
2
3
Calories
Protein
Vitamin A
Vitamin C
Thiamin (B1)
Riboflavin (B2)
Niacin
Calcium
Iron
Contribution to U.S. Recommended Daily Allowances

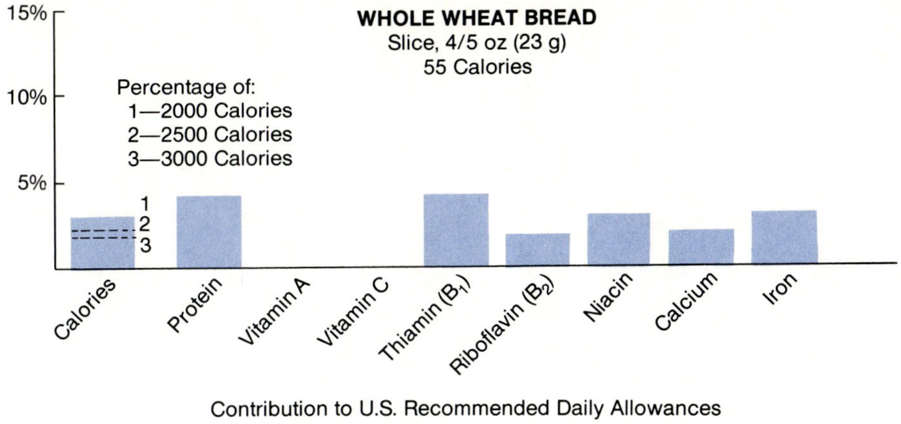

15%
10%
5%
WHOLE WHEAT BREAD
Slice, 4/5 oz (23 g)
55 Calories
Percentage of:
1—2000 Calories
2—2500 Calories
3—3000 Calories
1
2
3
Calories
Protein
Vitamin A
Vitamin C
Thiamin (B_1)
Riboflavin (B_2)
Niacin
Calcium
Iron
Contribution to U.S. Recommended Daily Allowances

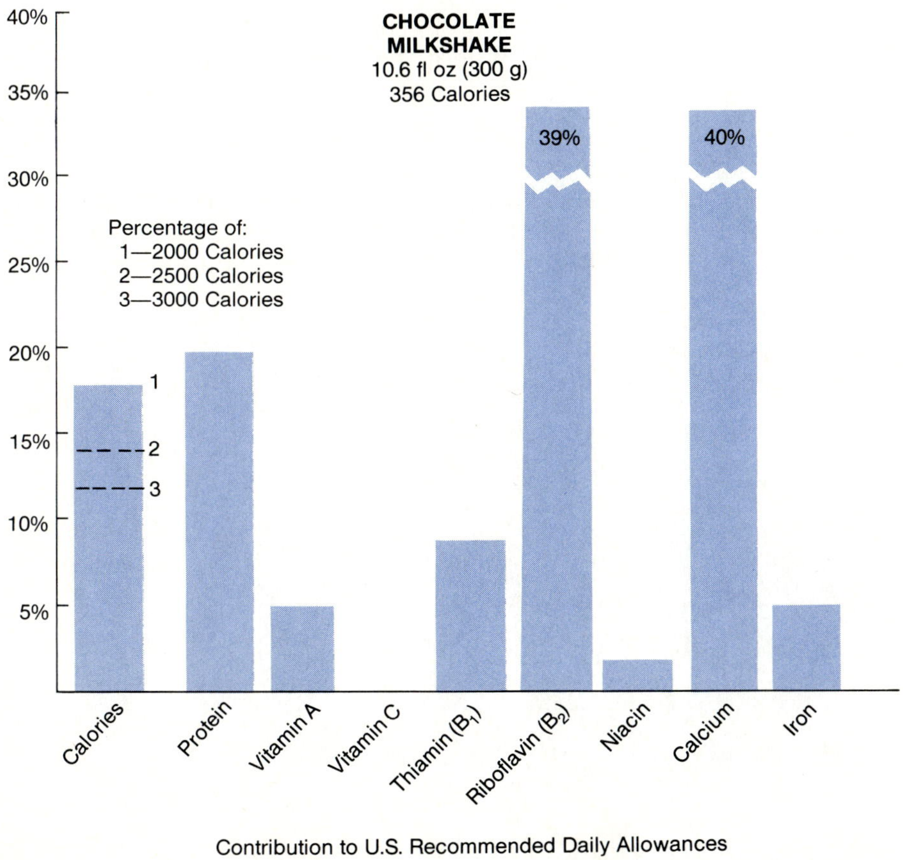

40%
35%
30%
25%
20%
15%
10%
5%
CHOCOLATE MILKSHAKE
10.6 fl oz (300 g)
356 Calories
39%
40%
Percentage of:
1—2000 Calories
2—2500 Calories
3—3000 Calories
1
2
3
Calories
Protein
Vitamin A
Vitamin C
Thiamin (B_1)
Riboflavin (B_2)
Niacin
Calcium
Iron
Contribution to U.S. Recommended Daily Allowances

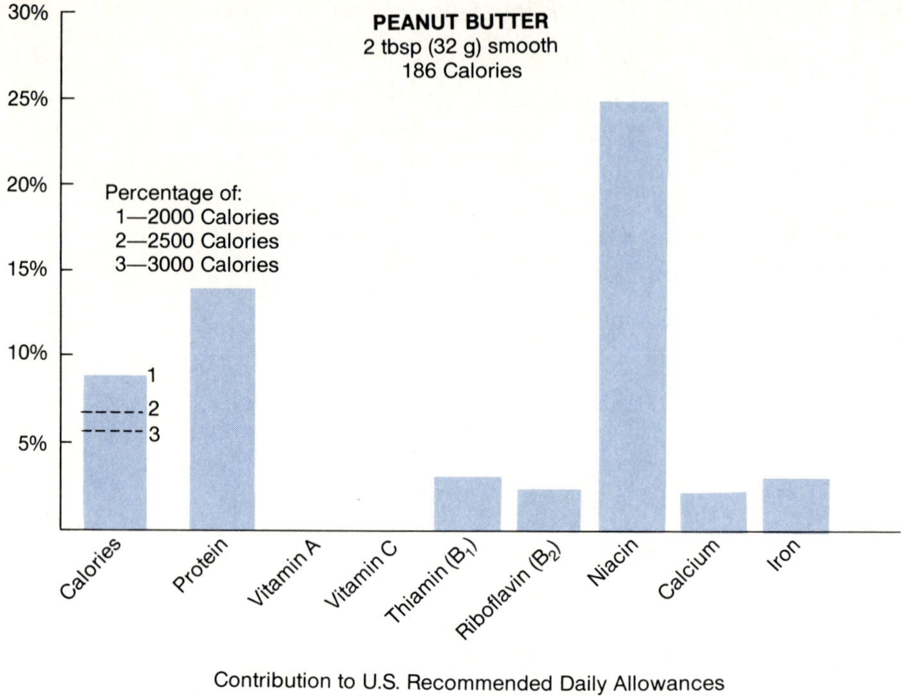
PEANUT BUTTER
2 tbsp (32 g) smooth
186 Calories
30%
25%
20%
15%
10%
5%
Percentage of:
1—2000 Calories
2—2500 Calories
3—3000 Calories
1
2
3
Calories
Protein
Vitamin A
Vitamin C
Thiamin (B_1)
Riboflavin (B_2)
Niacin
Calcium
Iron
Contribution to U.S. Recommended Daily Allowances

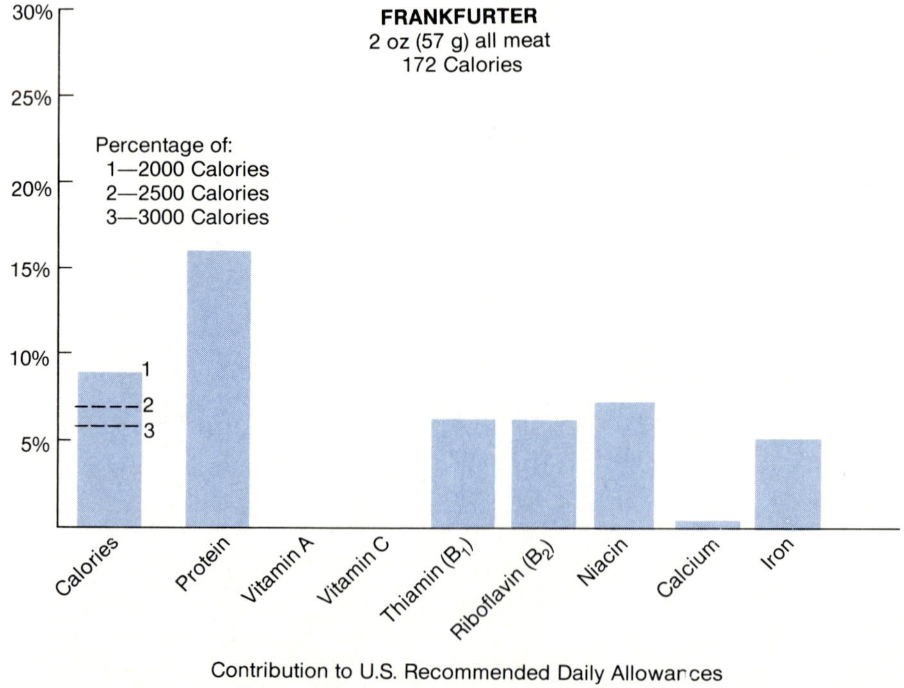
FRANKFURTER
2 oz (57 g) all meat
172 Calories
30%
25%
20%
15%
10%
5%
Percentage of:
1—2000 Calories
2—2500 Calories
3—3000 Calories
1
2
3
Calories
Protein
Vitamin A
Vitamin C
Thiamin (B_1)
Riboflavin (B_2)
Niacin
Calcium
Iron
Contribution to U.S. Recommended Daily Allowances

The following pages discuss several nutritional topics of interest.

Weight control

A problem frequently encountered is dieting to lose weight or eating to maintain a desirable weight. As you will no doubt remember from Chapter 14, combustion of different materials produces different amounts of heat energy. The complete metabolism of biochemicals to CO_2 and water in the living system produces the same total energy as if the same material were burned in a flame. Carbohydrates produce about 4.5 Calories, or kilocalories (kcal), of energy per gram; proteins produce 4.5; and fats produce 9:

$$\underset{\text{Glucose}}{1\text{ g carbohydrate}} + O_2 \xrightarrow{\text{enzyme}} CO_2 + H_2O \qquad \Delta H = -4500\text{ cal or } -4.5\text{ kcal or } -4.5\text{ Calories}$$

$$\underset{\text{Amino acids}}{1\text{ g protein}} + O_2 \xrightarrow{\text{enzyme}} CO_2 + H_2O + \text{other products} \qquad \Delta H = -\ 4.5\text{ Calories}$$

$$\underset{\text{Fatty acids}}{1\text{ g fat}} + O_2 \xrightarrow{\text{enzyme}} CO_2 + H_2O \qquad \Delta H = -9\text{ Calories}$$

The principle of weight control is simple. One takes in a certain amount of energy stored in the chemical bonds of food. That energy is released when this food is digested and metabolized. This energy is then used for driving energy-requiring synthesis reactions, moving muscles, and producing heat; or it is stored, primarily as fat. If more energy is consumed than body processes, heat, and exercise require, fat builds up. This is diagramed in Figure 15-7.

Body processes use a relatively small and fairly constant amount of energy. The heat needed to maintain normal body temperature under normal conditions is also fairly constant. If one lives outside in cold weather, the amount of heat required goes up considerably, however. The main controls we have over body fat are the amounts and kinds of food we eat and the

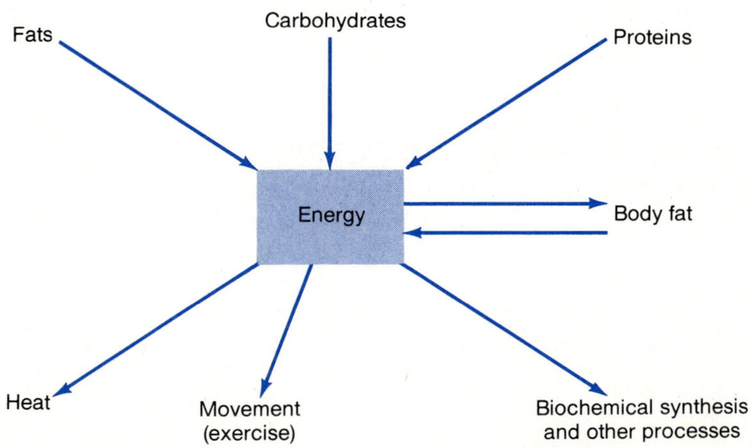

FIGURE 15-7
Fate of dietary energy.

exercise we do. Since fats have a higher energy content, lowering the proportion of them in the diet will lower energy intake; but it is not desirable to eliminate them completely, since they are part of our overall nutritional needs. Since fat takes longer to digest than carbohydrates and proteins, it stays in the digestive system longer and delays hunger. This can help one to eat less during a given day. Intake of carbohydrates likewise can be lowered *but should not be eliminated*. Since carbohydrate metabolism is required for the body to use up stored fat, one who is trying to lose weight should have some carbohydrates. The best and safest way to lose weight is to eat a *little less* than usual and exercise a *little more*. This is better than "crash diets." Weight-reducing medications may help to reduce appetite and aid in eating less, but their value is marginal. There is no such thing as "pills that melt fat away"—fraudulent advertisements notwithstanding.* Being aware of the caloric values of foods and applying the principles just discussed (using will power!) are the best approach.

There are several genuine aids to the approach just discussed (but they are not substitutes for it). Encouragement from friends or colleagues is the basis for the success of commercial weight-loss agencies. The staff at these establishments also (if the agencies are reputable) apply the principles outlined above. Some foods contain nutrients which reduce the appetite—niacin (vitamin B_3) is reported to be one of these, so that meats which are high in this vitamin may help one to eat less. When dry fibers such as cellulose or dry algae are ingested, they swell when they come in contact with the water in the stomach and give a "full" sensation, which also helps one to eat less. Note, however, that all these aids are only helping one to consume fewer calories. They have no specific medical weight-reduction value. It is also claimed by some that regular vigorous exercise not only consumes energy during the activity but also raises the general level of metabolism and helps a great deal with weight control.

Evaluating your own diet

Table 15-8 shows a personal dietary chart. One can determine the nutrient value of each food listed from package labels or from Tables 15-5 and 15-6. (More complete information about nutrient values is available from many sources, including the *CRC Handbook of Chemistry and Physics,* which is published and updated periodically by the CRC Press, Inc., Cleveland, Ohio.) The total values per day of each nutrient can be compared with the standard values shown, and any gross deficiencies can be corrected. This is somewhat time-consuming and should be done over a several-day period of normal eating. It should also be observed that not all the nutrients are included on the chart.

* The author once heard reports of a pill that (when chemical analysis was attempted) was found to contain a tapeworm. This would have aided in weight reduction, but its use would be quite dangerous, to say the least.

TABLE 15-8
Diet evaluation chart

MEAL	FOOD ITEM	WEIGHT	CALORIES, kcal	PROTEIN, g	CARBOHYDRATE, g	FATS, g	MINERALS, mg		VITAMINS							
							Ca	Fe	A, IU	D, IU	B_1, mg	B_2, mg	B_3, mg	B_{12}, μg	C, mg
Breakfast															
Lunch															
Supper															
Daily total															
RDA			3000	50	50	50	800	10	5000	400	1.5	1.7	20	6	70
Conclusion															

Food additives

Most commercial foods have a number of chemicals added to them. Some of these chemicals have been implicated as contributing to cancer, hyperactivity, and general poor health. Used in large quantities, these materials do cause problems. It is not necessarily true, however, that if a certain amount of a substance is harmful, a smaller amount is also harmful (that is, a proportionally smaller amount). A classic example is the element selenium (Se), which is toxic at relatively small doses. It is not only nontoxic but actually essential in trace amounts. The evidence available at the present time indicates that the *additives used today are beneficial*.

In addition, without these additives, even more serious problems would arise. A good example is the use of sodium nitrite ($NaNO_2$) in processed meats such as hot dogs, bologna, and bacon. One of the most toxic materials known is the toxin produced by the bacterium *Clostridium botulinum*. Meats are particularly prone to contamination with this organism, and the only good way of controlling it is to add small amounts of nitrite. Many people would die each year if no measures or less effective measures were used against *C. botulinum*. The fact that nitrite *may* slightly increase the risk of encountering other health problems is overshadowed by the great benefit derived in controlling botulism. A similar case can be made for some but not all additives.

Natural foods often contain some of the same compounds which are added artificially to processed ones. Fresh vegetables such as beets contain high concentrations of nitrites, even higher in some cases than the concentration of nitrites added. For example, hot dogs contain 150 parts per million (ppm) $NaNO_2$, while beets contain 220 ppm. Sodium nitrate, which is found in natural foods, additives, and drinking water (from farmland runoff), is converted to nitrites by human enzymes in saliva. Cured meats, such as ham or bacon, contribute about 20 percent of the nitrite in our diets, while saliva contributes about 75 percent (this nitrite comes mostly from the nitrates in vegetables). Thus it makes little sense to worry about relatively small amounts of ubiquitous materials just because they are "artificial additives."

One other point that should be discussed in connection with the use of food additives is the tendency of certain groups of people to decry the use of anything artificial and to raise a danger sign when a substance is called *a chemical*. Remember that *all physical substances* are composed of atoms and molecules of the 92 naturally occurring elements. Water which occurs naturally on earth is identical *in every respect* with that produced synthetically in a laboratory (some water has other materials dissolved in it, however). It is true that some substances, both natural and synthetic, are harmful and should be avoided, and others are less than helpful and should be used in moderation. On the other hand, however, the fact that something is produced synthetically does not necessarily make it substantially different from a natural material.

Table 15-9 lists some common food additives along with their purposes.

TABLE 15-9

Common food additives

SUBSTANCE	PURPOSE	TYPICAL AMOUNT USED,* PERCENT OF TOTAL	SAFE LEVEL OF INTAKE,* mg/kg BODY WEIGHT
Amino acids	Nutritional supplement		
Vitamins and minerals	Enhancement of the nutrient value of foods		
Salt	Flavor enhancer and preservative		
Monosodium glutamate (MSG)	Flavor enhancer		<100
Acids (acetic, citric, phosphoric)	Adjustment of pH (for stability) and addition of tartness (for flavor)	1	
Bases	Adjustment of pH		
Magnesium carbonate	Anticaking agent		
Propylene glycol	Moisturizer	0.05	<100
Synthetic flavorings (esters and aldehydes)	Improvement of taste	0.0003	<2
Saccharin	Sweetener	0.001	<5
Sodium benzoate	Preservative (inhibits bacterial and mold growth)	0.1	<5
Sodium or potassium propionates (or propionic acid)	Preservative	0.15	<10
Butylated hydroxytoluene (BHT) and butylated hydroxyanisole (BHA)	Antioxidants (prevent oils from becoming rancid)	0.005	<0.5
Monoglycerides and diglycerides	Emulsifying agents (prevent separation of oil and water layers)	0.1	
Vegetable gum	Emulsifying agent	0.5	
Propyl gallate	Antioxidant	0.003	<0.2
Sodium nitrate	Antibotulism agent in meats	0.00015	?
Silicon dioxide (powdered sand)	Free-flowing agent		

*If no entry appears for an item in this column, unlimited use is permitted.

Natural and commercial foods

The nutrition industry, of course, is not the group to ask for an unbiased opinion on food additives. The large food companies, which use a number of additives, will defend their use if it increases profits and is not grossly harmful. The natural food or health food segment of the industry is likewise interested in monetary gain. There is nothing wrong with profit, but it does tend to blur objectivity.

Natural foods can be defined as those which are produced by plants and animals not given any drugs, synthetic fertilizers, or pesticides during their growth and to which nothing has been intentionally added during processing. Some people claim that such foods are vastly superior to ordinary processed foods. All other things being equal, this claim may have some validity, but it is greatly overplayed. The healthiest situation would be to live on a farm and to raise all your own fruits, vegetables, and animals. A balanced diet of fresh produce is, no doubt, ideal.

A number of problems are encountered, however, with some foods. First, without pesticides and fertilizers most of the food would be of poor quality, since the growth of food year after year takes materials from the soil. After all, that is where many nutrients come from. Future plants grown on that land would be inferior. Insects would consume a significant portion of the produce, so that more land would be needed to grow enough food for a given family. (But nonchemical pest controls—traps, for example—are being developed and could at some future time be advantageously employed.) The health of cattle, hogs, and poultry would decline over a period of several years; if these animals lived, the quality of their products would not be high. In addition to these problems, it is not practical in our culture to have each family raise its own food. This means that safe shipping and processing become essential, and the role of many additives is to make this possible.

Natural foods are also more expensive. Whole-wheat breads that do not contain additives cost a good deal more than ordinary white breads. They are more difficult to manufacture, since some additives facilitate production, and the loss due to spoilage is greater. This means that a higher profit margin is required on the items sold. It should also be added that some less expensive bread products are simply colored with a brown dye to make them "look healthy"; actually these breads are the same as white breads. This practice is also used sometimes by doughnut shops to make their "whole-wheat" items, even though no whole-wheat flour is used. Moreover, it is not practical for many families whose budgets are limited to consume the more expensive natural foods. Likewise, it would not be possible to produce *enough* "natural" food to feed the entire population.

We can thus conclude that a diet consisting only of natural foods would be good but is not feasible. Fresh vegetables, fruits, meats, and other nonprocessed foods can and should be part of the diet, but they have most likely been produced in good quantity and with good quality by using fertilizers, pesticides, etc.

Hyperactivity

There is one problem frequently encountered today that may well be aggravated by small amounts of natural *and* synthetic materials found in food. This is the problem of hyperactivity, or hyperkinesis. While hyperactivity is most often found in children, it can also be found in people of all ages. Its symptoms include a high level of mental and physical random activity and the inability to sit still for more than a few minutes or to concentrate on one subject long enough to learn. Children with this condition appear to be undisciplined and have trouble learning in school. This is a pathological condition, and if severe enough, requires medical attention. The root causes are physiological, and no cure is presently available. However, the symptoms can be greatly reduced in less severe cases by modification of the diet or in more severe cases by treatment with certain drugs. The person can function at a more nearly normal level, and learning disabilities can be greatly reduced. Supposed culprits in the diet are preservatives such as butylated hydroxytoluene (BHT) and butylated hydroxyanisole (BHA) as well as synthetic coloring and flavoring materials. The brown "caramel color" added to give a "rich" look to syrups and colas has been suspect. Medications for children are often colored and flavored to make them more palatable; such coloring and flavoring may increase hyperactivity in those who are susceptible. In cases where these symptoms are encountered, the use of foods that contain none of the materials mentioned above often solves the problems. Their presence in prepared foods is so common that a "natural food" diet is a good approach, although it is difficult to maintain and requires careful reading of ingredient labels. More recent work, however, has shown that food additives are probably not major factors, and the special diets that do work do so for other reasons. Greater attention given a child on such a diet may account for part of the improved behavior; lowering of sugars in the diet may also be important. Over half the hyperactive children exhibit abnormal sugar metabolism, and diet can control whatever hyperactivating effects this may cause.

Dietary controversies

The public is constantly bombarded with claims by people with medical or other scientific training that a given type of diet is particularly harmful, or helpful, as the case may be. Some of these claims are difficult to evaluate, since gathering solid scientific data is very difficult. Each person's diet is complex, and isolating the effect of only one component is virtually impossible. For example, what are the long-term effects of drinking coffee? Some people who drink coffee do so with doughnuts, others with scrambled eggs and toast. Some have watermelon for dessert after the coffee while others eat chocolate cake. These other components of the diet most likely change the effects of the coffee itself. To isolate the effects of coffee, it would be necessary to feed two large groups of people with similar heredities identical diets (except for coffee) over a *long period of time*. One group would get coffee, and the other group would not. This is, practically speaking, impossible, and thus a firm conclusion is elusive. Several controversial claims will be discussed briefly in the section that follows, and a conclusion will be drawn about the evidence currently available.

One such claim is that refined carbohydrates such as white flour and table sugar are responsible for many health problems. The claim has been made that ordinary table sugar (refined sugar) greatly reduces one's muscular strength. Such statements are usually to the effect that a few grains placed in the mouth will cause a strong person to lose most of his or her strength. These are *not* the results obtained when this experiment is tried by objective investigators, and these statements should not be taken as fact. Nothing is added to crude (or natural) sugar during the refinery process; all that happens is that impurities are removed. It is unlikely that removing impurities causes refined sugar to become harmful. It is true that large excesses of sugar may cause a dietary imbalance, and it is also true that the average American consumes about 100 lb of sugar annually. We as a group do overconsume sweets, and each person should be aware of this trend, but it is *doubtful* that relatively small amounts of sucrose are harmful. It is probably true that an unbalanced diet rich in carbohydrates and very low (or lacking) in fats and proteins causes docility and lethargy in addition to other nutritional problems.

The author has encountered people who made rather negative claims about any chemical additives and any commercial processing of foods. The ideas that vegetables must be used within a few hours (or even a few minutes) if they are to keep their nutritional value, that heating destroys most vitamins, and that pasturizing destroys most of the protein in milk have all been presented to the public as fact by certain members of the medical community. Some people claim also that anything with a white color should be avoided. This includes white bread, white (or refined) sugar, and even milk. As far as can actually be determined, there is *no basis* for such claims if they applied to the general population as a whole. Of course, large excesses of any one food or the exclusion of any one of the food groups from one's diet can be detrimental to one's health. There are, moreover, people with allergies or special medical problems who must limit or eliminate certain foods. The idea to be recognized here is that there is no basis in fact to the general claims of adverse effects on human nutrition from processing, adding "chemicals," or being a certain color.

As an example of a mistaken idea, consider the belief that heating destroys vitamin C. The author has personally analyzed orange juice both before and after boiling and found that it contained the same amount (about 30 to 35 mg per 100 mL) of vitamin C after boiling.

Another mistaken idea is that pasteurizing destroys protein in milk. While the processing of milk may change the exact nature of the protein (casein) in the sense that the overall folding of the long chain is different, it does not change the amino acids composing it. Once the milk is ingested, digested into the individual amino acids, and absorbed into the blood stream, these amino acids are *identical* to those from unprocessd milk. Thus there is *no nutritional difference*.

The claim that dietary sugar *causes* tooth decay, while not *totally* invalid, is misleading. It has been shown that when sugar is consumed with other foods during a meal, it has an insignificant effect on the incidence of dental caries (decay). Heredity and general nutrition, which can affect the hardness of tooth enamel, are the two most significant factors in determining one's *tendency* toward developing cavities. Long exposure of the teeth to "chewy" sweets contributes to decay, as does eating sticky high-sugar foods such as raisins, while eating an equal amount of sugar on cereal with milk probably does not. Good dental hygiene habits, such as regular brushing and fluoride treatment, help prevent tooth decay.

The use of large doses of certain vitamins to cure a number of ailments has also been reported. The best-known claim is that of Linus Pauling that megadoses of vitamin C will cure or prevent the common cold and other viral infections. Linus Pauling is a very fine scientist who did pioneering work in bonding and discovered the α- (alpha-)helix structure of proteins. He has earned *two* Nobel Prizes (which is extremely rare). His opinion of vitamin C is therefore not to be dismissed lightly. Other scientists have confirmed his claim, but most experiments done by objective workers show no difference between groups to which vitamin C has been administered and control groups. This is a genuine controversy, with different people getting different results. Sufficient data (remember that there is no such a thing as a final answer in science) are not available yet.

Another dietary claim is that one would be more healthy on a diet consisting totally of fruits and vegetables. Those who adhere to only these foods can certainly be healthy, since it is possible to get all the essential nutrients from vegetable sources. On the other hand, with the possible exception of cholesterol content, it has not been shown that meat and other animal products are in any way harmful. Meat is more expensive than vegetables in term of both its purchase price and the resources used in its production. A world facing a possible future food shortage and those countries currently having difficulties producing enough food would be wise to emphasize vegetarian diets. One thing to keep in mind, however, is that animal protein is more complete than plant protein (i.e., it has a better balance of amino acids).

The final dietary controversy to be considered here is dietary cholesterol. Cardiovascular disease is caused by the following general process:

1 A small injury or break in the inner wall of a blood vessel occurs. This injury can be caused by tension, high blood pressure (caused by several factors), or physical external injury.

2 This injury heals, leaving scar tissue and a material called *plaque*. Plaque is largely deposits of cholesterol, which cause the diameter of the vessel to become smaller. (See Figure 15-8.)

3 It is possible for a blood clot or a piece of plaque flowing through the vessels near the heart to become lodged in a vessel so that the blood supply to the heart muscle is partially cut off. That part of the heart with an inadequate blood supply dies. A blood vessel that is already badly clogged by plaque can be completely closed off by nicotine (a vessel constrictor) or other materials. A heart attack results.

4 If a small part of the heart dies as a result of a heart attack, it can be repaired by the body; if a large portion dies, the body cannot repair all the damage, and the person dies.

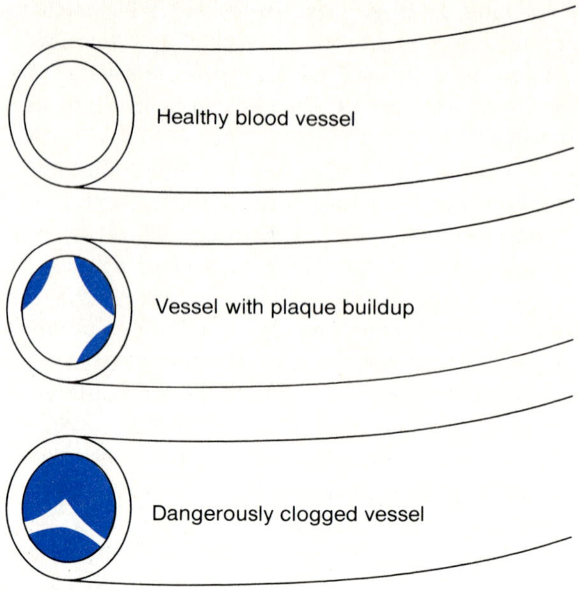

FIGURE 15-8
Plaque buildup in blood vessels.

The vessel injuries and subsequent plaque buildup are a result of at least five major factors:

1 *Heredity*. One's blood pressure and the *tendency* to develop plaque buildup are largely genetic.

2 *Exercise*. A regular, vigorous exercise program helps to prevent cholesterol from coming out of solution in the blood and slows down plaque formation. This raises the level of high-density lipoproteins (HDLs), which solubilize cholesterol.

3 *Dietary cholesterol*. Regularly eating large amounts of food containing high concentrations of cholesterol probably causes plaque to form faster. The daily intake of cholesterol should be kept below 200 mg.

4 *Dietary fats*. The consuming of unsaturated fats (oils) aids in reducing the tendency to develop cardiovascular disease and probably also raises the HDL level.*

5 High levels of glucose stimulate the body's system to synthesize cholesterol. This occurs because the insulin needed for glucose metabolism also causes cholesterol to be produced. Fructose, which is available in grocery stores and which is the main sugar found in fruit and honey, does not have this effect. Regular use of fructose and keeping sucrose levels low will, over a period of several months, help to lower serum cholesterol. (Sucrose is a disaccharide containing glucose.)

---

*Some studies suggest that high oil diets may contribute to cancer incidence. Thus, the balance of fats in a person's diet should be controlled by whether that person has a family history of heart disease or cancer.

TABLE 15-10

Foods with high cholesterol content

FOOD	SERVING SIZE	AMOUNT OF CHOLESTEROL, mg
Brains	¼ lb	1000
Liver	¼ lb	450
Oysters	¼ lb	325
Egg	1 large	252
Shrimp	¼ lb	125
Beef	¼ lb	100
Pork	¼ lb	90
Chicken	¼ lb	90
Fish	¼ lb	60
Whole milk	1 glass	35
Skim milk	1 glass	5
Ice cream	½ cup	50
Butter	1 teaspoon	12
	¼ lb	300

Table 15-10 shows foods that are particularly high in cholesterol. For those people who have a family history of heart disease, it would be wise to keep the level of cholesterol in the diet relatively low, but since it is needed by the body for normal functions and is synthesized by the body, there is no need to attempt to reduce its intake to near zero (except if directed by one's physician, of course). A serum cholesterol level below 200 mg per 100 mL of blood is desirable, although the average is 230 to 250.

The best approach for the prevention of cardiovascular disease is to eat a relatively low-cholesterol diet, to eat most of your lipids as oils, to get plenty of exercise, not to smoke, to eat a low-sucrose diet, and to choose your parents carefully.

Artificial sweeteners

Several compounds are useful as artificial sweeteners in products for people who must not consume sugar, either because they are trying to reduce their calorie intake or because they have other health problems.

Saccharin has been used for a long time in this capacity:

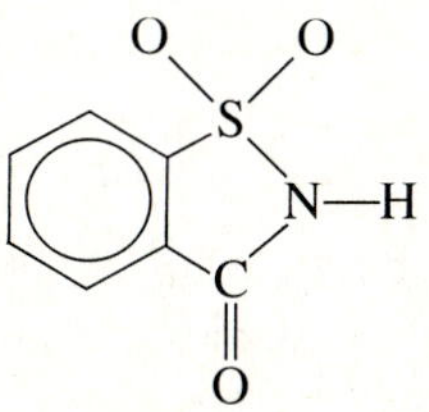

Saccharin

However, it has several disadvantages. It leaves a bitter aftertaste; it does not taste exactly like sugar; and it is weakly carcinogenic.

As was noted in Chapter 3, cyclamates were previously used but now are banned. Calcium cyclamate is shown below:

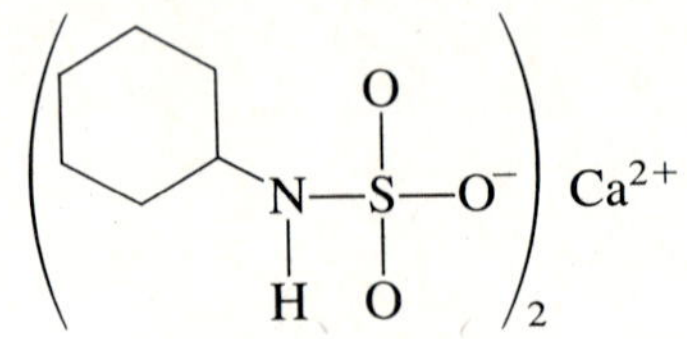

Calcium cyclamate

The only nonsugar sweetener besides saccharin now approved by the U.S. Food and Drug Administration is aspartame. Its structure is shown below:

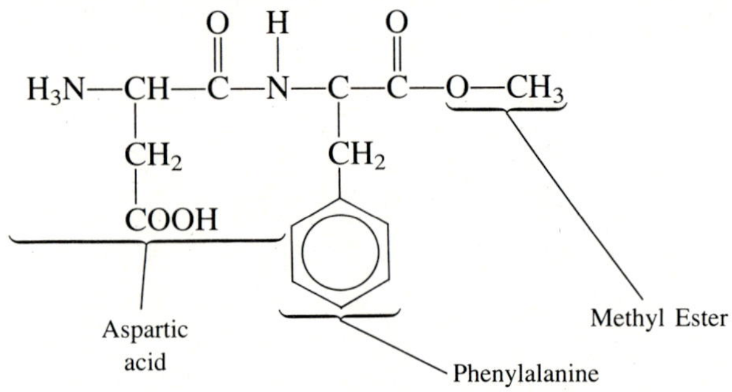

Aspartame

As can be seen, it is composed of an aspartic acid moiety along with the methyl ester of phenylalanine. A dilute solution of this material is many times sweeter than sucrose (table sugar)—up to two times sweeter. Since aspartame is composed of two natural amino acids, it is not surprising that tests have shown no medical problems, and it is presumed to be completely safe.

Another material, acesulfame (the potassium salt), is expected to be marketed soon. Its structure is shown below:

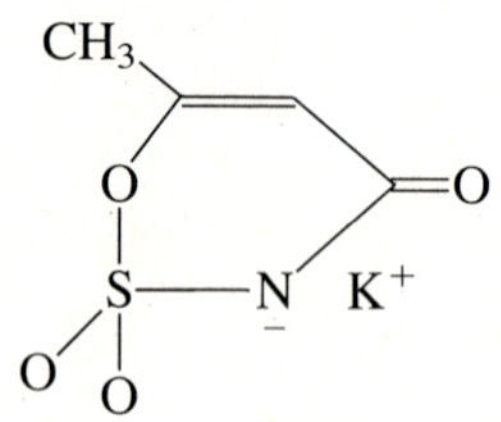

Acesulfame-K

Acesulfame is over 100 times sweeter than sugar, and tests have shown this unusual molecule to be safe for human consumption.

A protein called *Talin,* isolated from an African plant, is several thousand times as sweet as sugar; it has a complex molecular structure with 20 amino acids. Its applications and safety are currently being evaluated in several countries, and it is actually in use in some places.

It is not known why substances such as these are sweet. Most of them do not have similar structures, and indeed some materials with structures similar to those that are sweet are not sweet at all.

Another approach to "artificial" sweetening is sorbitol:

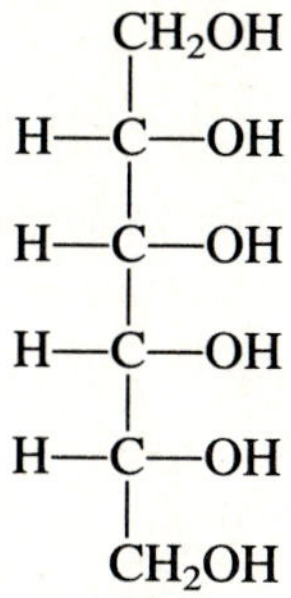

Sorbitol

Sorbitol is made from glucose; but it cannot be broken down by bacteria to cause tooth decay, nor can it pass through cell walls. Because it cannot get into the cells, it cannot be metabolized; therefore, it does not contribute to the caloric value of foods (such as "diet candy") in which it is found.

Table 15-4 (refer back to page 398) shows the relative sweetness of several artificial sweeteners as compared with natural sugars.

STUDY GUIDE

Terms to learn

Amide bond
Amino acid
Aspartame
Carbohydrates
Cellulose
Essential amino acid
Essential fatty acid
Enzyme
Fat
Food additive
Glucose
Glycogen
HDL
Hyperactivity
Lipid
MDR
Mineral
Nucleic acid
Oil
Plaque (in blood vessels)
Protein
RDA
Starch
Sugar
Vitamin

Skill practice problems

1 True or false? A high-cholesterol diet is the main reason for plaque buildup in blood vessels.

2 How much vitamin C is in a 1-lb potato? (See Table 15-6.)

3 Amino acids are the simple molecules from which ______________ are made.

4 Starch is changed in the digestive system to ______________.

5 Enzymes in the human digestive system *cannot* break α- or β-glucosidic bonds in polyglucose materials.

6 Enzymes are needed for which of the following?
(a) Breaking down food
(b) Synthesis of biomolecules
(c) Energy production
(d) All of the above

7 Which of the following is *not* composed primarily of protein?
(a) Oils
(b) Enzymes
(c) Muscle
(d) Hair

8 Which of the following is an essential amino acid?
(a) Alanine
(b) Leucine
(c) Glycine
(d) Proline

9 What is the R group in valine?

10 What are the proteinlike compounds containing *two* amino acids that one could prepare by using glycine (Gly) and leucine (Leu)?

11 What are the proteinlike compounds containing *three* amino acids that one could prepare by using glycine (Gly) and leucine (Leu)?

12 Write the equation for digestion of fats.

13 How many Calories (kcal) are there in 1 oz of sugar?

14 For a rapid weight loss during a diet, one should not eat either fats or carbohydrates. True or false?

15 The RDA for adult humans of vitamin A is ______________.

16 The term RDA means ______________.

17 Why is it difficult to determine the long-term effect of a given food on human health?

Goals

To understand the important role of drugs in our culture

To gain practical consumer information about over-the-counter drugs and their use

To develop principles of personal responsibility about drug use

To gain correct information about the effect of nonmedicinal drugs on our bodies

To understand the need for safety in drug use

To understand the great effort required to produce a useful drug

To learn about current areas of pharmaceutical research

16

MEDICINAL CHEMISTRY FOR THE CONSUMER

If you visit one of the reconstructed colonial American towns (such as Sturbridge Village in Massachusetts), you will probably be impressed by the frequency of early death during that period. Plaques on the walls of many houses tell of the deaths of all the children in a family within a span of a few weeks from infectious disease. As late as the 1890s and the early twentieth century, whole families died of cholera. Some people who are still alive were crippled with polio before the development in 1955 of the Salk vaccine, which effectively stopped polio epidemics. Of all the areas in which science has contributed to human well-being, probably health care is the most dramatic and important. This is the result of two major factors. One is the understanding that many (though not all) illnesses are caused by microorganisms. The second is the development and use of chemical substances to cure or control sickness.

DRUGS AND THEIR USE

Let us define a *drug* as any substance taken into the body or used on the body to produce a specific desired effect. A *medicine,* or *medicinal drug,* is used to control or relieve an abnormal condition such as pain or disease. A *nonmedicinal drug* is usually used to produce abnormal conditions such as a "high," hallucinations, relief of inhibitions, and wakefulness when sleep would be normal. A systematic presentation of various types of drugs is given later in this chapter, in Table 16-3.

As we become more and more comfortable with our high standard of living, pain (even mild pain) or discomfort seems to be intolerable. As a result, the use of both prescription drugs (those obtained only as a result of a physician's orders) and over-the-counter (OTC) drugs (those that can be purchased in any quantity at a drugstore any time one wishes) has proliferated in recent years. (See Figure 16-1.) At the slightest sign of a headache many of us go to the medicine cabinet and "take two aspirin." Likewise, we indulge our appetites to the point of abusing our bodies (eating until we are uncomfortable) and then seek relief by taking an over-the-counter formulation. Part of this is a result of the hedonistic philosophy which we subconsciously embrace—the idea that we should always feel good and experience no pain. Another reason for the intense use of drugs is the constant bombardment of our minds with advertising propaganda.

Not only do we take more nonprescription drugs than necessary but we have placed subtle pressure on our personal physicians to "prescribe something" or give an injection to "make me well instantly." This has led at times to the overprescribing of drugs.

The other side of the drug picture is that when we really do have medical problems, chemical formulations which can be used to solve most of them are available at all pharmacies. These range from headache remedies, to suntan lotion, to antibiotics and medication to help a sick child to sleep, to treatments for people with epilepsy and those who would die from pneumonia or even cancer. Except for cancer patients and accident victims, very few young people die today. This is the result of medical research, chemical synthesis, and

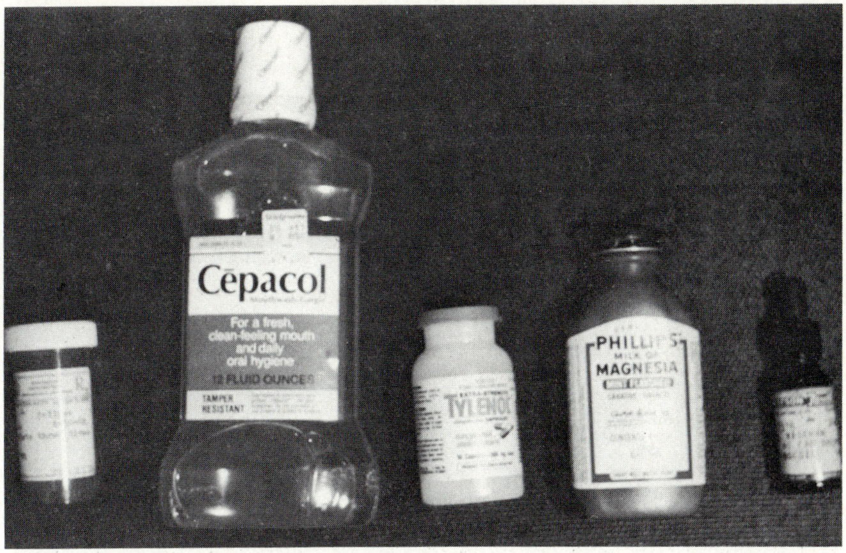

FIGURE 16-1
Some of the many drug products we use regularly.

the excellent training of an abundant supply of physicians and pharmacists. This situation has gradually come about over the past century, but many of the major advances have occurred in the last 50 years. Sulfa drugs were first used in the 1930s. Penicillin was discovered in 1928 and first used during World War II. It has been produced in great abundance since 1950 by a partial synthesis method (that is, part of the molecule is isolated from natural sources and part is prepared using chemical synthesis techniques). Tuberculosis, polio, and smallpox have been almost eradicated by the use of drugs and vaccines.

We have available a plethora of drugs which can be either used advantageously or misused. We will consider some personal guidelines for drug use at the end of this chapter, but to begin with we will look at some of the common medicinal drugs. Both prescription and over-the-counter drugs will be examined in order to point out some principles of drug use and effects. Practical aspects of buying and using nonprescription drugs will also be covered. Much of medicinal science uses the body's natural defenses, surgery, or vaccination, but we will limit our focus to the use of drugs.

COMMON MEDICINAL DRUGS

Aspirin

One of the oldest and most commonly used drugs is aspirin. It has been used since 1899 and has three primary physiological properties: it is (1) an analgesic (a general painkiller), (2) an antipyretic (it lowers fever during an infectious illness), and (3) an anti-inflammatory agent (for example, it relieves arthritic inflammation). Aspirin is the best antipyretic known, and it is better than narcotics for relief of aches. It also is useful in treating victims of heart attacks and a number of other serious ailments.

Chemically, aspirin is acetylsalicylic acid, which is prepared as shown below from salicylic acid and acetic anhydride:

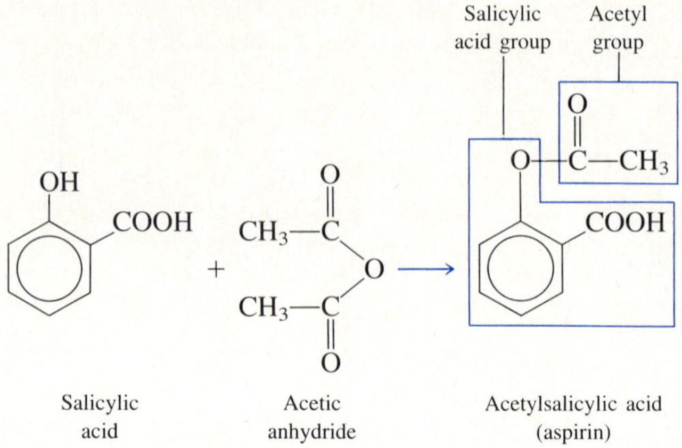

A principle of drug development can be illustrated by this synthesis. Actually salicylic acid itself is the material which possesses the desired physiological effect, but the —OH group causes the salicylic acid to burn the mouth and throat lining. (A compound with an —OH group on an aromatic ring is called a *phenol*.) If the —OH group is modified by adding an acetyl group to it, this undesirable property is eliminated and the aspirin can be taken safely. Once in the digestive system, aspirin changes back to salicylic acid and acetic acid, which are absorbed into the bloodstream. The acetyl group is called a *blocking group,* since it blocks (changes) the —OH functional group and eliminates the undesired side effect.

Contrary to popular opinion, there is no such thing as a perfectly safe drug. All materials have some undesirable side effect in addition to the desired therapeutic effect. Aspirin causes irritation of the stomach walls and the loss of several milliliters of blood every time a tablet is taken. This irritates ulcerated stomachs or can cause them. It also causes hyperactivity in people with a predisposition (usually children). Slow blood clotting (clotting takes about twice as long) and resistance to infection also result. In very large doses (60 to 90 tablets) it can even cause death from salicylate poisoning. The toxicity of a compound is measured by a value called the LD_{50} (lethal dose for 50 percent of the test animals). The *higher* the LD_{50} value, the *less* toxic the material. For aspirin the LD_{50} is 1.75 g of aspirin per kilogram of body weight in rats and only 0.15 g/kg in humans. This means that 10 tablets can be fatal to a 45-lb child.

The *therapeutic index* (TI) of a drug is the amount required to cause poisoning compared with the amount needed for medical purposes:

$$\text{TI} = \frac{\text{amount for toxic levels}}{\text{amount for therapeutic levels}}$$

Aspirin has one of the highest TI values, about 20. This means that it takes 20 times as much aspirin to poison as to stop pain or lower fever, which is quite a safe margin. (By comparison, the TI of diphenylhydantoin, which is used to control seizures, is only about 2.)

A question may come to mind when we consider common usage and advertising: Is there a difference in the effectiveness of different brands of aspirin? Some companies, as you probably already know, claim that their product is superior. All aspirin tablets contain 5 grains (325 mg) of acetylsalicylic acid and have a high purity as defined by the Food and Drug Administration (FDA). The FDA periodically checks to make sure that *all* companies adhere to the same high standards of weight, purity, and accuracy of tablet count (±5 percent). There are some differences among brands, such as attractiveness of packaging, volume of advertising, and—perhaps most important—the tightness to which the tablets are compacted or other formulation techniques. Tightness, along with other aspects of formulation, determines how fast a tablet disintegrates in the stomach and is absorbed into the bloodstream. The difference between the fastest and slowest aspirin, however, is a matter of several minutes at most. Since it takes about 50 min for the desired effect to take place, a difference of 1 to 2 min in dissolving time is insignificant. Some brands of aspirin cost up to five times as much as others. But it can be said, in light of the facts we have just noted, that *all* regular aspirin tablets are alike for all practical purposes. (In fact, this is true for most drugs, both over-the-counter and prescription. A few companies may formulate a given drug so that it is slightly more effective, but that is unusual.)

There are a number of multicomponent formulations that contain aspirin. *Buffered aspirin* contains a slightly basic material such as $CaCO_3$ or $NaHCO_3$ to help neutralize acidity and reduce stomach irritation. This has no effect on relief of pain, however. *Effervescent aspirin* (Figure 16-2) contains a solid acid and sodium bicarbonate ($NaHCO_3$), so that when it is added to water, the bubbling of the CO_2 causes the aspirin to dissolve. When this solution is ingested, the aspirin is absorbed almost immediately. The solid acid does not

FIGURE 16-2
Different dosage forms of aspirin. *Left,* effervescent aspirin; *right,* two regular aspirin tablets with hot water.

react with $NaHCO_3$ when the aspirin is in tablet form; but when the aspirin is added to water, a vigorous change occurs:

$$\underbrace{\begin{array}{c}\text{Aspirin (solid)}\\ + \\ NaHCO_3\text{ (solid)}\\ + \\ \text{tartaric acid (solid)}\end{array}}_{\begin{array}{c}\text{Solid effervescent}\\ \text{aspirin tablet}\end{array}} \xrightarrow{H_2O} \begin{array}{c}\text{aspirin (solution)}\\ + \\ CO_2\text{ (gas bubbles)}\\ + \\ \text{sodium tartrate (solution)}\end{array}$$

Aspirin itself cannot be shipped or stored in solution form, because it would decompose to salicylic acid and acetic acid in a few days or weeks. Tablets also decompose slowly when exposed to moist air, but unless they are many years old, this is not a serious problem. One usually swallows a tablet rapidly with water, and thus there is little contact with mouth and throat.

Effervescent formulations do relieve pain a little faster (in 30 to 40 min compared with about 50 min for regular tablets). The same effect can be obtained, however, by taking two of the least expensive regular tablets with hot water (to facilitate dissolving) on an empty stomach (to facilitate absorption into the bloodstream); see Figure 16-2. The so-called double-strength formula simply has 650 mg of aspirin, the equivalent of two tablets. No aspirin formulation acts faster than the method just described. The blood level of salicylic acid 30 min after taking aspirin this way is about twice as high as that using the usual method of ingestion, and no commercial formulation surpasses that level. Some tablets combine caffeine with aspirin, since caffeine is a stimulant and gives a sense of well-being. It does not actually reduce pain or increase the speed of relief, however.

Liquid ''aspirin'' used by children who are too young to take tablets is a suspension of salicylamide:

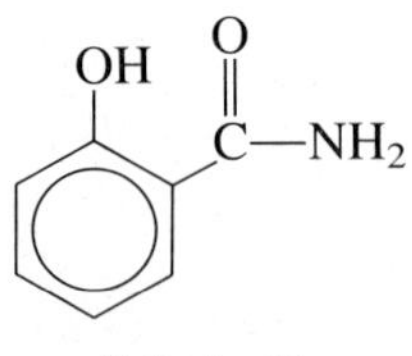

Salicylamide

In salicylamide the —OH group is not blocked as it is in aspirin; the —COOH group is changed instead. Apparently the irritating effect of salicylic acid itself requires both the —OH group and the —COOH group. Salicylamide is stable enough to remain unchanged in water but can still be changed to salicylic acid in the child's digestive system. It is a little slower to convert and is slightly more toxic than aspirin (its LD_{50} is 1.4 g/kg in mice), but it is often critical to a child's health to reduce a high fever rapidly to avoid possible brain damage. This is another example of modification of a basically useful molecule by synthetic chemistry to produce a material to meet a specific medical need.

Another related compound is methyl salicylate, commonly called *oil of wintergreen;* it is a liquid and is used in rubbing compounds for sore muscles:

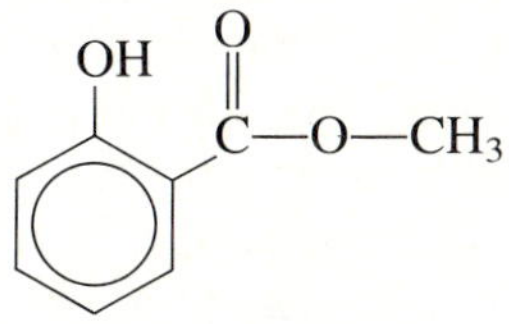

Methyl salicylate

Acetaminophen

Acetaminophen is a member of a class called *aminophenols*. The three members of this group are shown below:

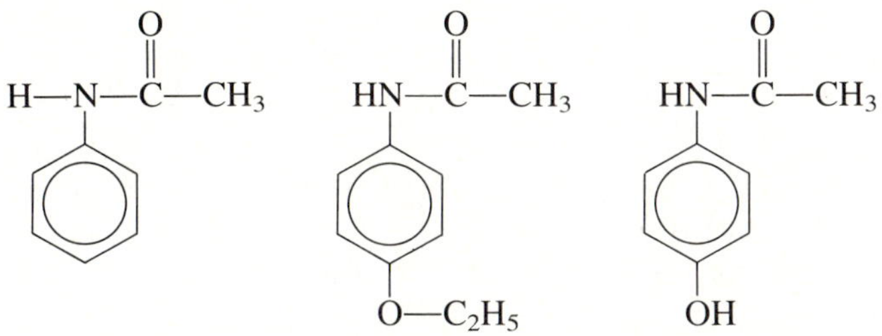

Acetanilide Phenacetin Acetaminophen

Acetanilide is somewhat toxic and is no longer used. It has an LD_{50} of 0.8 g/kg in rats, which is about twice as toxic as aspirin. Phenacetin is less toxic and has been used in "APC" tablets (APC stands for aspirin, phenacetin, and caffeine). Because it is suspected of causing liver damage, it is no longer used. Acetaminophen is the best of the three and is sold under various brands such as Tylenol and Datril. It is an analgesic and antipyretic agent, like aspirin, but it is not an anti-inflammatory agent. The dose of acetaminophen in a tablet is the same as aspirin, 325 mg per tablet, and its rate of action is also about the same as aspirin. It is slightly less toxic than aspirin (its LD_{50} is greater than 4 g/kg in rabbits) and does not cause stomach bleeding. Some people become nauseated when taking it, however, so that the choice between acetaminophen and aspirin is largely a personal preference. Most acetaminophen is more expensive than aspirin.

FIGURE 16-3
Alexander Fleming, who first isolated penicillin. (Photograph from the Smithsonian Institution.)

Penicillin

We now turn from over-the-counter drugs to one of the most commonly used prescription antibiotics. Penicillin was discovered by Alexander Fleming (see Figure 16-3) in 1928 as a product of bread mold, but its medical value was first recognized in 1939 by Flory and Chain. It was one of the most valuable medicines used during World War II to counteract infections due to wounds (see Chapter 1). At that time it had to be isolated by extraction from the culture medium on which the mold was grown. This was slow, expensive,

FIGURE 16-4
Kenneth Henry-Logan, who first proved the molecular structure of penicillin. (Courtesy of Dr. Henry-Logan.)

and limited; and even though there were several kinds of penicillin, these were all natural ones. A major effort by a large number of British and American chemists during and after the war failed to determine conclusively the molecular structure of penicillin. But in 1950, Kenneth Henry-Logan (see Figure 16-4) finally synthesized one of the suspected structures. (Previous work by other chemists was known to him.) It had all the properties of natural penicillin and thus had to have the same molecular structure. Since Henry-Logan knew what structure he had produced, he had proved the structure of the naturally produced material.

That structure is shown below:

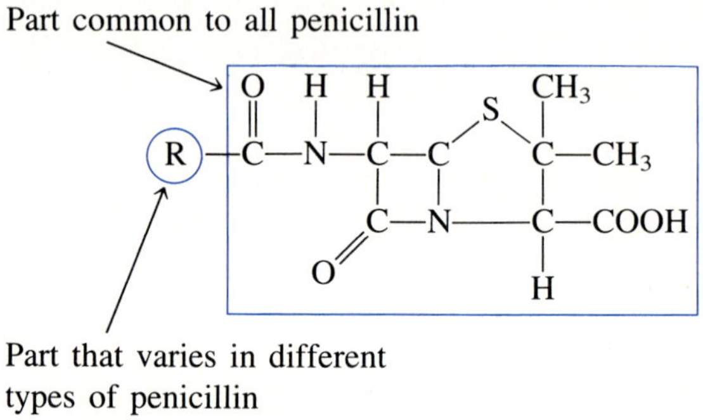

As can be seen, it is a complex molecule with several functional groups. The symbol R in the structure stands for any one of several molecular groups that are present in the various types of penicillin. The common ones are:

R	NAME
 $—CH_2—$	Penicillin G
$CH_3—CH_2—CH_2—CH_2—CH_2—CH_2—CH_2—$	Penicillin K
 $—O—CH_2—$	Penicillin V
$CH_2{=}CH—CH_2—S—CH_2—$	Penicillin O

Although the entire molecule can be synthesized, that is quite expensive. It is now possible to grow natural penicillin and chemically change the R group to the desired one. Different bacteria respond to different types of penicillin, and this gives physicians more tools with which to work. When penicillin was first isolated from mold cultures, it cost about $100 per gram. After the structure was proved, a partly synthetic product became available and the techniques of isolation were refined, which reduced the cost to a few cents per gram. The research seemed long and was expensive, but now everyone, no matter how poor, can afford and benefit from penicillin. Of all the drugs this class is probably the safest (except for those with allergies to it). The therapeutic index is very high (several hundred), and at times very large doses have been given with no toxic effect.

Because of the frequent use of penicillin for virtually every infection for a number of years, some bacteria have developed penicillin-resistant strains. This is especially true of the bacteria that cause venereal disease. The use of penicillin in the treatment of venereal disease is an excellent example of the abuse of drugs. People assumed that venereal disease could easily be cured by taking penicillin and became careless in their sexual behavior. (While this was probably not the main factor in the change in sexual patterns, it was a contributing factor.) This produced a great increase in penicillin use and increased the probability that resistant strains would develop. Some of the resistant strains are not killed by presently available antibiotics and are thus incurable. Thus, by abusing the tools available to medical science, a serious problem has been created.

While the exact biochemical action of some older drugs, such as aspirin, is not fully understood, there is a body of evidence which allows us to explain, with reasonable certainty, how penicillin stops bacterial growth. During the initial development of a bacterial cell, the cell wall is also growing. The wall is like a net; and as it grows, long chains of atoms form which have "tails" of atoms hanging from the chain. The ends of these tails are then linked together to form the net. This is shown in Figure 16-5. Each step

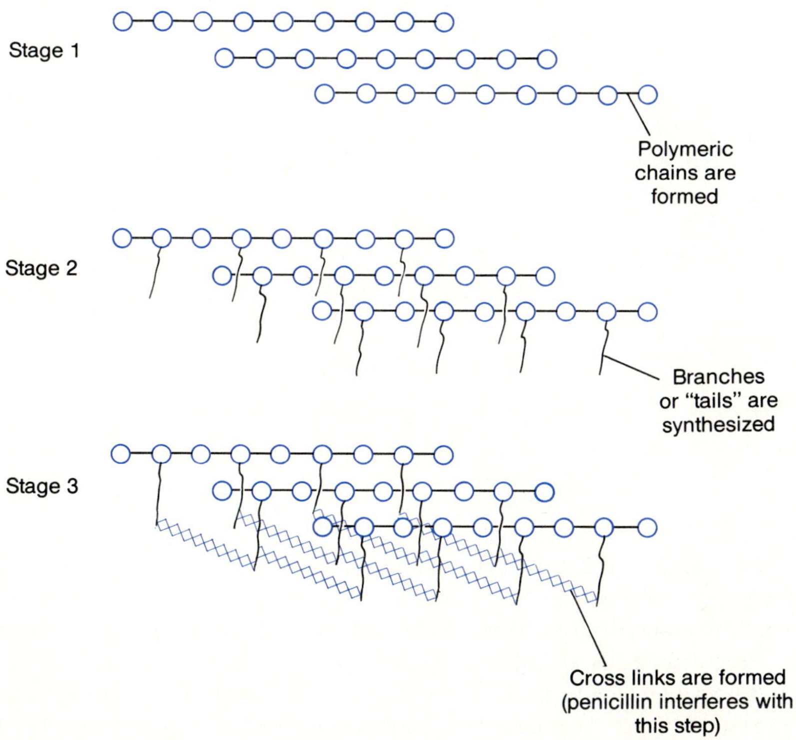

FIGURE 16-5
Cell wall development and the mechanism of the action of penicillin.

results because enzymes cause new chemical bonds to form. Penicillin interferes with (inhibits) the action of the cross-linking enzyme, so that the last step is never completed. The net is not formed, and the young bacteria die, never growing old enough to reproduce. The infection is thus stopped.

Sulfa drugs

Another important class of antibiotics is the sulfa drugs. The synthesis of sulfanilamide is used as an example of a complex organic synthetic sequence in Chapter 11. Other organic groups can be added to the

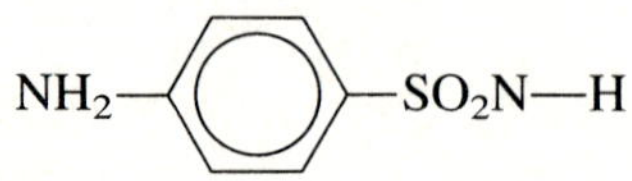

group to produce any one of several drugs with similar yet distinct activities. Sulfathiazole is an example. About 5000 sulfa drugs have been prepared, but only 10 to 15 are in use as antibiotics today. The structures of sulfanilamide and sulfathiazole are shown below:

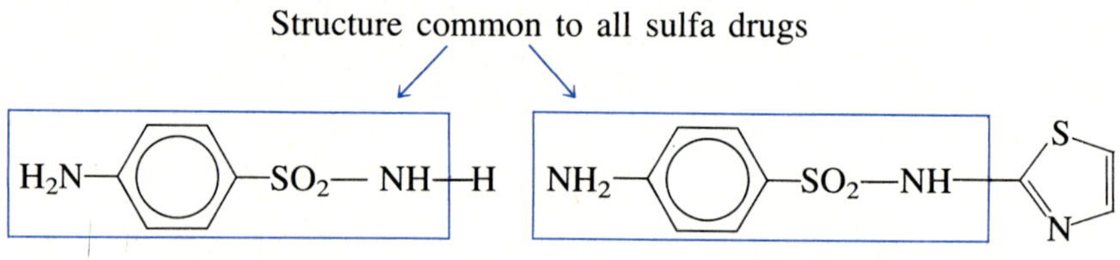

Sulfanilamide　　Sulfathiazole

Benzocaine

We now turn to a compound which is part of many nonprescription formulations. Benzocaine is a local anesthetic (painkiller) that is used to numb the pain of sore throat, sunburn, cuts, etc. It is a simple compound and can easily be prepared by a beginning chemistry student. Its structure is shown below:

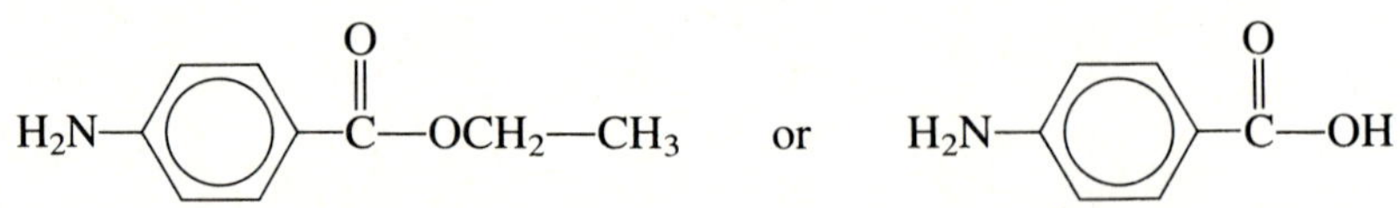

Ethyl ester of *p*-aminobenzoic acid (benzocaine)　or　*p*-aminobenzoic acid (PABA)

Not only does it kill pain when applied topically (on the surface of the skin); along with a similar compound, *p*-aminobenzoic acid (PABA), it also has the ability to absorb the ultraviolet rays of the sun. These are the rays which cause burning; thus PABA is called *Sunscreen* by the suntan lotion industry. As a consumer one should look for the presence of one of these materials in any painkilling formulation such as sore throat medication, sprays for cuts, or suntan lotion.

Suntanning is such a popular activity that it will be discussed briefly in the context of medications. The ultraviolet (UV) rays of sunlight cause both burning and tanning. These are two separate processes. Contrary to the general belief, a red sunburn does not "turn into a tan." The longer UV rays—those longer than 320 nanometers (nm)—penetrate the skin and stimulate the formulation of *melanin,* and once formed, the melanin becomes darker on exposure to more UV light. The shorter-wavelength UV light (less than 320 nm) simply destroys the surface skin cells. This is called a *sunburn* and is red or pink in color. During the summer, the initial redness heals while melanin is slowly forming from continuing exposure to the longer UV rays of the sun—independently of each other. The sun's rays can be blocked by clothing or an umbrella or such opaque materials as ZnO_2 (white) in a cream formulation. Clouds do not stop most of the UV rays, and while one cannot feel the heat (infrared rays) or see the bright light from the sun on a cloudy day, a bad burn can occur at such a time. A wet cotton T shirt will stop only part of the UV rays, while a dry one essentially blocks all the UV rays.

PABA absorbs the short UV rays and prevents burning while allowing the long ones to pass through to the skin to form melanin. Some suntan lotions are simply oils to prevent drying of the skin. They are no better, and usually more expensive, than the least expensive drugstore skin cream. Those which actually prevent burning contain PABA, and the best are those that have the most PABA and cost the least. Some products also have a brown or yellow substance such as iodine (I_2) to give the *illusion* of quicker tanning. Some lotions have numbers indicating the amount of PABA. Those having larger numbers contain higher concentrations of PABA and thus protect more effectively against sunburn.

Dextromethorphan

In our exploration of a wide range of common drugs of interest to consumers, we now come to one that has only one specific use. Dextromethorphan (often used as its hydrobromide salt) is a nonnarcotic over-the-counter cough suppressant—probably the only really good one other than codeine (which is a narcotic obtainable by prescription only). Any cough medication should contain it; the usual oral dose is 15 mg. (See Figure 16-6.)

The structure is shown below:

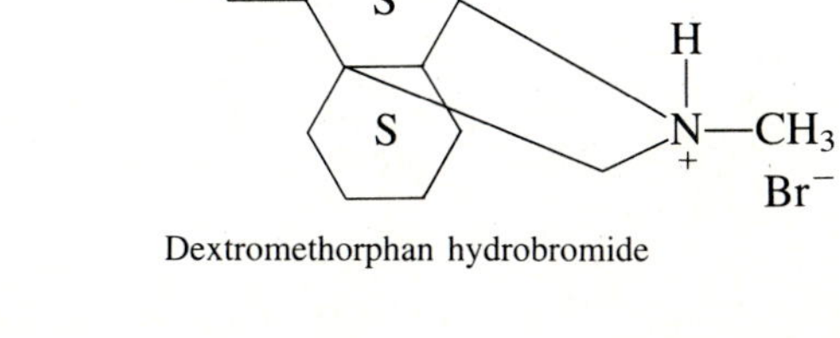

Dextromethorphan hydrobromide

FIGURE 16-6

Typical ingredient label for an over-the-counter drug.

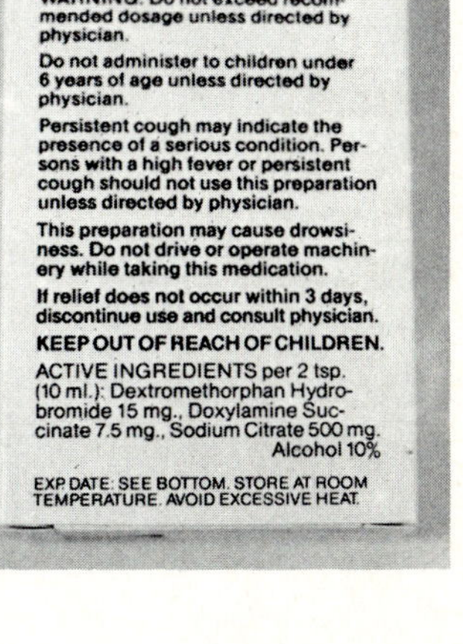

Cetylpyridinium chloride

Cetylpyridinium chloride belongs to a class of topical bactericides called *quaternary ammonium salts* (*quats* for short). The structure is shown below:

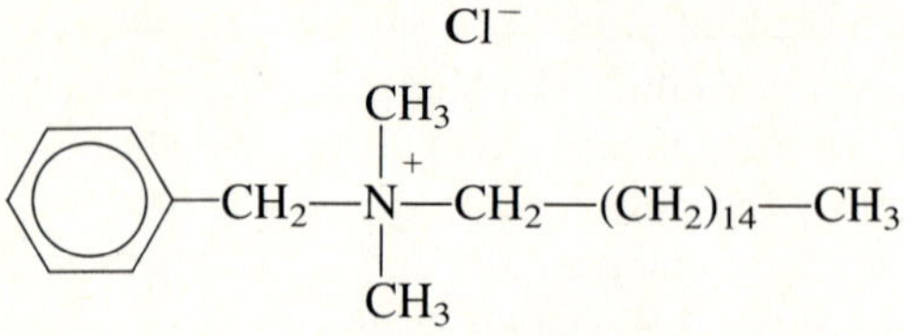

Cetyldimethylbenzyl-
ammonium chloride

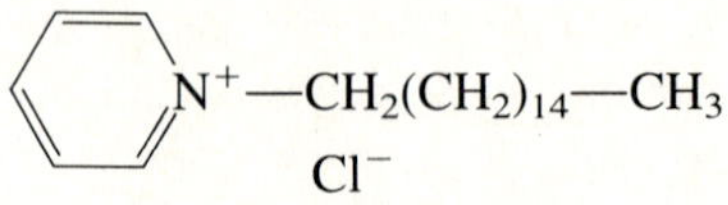

Cetylpyridinium chloride

There are several similar materials used whose structure can be recognized by the presence of a nitrogen atom with four bonds and a positive charge. Their names can be recognized on labels because they all end in *-ium chloride* or *-ium bromide*. The quats are somewhat toxic and should not be ingested, but they are most commonly found in mouthwashes and special skin cleansers. For a mouthwash to be generally effective, it should destroy most of the bacteria present. Ethyl or isopropyl alcohols are not good fast-acting bactericides in low concentrations. They work well, however, on long exposure or at high concentrations (such as 70 percent alcohol in water). Some mouthwashes contain 15 to 25 percent alcohol, but the short exposure time during gargling is not sufficient to kill most of the bacteria. You should look for a quat in the mouthwash you are considering.

Topical antiseptics

Four of the most common materials used for preventing infection in open cuts are merthiolate, tincture of iodine (I_2), mercurochrome, and hydrogen peroxide. Iodine is the oldest and is still used effectively. It is not very soluble in water (the I_2 molecule is nonpolar) but does dissolve in alcohol to give a brown solution (the term *tincture of* means *an alcohol solution of*). The alcohol causes an open wound to sting, and if a large area has been scraped or burned, the application of tincture of iodine can be quite painful.

Mercurochrome, whose structure is shown below, has two negatively charged groups, which makes it polar and therefore soluble in water:

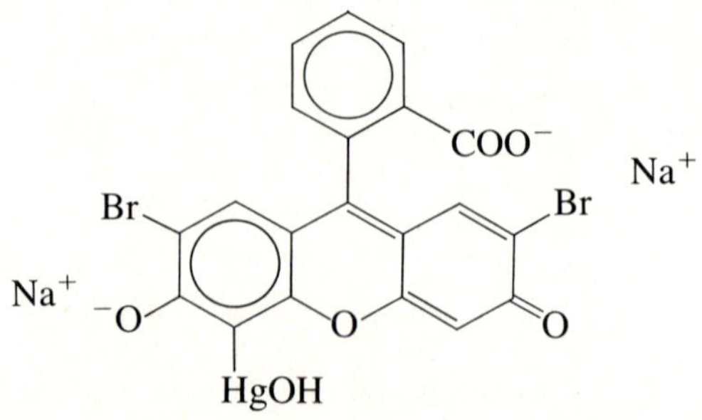

Mercurochrome

It has an intense red color (the molecules absorb green light), and since the water in which it is dissolved does not burn an open cut, it is useful for small children and in other applications where pain is a problem.

Another common topical antiseptic is merthiolate, whose structure is shown below:

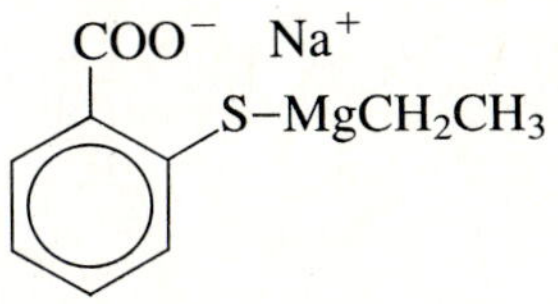

Merthiolate

Merthiolate is almost colorless, and because of the polar COO^-Na^+ group it is soluble in water. Preparations of it are colorless, and since it is dissolved in water, it does not burn. With no stain and no pain, it is an ideal topical antiseptic. When it was first marketed in this form, it did not sell well, however. People were used to iodine and mercurochrome, which were colored; therefore, for an antiseptic to be useful it was thought to need color. There is also a mentality which thinks that for a medication to be useful and helpful, it must either be painful or taste terrible (there is apparently a certain amount of self-satisfaction in pain). An inert red dye was added, and a little alcohol was put into the solution to make it sting; it has been selling quite well ever since. This is a good example of how fallacious ideas are sometimes accepted by the general public and how companies sometimes pander to them, as long as no harm is done, in order to sell a product.

The fourth antiseptic is hydrogen peroxide (H_2O_2), which comes as a 3 percent solution. It reacts with many materials and can be either an oxidizing or a reducing agent. In a cut, it is changed to water and gives off pure O_2 gas bubbles which kill the bacteria:

$$2H_2O_2 \longrightarrow 2H_2O + O_2$$

Since the peroxide slowly decomposes on standing, its concentration should be checked periodically using the method discussed in Chapter 17 under "Household Bleach Formulations."

Birth control pills

The last type of medicinal drug we will consider is contraceptives. Many people, for financial or other reasons, want to limit the number of children born to them. Others, for various reasons—including personal preference and considerations of health—decide to have no children. Still others are concerned with excessive population growth and decide to do their part to limit it by having only one or two offspring. Sexual activity is desirable to satisfy the physical needs of both partners, and it is important in building and maintaining a satisfying relationship. Thus, since engaging each other sexually is desirable and limiting conception is also considered important by many people, contraceptives are widely used. Some methods are physical; these include the intrauterine device (IUD), the diaphragm, and the condom. But because IUDs are not always safe and the condom can interfere with sexual pleasure, many couples select a chemical (or drug) method.

In nature, once a female becomes pregnant, hormones are secreted which prevent further union of sperm and egg cells. This keeps two out-of-phase conceptions from occurring at the same time, since this would be disastrous to both fetuses. "The pill" is a drug formulation containing female hormones which chemically simulate pregnancy in the female reproductive system. This means that from a biochemical point of view, the woman using birth control pills is in a continual state of "pregnancy," and a real pregnancy cannot occur. Use of birth control pills in no way interferes with sexual activity and is quite reliable (90 to 95 percent). The desirability of long-term artificially induced hormone levels is questionable, but it appears to be reasonably safe. The compounds used are derivatives of the hormone estrogen. The structures of two of the most commonly used are shown below:

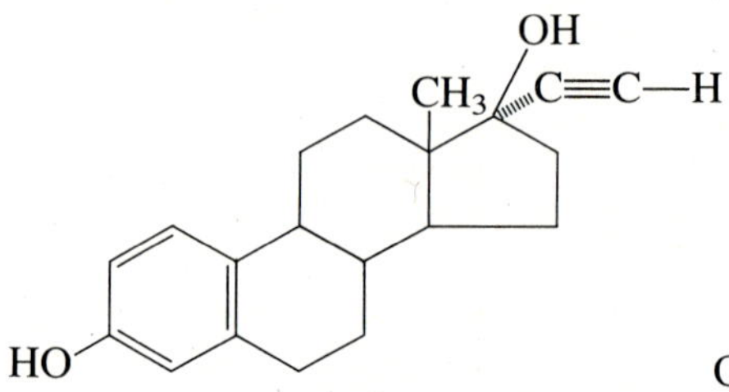

Ethynyl estradiol

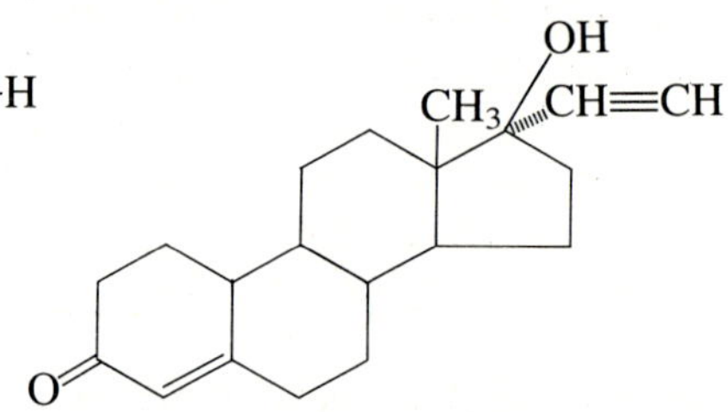

Norethindrone

Spermicidal foams and jellies, used intravaginally, kill sperm cells before fertilization. A recent development is a polyurethane foam sponge containing a sperm-killing agent. This sponge is inserted before intercourse and serves as a physical barrier to sperm mobility as well as being toxic to sperm. This product is advertised as being highly effective and convenient.

Other types of chemical contraceptives are in the research stage, but they are not being marketed yet. They include compounds that would prevent a fertilized egg from becoming implanted in the uterine wall. The fertilized egg would simply be excreted. Such a compound would be taken *after* intercourse and is called the *morning-after pill;* it actually causes a very early abortion. Another concept being worked on is a male pill. In this case the man would take the drug rather than the woman and the sperm would be inactivated. It is not known with certainty if the morning-after pill and the male pill will be marketed someday, but it is likely that they will.

COMMON NONMEDICINAL DRUGS

Nonmedicinal drugs were defined earlier as those used to produce an abnormal effect. This section will not give an exhaustive treatment of all drugs of abuse, but it will give some information on those that are most commonly used.

Ethyl alcohol

The alcohol in alcoholic beverages is ethyl alcohol (ethanol, grain alcohol). This is the most commonly used nonmedicinal drug in our culture. It is the least toxic of all alcohols. The other commonly available alcohols are methyl alcohol, which is lethal at 100 mL and can cause blindness at 10 mL, and isopropyl alcohol, which is also deadly in relatively small doses.

Ethyl alcohol is produced synthetically by adding water to ethylene, and most produced this way is used for chemical purposes and solvents:

$$\underset{\text{Ethylene}}{H_2C{=}CH_2} + H_2O \xrightarrow{H_2SO_4} \underset{\text{Ethyl alcohol}}{H_3C{-}CH_2{-}OH}$$

Beverage alcohol is produced by enzymatic fermentation of starch from fruit or grains:

$$\underset{\text{Starch}}{(C_6H_{10}O_5)_n} \xrightarrow{\text{enzyme}} \underset{\text{Glucose}}{C_6H_{12}O_6} \xrightarrow{\text{enzyme}} \underset{\text{Ethyl alcohol}}{CH_3CH_2OH}$$

Because over 15 percent alcohol in the fermenting medium inactivates the enzymes, once the alcohol content is at that level, fermentation stops. In order to reach a higher concentration of alcohol, the mixture must be distilled. The alcohol boils at 80°C, while water boils at 100°C. When the fermentation mixture is heated, the alcohol boils first, and the distillate (the material being distilled) is richer in alcohol. Figure 16-7 shows the distillation of an alcoholic beverage, and Table 16-1 (page 445) lists several common alcoholic beverages along with the source of starch and the percent alcohol present. The term *proof* has been carried down from antiquity and is still used by the beverage industry to designate alcohol content (see Table 16-1 for examples):

$$\text{Proof} = 2\ (\text{percent alcohol content})$$

The different flavors and colors of beverages are due to small amounts of other alcohols, aldehydes, and ketones which form during fermentation. These are called *fusel oils*.

The federal government places a heavy tax on alcoholic beverages—about $20 per gallon. The production cost is only about $1 per gallon, and much alcohol is used for chemical purposes. To render this alcohol undrinkable and therefore legally avoid paying the beverage tax, most alcohol has small amounts of toxic, unpalatable materials added. Ethyl alcohol thus treated is called *denatured alcohol,* or *proprietary solvent*. It may contain gasoline, methyl alcohol, acetone, or any one of nearly 100 approved denaturation mixtures. These materials cannot be removed chemically or physically, and so the denatured material cannot be changed back into drinkable alcohol. Drinking denatured alcohol can cause blindness or death.

Medically, ethanol is an antiseptic and hypnotic (sleep inducer), and in general use it is a tension reliever, inhibition lowerer, and intoxicant. No activity, either mental or physical, is enhanced by the ingestion of alcohol—people just think they are doing something better. Alcohol is not only a

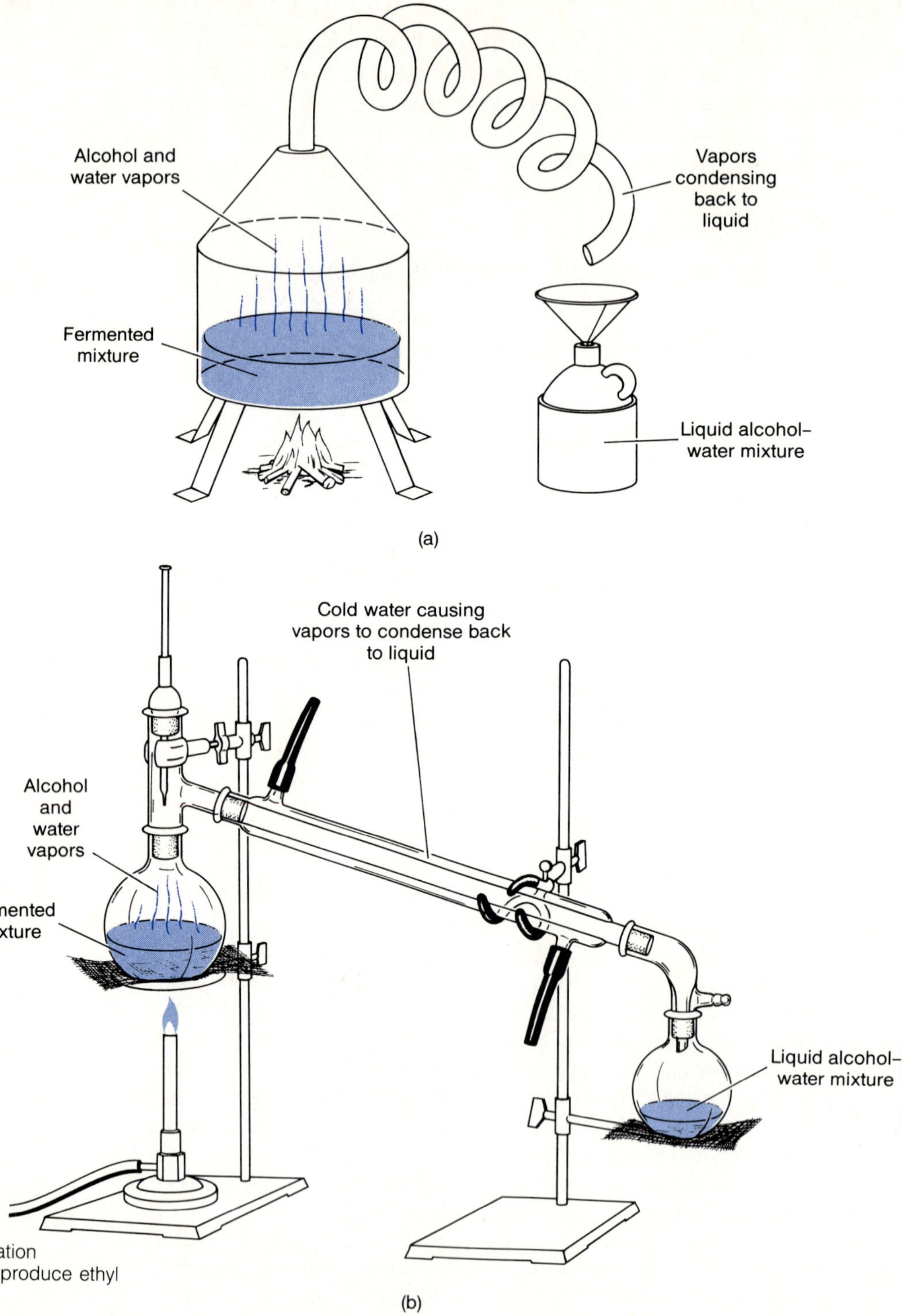

FIGURE 16-7
Two types of distillation apparatus used to produce ethyl alcohol. (a) Crude. (b) Laboratory.

TABLE 16-1

Common alcoholic beverages

NAME	SOURCE OF STARCH	TREATMENT	PERCENT ALCOHOL	PROOF
Beer	Barley	Fermented	5	10
Wine	Grapes	Fermented	10–15	20–30
Whiskey	Corn	Distilled	40–50	80–100
Rum	Molasses	Distilled	40–50	80–100
Vodka	Potatoes	Distilled	40–50	80–100
Gin	Rye (flavored with juniper berries)	Distilled	40–50	80–100

socially acceptable drug but also one which there is frequently considerable social pressure to use. At present there is considerable effort to change this attitude, especially with regard to drinking and driving.

Alcohol is metabolized by enzymes produced by the liver, and it is the size of one's liver (often roughly proportional to body size, but not necessarily so) that determines the rate of consumption that can be tolerated. The average person can metabolize about 30 mL of pure alcohol every hour. This is about 1 oz of alcohol or 2 oz of 100 proof (50 percent) liquor—two drinks. If no more than that is consumed, no obvious physical or mental incapacity usually occurs. At a blood level of 0.05 percent alcohol, drunkenness begins to be exhibited—a person feels carefree and uninhibited and judgment is hampered. This usually occurs at a consumption rate of 2 oz of alcohol per hour (4 oz of liquor). A person with a blood level of 0.10 percent is considered to be drunk and has trouble walking, talking, etc. Although alcohol does induce sleep, the sleep is not as restful as natural sleep. Contrary to popular opinion, sexual activity is not enhanced by alcohol. Some inhibition may be reduced at low blood levels, but otherwise the ability to perform is reduced or eliminated, depending on the blood level.

Some chemicals when taken together simply act independently—each does what it would do if the other material were not present. On the other hand, some materials when used together show effects much greater than the two effects added together. The combination acts like a new drug, and the cooperativeness is called the *synergistic effect*. Alcohol acts synergistically when taken while any one of the following is already in the bloodstream: barbiturates, antihistamines, narcotics, tranquilizers, or other sedatives. These, especially barbiturates, should never be used with alcohol. Most people can tolerate a few drinks with few ill effects. However a person who is taking prescription phenobarbital might go into a fatal coma from those same few drinks.

The rate at which alcohol is metabolized or at which a hangover disappears is dependent on the level in the blood of the liver enzyme that metabolizes alcohol. Such external activities as cold showers and fresh air have no effect, nor does ingestion, after the fact, of coffee, raw eggs, etc. It is true, however, that food in the stomach *before* or *during* drinking slows the absorption of alcohol into the blood and helps reduce the alcohol blood level.

The effects of very moderate alcohol consumption, such as a glass of wine with a meal or a beer once in a while, have not been shown to be a health hazard. Indeed there is evidence that moderate consumption may be quite beneficial. The effects of drinking to the point of frequent drunkenness, however, are all too familiar to anyone in our culture. Relationships can be broken, and physical harm done to other people. Many domestic murders, and half the fatal automobile accidents, are caused by people who are intoxicated. Heavy drinking over a number of years causes serious health problems as well. These include damage to the liver, brain, and heart as well as stomach ulcers and increased incidence of cancer of the mouth and throat. Lower resistance to infection, especially pneumonia, and a higher frequency of mental problems are seen in those who drink heavily. Six drinks per day consumed by a pregnant woman can cause fetal deformity and increase the chances of stillbirth. (A pregnant woman should not drink more than a little wine with meals.)

Barbiturates

Barbiturates are a class of central nervous system depressants (''downers'') used medically as sedatives and anticonvulsants for the treatment of epilepsy. The general structure and the structures of two commonly used members of this class are shown below:

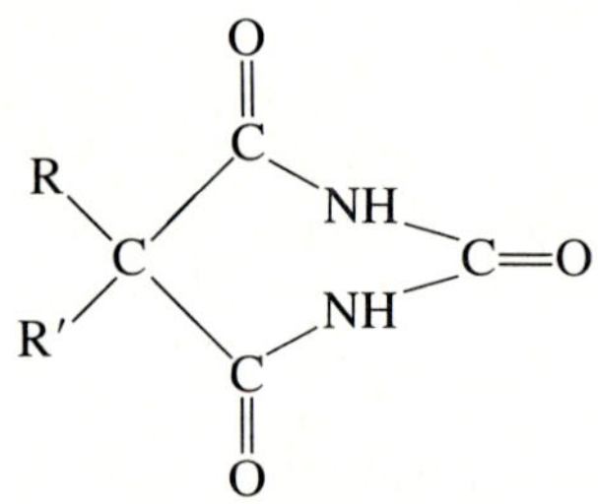

General barbiturate structure (R and R′ can vary but are hydrocarbon groups)

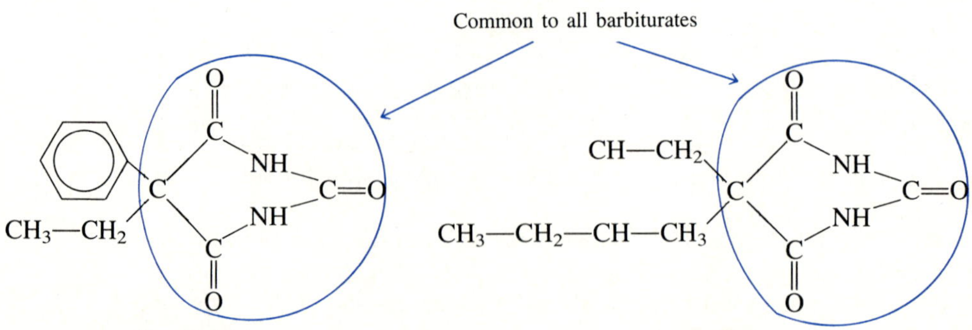

Phenobarbital

Secobarbital

Barbiturates are used by some people to counteract the undesirable side effects of stimulants such as amphetamines. The undesirable synergistic effects of barbiturates with alcohol have already been discussed, and people taking barbiturates should be aware that they can become physically dependent on (addicted to) them.

Amphetamine

Dextroamphetamine (sometimes administered as the sulfate or hydrochloride) is a legally controlled substance that is found in some prescription medication and used illegally by some people as an "upper." It is a central nervous system stimulant and produces a feeling of euphoria (a "high"). While the high is the desired effect for those who use it as a "recreational" drug, it also increases the heart rate and most other body functions except appetite. Since it also produces nervousness and insomnia, barbiturates are needed to help with sleep, but they cause depression. Thus more amphetamine is needed to get over the depression, which produces more insomnia, etc. This vicious circle is very hard to break once it is started, and one can become physically and emotionally dependent on both amphetamines and barbiturates. The withdrawal symptoms experienced by an addicted person are unpleasant.

The structure of amphetamine is shown below:

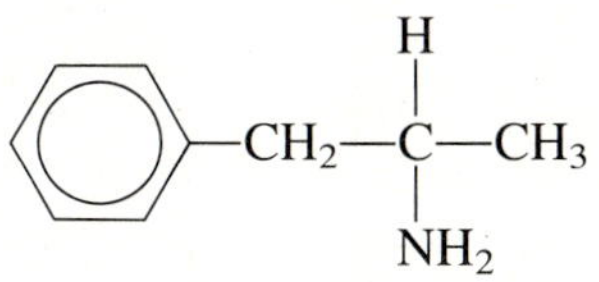

Amphetamine

Lysergic acid diethylamide

Lysergic acid diethylamide (LSD) is one of the most powerful drugs known; the amount needed to cause the effect is extremely small. It is in the class of compounds called *mind modifiers* or *hallucinogens*. This material, commonly called simply *acid*, causes the person using it to perceive as reality things that are not real. Persons using it may think that they can fly or that they are strong enough to stop moving vehicles, and upon acting accordingly do harm to themselves or others. It is also reported that persons using this drug see odors and smell colors and have other experiences indicative of scrambling of patterns in the brain. These experiences can occur long after the drug is used (even years after use). LSD is a very dangerous material to use because it can destroy a person's mind and personality.

The structure of LSD is shown below:

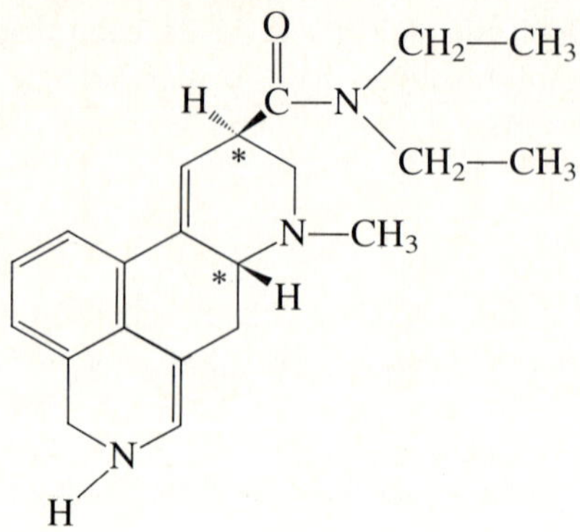

Lysergic acid diethylamide

The hallucinating dose is only 0.000025 g, and since this is too small to be seen, much less handled, it is taken on a sugar cube or on another "filler." The liver enzymes quickly metabolize it, so that probably only 0.000005 g (one-fifth of that taken in) actually reaches the brain and causes mind modifications. This points out the extreme sensitivity of our minds to minute amounts of chemicals. It also illustrates a biochemical principle: all drugs, both medicinal and nonmedicinal, are metabolized (changed) to inactive or nontoxic materials in the body. The actual body level of any drug is determined by a balance between how fast it is absorbed and how fast it is destroyed by normal body enzymes.

Another biochemical principle that is illustrated by LSD, which also holds for many other drugs, is that not only an exact structure but also a precise geometry is critical for biological activity. In the structure of LSD there are asterisks at two of the carbons. At the lower of these carbon atoms, the hydrogen atom is shown coming out of the paper in front of the plane of the main ring structure; at the upper carbon atom, the hydrogen is shown going behind the plane of the main ring, and the $-\overset{\overset{\displaystyle O}{\|}}{C}-N-(CH_2-CH_3)_2$ group is shown in front of the plane of the ring. This exact arrangement (or steriochemistry) must be present for LSD to be biologically active. If either one of these arrangements at these two carbon atoms is reversed, the resulting substance is *totally inactive*.

While the mechanism of action of most drugs is not known, it is almost certain that they all work by lowering or raising the activity of enzymes. Since enzymes themselves have a carefully defined three-dimensional structure, the drug, likewise, must have a certain definite shape if it is to interact either constructively or destructively with an enzyme and exhibit its physiological activity. The study of the multitude of complex biochemical processes that occur in our bodies daily is a fascinating subject and can cause one to stand in awe of the physical bodies we possess and daily take for granted.

TABLE 16-2

Caffeine in foods, beverages, and over-the-counter drugs

ITEM	AMOUNT OF CAFFEINE, mg
Percolated coffee (1 cup)	100
Instant coffee (1 cup)	65
Coffee, decaffeinated (1 cup)	3
Tea, regular brew (1 cup)	35
Hot chocolate (1 cup)	5
"Sun-tea," brewed by soaking tea bags in water warmed in the sun (1 cup)	15
Cola (1 cup)	35
Milk chocolate candy (1 oz)	10
Alertness tablets	30–100

Caffeine

Along with aspirin, alcohol, and tobacco, caffeine is one of the most commonly used drugs. Coffee has 60 to 100 mg of caffeine per cup, and tea has about half that level. A chocolate candy bar contains 10 mg, and many cola and even noncola beverages contain caffeine. Other examples are shown in Table 16-2. Caffeine is an "upper"—that is, a central nervous system stimulant (see Figure 16-8)—and also a cardiac and respiratory stimulant. A fairly mild physical dependence can be developed, and many people need a cup of coffee before they feel good each morning as a result of this dependence.

Let us look at Figure 16-8 more closely to see how caffeine acts on the central nervous system. Like all stimulants, caffeine causes an increase in the release of biochemical substances (A in Figure 16-8) in the nervous system (these substances are called *neurotransmitters*). Other biochemicals (B in the figure) are automatically produced in response to this to counteract the extra stimulation. These other biochemicals reduce the ability of nerves to function. When a person who has developed a dependence on caffeine gets up in the morning, biochemicals of the "B" type are present in excess of the normal amount, and the nervous system is sluggish. When caffeine is ingested, substances of the "A" type are produced, and things begin to function normally. Once that person stops using caffeine for a period of time, the amount of "B" is slowly reduced, since no excess "A" is present to stimulate the production of "B." Once the level of "B" is back to normal, the normal level of "A" is quite sufficient to make the person feel good. In part 1 of Figure 16-8, the normal nervous system has the correct amounts of "A" and "B" for proper functioning. When caffeine is first ingested (part 2), more "A" is present as a result of the stimulatory effect of caffeine on its production, and a high is experienced. After a person uses caffeine in large amounts for a while, more "B" is formed to counteract the higher level of "A," and the system functions normally (part 3). The system has become physically dependent on caffeine. When such a person has not had caffeine in the morning, the level of "A" remains low, but "B" is still at a high level, having been stimulated to this high level by previous doses of caffeine. This causes sluggishness, as

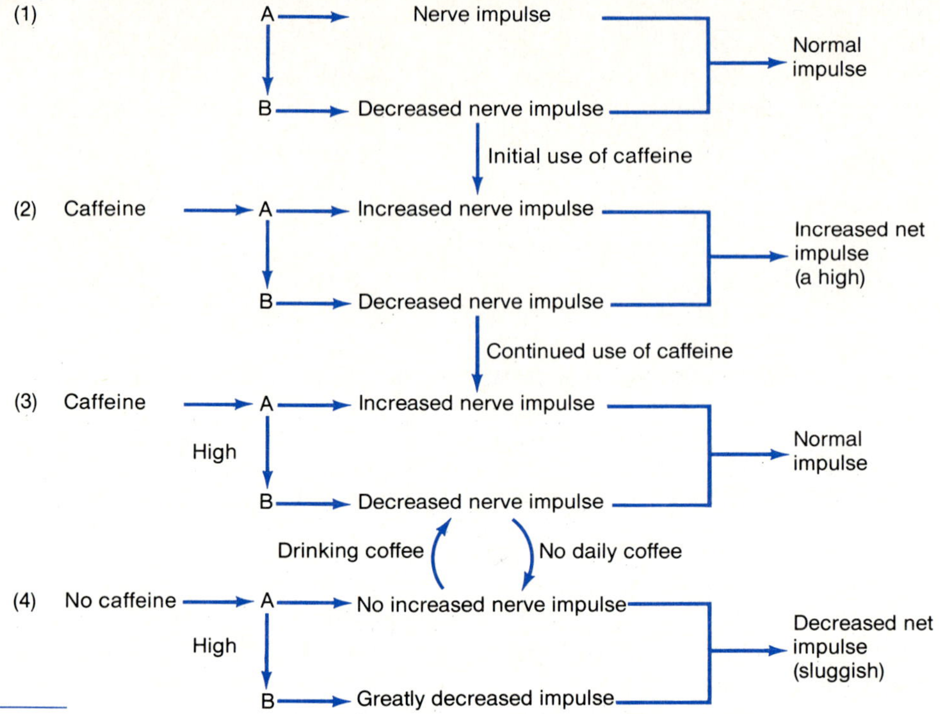

FIGURE 16-8
How caffeine acts on the nervous system.

shown in part 4. Once coffee has been ingested, the person moves back to part 3 and feels good. Prolonged abstinence from caffeine will eventually return the system to part 1.

All drugs that cause physical dependence or addiction work in much this same way. Different types of "A" and "B" are produced with different drugs; the effects experienced are more or less serious; and the withdrawal symptoms are quite different for different drugs. For example, narcotic addiction works in essentially the same way as dependence on a more commonly used drug such as caffeine; however, the effects experienced are more serious, and the withdrawal symptoms are more severe. Symptoms of too much caffeine include nervousness, insomnia, and rapid, irregular heartbeat (tachycardia). These symptoms are similar to those experienced from too much amphetamines.

The structure of caffeine is shown below:

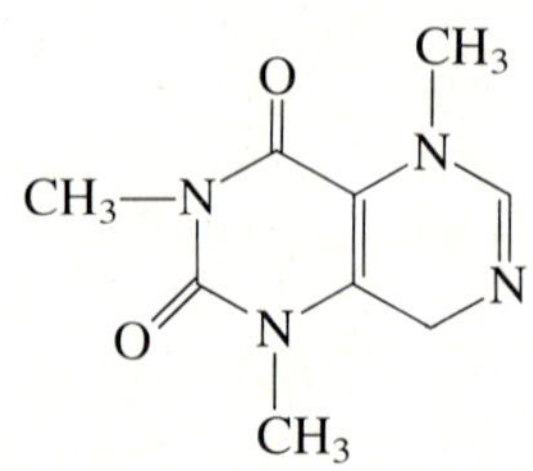

Caffeine

Tobacco

There are two compounds of interest in tobacco smoke: nicotine and benzpyrene. Nicotine is the stimulant that produces the mild euphoria associated with smoking (especially cigarettes), and benzpyrene is the carcinogenic (cancer-causing) material:

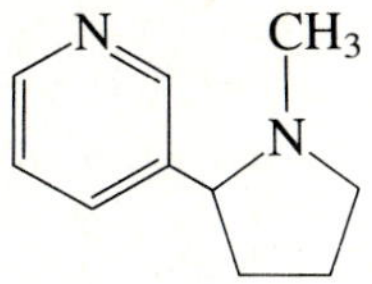

Nicotine

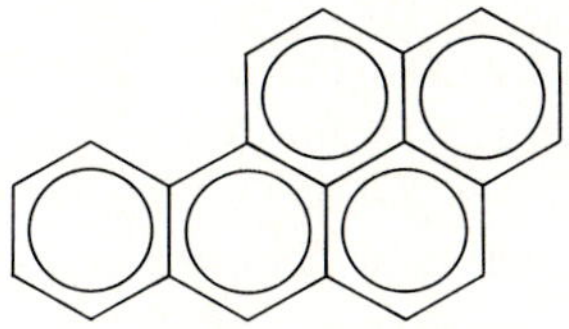

Benzpyrene

Nicotine is a highly toxic material with an LD_{50} in rats of 0.05 g/kg. As little as 0.001 g/kg can kill a rat, and only 0.04 g can kill a human being. It is used commercially as an insecticide. Each cigarette contains 0.0015 to 0.0005 g (1.5 to 0.5 mg) nicotine.

Benzpyrene (compound I in Figure 16-9) is formed when tobacco is burned and is the most dangerous component of "tar," a dark, sticky material found not only in tobacco smoke but also in burning leaves and grass and burned and charcoal-broiled foods. A cigarette yields 5 to 20 mg of tar. Enzymes (ENZ_1) in the body convert benzpyrene to the alcohol epoxide shown in Figure 16-9. Compound II is the "primary carcinogen"—the material that actually changes the DNA. The mutant (abnormal) DNA causes the cells to grow in a grotesque and rapid manner, and a malignant tumor results.

There is another enzyme (ENZ_2) that can destroy compound II and prevent it from initiating cancer (see Figure 16-9). Current evidence suggests that

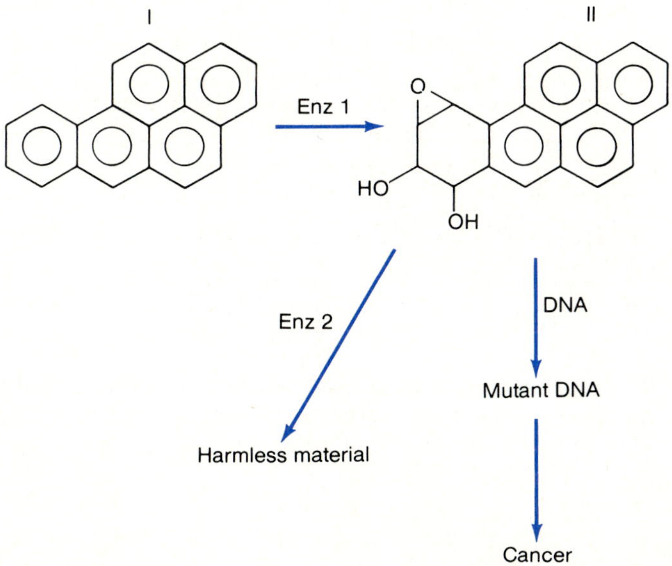

FIGURE 16-9 How benzpyrene acts to induce cancer.

it is the relative amounts of these two enzymes that determine if cancer will actually occur in a person who smokes. The hypothesis is that if a person has large amounts of ENZ_1 and little of ENZ_2, much more compound II is formed than is destroyed, and a high level of compound II remains to react with DNA. If a person, apparently because of hereditary factors, has about equal amounts of ENZ_1 and ENZ_2—which means that compound II is destroyed about as fast as it is formed—the chance of cancer is smaller. This is the best explanation to date for the observation that some people who smoke do not get cancer, while some who smoke do. A series of studies have shown that about 45 percent of the population have a favorable enzyme ratio and are at low risk. Another 45 percent have about 15 times as much ENZ_1 as ENZ_2 and have a significant risk of getting cancer if they smoke. The remaining 10 percent have 30 times as much of ENZ_1 as ENZ_2 and are at high risk. Chemical tests for ENZ_1 levels are available only for research purposes at this time, and they are not very reliable. The best way to find your probable level is to consider the smoking habits of, and cancer incidence in, your older relatives.

Table 16-3, below, gives a brief summary of medicinal and nonmedicinal drugs.

TABLE 16-3

Classification of drugs

DESIGNATION	DESCRIPTION	EXAMPLES
Over-the-counter (OTC) drugs	Are available to anyone at any time	Aspirin Vitamin tablets Mercurochrome
Unregulated prescription drugs	Are available only to those for whom a physician has prescribed it	Sulfa drug Penicillin Phenytoin (Dilantin)
Unregulated nonmedicinal drugs* (recreational drugs)	Are available in beverages or foods or in other easily obtainable sources	Alcohol Caffeine Nicotine
Controlled substances:†		
Schedule I	Are commonly abused—no medical use	LSD Mescaline
Schedule II	Are commonly abused—also have important medical uses	Morphine Cocaine Amphetamines
Schedules III, IV, V	Have possibility of abuse—have primarily medical uses.	Diazepam (Valium) Phenobarbitol

*These are frequently taxed, however.

†Materials whose sale, distribution, and possession are controlled by the Drug Enforcement Administration (DEA) of the U.S. Department of Justice. Abuse of these drugs is penalized.

MEDICINAL CHEMICAL RESEARCH

The chemical compounds discussed in the section on medicinal drugs were first synthesized, and their medicinal properties were discovered, as a result of research by professional scientists (chemists, biologists, and physicians). While most research is tedious, sometimes discouraging for the experimenter, and expensive, it is the only way progress can be made in lowering the suffering due to sickness. The other main causes of suffering—war and accidents—can be reduced by human will and care. While infectious diseases have largely been brought under control, and some sicknesses are brought about knowingly (some forms of cancer, alcoholism, etc.), there are a number of areas where research is still needed. These include cancer treatment, cardiovascular diseases such as high blood pressure and heart attacks, central nervous system disorders such as epilepsy and mental illness, and degenerative disorders such as muscular dystrophy and Alzheimer's disease.

There are three approaches one can take to such problems.

The first and most fundamental, but also the most difficult, is to try to *understand the cause* of the malfunction. Since our knowledge of the normal human biosystem is limited and techniques for studying it are still being developed, there have been no major successes using this approach to date. It must be continued, however, because some problems will, no doubt, be solved eventually only by this method. This does not mean that we have had no success with the rational approach. Once it was realized that diabetes is caused by a deficiency of insulin, animal insulin was used to treat it. However, although it became possible to treat diabetes, no one understood the biochemical error in the body that prevents insulin from being produced.

The second approach is to attempt to produce new compounds which are similar in structure to those already known to solve a problem. Once natural penicillin was found to stop staphylococcus and streptococcus infections and its structure was determined, different compounds with different R groups could be synthesized and tested to see if they were more or less effective than natural penicillin. The R groups could be varied by varying the numbers of carbon atoms and by adding benzene rings, double bonds, halogen atoms, etc. Once the medicinal chemists had prepared each material, biologists and laboratory technicians could test them to see if they inhibited the growth of bacteria in a culture medium. This approach was used to prepare many different members of the penicillin, sulfa drug, and barbiturate classes. Those compounds which proved effective—a small number compared with those originally synthesized—are the ones in use today. A similar approach is also being tried at the present time to find better anticonvulsants. Chemists are preparing compounds which have molecular structures similar to the 15 known anticonvulsants, and these compounds are being tested by the National Institutes of Health (NIH) on rats.

It is known that phenobarbital (shown below) is a good anticonvulsant (it stops seizures in people with epilepsy). It has also been shown that dimethyl maleic hydrazide (see below) stops seizures in test animals but is too toxic to be used as a drug:

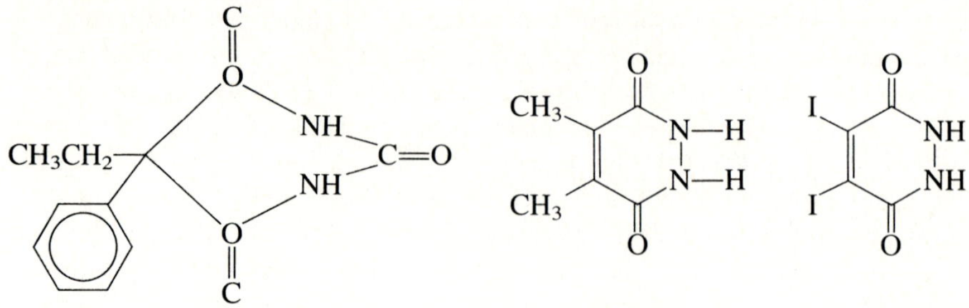

Phenobarbital

Dimethyl maleic hydrazide

Diiodo maleic hydrazide

This information was determined as follows. Several compounds are known to cause seizures in animals. If a known anticonvulsant such as phenobarbital is given to the animal before administering the seizure-producing stimulus, no convulsions occur. In the screening method, the suspected anticonvulsant drug is given to several rats, and a little while later they are given one of the seizure-producing drugs. If no convulsions are observed, then the compound is considered to have anticonvulsant properties. Dimethyl maleic hydrazide was synthesized because it contained a ring with nitrogen atoms (heterocyclic ring) and C=O groups and alkyl groups attached to the ring. These are some of the same molecular features found in phenobarbital and most other known antiepileptic drugs. When the dimethyl compound was tested, it did indeed stop seizures, but the rats became sick, indicating toxicity.

With this information, a medicinal chemist proceeded as follows. Since an iodine atom has about the same size (diameter) as a CH_3— group, diiodo maleic hydrazide (see above) was synthesized. The reasoning was that if it had the same shape and size as the dimethyl compound, it might have the same anticonvulsant properties but, by chance, might not be toxic. Testing of the diiodo compound showed that it had the same seizure-stopping abilities as the dimethyl compound but also was toxic. While this approach did not lead to a new drug in this case, it does illustrate the kind of reasoning that does lead to success in some cases. It also illustrates the patience a chemist must have and the expense of carrying out such research. The percent of successful compounds is small, the test animals are expensive to keep, and many salaries must be paid over a number of years. Figure 16-10 shows a typical report form and indicates that the dimethyl maleic hydrazide did stop convulsions but was also toxic.

The third approach is to test any newly prepared material for many desirable physiological properties. For example, when a chemist working for a pharmaceutical company prepares a compound for testing as an anticancer drug, the company will also test it for antibacterial, antiviral, and anti-

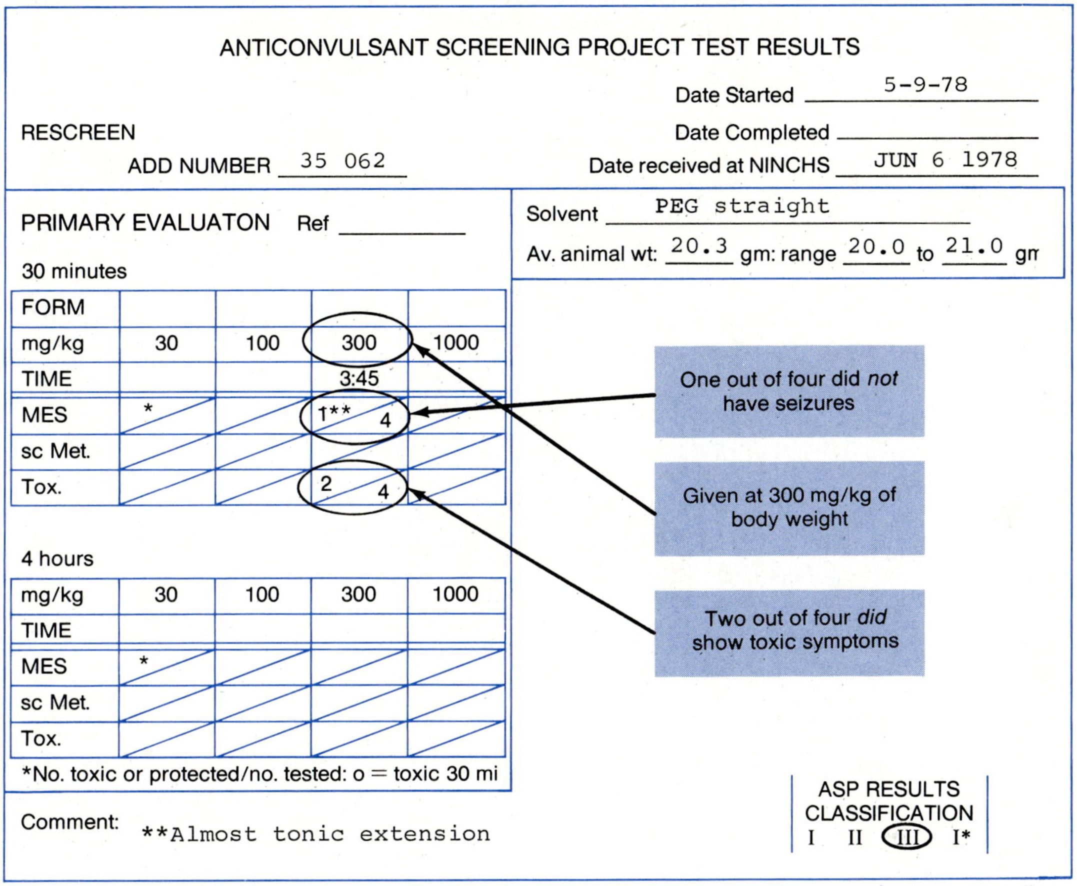

ANTICONVULSANT SCREENING PROJECT TEST RESULTS

Date Started 5-9-78

RESCREEN

Date Completed

ADD NUMBER 35 062

Date received at NINCHS JUN 6 1978

PRIMARY EVALUATON Ref

Solvent PEG straight

Av. animal wt: 20.3 gm: range 20.0 to 21.0 gm

30 minutes

FORM				
mg/kg	30	100	300	1000
TIME			3:45	
MES	*		1** / 4	
sc Met.				
Tox.			2 / 4	

4 hours

mg/kg	30	100	300	1000
TIME				
MES	*			
sc Met.				
Tox.				

*No. toxic or protected/no. tested: o = toxic 30 mi

Comment: **Almost tonic extension

ASP RESULTS CLASSIFICATION

I II III I*

FIGURE 16-10

A typical drug screening report, on dimethyl maleic hydrazide.

fungicidal properties as well as for as many as 30 or 40 other possible properties. This is to help ensure that the company will not miss a marketable material which could help solve a medical problem. Academic chemists can also submit newly prepared compounds to these screening programs.

Only 1 compound out of about 8000 tested is finally marketed. Not only must a chemical be effective in solving a problem, it must also have a low toxicity for both long-term and short-term use. Toxicity is tested first in test animals and finally, if considered safe on the basis of these animal tests, in humans. Determining the effectiveness and safety of a drug is expensive and time-consuming. In order to recover the costs of development and testing, a drug company, therefore, has to have a high profit margin on the drugs eventually sold. It takes either company profits or tax money to pay for the discovery and marketing of drugs. The public, if it wants the benefit of good medical care, must be willing to bear this cost. Figure 16-11 shows the steps in preparing a drug and the approximate times and amounts of money required.

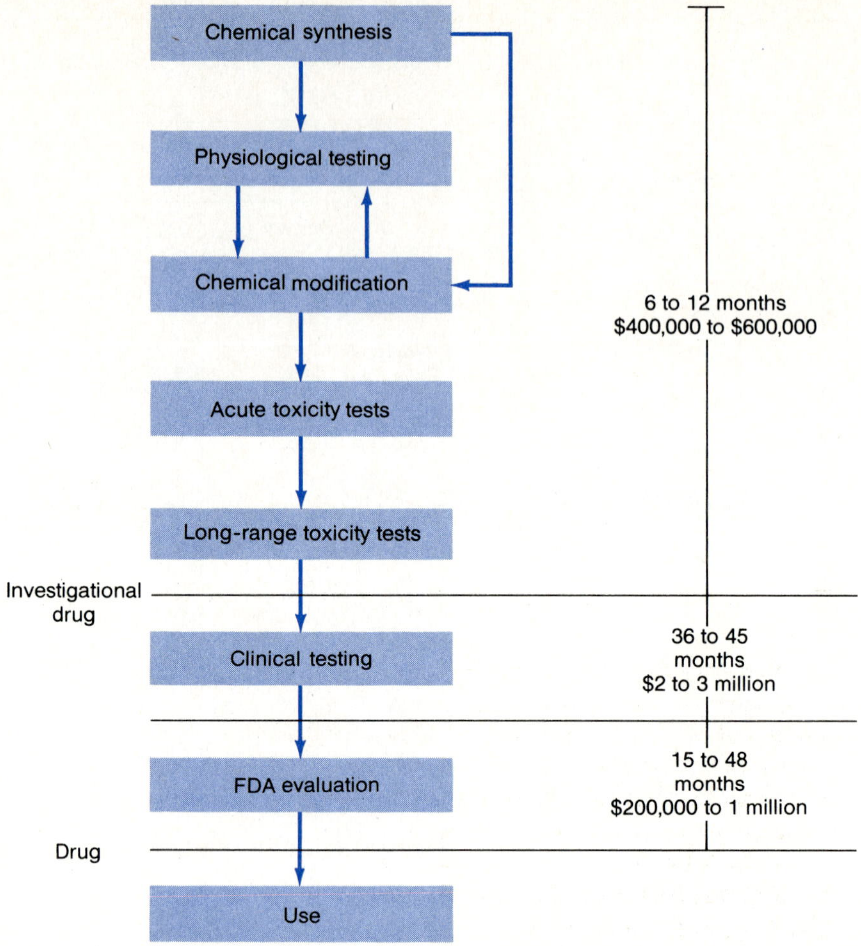

FIGURE 16-11
Overall scheme for development of a drug.

SALE AND ADVERTISING OF DRUGS

Brand names and generic names

Each drug has a proper chemical name. Because many chemical names are complex, however, commonly accepted shorter names are often used; these are called *generic names*. In addition, a particular company will use a *brand name* which may be used only by that company. Table 16-4 gives examples of the three levels of names. The company that first markets a given drug usually has high research costs and must charge a fairly high price. That company may or may not have a patent on the compound, but even if it does, the patent is for a limited duration. At some point other companies can simply begin to produce the same material, and since they have only production and marketing expenses (no research costs), they can sell it at a lower price. These

TABLE 16-4

Chemical, generic, and brand names of several common drugs

CHEMICAL NAME	GENERIC NAME	BRAND NAME
Acetylsalicylic acid	Aspirin	Bayer aspirin Norwich aspirin Empirin
5-Ethyl-5-phenylbarbituric acid	Phenobarbital	Luminal Phenobal Somonal
5,5-Diphenyl-2,4-imidazol-idinedione	Diphenylhydantoin or phenytoin	Dilantin Lepitoin Diphentoin
Ethyl-4-aminobenzoate	Benzocaine	Anesthesin Americaine
4-Hydroxyacetanilide	Acetaminophen	Tylenol Tempra
———	Propoxyphene	Darvon
	Diazepam	Valium
	Phenylephrine	Neo-Synephrine

lower-priced drugs are sometimes called *generics,* since the generic name of the lower-priced drugs is the same as that of the more expensive brands. The least expensive drugs are usually the brands bearing a house name—White Drug Stores will sell White brand aspirin, for example.

Prescription drugs also vary in cost for the same reason, and some pharmacists prefer the more expensive brands, claiming that they are better. As stated previously, except in rare cases where formulation affects the drug's usefulness, the generics are identical in their medicinal properties. Some states require that a pharmacy dispense the lowest-priced generic drug unless a customer or physician specifically requests otherwise. This law has both good and bad effects. It is usually to consumers' advantage in the short term—they save money and get a good product. On the other hand, the companies with large research and development costs are less able to recover that money in the form of profits and so are less likely to invest in future drugs. In this case society is the loser in the long run. The pros and cons of this legislation are still being debated.

When deciding on an over-the-counter medication, you should carefully read the ingredient list on the label and look for the best compounds for a given purpose. *The Handbook of Non-Precription Drugs,* 6th ed., American Pharmaceutical Association, Washington, D.C., 1979, and *The Merck Index,* 9th ed., Merck and Co., Inc., Rahway, New Jersey, 1976, are available sources of names and physiological properties of the ingredients listed. The former is the best publication available from which consumers can get valuable advice on over-the-counter products. The latter is a very complete listing of all commonly used medicinal compounds. A pharmacist can also help you to decide on the best formulation.

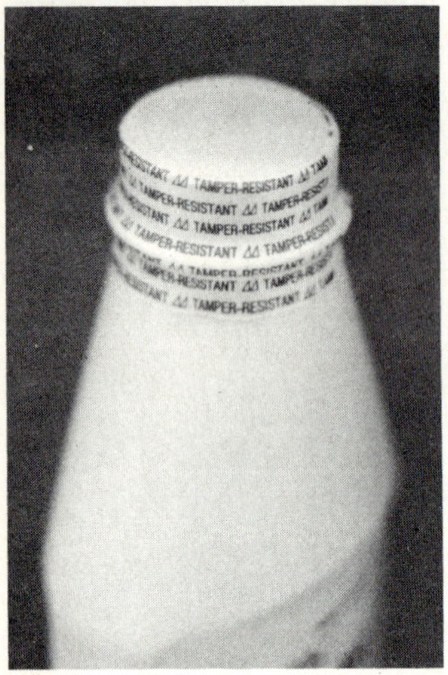

FIGURE 16-12
Tamper-resistant packaging for a drug.

The general principles are listed below:

1 A product should have an active ingredient that has been shown to help the problem you are trying to solve.

2 A low-priced product with equal amounts of the same compounds is just as good as a more expensive brand name.

Safety in packing

In 1982 a frightening event occurred. Some person or group placed potassium cyanide (KCN) powder inside capsules originally filled with Tylenol—having first obtained Tylenol from a drugstore, then replaced the white acetaminophen with the KCN (which is also a white powder), and finally placed the containers of poisoned capsules on a store shelf in the Chicago area. The cyanide ion is a very toxic material, and several people died after taking what they believed was medication for a headache. Johnson and Johnson, the manufacturers of Tylenol, recalled the unused bottles of this product and lost a great deal of money as well as incurred a damaged reputation.

Because the packages of Tylenol were normally not sealed in such a way that a consumer could easily tell if they had been previously opened, this tampering was not noticed by the unsuspecting consumers. Since that time, all drug companies have taken great precautions to ensure that if a package has been tampered with, the tampered package can be easily recognized by the store or the consumer. One technique used to make a bottle tamper-resistant is to place a formed plastic membrane over the screw cap and then place a paper seal over the top of the bottle under the screw cap. (See Figure 16-12.) If someone has opened the bottle and placed poison in it, both seals would be broken, and a consumer observing this would not use the medication but would return it for an unbroken replacement. Another approach is to place the bottle of medication inside a metal can whose top must be torn off. This is similar to cans used for tennis balls, potato chips, puddings, etc. (See Figure 16-13.)

FIGURE 16-13
A drug sealed in a metal can.

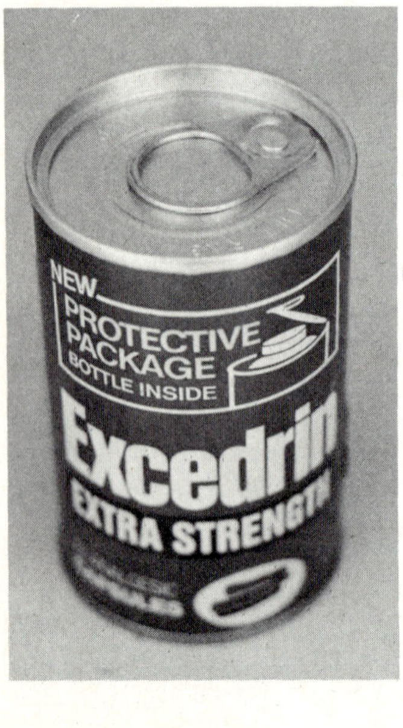

PERSONAL USE OF DRUGS

In addition to choosing the correct product from the many of each type available, each person has to make decisions about the use of all types of chemicals. In this chapter, an attempt has been made to point out the chemical principles involved in the use of drugs and explain in some detail what happens in certain more common cases such as the use of aspirin, caffeine, and alcohol. Some misconceptions have also been pointed out.

It is the author's belief that drugs of all kinds—over-the-counter, medicinal, or nonmedicinal—should be used with careful thought and with full knowledge of the facts. To continually ingest a compound just for the sake of habit or convenience is to play ''fast and lose'' with the delicate, yet marvelous, biochemical system in which we live (our bodies). You should do a cost-benefit analysis in each case. For example, if you have a slight headache, it is not too difficult to live with it and allow it to be healed by natural body processes. This is especially true in view of the side effects of taking an aspirin. On the other hand, if the headache becomes so bad that you cannot function at work or cannot study, or if it causes you to be so irritable that friends and relatives are alienated, then the side effects are not as important, relatively speaking, and the aspirin should be taken. Ask yourself, ''Which is worse, the side effects of the medication or the effects of staying sick?''

A common cold can most likely be helped as much by a hot bath, rest, and plenty of liquids as by the myriad cold ''remedies'' available that cause drowsiness and make one feel almost as bad as the cold itself. For more severe cases, of course, one should use the medications available, but no drug should be taken flippantly.

Another factor to consider is the habit-forming properties of the material. This applies most frequently to coffee and soft drinks containing caffeine and to alcoholic beverages and tobacco, although many drugs, especially those that act on the central nervous system, tend to be habit-forming, including most ''street'' drugs. As many people can testify, it becomes a personal burden and an expense to need 6 to 10 soft drinks *every* day and 2 cups of coffee *every* morning, or not to be able to interact socially before a few drinks. All these things reduce the quality of life that we all strive for even though they sometimes give the illusion of increasing happiness.

Another factor which any drug user should be aware of is the mind- and personality-altering effects of many drugs. Those who find it necessary for medical reasons to be on drugs constantly can become irritable, can become unable to think clearly, and can actually develop a different temperament. It is important that such persons realize this and face the reality of it in adjusting to their situation. Those around these people, likewise, must realize that the personality change is a natural side effect, like nausea or sleepiness. Patience is required by all in such circumstances.

One must also ask the very important question, ''Am I taking this drug to *solve a problem* and add quality to life, or to *escape reality* so that I will not have to face my problems?'' There is no fundamental difference between one who takes Valium, one who needs alcohol in order to cope with life, and one who takes heroin. The side effects of heroin are more devastating, to be sure, but in all three situations the motive is questionable and the action is usually counterproductive in the long run. Problems are not solved by ignoring them. One should keep all these facts and questions in mind—then make an intelligent decision, being fully responsible for one's own decision and the future consequences of it.

There are several misconceptions that should be mentioned about the personal use of drugs. There is at this time no such thing as a drug to improve intelligence or facilitate learning. It has been claimed that those who study under the influence of caffeine learn more effectively, but that has not been proved. Research is being done in this area, but so far with no success; you will have to continue to learn chemistry the old-fashioned way.

It is also doubtful that any drug can specifically relieve menstrual cramps.

This pain is due to cramping of the uterine muscle, and it is claimed that cinnamedrine (used in some products for menstrual relief) relaxes the muscle, relieving the pain. This has not been proved, however, and many people believe that a small amount of caffeine as a mood elevator (upper) and aspirin or acetaminophin for pain relief are the best approach. Most products for this purpose have both of these components.

Pheromones play an important role in the mating of insects and various animals. Pheromones are usually chemical signals (smells) given off by the male or female that draw members of the opposite sex to each other as well as stimulate them to action. There are claims that certain materials cause a sexual response in humans. Such materials are called *aphrodisiacs* and include powdered rhinoceros horn, Spanish fly, alcohol, and various perfumes such as musk oil. As already mentioned, alcohol may lower inhibitions but does not increase the sex drive and actually reduces the ability to perform. Rhinoceros horn is apparently a placebo—a compound that causes the expected effect only because of the power of suggestion. Spanish fly is prepared from the powdered bodies of green flies. The active ingredient is called *cantharadine* (it has recently been prepared synthetically), and it is an irritant. When it is taken internally, it irritates the blood vessels; and since the genitals have a high blood supply, they are irritated more than the rest of the body and one's attention is drawn to sexual activity. Perfume probably works by the power of suggestion, including reminding one of past pleasant experiences. Thus, at present there is *no known human pheromone*—nothing that chemically causes arousal or increases sexual activity. It could be argued that Spanish fly is an aphrodisiac because it causes physical stimulation. However, this is apparently a different phenomenon from the uncontrolled urge that pheromones produce in animals.

The final area to be considered is that of the placebo. When people are given a ''sugar pill'' and are told that it will relieve pain, for example, they report a lowering of pain almost as much as those taking a narcotic. It was commonly thought that this effect was only in the mind of the person—that it had a purely psychological basis. It has been shown, however, that the thought that one is getting a painkiller actually triggers the production of natural opiates (painkillers) called *enkephalins* and *endorphins*. Thus the effect is not just in one's mind; it has a basis in physical reality. The placebo effect may also be real for areas other than pain relief as well.

STUDY GUIDE

Terms to learn

Acetaminophen
Acetylsalicylic acid
Amphetamine
Aphrodisiac
Barbiturate
Benzocaine
Benzpyrene
Blocking group

Caffeine	Melanin
Denatured alcohol	Nicotine
Dextromethorphan	NIH
Drug (medicinal)	OTC
Drug (nonmedicinal)	PABA
Ethyl alcohol	Penicillin
FDA	Physical dependence
Generic name	Quat
LD_{50}	Synergistic
LSD	Sulfa drug
Medicine	TI (theraputic index)

Skill practice problems

1 List the side effects of taking aspirin.

2 If the LD_{50} for salicylamide is 1.4 g/kg, how much would have to be given to rats weighing 150 g in order to kill half of them?

3 Are high profit margins or drug products really needed?

4 Which has more caffeine, a cup of coffee or three candy bars? (See Table 16-2.)

5 What government agency is responsible for monitoring the safety of all drugs?

6 What is the advantage of buffered aspirin?

7 How should one take aspirin for fastest relief?

8 What two factors control the blood level of any drug?

9 What is the complete structural formula for penicillin V?

10 What is the advantage of acetaminophen over aspirin?

11 How does penicillin kill bacteria?
(a) It keeps them from using nutrients.
(b) It stops them from dividing.
(c) It prevents cell wall development.
(d) It destroys the cell membrane.

12 Which of the following will stop the burning UV rays of the sun?
(a) Clouds
(b) Dry clothing
(c) Wet clothing
(d) PABA
(e) An umbrella

13 What is the best topical anesthetic?

14 What is the best type of topical bacteriocide?

15 What is the percent of alcohol in an 80-proof mixture?

16 Rum is prepared by fermenting ______________.

17 Ethyl alcohol that is made undrinkable is called ______________.

18 A person who has had ten 1-oz drinks of 100-proof whiskey in a short time will take how long to metabolize the alcohol?

19 Which of the following show a synergistic effect with alcohol?
- **(a)** Aspirin
- **(b)** Caffeine
- **(c)** Sleeping pills
- **(d)** Barbiturates

20 What is the structural formula of aspirin?

21 What is the structural formula of benzpyrene?

22 What is the generic name of the material found in Valium?

23 Do all who smoke heavily have an equal chance of getting cancer?

24 Are generic brands usually as effective as more expensive name brands?

25 Which of the following statements are true? Which are false?
- **(a)** Spanish fly is a human sexual stimulant.
- **(b)** "Sugar pills" (placebos) have no biochemical effect.
- **(c)** Development of new drugs is generally costly.
- **(d)** Caffeine can cause physical dependence.
- **(e)** Methyl alcohol is commonly called *grain alcohol*.

Questions for discussion

1 Write a personal statement of your own philosophy of drug use. Are your actions consistent with this philosophy? Does reason dictate a change in any of your statements of philosophy?

2 Do you think the generic-drug laws are good or bad? Why?

Goals

To understand that ordinary household materials are chemical in nature and that their usefulness is a result of their chemical and physical properties

To gain practical information about applying chemical knowledge to solve commonly encountered household problems

17

HOUSEHOLD CHEMISTRY

We are using chemical substances more and more frequently in our daily lives. Cooking, cleaning, personal hygiene, automotive care, and general household maintenance involve applied chemistry. In this chapter we will examine the chemical nature of household products, the way in which they work, and the safety and economics involved in their use. We will also consider some misconceptions about household products and practical ways of solving some everyday problems.

CLEANING PRODUCTS

Soaps and detergents

One of the most commonly used and oldest chemicals is soap. The formula is:

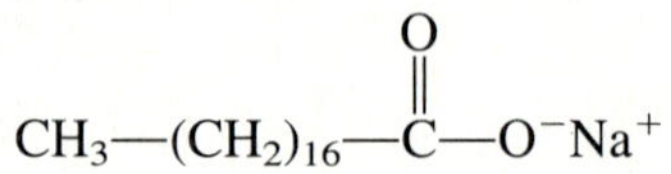

Sodium stearate (soap)

Soap is prepared by reacting animal or vegetable fats with a strong base such as sodium hydroxide (NaOH). Fat (see Chapter 15) is composed of three ester groups joined to three long-chain fatty acids and a glycerin backbone. The NaOH breaks the ester bonds, producing a trialcohol (glycerin) and the sodium salt of the fatty acid (soap).

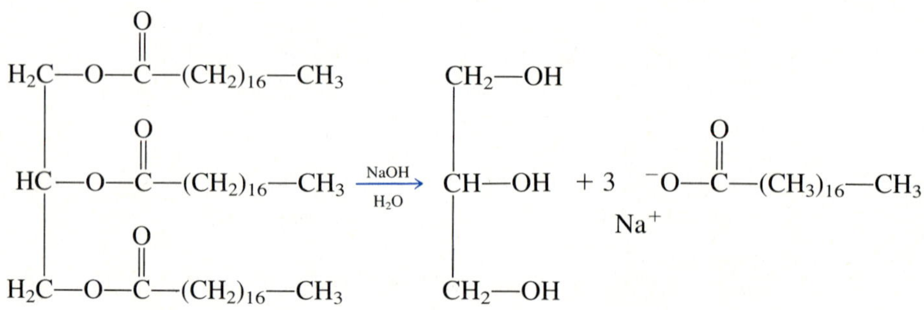

Fat — Glycerin (trialcohol)

In ancient times, and up until about 1940 in this country, it was a common practice to make soap at home. The practice died out slowly, and by 1960 few people were still making their own soap. Originally the base was obtained by letting rainwater run through wood ashes. The water was evaporated to produce a concentrated solution, which was then boiled with waste animal fat over an open fire. In the early and middle 1900s, the NaOH was purchased as lye and boiled with fat cut from pork and beef. This rather crude soap was harsh on hands because it contained some NaOH, which reacted with, and thus removed, the natural oils from the skin. Figure 17-1 diagrams this process, which a few people still carry out today.

Once modern commercial soap is prepared, it is formulated in various

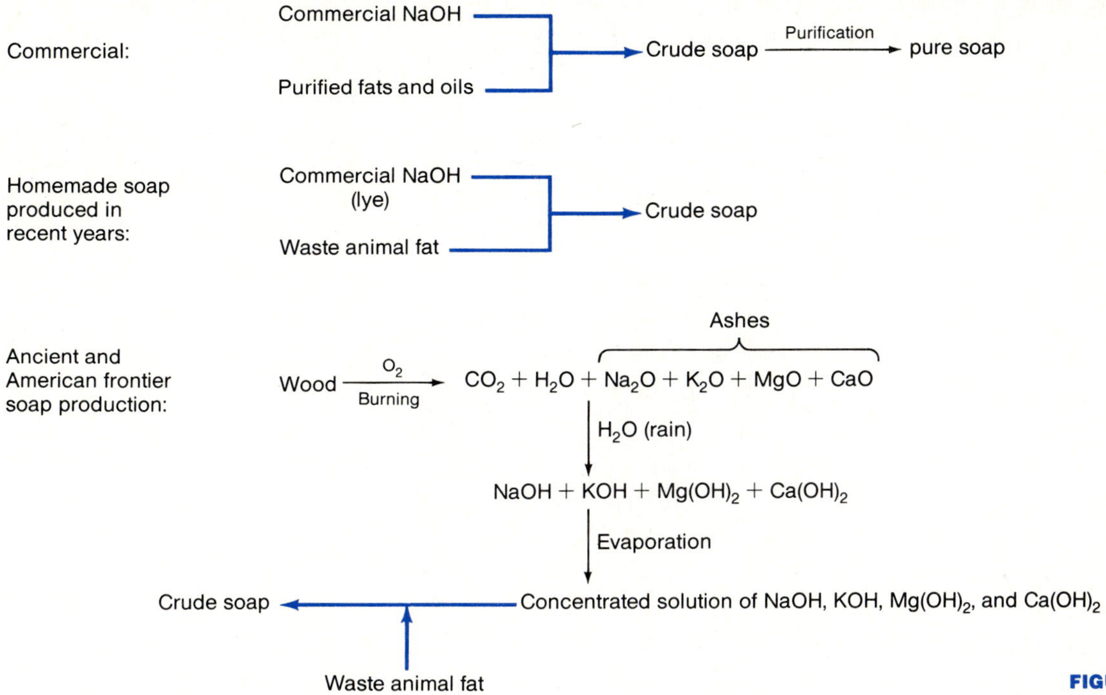

FIGURE 17-1
Production of soap.

ways. Some is simply pressed into bars for sale to the public. Some has color and perfume added to make a more pleasant soap, while some has deodorant or a topical bactericide.

Liquid soaps for hands or dishes are composed of soap suspended in water to make a thick concentrated solution, which can be used easily from a plastic squeeze bottle or pump container. For use on really dirty hands, soap is mixed with sand or powdered pumice, which serves as an abrasive; and for use as a scouring powder both abrasives and bleaches are added to powdered soap.

Most dirt on skin, clothing, dishes, or floors is combined with grease and is not soluble in water (grease is nonpolar and water is polar). Soap molecules have two rather well-defined regions and thus a dual nature. The charged portion with the carbon-oxygen bonds is polar and tends to dissolve in water. The long hydrocarbon end is nonpolar and tends to be insoluble in water but soluble in the nonpolar grease (see Figure 17-2*a*). If one attempts to dissolve fat or oil in water, small particles of the nonpolar phase become suspended in the water, but no dissolving occurs (see Figure 17-2*b*). When soap is added, the soap molecules align themselves so that the hydrocarbon "tails" are in the oil droplets and the polar "heads" are outside the oil and in the water. The oil droplets become surrounded by negatively charged $—COO^-$ groups and can be transported by the water. The water molecules perceive only the outer negative layer and readily associate with it. The grease (along with any other

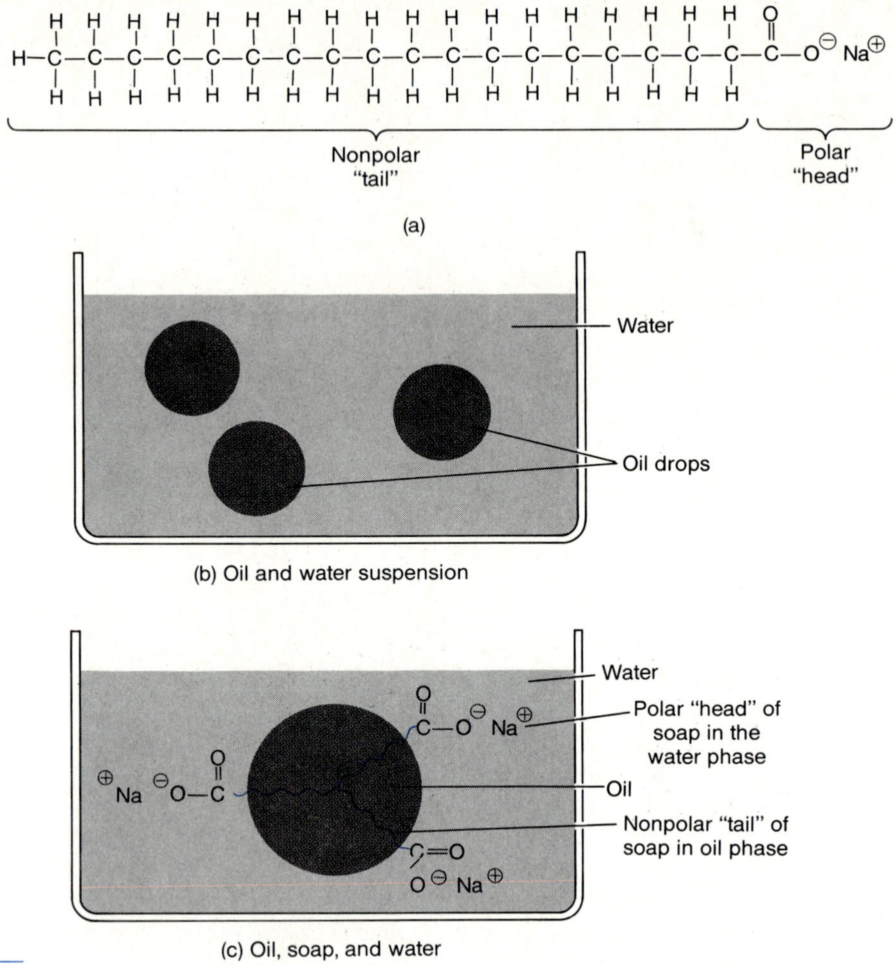

FIGURE 17-2
Cleaning action of soap.

dirt associated with it) is now soluble (or at least well suspended) in the water and can be washed away. This is shown in Figure 17-2*c*.

Ordinary soap belongs to a large class of molecules called *surfactants* or *emulsifiers*. These are compounds that break down the incompatibility barriers (surfaces) between polar and nonpolar materials and allow them to form suspensions that will not separate into two layers. Many skin creams contain both oil and water, and thus surfactants are needed to give the skin creams a smooth consistency.

When soap itself dissolves in water, it does so by forming *micelles*. These are special clusters of molecules positioned in such a way as to have the nonpolar tails together in the center of the cluster and the polar heads outward toward the continuous water phase. Since these micelles are actually much like very small oil drops and are larger than simple dissolved molecules, they interfere with light transmission, and the mixture looks cloudy. The micelles are too small to settle out; they stay suspended as a result of the natural

FIGURE 17-3
The Tyndall effect. *Left:* Light beam shining through a true solution. *Right:* Light beam shining through a colloidal suspension (soap in water).

thermal motion of the water molecules. Such mixtures are called *colloidal suspensions,* and the visualization of a light beam passed through them is called the *Tyndall effect*. True solutions are homogeneous (the same throughout—see Chapter 1) and are transparent to light (they are clear, not cloudy), although they are often colored. Figure 17-3 shows a light beam shining through a true solution and a colloidal suspension. The Tyndall effect can be seen in the colloidal suspension, but in the true solution the light beam is not visible. Figure 17-4 (page 468), a diagram of soap micelles, shows why soap-water mixtures are cloudy.

When soaps are used for cleaning, there are two problems that can be encountered. First, when acids are present, they react with the soap to produce the corresponding carboxylic acids, which are almost totally insoluble in water. This reaction renders the soap inactive (unable to clean), since there is now no ionic portion to dissolve in the water. In addition, the insoluble acid forms a solid scum that is undesirable. Acids on skin or clothing result from decomposition of perspiration. The reaction of soap and acid is:

$$\underset{\text{Sodium stearate (soap)}}{CH_3\text{—}(CH_2)_{16}\text{—}\overset{\overset{\displaystyle O}{\|}}{C}\text{—}O^-Na^+} + \underset{\text{Acid}}{H^+} \longrightarrow \underset{\text{Stearic acid (solid scum)}}{CH_3(CH_2)_{16}\text{—}\overset{\overset{\displaystyle O}{\|}}{C}\text{—}OH} + Na^+$$

The second problem occurs when soap is used in water containing Ca^{2+} or Mg^{2+} ions. Such water is called *hard water,* and the ions of alkaline earth metals react with the soap to form insoluble calcium or magnesium salts (this also causes a solid precipitate, or scum, to form in bathtubs, on laundry, etc.):

$$\underset{\substack{\text{Sodium stearate}\\ \text{(soluble)}}}{2CH_3(CH_2)_{16}\text{—}\overset{\overset{\displaystyle O}{\|}}{C}\text{—}O^-Na^+} + Ca^{2+} \longrightarrow \underset{\substack{\text{Calcium stearate}\\ \text{(insoluble)}}}{(CH_3\text{—}(CH_2)_{16}\text{—}\overset{\overset{\displaystyle O}{\|}}{C}\text{—}O)_2\text{—}Ca} + 2Na^+$$

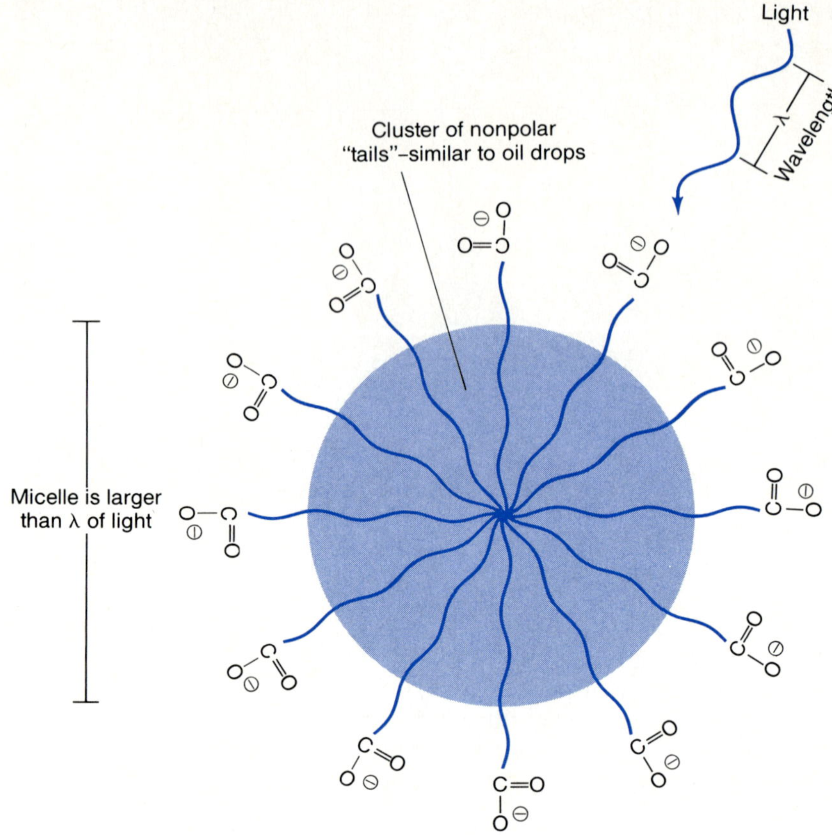

FIGURE 17-4
Soap micelles.

This second problem can be partially solved by the use of *detergents*. Detergents are similar to soaps in that the molecules have a small polar part and a long nonpolar hydrocarbon part. There are three types of detergent materials shown below. The *anionic* type (in which the organic molecule contains a *negative* charge) is the most common household laundry detergent. The *cationic* type (in which the organic molecule has a *positive* charge) is used in restaurants where antibacterial action is important. (All soaps and detergents have antibacterial properties.) The *neutral* type, which has no charge but does have polar bonds in part of the molecule, is used as a surfactant in cosmetic formulations.

Anionic detergent

$$CH_3—(CH)_{11}—O—\underset{\underset{\displaystyle O}{\|}}{\overset{\overset{\displaystyle O}{\|}}{S}}—O^- \quad Na^+$$

Sodium lauryl sulfate
(common laundry detergent)

Cationic detergent

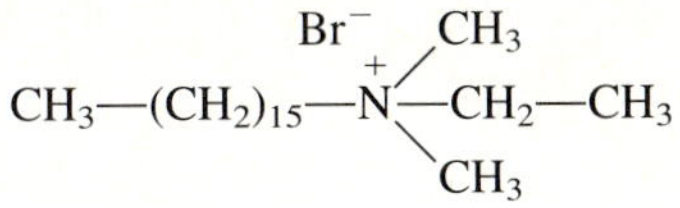

Dimethylethylhexadecylammonium bromide
(a quaternary ammonium salt—a quat)

Neutral detergent

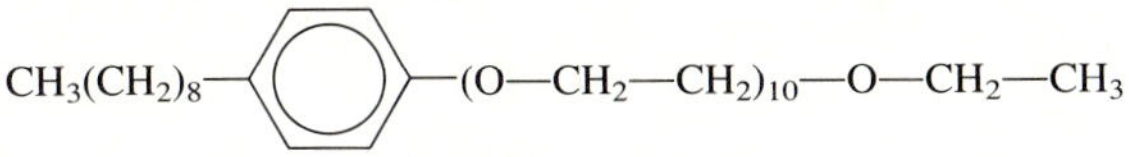

The cationic and neutral detergents do not react with either acids or metal ions and are thus not inactivated by these materials. The anionic detergents are less expensive and more powerful and are therefore more commonly used. They are less affected by metal ions than soap is, since their calcium salts are soluble.

Detergents act by the same mechanism as soap (shown earlier, in Figure 17-2). Those soaps and detergents which have *straight-chain* nonpolar hydrocarbon parts are degraded by natural bacteria in waterways, and if they are used, no foam is formed on the rivers and lakes into which household sewage is dumped. These are called *biodegradable* and should be used exclusively to help solve the water pollution problem. In some molecules of detergent, if a benzene ring is present (as in the neutral detergent shown above), the material stays in the water virtually unchanged and causes deterioration of biosystems as well as the formation of unsightly foam billows at rapids.

Soap additives

To make a cleaning product more attractive, convenient, or useful, a number of additives are used, especially in laundry formulations.

Carboxymethylcellulose (CMC) Carboxymethylcellulose is used as a suspending or thickening agent. Particles of dirt that are loosened tend to settle back onto the clothing. CMC helps to prevent this from happening, so that when the water is drained from the clothes, the dirt is carried along with it.

Fluorescent dyes A fluorescent material is one that absorbs light of one wavelength and gives off light of a different wavelength. Clothes tend to look whiter if they are actually slightly blue, and they tend to look dirty if they are slightly yellow. The dyes used in the soaps are absorbed onto white clothes, and when white light shines on the clothes, they radiate a faint blue glow. They are not any cleaner than if washed in ordinary soap, but they *look* "whiter and brighter than new." This dye is especially noticeable at a dance, bar, or skating rink, where fluorescent ("black") lights are on and the regular white lights are turned low. Under the black light all white pieces of clothing washed in such a detergent actually have a bluish-purple color. These materials are sometimes called *optical brighteners*. The molecular structure of one in common use is given below:

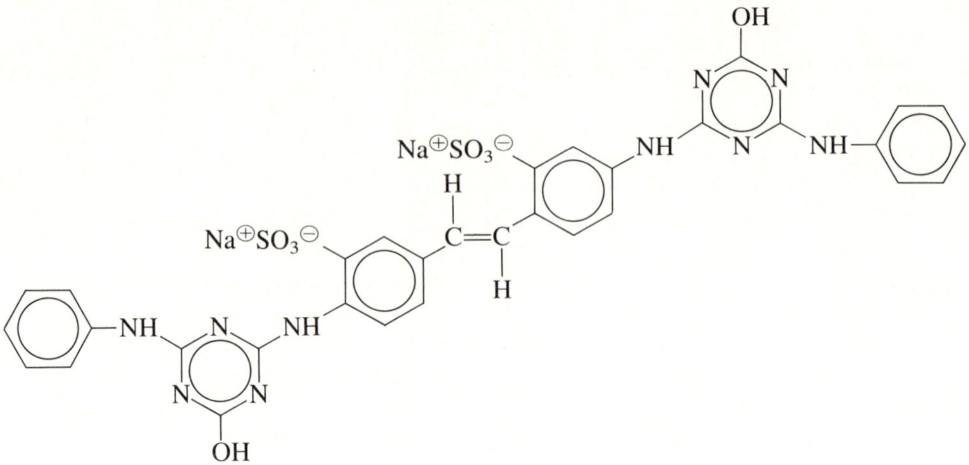

An optical brightener

Bleaches Bleaches are compounds that change the colored molecules in a stain to colorless ones. Most colored materials absorb a characteristic wavelength of visible light because of a series of carbon-carbon double bonds in the molecules. Oxidizing agents such as sodium hypochlorite (NaClO) or hydrogen peroxide (H_2O_2) destroy the double bonds. The modified molecules do not absorb light and are, thus, colorless. They are still on the clothes but are not visible. The process is shown below:

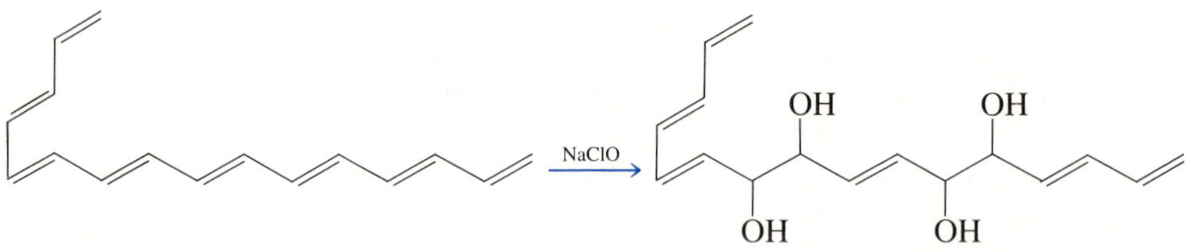

8 double bonds in a series (orange-colored)

Some of the double bonds have been changed (colorless)

Enzymes As we saw in Chapter 15, enzymes are protein catalysts. Some enzymes are used by the human digestive system to break down proteins into amino acids, carbohydrates into simple sugars, and fats into glycerin and fatty acids. Most of these digested materials are soluble in water, while the foods themselves are not. Much of the difficult-to-remove dirt on clothing is from food. Eggs, ketchup, gravy, grease, blood, etc., are not readily removed by soap; but enzymes such as papain react with the food stains, and the resulting digested materials dissolve and are rinsed away. Think of a washing machine in which an enzyme-containing detergent formulation is used as a ''giant stomach'' digesting the food off the clothes. Some people, however, are adversely sensitive to traces of enzymes left on the clothes.

Borax Borax, whose formula is $Na_2B_4O_7 \cdot 10H_2O$, is a slightly basic material (it is mined from large deposits in California). It reacts with acids that weaken the cleaning ability of soaps or detergents and thus helps the soaps or detergents to be more effective and reduces scum formation. Clothing that has the last traces of odor-causing acidic materials removed or neutralized has a more pleasant, ''cleaner'' smell. Borax helps accomplish this, since it is a base and reacts with the odor-causing acids.

Phosphates Sodium triphosphate ($Na_5P_3O_{10}$) is present in fairly large amounts in most detergent mixtures. It serves three useful purposes and causes at least one serious problem. The phosphate is a base and thus serves the same purpose as borax (see above). It also complexes the Ca^{2+} and Mg^{2+} ions so that they do not react with, and inactivate, the soap or detergent. Since detergents are much more effective cleaning agents than soaps, it takes less detergent than soap to wash a load of clothes. Detergents are also more expensive to manufacture than soap is, but since it takes less detergent, the cost per load is about the same as for soap (perhaps even less). When detergents were first marketed, people were unwilling to buy a small box of detergent at a high price even though it would last just as long as (or longer than) a large box of soap. The solution to this marketing problem was the addition to the formulation of an inexpensive filler so that the same amount of detergent would fill a larger box. Phosphate is added as an inexpensive filler and also for the two other purposes mentioned above. It is called *builder* instead of a *filler* for promotional reasons (*builder* sounds better than *filler*).

Phosphates serve as fertilizers (as you may remember from Chapter 10); and when sewage (which contains laundry water) is dumped into rivers and lakes, the phosphate causes water plants to grow in profusion, which chokes or fills the water. The lower-level plants get no sunlight and die, and the rotting plants use up all the O_2 in the water. The body of water cannot support life and eventually becomes badly polluted and useless. This is called *eutrophication*. This problem has been partly solved by laws that limit the amount of phosphate in detergents. The phosphorus (P) content is now limited by law to not over 0.5% P or 2.0% $Na_5P_3O_{10}$.

Household bleach formulations

Bleaches are materials that remove unwanted color from a fabric or surface. Most of them are oxidizing agents which destroy the double bonds that cause light absorption. The mildest bleach is hydrogen peroxide (H_2O_2); it is especially useful for removing fruit stains from colored cloth. It is mild enough that it does not bleach the cloth itself but strong enough to oxidize the fruit stain. It is ideal for removing a strawberry stain from a yellow shirt, for example. It usually is purchased as a 3% solution in water and should be kept in a dark bottle. It decomposes fairly rapidly, and if a bottle has been around the house for more than 9 months to a year, the contents may be inactive. The activity of a bottle of H_2O_2 can be checked by adding some chlorine bleach

(NaClO) to a small sample of it. If it bubbles vigorously, then it is still good. The reaction occurring during this test is:

$$NaClO + H_2O_2 \longrightarrow Na^+ + Cl^- + H_2O + O_2 \uparrow$$

The most commonly used bleach is liquid chlorine bleach (Clorox, for example). These bleaches are usually 5.25% sodium hypochlorite solution (NaClO), but some brands are as high as 6.0%. These solutions are fairly strong oxidizing agents and are appropriate for stains on white cloth but will frequently bleach the dye in colored cloth. When strong solutions (that is, undiluted commercial bleaches) are used and clothes are allowed to soak for a period of time, the fibers of the cloth can be greatly weakened. These bleaches should be used in diluted form or for a short time for most applications.

A third type of bleach has one particular application. It is oxalic acid ($H_2C_2O_4$), which can be purchased at a drugstore in solid form. A solution of it in warm water removes iron stains. Either red-brown rust stains or some ink stains (from ink which contains an iron compound) can be bleached from cloth, rugs, porcelain, or wood. The oxalate ion ($C_2O_4^{2-}$) complexes with the Fe^{2+} or Fe^{3+} ion and renders it both water-soluble and colorless. The oxalic acid crystals should be dissolved in warm water, and the solution rubbed on the stain. After standing overnight, the material can be laundered or scrubbed with soap and water. Caution should be taken, since oxalic acid is extremely toxic if ingested.

Window cleaners

Glass cleaners should contain only volatile materials (ones that evaporate) so that no solid residue is left on the glass. No abrasives can be used, since glass is softer than pumice or sand and is scratched by them. Most window-cleaning formulations contain a mixture of water, ethyl or isopropyl alcohol, and ammonia. The alcohol is used to dissolve somewhat nonpolar materials and to help the water to evaporate faster, the water is used to dissolve polar materials, and the ammonia is a surfactant and allows any grease to become at least partly soluble. All three evaporate in a very short time, allowing one to see almost immediately if the window is clean. These preparations can easily be prepared from water, rubbing alcohol, and household ammonia in the proportions shown below:

	COST, CENTS	PERCENT OF TOTAL
Window cleaner:		
Water		45
Rubbing alcohol	30	45
Ammonia	10	10
Total cost	40	
Commercial window cleaner (including spray bottle)	90	
Savings	50	

The exact amounts are not critical. It is less expensive to prepare a window cleaner than to buy one. Although one is not going to get rich by making one's own window cleaner, this example does point out that there is nothing mysterious about special-purpose products—they are mixtures of ordinary materials with a little color added and perhaps a fictitiously "special" ingredient such as "ammonia-D" or "ammonia +" to make the consumer think that there is some special magic ingredient in a particular brand. This is simply an advertising gimmick.

Drain cleaners

Bathroom drains tend to accumulate hair, and kitchen sink drains gradually become clogged with grease. In both cases, a concentrated solution of NaOH will solve the problem. In kitchen drains the NaOH reacts with some of the grease to form soap (see the section in this chapter on soaps and detergents), which is easily washed away with water and emulsifies the rest of the grease. Since many drain cleaner formulations contain solid NaOH, which when added to the water in the drainpipe trap "cakes up" and forms a hard mass, something has to be added to cause agitation. Aluminum reacts with NaOH to produce hydrogen gas bubbles according to the reaction below:

$$2Al + 6NaOH \longrightarrow 2Na_3AlO_3 + 3H_2$$

Thus a little aluminum (in the form of a powder or small chunks) is added to the NaOH for this purpose. The dissolving of NaOH in water is a highly exothermic process—the solution becomes very hot (80–100°C). As shown in Figure 17-5, the solid NaOH is added to the grease-packed drain. The NaOH dissolves, producing heat which melts the grease. This process is aided by the H_2 bubbles produced by the aluminum. The heat and agitation cause the grease to react fairly rapidly with the NaOH, converting it to soap. After a

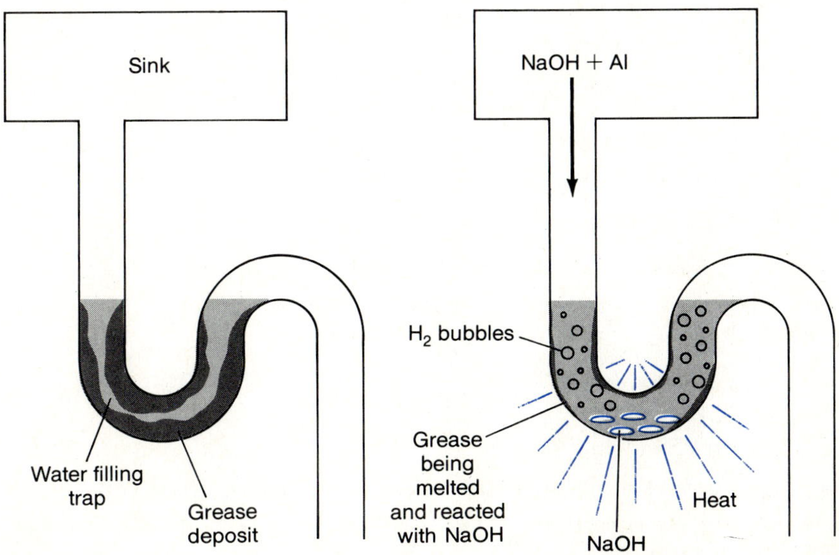

FIGURE 17-5
Action of drain cleaners.

few minutes, the soap along with the grease it has emulsified is washed away with water.

Hair in bathroom drains is composed of protein (see Chapter 15). NaOH will cause the protein to break down into amino acids, which are quite soluble in water. The reaction is shown below:

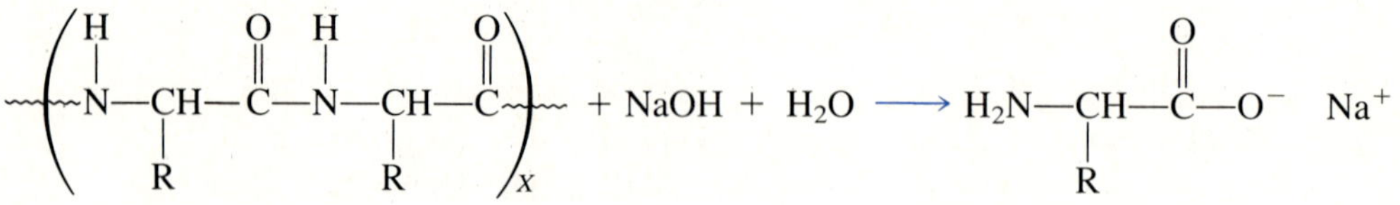

General hair structure (protein) — Individual molecules of amino acids

Toilet bowl cleaners

The discolorations and tight scale buildup that occur in toilet bowls are mostly calcium carbonate ($CaCO_3$) deposits from hard water. The materials used for cleaning contain some kind of strong acidic material, since the $CaCO_3$ reacts with acid to form water-soluble materials. The solid-type cleaners are mostly $NaHSO_4$, and the liquid type contain HCl. Both materials are acids and react with the $CaCO_3$ scale as follows:

$$2NaHSO_4 + \underset{\text{Insoluble scale}}{CaCO_3} \longrightarrow 2Na^+ + Ca^{2+} + 2SO_4^{2-} + H_2O + CO_2\uparrow$$

$$2HCl + \underset{\text{Insoluble scale}}{CaCO_3} \longrightarrow Ca^{2+} + 2Cl^- + H_2O + CO_2\uparrow$$

In addition to the acidic material, these formulations sometimes contain a carbonate such as $NaHCO_3$ to cause bubbling and therefore agitation to help break the scale loose and give the illusion of being a "powerful cleaner." Some also contain a blue coloring material, since blue is supposed to look "clean and pure." This is purely the result of psychological conditioning by the advertising industry; but people apparently do tend to buy more products—from mouthwashes and laundry detergents to toilet bowl cleaners—that are blue.

A precaution: One should *never* use toilet bowl cleaners with bleaches. There is a temptation to bleach and clean scale at the same time, but the two formulations react to produce the very toxic chlorine gas. The reaction of bleach with liquid toilet bowl cleaners is as follows:

$$\underset{\text{Bowl cleaner}}{2HCl} + \underset{\text{Bleach}}{NaClO} \longrightarrow Na^+ + Cl^- + H_2O + \underset{\text{Toxic}}{Cl_2\uparrow}$$

But Cl_2 is also formed with solid toilet bowl cleaners. If the toilet bowl cleaner does not also remove the stain, one can use NaClO bleach at a *separate time,* when no toilet bowl cleaner is present. Since most stains of this type are from iron, oxalic acid (see the section in this chapter on bleaches) can be used. Because of the toxic nature of oxalic acid, most of the water should probably be removed and disposed of in a place other than the sewage system, since this eventually goes to the waterways from which drinking water is drawn. While oxalic acid is good for removing iron stains in other situations, it is impractical for cleaning toilets.

Tub cleaners

As was mentioned in the section in this chapter on soaps, sodium stearate reacts with hard water to form calcium stearate, which is a solid material. Calcium stearate is present in almost any used bathwater and tends to collect, along with body oils (or bath oils), on the sides of the bathtub. This also happens to a smaller extent on shower walls. It can be removed by scouring powder and a little work, but there are also commercial cleaners available that specifically remove the calcium.

The best agent for this is the sodium salt of ethylenediaminetetraacetic acid (EDTA), whose structure is shown below:

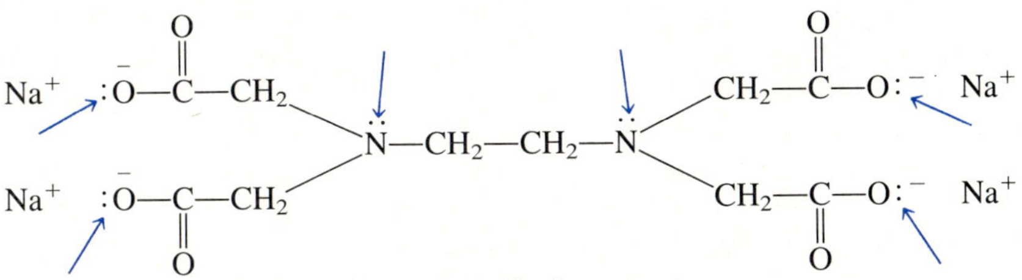

Sodium ethylenediaminetetraacetic acid (EDTA)

The action of this compound involves a phenomenon called *chelation*. The *metallic* elements in the fourth period on the periodic chart from Ca to Cu have empty or only partly filled *d* orbitals (see Appendix D). EDTA has several unshared pairs of electrons (see arrows in formula), and these can be shared with the empty *d* orbitals of Ca^{2+} ions, forming a complex sometimes called a *coordination complex*. The compound with the originally unshared electron pairs is called a *ligand*; and when a ligand has several pairs of electrons to share, so that it totally surrounds the metal when the complex is formed, it is called a *chelating agent*. When the cleaner is sprayed on the tub or shower walls, the following reaction occurs, and both the soap and the Ca-EDTA coordination complex are easily washed away with water:

$$(CH_3(CH_2)_{16}{-}\overset{\overset{\displaystyle O}{\|}}{C}{-}O)_2Ca + \text{Na-EDTA complex} \longrightarrow$$

Calcium stearate (bathtub scum, insoluble)

$$2CH_3{-}(CH_2)_{16}{-}\overset{\overset{\displaystyle O}{\|}}{C}{-}O^-Na^+ + \text{Ca-EDTA complex}$$

Soap (soluble) — Soluble complex

This is the same process by which oxalic acid removes iron stains—it forms a soluble chelate. Medically this concept is used in the treatment of heavy-metal poisoning. If one has taken in toxic levels of lead (Pb), for example, EDTA or other chelating agents are given in small doses over a period of time and the metal is chelated and removed from the body. This is called *chelation therapy*.

Shampoo

Hair, like most materials, becomes "dirty" when it collects grease that may or may not have dust and other dirt associated with it. Most of the oil is from natural oil secretion at the hair root. Formulations of soap or detergent have been prepared for washing hair, although in many cases ordinary soap would suffice. These shampoos have a dozen or more compounds in them to make an attractive, pleasant, easily usable product. First, detergents such as sodium lauryl sulfate (see the section in this chapter on detergents) are used to prevent the calcium in the water from forming a hard-to-remove solid. (You would not want bathtub scum all over your hair, would you?)

Since natural suspensions of soap or detergent are cloudy because of micelle formation, materials such as alcohols are sometimes added to dissolve the micelles and make a clear solution (which makes the solution look as though it is cleaner and therefore "purer"). Other companies do not bother with this and leave a cloudy or even opaque suspension. There is no difference in cleaning power between the two types.

Soaps and detergents, when placed in water, form mildly basic solutions. This base can burn eyes and, if used regularly, can cause hair, which is protein, to deteriorate slightly. Refer to the section in this chapter on drain cleaners for the reaction of hair with a strong base such as NaOH. Although soap does not react nearly as fast with hair as the NaOH does, the strands of hair are split (see the section in this chapter on wool, which is a natural fiber), and the hair becomes unsightly. For both these reasons, acidic materials are added to shampoos to lower the pH of the shampoo (see Chapter 10). Any pH value between 5 and 8 will prevent damage to the hair. To prevent eye irritation the shampoo should have a narrower pH range of about 6.0 to 7.0, as all shampoos formulated for small children ("baby shampoos") have. Table 17-1 (opposite page) shows the pH values of several shampoo formulations.

Stain and material removers

We now come to a common problem—removing marks or stains from various materials or surfaces. There are two approaches that can be used. The first is to bleach or discolor the material using one of the materials already discussed. They include H_2O_2 for fruit stains, NaClO for many stains on white cloth, and $H_2C_2O_4$ for iron rust stains on most materials.

The second approach is to dissolve the unwanted material in an appropriate solvent. This approach applies the principle that the material to be dissolved will be soluble in a solvent of similar polarity, that is, like dissolves like. Sometimes one can estimate the polarity of a material from known solubilities. For example, if a marker is known to be "waterproof," that means it

TABLE 17-1

pH values of some shampoos*

BRAND	pH VALUE
Selson Blue	3.7
Alberto VO 5	4.4
Suave	6.1
Johnson and Johnson's (for babies)	6.9
Whella Balsam	7.3
Prell Concentrate	7.3
Dial	7.7
Head and Shoulders	7.7
Pharmacy Lemon Scented	8.0
Pet Shoppe (for pets)	8.7

*John Griffin, Robert Corcoran, and Kenn Akana, "The pH of Hair Shampoo," *Journal of Chemical Education,* **54:**553 (1977).

does not dissolve in water. This indicates that it is nonpolar (since water is polar), and to remove it a nonpolar solvent is needed. Likewise, asphalt or tar spots on an automobile can be removed only by nonpolar solvents, since one knows from experience that paved roads made of asphalt or tar do not dissolve when it rains. At times, the polarity and, therefore, the solubility in various solvents are not known. In this case one tries solvents of gradually changing polarities. Table 17-2 lists common solvents, their usual sources, and their relative polarities. Polarity *decreases* down the list, which means that less polar stains are removed best by the solvents near the bottom of the list. One of these five kinds of solvents will dissolve most materials and can be tried in order. If water does not dissolve something, try alcohol; and so on.

TABLE 17-2

Common solvents and their polarities

SOLVENT	SOURCE	POLARITY DESIGNATION
Water		Very polar
70% alcohol and 30% water	Rubbing alcohol	Polar
95% ethyl alcohol	Denatured alcohol Shellac thinner	Somewhat polar
Acetone or ethyl acetate	Fingernail polish remover	Slightly polar
Ethylene dichloride	Drugstore	Very slightly polar
Hydrocarbons*	Gasoline Upholstery cleaner Charcoal lighter fluid Paint thinner Varsol Turpentine	Nonpolar

**Caution:* These hydrocarbons are flammable and should always be used with plenty of ventilation and away from sources of ignition.

Table 17-3 lists a number of stains and other unwanted materials along with the best method of removing each of them.

Good reference books for stain removal and home preparation of commercial formulations are:

Norman H. Stark, *The Formula Manual,* 3d ed., Stark Research Corporation, Tucson, Arizona, 1975.

Kenneth M. Swezy and Robert Sharff, *Formulas, Methods, Tips and Data for the Home and Workshop,* Popular Science, Harper & Row, New York, 1979.

TABLE 17-3

Methods of dissolving unwanted materials or removing stains

MATERIAL OR STAIN	METHOD OF REMOVAL
Coffee	NaClO
Soft drink	NaClO
Tobacco	NaClO
Mustard	NaClO
Grass	NaClO for white cloth or denatured alcohol for colored cloth
Berry	NaClO for white cloth or H_2O_2 for colored cloth
Blood	NaClO for white cloth or enzyme detergent for colored cloth
Scorch	H_2O_2
Lipstick	Ethylene dichloride
Chewing gum	Ethylene dichloride (softens and dissolves it slowly) or cooking oil or peanut butter (also softens it) followed by washing in soap and water
Ink	Oxalic acid solution or H_2O_2 or $(NH_4)_2O_2$ (ammonium peroxide)
Rust	Oxalic acid solution or citric acid solution
Asphalt (tar)	Hydrocarbons*
Grease or oils	Hydrocarbons or ethylene dichloride
Iodine	Denatured alcohol
Mildew†	4:1:2 mixture of H_2O_2:NH_4Cl:alcohol in 15 parts H_2O‡
Perspiration	Household ammonia
Paraffin wax (candle) or crayon	Hydrocarbons or boiling water§
Urine†	Citric acid solution

*Lighter fluid works quite well.
†N. H. Stark, *The Formula Manual,* 3d ed., Stark Research Corporation, Tucson, Arizona, 1975.
‡These are volume ratios. Material should be soaked in this mixture.
§Scrape the excess, then dissolve the residue with a hydrocarbon. If the stain is on a cloth, boil the cloth in water to melt the wax (it will float to the top).

CHEMISTRY OF FIBERS

Both natural and synthetic fibers are composed of very long molecules called *polymers*. These polymers are prepared either naturally or synthetically by the linking together of many small molecules that are usually referred to as *monomers*. Several natural polymers are discussed first; then several synthetic polymers are described.

Natural fibers

Cotton Cotton, for example, is a form of cellulose (see Chapter 15), and is therefore composed of many glucose units linked by β-glucoside bonds (beta-glucoside bonds). Its structure is shown below:

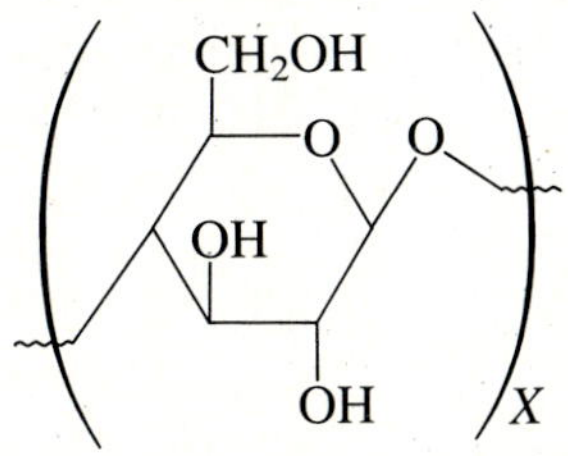

This is a linear polymer with hundreds of glucose units bonded together in a continuous sequence. The *X* signifies an indefinitely large number.

Silk Silk is likewise a polymer, but it is composed of amino acids linked by amide bonds. It contains a repeating sequence of six amino acids (see Chapter 15):

$$\sim\!\!(\text{gly-ser-gly-ala-gly-ala})_X\!\!\sim$$

These chains are bound to each other by *hydrogen bonds,* indicated below by the short dotted lines:

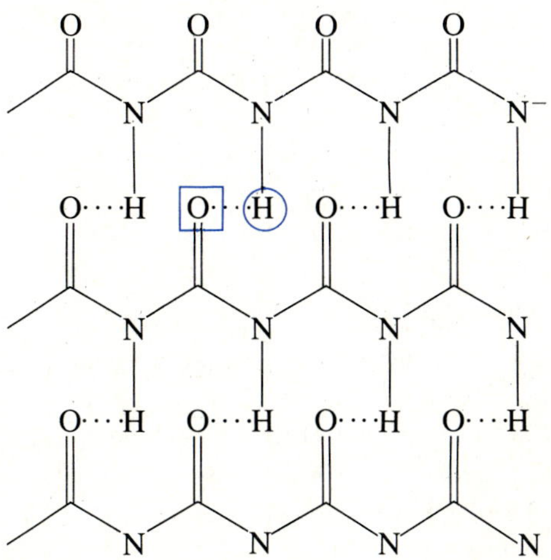

Hydrogen bonds are strong forces that hold molecules together. Hydrogen bonds are *intermolecular forces*—forces of attraction between neighboring molecules. (*Intramolecular forces,* by contrast, are bonds within a molecule.) Although intermolecular forces are weaker than covalent bonds, they account for the great strength of silk. Hydrogen bonding occurs between hydrogen which is covalently bonded to either oxygen, nitrogen, or fluorine in one molecule and another atom of nitrogen, oxygen, or fluorine in another molecule. Thus the hydrogen bonded covalently to a nitrogen (circled hydrogen above) can form a hydrogen bond with an oxygen in another polymer chain (C=O in the box above). This hydrogen bond holds the two polymer molecules together.

Wool Wool is also a proteinlike material, but in wool the polypeptide chains are in a helical (spiral) form and several of these spirals are held together by covalent disulfide bonds. The helices themselves are held together by hydrogen bonds between adjacent loops within a single polymer:

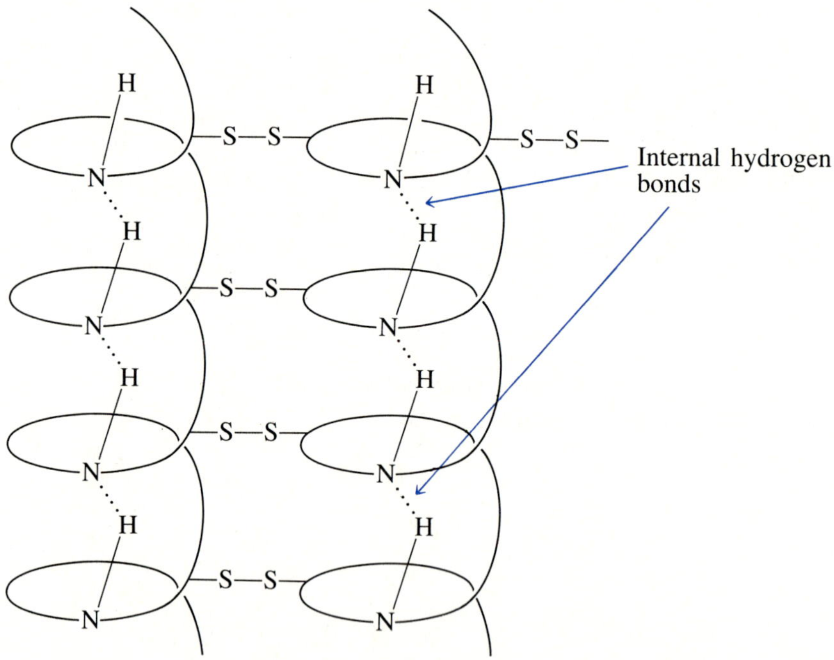

Silk does not stretch, because the molecules are already extended. Wool, on the other hand, contains spiral polymers and does stretch. When wool gets wet, hydrogen bonds form with external water molecules rather than with another part of the same spiral. With little to hold the spiral together, the wool fiber can be stretched much like a spring. When it drys, the original hydrogen bonds reform and the wool shrinks again.

Wet stretched wool is shown below from a molecular point of view:

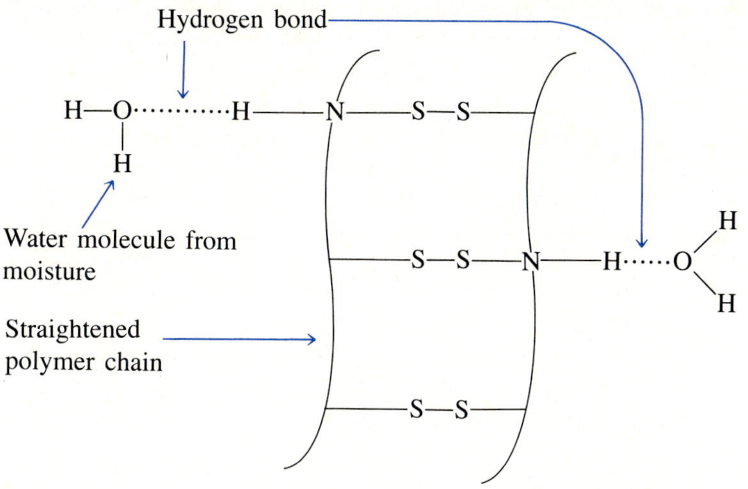

This same process occurs when human hair is wet. It stretches and can be shaped (curled or straightened) and allowed to dry in that shape. The new shape is maintained by the newly formed internal hydrogen bonding until the hair becomes wet again. This illustrates a principle mentioned earlier. The behavior of all materials—both the physical and the chemical properties of synthetic or natural materials and everyday materials or materials used mainly in the research laboratory—is a direct consequence of the molecular structure.

Synthetic fibers

Synthetic fibers are made by bonding various monomers together to form long molecules that act somewhat like natural fibers.

Polyester Certainly the most common synthetic fibers are those of the polyester type. These fibers are used alone (Dacron), used mixed at a ratio of 60 to 40 with cotton, or used in ribbon form to make audiotape and videotape (Mylar). Chapter 11 presented the ester functional group and indicated that esters are prepared from carboxylic acids and alcohols as shown below:

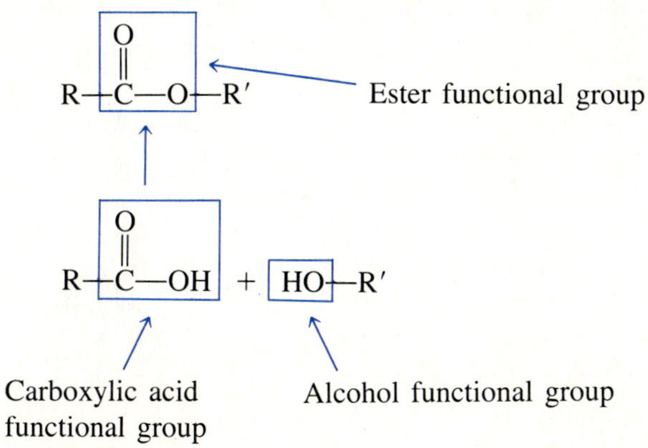

If one starts with a difunctional acid (one containing two —COOH groups) and a difunctional alcohol (one containing two —OH groups), continuous polymer chains can be produced. The two compounds used are shown below:

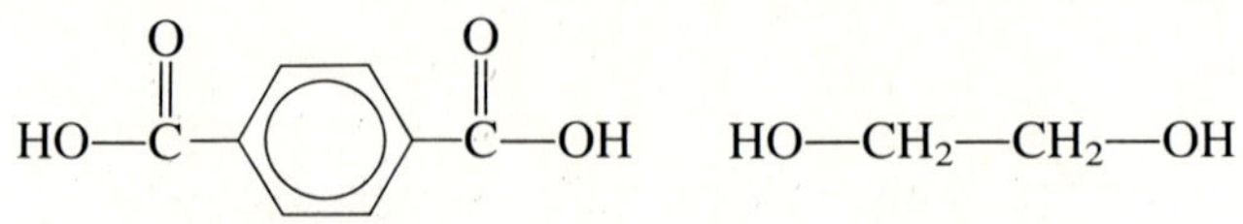

Terphthalic acid (difunctional)

Ethylene glycol (difunctional)

When the two compounds react and give up water molecules, continuous chains are formed. The polymeric *copolymer* (one having two different monomer units) is called *polyethyleneterphthalate:*

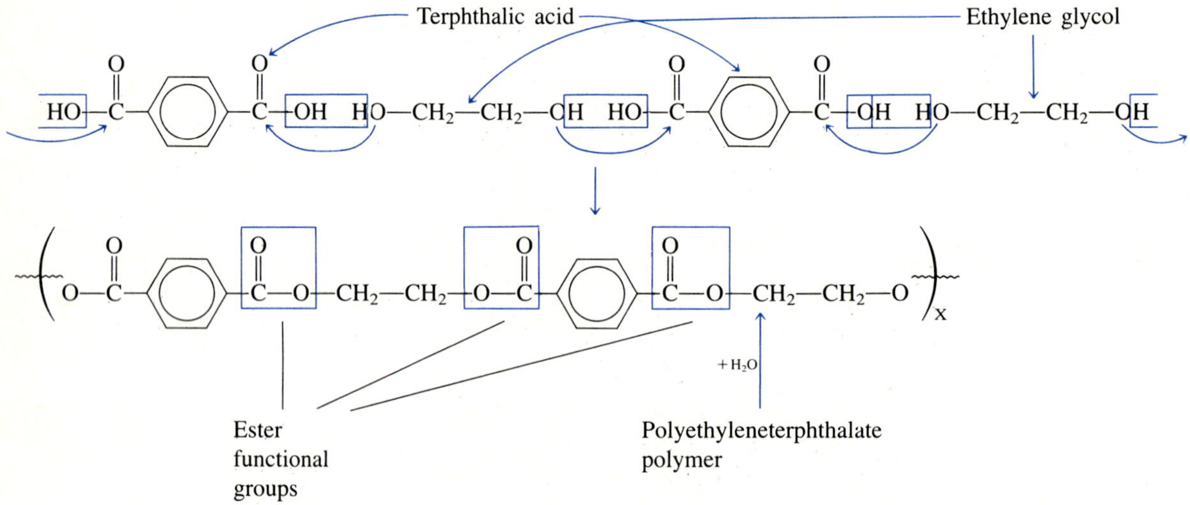

Nylon Another important synthetic fiber is nylon. It is a polyamide prepared by reacting, for example, the diacid and diamine shown at the top of the opposite page.

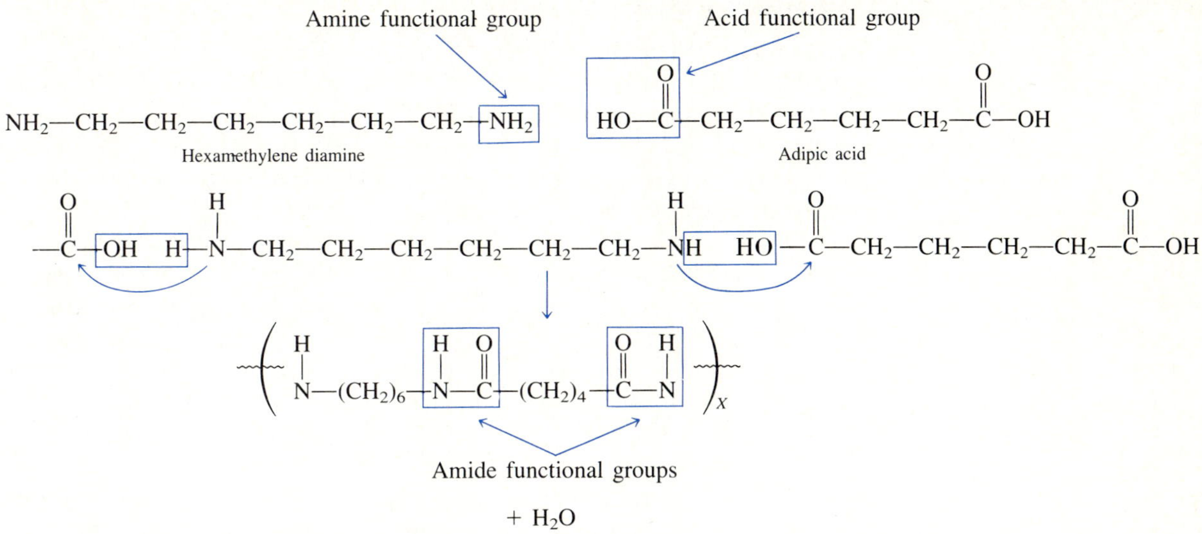

Remember that the amide functional group is:

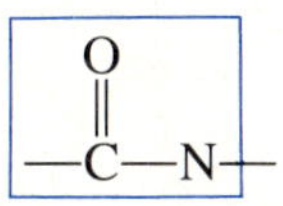

The polyamide shown above is called *nylon-6,6* because there are six carbon atoms in each part (in the acid and in the amine). Other types of nylon such as 5,5; 5,6; 6,7, can be prepared; each has slightly different properties.

Spandex Spandex is an elastic fiber and is used along with other fibers to produce snug-fitting clothing. It is a polyester with two different dialcohols used as monomers. Polymerization is carried out so as to produce *short* segments with each type of alcohol. These are called *blocks*. These two blocks, along with a third polyamide block, are mixed to form a polymer with the different segments arranged in random order. This type of polymer is called a *block copolymer*. It is shown below:

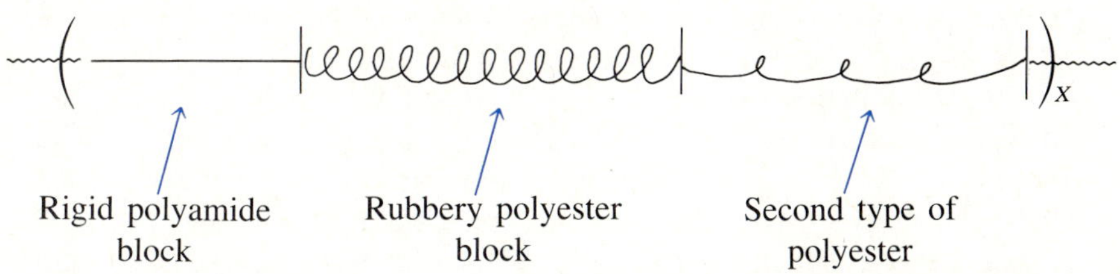

The fiber stretches to the limit of the polyester (until the segments are straight). When the fiber is released, however, it tends to go back to the natural folded or springlike shape and pulls the cloth tightly against the body.

Properties of fibers

As a way of reemphasizing the idea that the properties of substances are a result of their molecular structure, let's consider the behavior of polyester and cotton—why they vary in both comfort and resistance to wrinkles. Cotton has three OH groups on each monomer unit in the polymeric chain. Since these are polar groups and can form hydrogen bonds, the cotton fiber can absorb water molecules. This removes the water of perspiration from the skin by soaking it into the fiber rather than having it evaporate. The cotton does not feel wet, because the water is inside the fiber. Polyester, on the other hand, has no OH groups and is nonpolar. Since the water of perspiration cannot be absorbed, it wets the outside of the polyester fiber, causing it to feel cold and clammy. The 60-to-40 mixture of cotton and polyester helps to solve this comfort problem. The cotton in the mixture absorbs the water, adding comfort, while the polyester contributes other desirable properties, such as wrinkle-resistance.

To explain why cotton wrinkles easily while being worn but polyester does not, we will have to examine a phase of matter we have thus far ignored. Most matter can exist only as a gas, liquid, or solid. Solids are completely rigid, and liquids have no structural integrity at all. For most pure materials, there is no state between a solid and a liquid. Water, for example, goes directly from rigid ice to fluid liquid water. Polymeric materials, such as the fibers we have been examining, have a flexible solid or rubbery state, similar to glass. Consider the common material polyethylene. At very cold temperatures it is brittle like other organic solids. As the temperature is raised, it reaches a point where it changes to a rubbery state. This is called the *glass transition temperature* and is symbolized by T_g. The T_g of each polymer—including polyester, nylon, cotton, and silk—is characteristic of each material just as melting and boiling points are. Eventually these materials melt if the temperature is high enough.

The T_g of cotton is about 50°C, and that of polyester is 100°C. As long as a fiber is *below* its T_g temperature, it does not wrinkle. If it is bent, it springs back to its original shape when released. When it is above its T_g (but below the melting temperature), it is flexible, and when bent, stays in the new shape. In this state it acts like a copper wire or piece of clay. The T_g of fibers is lowered by moisture, and the T_g of moist cotton is about 20°C, while that of polyester is a little below 100°C (about 80°C).

When one is wearing a cotton shirt or blouse and leans against the back of a chair, perspiration lowers the T_g to 20°C. Since this is below body temperature (37°C), the cotton goes into the rubbery state and the natural folds become bends. When one stands up, the cloth cools down and the water evaporates. The fiber is now below its T_g temperature,* and the bends are permanent (i.e., the shirt is wrinkled). When one wears a piece of polyester clothing, the perspiration only lowers the T_g to 80°C, which is well above body temperature. When one leans against the chair this time, the fibers

*The T_g is back up to about 50°C, since the moisture is gone and the cloth has cooled down to about 30°C.

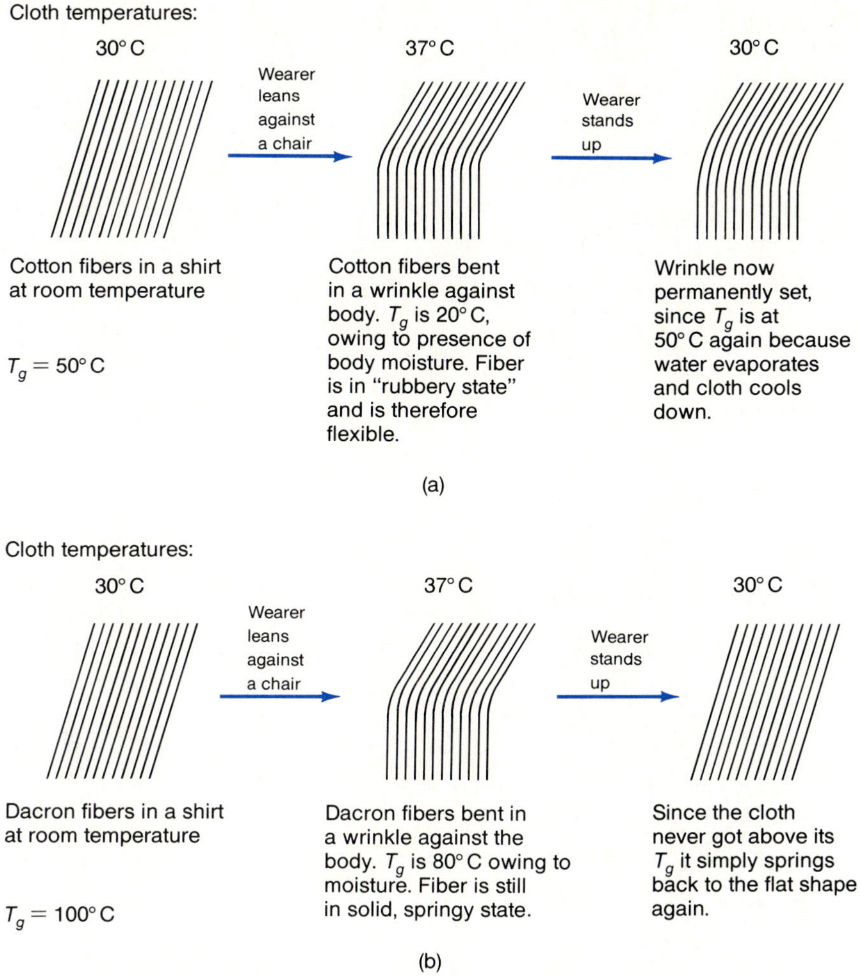

FIGURE 17-6
Wrinkle-resistance of polymers. (a) Cotton. (b) Dacron.

remain in the solid state, and they spring back into their straight shape when one stands up. (See Figure 17-6.)

Ironing cotton cloth also involves this same principle. To flatten cotton one has to get it above its T_g so that it can be forced into a flat shape. This can be done with a hot clothes iron, but is easier if the cloth is moistened first or steam is used during ironing. This moisture lowers the T_g and therefore lowers the temperature needed to remove wrinkles.

PREVENTING AND EXTINGUISHING FIRES

Fireproofing

The presence of combustible materials in the home adds to the potential of a serious home fire. This problem has not been as serious in recent years as it was formerly, because fewer people use open flames for heating and cooking—we have used electricity for lighting and cooking for many years. With the increasing expense of energy consumption, however, more people are using wood stoves or fireplace inserts as supplemental energy forms; this increase, along with the traditional use of candles, presents an increased risk of an accidental home fire. The fabrics used in decorations such as curtains or burlap false ceilings, as well as Christmas trees, are among those combustible items that can be rendered resistant to fire.

Burning is a chemical process which involves the combining of a carbon-containing (organic) material with O_2:

$$\text{Organic material} + O_2 \longrightarrow CO_2 + H_2O + \text{heat}$$

Fires start when sufficient heat is applied to a portion of the material to cause a few molecules to undergo this reaction. The flame then spreads by the formation of *free radicals*—that is, atoms and molecules with a single unpaired electron:

$H_3C^{\bullet}$ ← Free radicals
$O^{\bullet}$ ← Free radicals

These free radicals, which are very reactive and form in the beginning stages of ignition, react with other molecules, producing an increasing number of free radicals. Thus, the reaction becomes even more intense, giving off more heat until finally a *cloud of free radicals* giving off large quantities of energy is visible—this is called a *flame*.

There are several chemicals that can be added to combustible materials which react with the free radicals formed initially and inactivate them. This stops the ignition process from spreading:

$$H_3C^{\bullet} + \text{inorganic compound} \longrightarrow \text{nonreactive species}$$

Solutions of materials such as ammonium sulfate, boric acid, and ammonium chloride can be soaked into a flammable fabric. Following drying, this treated fabric will be resistant to combustion. For example, if one soaks a piece of burlap in a solution of ammonium chloride for 10 min and dries it overnight, it can be held in the flame of a gas stove and it will not burn. Some charring (blackening) will occur, but it will not actually flame. Likewise, a freshly cut Christmas tree can be watered with a solution containing boric acid. The tree will then be less likely to burn—it will take more heat to ignite it, and the burning will be less vigorous. A reference book already cited, *The Formula Manual* by Norman Stark, gives more information about this subject.

Fire extinguishing

Fires, like all chemical reactions, require a continuous supply of reactants in order to be sustained. In the case of combustion, the combustible material (fuel) and oxygen (from the air) must be in continual supply if the process is to continue or spread. Fires can be extinguished by (1) removing the fuel, (2) removing the air (preventing the two from coming into contact), or (3) cooling the system so that the heat is not sufficient to maintain the flame.

Water is placed on fires primarily as a coolant, and bed blankets or towels can be thrown over a small fire to decrease the air supply. Solid materials such as sand and salt can also be used to prevent air from getting to the flames.

Barriers of noncombustible vapor such as CCl_4 or CO_2 can also be used. Commercial fire extinguishing devices found in most public places are usually one of two types. The first type contains liquid CO_2 under pressure; when released, the CO_2, which is heavier than air, will form a blanket and separate the oxygen from the flame long enough to quench the fire. The second type contains solid $NaHCO_3$ (sodium bicarbonate) in one compartment and an acid in another. When the device is inverted, the two mix, forming a CO_2-containing foam that serves the same purpose as the CO_2 from the first type of fire extinguishing device. The following reaction occurs:

$$\underset{\text{Sodium bicarbonate}}{NaHCO_3} + \underset{\text{Acetic acid}}{HC_2H_3O_2} \longrightarrow \underset{\text{Carbon dioxide gas}}{CO_2} + NaC_2H_3O_2 + H_2O$$

For kitchen fires, solid baking soda ($NaHCO_3$) can be thrown on the flame. The heat from the fire decomposes it as follows, producing a blanket of CO_2 which stops the fire:

$$2NaHCO_3 \longrightarrow CO_2 + Na_2CO_3 + H_2O$$

This is especially good for grease or wax fires.

Some fire extinguishers contain CCl_4 (carbon tetrachloride). CCl_4 is not combustible and evaporates rapidly to form a dense vapor blanket around the flame. (This vapor is hazardous, however; and these extinguishers should therefore be replaced by modern ones.) Combustion is an oxidation process, and since the carbon in CO_2 and the carbon in CCl_4 are both in the highest oxidation state possible, further oxidation is not possible and no combustion can occur with either one.

Oxidation in the combustion of methane takes place as follows:

$$CH_4 + O_2 \longrightarrow CO_2 + H_2O$$

CH_4: Oxidation state of -4

CO_2: Oxidation state of $+4$ (highest possible for carbon)

$-4 \rightarrow +4$: Oxidation has occurred

No oxidation is possible in CO_2 and CCl_4:

Oxidation state of $+4$:

$$CO_2 + O_2 \longrightarrow \text{no reaction}$$

$$CCl_4 + O_2 \longrightarrow \text{no reaction}$$

CHEMISTRY OF COOKING: SELECTED TOPICS

Many of the processes used in the kitchen to prepare foods involve chemical reactions or take advantage of the chemical and physical properties of the substances found in foods. In this section several food preparation situations will be looked at, and suggestions on proper cooking techniques or an explanation of *why* a standard procedure works will be given. For a very good and complete treatise on this subject the student is referred to *The Cook Book Decoder* by Arthur E. Grosser.*

Cooking meat

First let's consider the preparation of meats. Meat is composed of protein fibers and is about 70 to 80% water (by weight). To be tender enough to eat easily and to enhance flavor, the meat is cooked. The heat partially breaks the long protein molecules into shorter ones, making the meat tender. This tenderization can be enhanced in two ways. An external enzyme—papain, for example—can be added. Papain is the major component in commercial meat tenderizers and causes the protein strands to be broken at several places.

The ''aging'' of meat, especially beef, at temperatures just above freezing (5°C) also produces tender meat. Muscle cells contain natural enzymes that break down old protein in the living organism. These enzymes are still present in the meat and will slowly work at breaking some of the amide bonds in the meat protein if the meat is allowed to stand for several weeks. This is called *autoprotolysis*—the breaking of protein molecules by self-contained enzymes. The action of these proteolytic (protein-splitting) enzymes is similar to that of papain:

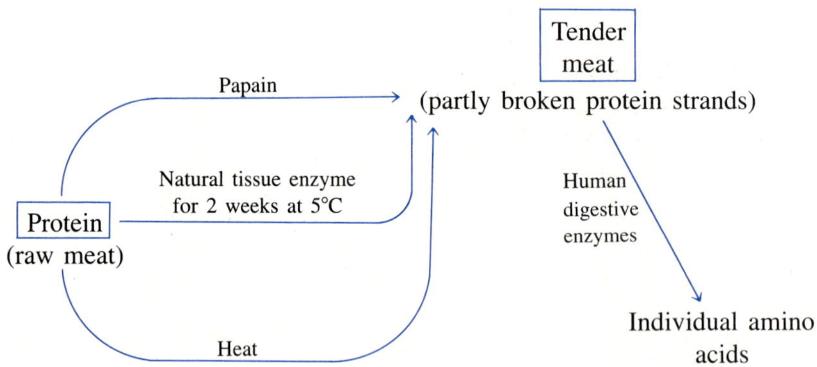

When meat is cooking, the loss of water from the tissue can cause toughening, which, of course, is undesirable. This can be prevented by quickly sealing the outside of the meat by subjecting it to high temperature such as by searing it in a hot skillet or hot oil. The meat can also be sealed by coating it with a batter before cooking it. Adding salt to meat and cooking it dry without sealing or coating it—for example, broiling a salted piece of steak—also can

*Beaufort Books, Inc., New York, 1981.

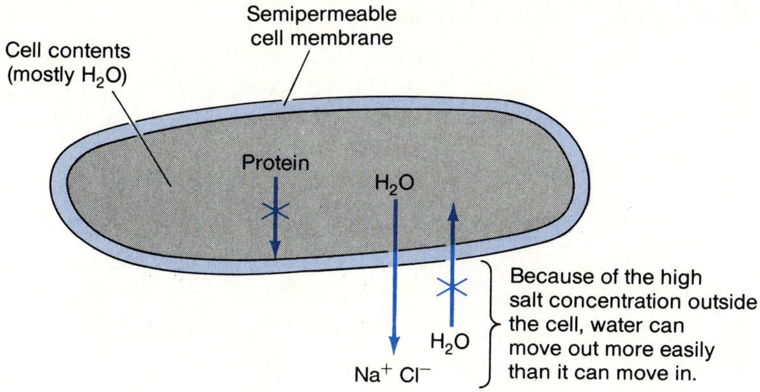

FIGURE 17-7
Osmotic drying of meat by salt.

toughen meat. The water in the tissue cells is drawn out by the salt through a process called *osmosis,* which causes it to be stringy and hard to chew. This is diagramed in Figure 17-7.

The cell is surrounded by a membrane through which certain molecules, including water, can travel; however, larger protein molecules cannot get through the membrane. The membrane is said to be *semipermeable* (permeated by some materials but not by others). A high salt concentration on the outside of the cell lowers the water concentration on the outside. The water from inside the cell can move out, but there is little water on the outside to move back in. This net migration of water out of the cell causes the cell to dry out and its protein to become tough. Thus one should not put salt on steaks until after they have been broiled.

Cooking fish

Another situation in which the principles of chemistry can be demonstrated is the cooking of fish. A "fishy" odor or taste is caused by amines (see Chapter 11). Amines are bases and, hence, will react with acids. In their unreacted state—that is, as amines—they are not very soluble in water and are volatile (they evaporate easily). Fish contains a small amount of amines which evaporate and diffuse into the cooking and eating area, permeating the air with their fishy odor. This can be largely controlled by soaking (marinading) the fish in an acidic substance such as vinegar or lemon juice. The amines react as follows:

$$\underset{\text{Amine}}{R{-}NH_2} + \underset{\text{From acid}}{H^+} \longrightarrow R{-}NH_3^+$$

The newly formed cation is water-soluble and nonvolatile. Since it cannot evaporate, it is not mobile either—it cannot get to your nose or anywhere else in the room. Washing a utensil or your hands with a weak acid will remove fishy odor because the $R{-}NH_3^+$ that results can be rinsed away quite easily.

Preventing fruits from darkening

Darkening of fruits during cooking, canning, or eating is a problem which sometimes faces a culinary artist. This is caused by the exposure of natural material in the fruit (phenols—see Chapter 11) to O_2 in air in the presence of enzymes which are also found in the fruit:

$$\text{Colorless phenols} + O_2 \xrightarrow{\text{enzymes}} \text{dark polymeric phenols}$$

This darkening can be prevented (for purely aesthetic reasons) by keeping O_2 away from the surface of the fruit or by inactivating the enzymes. Keeping oxygen away is best accomplished by totally immersing the fruit in water, especially previously boiled water which has very little oxygen dissolved in it. In preparing Waldorf salad this is accomplished by coating the apples with the dressing, which is an oil-based material such as mayonnaise. A longer-range and more effective approach is the permanent inactivation of the enzymes. Recall from Chapter 15 that enzymes are globular (ball-like) protein molecules. They can function as enzymes only if they retain their globular conformation. Any environment which causes them to unfold will inactivate or *denature* them; and since the forces holding them in the globular shape are rather weak, any one of a number of procedures can accomplish this. These procedures include:

1 Adding vinegar, citric acid, or lemon juice to change the pH of the medium

2 Cooking

3 Adding alcohol

4 Adding salt (an impractical procedure, however, because salt would change the flavor too drastically)

Curing meat

Curing or preserving meat is not very common, now that refrigeration is available, but it can be done as follows. This method works especially well for cuts of pork such as ham and bacon. One prepares a mixture of the following ingredients for 20 kg (40 to 50 lb of pork):

2.3 kg (5 lb) salt (flavor enhancer and preservative)

113 g (¼ lb) pepper (flavor enhancer)

57 g (2 oz) saltpeter (KNO_3) (prevents bacterial growth)

1.1 kg (2½ lb) brown sugar (flavor enhancer)

The salt dehydrates all bacteria by osmosis (in exactly the same way as it causes broiled steak to become tough), while the KNO_3 specifically prevents the growth of the bacteria that cause botulism (acute food poisoning caused by a deadly toxin produced by these bacteria). The other ingredients are primarily for flavor. One rubs this mixture on large pieces of fresh meat such as whole hams every day for several days while keeping the meat cold (outside in the winter) and laying it in the solid curing mixture. After about 6 to 8 weeks the excess material can be brushed off and the pieces hung in a dry place away from animals. This meat will keep indefinitely, but after a year or two it becomes extremely dry, tough, and strong-tasting.

Boiling eggs

Boiling eggs is a seemingly simple task that sometimes becomes annoyingly difficult. An egg is basically a fat-containing center (yolk) surrounded by a viscous water solution of globular protein (white) encased in a $CaCO_3$ shell. There is also some air in the egg, and the shell is permeable to air. The main problem with cooking eggs in their shells is cracking. Cracking is caused by putting cold eggs in hot water, which makes the shell crack from thermal shock just as many hard objects do. Cracking is also caused by the air inside the egg expanding faster than it can escape through the shell. Hot gases exert a higher pressure on a container than cold gases. Both these problems can largely be solved by initially placing the eggs in cold or room-temperature water. This allows the shell to change temperature gradually as the water heats up, and also gives the air time to escape. A little salt added to the water not only helps raise the boiling temperature of the water (which cooks the egg faster; see Chapter 9) but also aids in preventing any uncooked white from escaping into the water if the shell should crack. The salt causes the soluble globular protein in the egg white to extend to a linear form, which is an insoluble solid—the same insoluble linear form that results from the cooking process itself. This seals the crack and prevents further leaking:

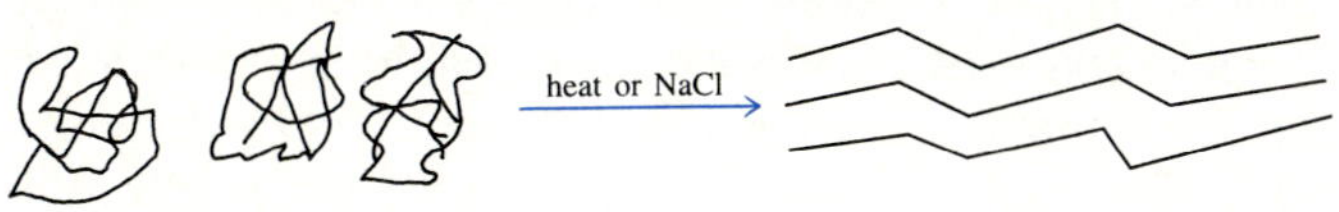

Soluble globular protein molecules (uncooked egg white)

Insoluble denatured protein (cooked egg white)

Leavening agents

The final area we will consider is leavening agents—those materials that cause a given food to be light and fluffy. These fall into several categories:

1 Chemicals that decompose to form a gas

2 Chemicals that react with each other to form a gas

3 H_2O that turns to steam under cooking conditions

4 Preformed foam which has air trapped in it

5 Formation of microbial CO_2

An example of *type 1* is baking soda—sodium bicarbonate ($NaHCO_3$). On heating the following reaction takes place:

$$2NaHCO_3 \xrightarrow{\text{heat}} Na_2CO_3 + CO_2 + H_2O$$

The CO_2 forms bubbles which lighten the pastry being prepared. Unfortunately, the Na_2CO_3, which is also formed, is quite basic and imparts a bitter taste to the food unless there is a lot of sugar to counteract it.

Type 2 is one of the most commonly used. An example is commercial baking powder, which is composed of $NaHCO_3$ and a dry solid acid such as tartaric acid. While the powder is in dry form, no reaction occurs between these components; but when the powder is mixed with liquid ingredients, the acid dissolves, producing a liquid with a low pH. The following reaction occurs:

$$NaHCO_3 + \text{tartaric acid} \longrightarrow CO_2 + \text{sodium tartrate} + H_2O$$

The $NaHCO_3$ and vinegar that are used in combination for making pancakes produce the leavening agent CO_2 gas by the same type of reaction. The acetic acid in the vinegar reacts with the baking soda as follows:

$$HC_2H_3O_2 + NaHCO_3 \longrightarrow CO_2 + NaC_2H_3O_2 + H_2O$$

In making buttermilk biscuits or pancakes, the lactic acid in the sour milk or buttermilk accomplishes the same task as the vinegar. In all these cases the by-product (sodium tartrate, sodium acetate, etc.) does *not* have a bitter taste.

An example of *type 3*—water as a leavening agent—is the addition of a small amount of water to scrambled eggs. Since water is present in most food preparations, it aids the other agents.

An example of *type 4*—preformed foam—is the use of whipped egg whites in omelets and waffles. When the egg whites are separated from the yolks, the globular proteins in the egg whites can be partly denatured by the physical force of beating. Beating also causes air to be incorporated into the denatured mixture, producing a foam. Since the yolk contains a lot of lipid material (including cholesterol) which acts as a defoaming agent, it is necessary to remove the yolk to get a good foam. Once the foam is formed, it can be mixed with the yolk to form a very light cooked egg commonly called an *omelet*. Likewise, the foam can be mixed with what is essentially a pancake batter to make a fluffy batter suitable for waffles.

Finally, *type 5*—the most complex leavening method—involves the use of yeast, a microorganism which contains an enzyme capable of converting starch and simple sugars, such as glucose, into CO_2 (this enzyme-catalyzed conversion is called *fermentation*):

$$\text{Glucose} \xrightarrow[\text{}]{\text{yeast enzymes}} CO_2 + \underset{\text{Ethyl alcohol}}{CH_3{-}CH_2OH}$$

This is the same process that is used to produce alcoholic beverages (see Chapter 16). Here, however, the CO_2 is a leavening agent, and the process is used to make ordinary breads. The alcohol is lost by evaporation during cooking. The procedure requires some skill, since the process is slow and the correct temperature must be maintained for several hours. If the temperature is too low, the reaction will not occur; if the temperature is too high, the heat will denature the enzymes (''kill the yeast'') and prevent fermentation.

We have examined several selected aspects of the chemistry of cooking. Most materials, other than flavorings, are used because of their chemical nature, and both mixing and cooking involve chemical processes. Some of these—such as gel formation in custards, gelatins, and gravies—are complex and are not well understood. Some, such as cooking green vegetables without a lid on the pot to allow acidic components to evaporate (to preserve color), are quite simple. Various acidic and basic materials are added to produce the correct pH so that a reaction can occur (or be prevented) or to impart a palatable color to the food. It is possible for people who are not professional chemists to become quite knowledgeable about basic cooking chemistry, which should aid them in improving their cooking skills.

IN CONCLUSION: A NOTE ON INGREDIENTS OF HOUSEHOLD PRODUCTS

Most household products—like the bottle of shampoo shown in Figure 17-8—have their ingredients listed on the label. Such lists of ingredients are often helpful to consumers because they make it possible to compare products. The discussions of properties and processes in this chapter can also be useful as background for comparing products.

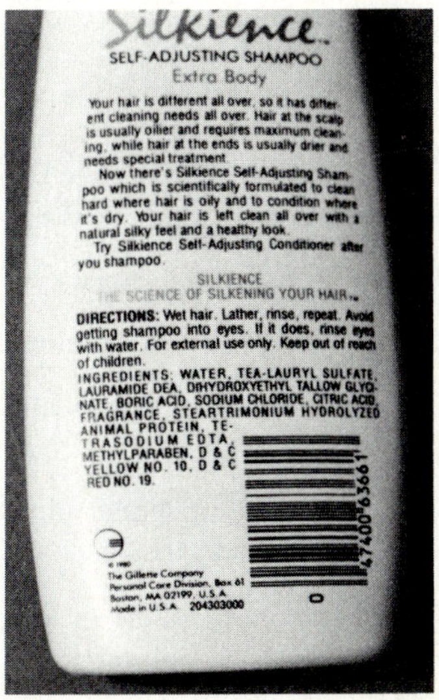

FIGURE 17-8
Most household products—like this shampoo—have their ingredients listed on the label. The consumer can sometimes use such lists to compare products.

STUDY GUIDE

Terms to learn

Biodegradable	Glycerin
Bleach	Ligand
Borax	Micelle
Chelation	Optical brightener
CMC	Polyamide
Colloidal suspension	Polyester
Coordination complex	Sodium stearate
Copolymer	Sodium triphosphate
Detergent	Surfactant
EDTA	T_g
Fluorescent dye	Tyndall effect
Free radical	

Skill practice problems

1 What is the formula for a typical animal grease?

2 What is the formula for soap?

3 Explain how soaps and detergents aid in removing dirt.

4 What simple chemical test can be done in the home to determine if peroxide is still good?

5 What is meant by the term *hard water?*

6 Why does hard water present a problem in the use of soap?

7 What is a ''quat''? What are two uses for ''quats''?

8 Why do white clothes ''glow'' in ''black light''?

9 How do enzymes remove food stains?

10 What is the active chemical in household bleach?

11 If a household uses 115 lb of laundry detergent each year, how many pounds of sodium triphosphate are put into the waterways in that year? (Assume 2% sodium triphosphate.) How much would a city of 100,000 families put into its river?

12 What would you use to remove each of the following materials or stains?

(a) Lipstick on a yellow shirt collar
(b) Strawberry stains on a white shirt
(c) Cola on a white dress
(d) Rust stains on a white rug
(e) Bluberry stains on a green dress
(f) Shellac spilled on a rug and allowed to dry
(g) Blood on a white handkerchief
(h) Grass stains on the knees of a child's good red pants
(i) Coffee stains on a white bikini
(j) Mildew on a bedspread
(k) Chewing gum stuck to a child's stomach
(l) Tar tracked across a sidewalk by someone who had walked on a newly paved driveway
(m) Paint that has dried on a rug

13 How many grams of O_2 are given off if a 200-g bottle of 3% H_2O_2 is reacted with NaClO? (See Appendix E for the method to use.)

14 What was the source of base for making soap on the American frontier?

15 Why should toilet bowl cleaners not be used with NaClO bleaches?

16 What material chelates and dissolves calcium stearate (bathtub ring)?

17 Which shampoos listed in Table 17-1 will burn eyes?

18 Which of the following pairs of materials form hydrogen bonds with each other?

(a) HF and C_2H_6
(b) $CH_3—O—H$ and NH_3
(c) $CH_3—O—CH_3$ and CH_4
(d) H_2O and $CH_3—CH_2—O—H$

19 What class of biomolecules does hair belong to?

20 Why do wool and hair stretch when wet?

21 Draw the structure of nylon-4,4.

22 Why does polyester fabric not wrinkle?

23 List 4 basic ways flames can be extinguished or prevented.

24 List at least 2 ways $NaHCO_3$ can be used in the home.

25 Why does lemon juice remove the odor of fish?

26 Why does citric acid prevent darkening of fruit? What other processes or substances accomplish the same task?

27 What is the purpose of tartaric acid in baking powder?

Appendix

A

METRIC AND UNITED STATES MEASUREMENTS

METRIC PREFIXES

WORD	NUMERICAL MEANING	SYMBOL
giga	10^9	G
mega	10^6	M
kilo	$1000 = 10^3$	k
deca	10	da
deci	0.1	d
centi	0.01	c
milli	$0.001 = 10^{-3}$	m
micro	10^{-6}	μ*
nano	10^{-9}	n

ABBREVIATIONS

Metric

meter	m
gram	g
liter	L
tonne	t (also called *metric* or *long ton*)

United States

foot	ft
inch	in
yard	yd
mile	mi
ounce	oz
fluid ounce	fl oz
gallon	gal
pint	pt
quart	qt
tablespoon	tblsp
teaspoon	tsp
ton	ton (also called a *short* or *English ton*)
pound	lb

* Greek letter mu.

CONVERSION TABLES

Length

United States–metric conversions

1 m = 1.09 yd = 39.37 in
1 yd = 0.91 m
1 in = 2.54 cm
1 km = 0.621 mi
1 ft = 30.5 cm

Conversions within the United States system

1 ft = 12 in
1 yd = 36 in
1 mi = 5280 ft

Volume

United States–metric conversions

1 L = 1.06 qt
1 qt = 0.946 L
1 gal = 3.76 L
1 fl oz = 29.6 mL
1 cup = 235 mL
1 tsp = 5 mL

Conversions within the United States system

1 gal = 4 qt
1 qt = 2 pt
1 pt = 2 cup
1 cup = 8 fl oz
1 pt = 16 fl oz
1 qt = 32 fl oz
1 fl oz = 6 tsp
1 tblsp = 3 tsp

Mass (weight)

United States–metric conversions

1 lb = 454 g
1 kg = 2.20 lb
1 oz = 28.5 g
1 ton = 0.9 t
1 t = 1.1 ton

Conversions within the United States system

1 ton = 2000 lb
1 lb = 16 oz

Conversions within the metric system

1 tonne = 1000 kg

Appendix

B

OPERATIONS USING SCIENTIFIC NOTATION

Often, we must not only express large and small numbers in scientific notation but also work with these numbers. This appendix examines multiplication and division.

MULTIPLICATION AND DIVISION

When multiplying numbers expressed in scientific notation, one *multiplies* the *decimal* portions in the usual way and *adds* the *exponents*. For example:

$$(2.4 \times 10^6)(2.1 \times 10^4)$$

Multiplying decimal numbers:

$$(2.4)(2.1) = 5.04$$

Adding exponents:

$$6 + 4 = 10$$

The answer is 5.04×10^{10}.

When dividing, one *divides* the *decimal* portion in the usual way and *subtracts* the *exponent* of the divisor (denominator of the fraction) from the exponent of the dividend (numerator of the fraction). For example:

$$\frac{2.7 \times 10^6}{1.3 \times 10^4}$$

Dividing decimal numbers:

$$2.7 \div 1.3 = 2.08$$

Subtracting exponents:

$$6 - 4 = 2$$

The answer is 2.08×10^2.

The following problems will allow you to master these operations. Even if you have a pocket calculator that handles exponential numbers, it is of some value to you to be able to perform these operations by hand.

MULTIPLYING AND DIVIDING EXPONENTIAL NUMBERS

Worked example

$$(2.71 \times 10^5)(1.3 \times 10^{-2})$$

1 Multiply 2.71×1.3:

$$2.71 \times 1.3 = 3.5$$

2 Add 5 and -2:

$$5 + (-2) = 5 - 2 = 3$$

Note The addition is *algebraic*.

3 The answer is 3.5×10^3.

Worked example

$$(1.5 \times 10^{-6})(3.1 \times 10^5)$$

1 $1.5 \times 3.1 = 4.6$

2 $-6 + 5 = -1$

3 The answer is 4.6×10^{-1}.

Worked example

$$(1.7 \times 10^8)(4)$$

1 $1.7 \times 4 = 6.8$

2 Since there is no exponential part to the second number, we simply use the 10^8 in our answer. Alternatively we could write $4 = 4 \times 10^0$. Then the exponent of the answer would be $8 + 0 = 8$.

3 The answer is 6.8×10^8.

Worked example

$$\frac{2.3 \times 10^8}{1.6 \times 10^{-6}}$$

1 Divide 2.3 by 1.6:

$$\frac{2.3}{1.6} = 1.4$$

2 Subtract -6 from 8:

$$8 - (-6) = 8 + 6 = 14$$

3 The answer is 1.4×10^{14}.

Worked example

$$\frac{2.3}{6.02 \times 10^{23}}$$

1 Divide 2.3 by 6.02:

$$\frac{2.3}{6.02} = 0.38$$

2 Since the numerator has no exponential term, simply change the sign of the exponent in the denominator:

$$\frac{1}{10^{23}} = 10^{-23}$$

Alternatively: $1 = 1 \times 10^{0}$, and so the exponent of the answer is $0 - 23 = -23$.

3 The answer is 0.38×10^{-23} or more properly 3.8×10^{-24}.

Note When moving the decimal to the right, lower the exponent (more negative); when moving the decimal to the left, raise the exponent (more positive).

Problem

$$(1.7 \times 10^{-8})(2.7 \times 10^{-6})$$

Solution

1 $1.7(2.7) = 4.6$

2 $-8 + (-6) = -14$

3 The answer is 4.6×10^{-14}.

Problem

$$(1.63 \times 10^{8})(8)$$

Solution

1 $1.63(8) = 13.0$

2 Since there is no exponent in the second number, write 13.0×10^{8}.

3 That is expressed more properly as 1.30×10^{9}.

Problem

$$\frac{3.61 \times 10^{-8}}{9.6 \times 10^{-2}}$$

Solution

1 $\frac{3.61}{9.6} = 0.38$

2 $-8 - (-2) = -8 + 2 = -6$

Note Remember that the subtraction is *algebraic*.

3 The answer is $0.38 \times 10^{-6} = 3.8 \times 10^{-7}$.

Problem

$$\frac{8.73 \times 10^{16}}{4}$$

Solution

1 $\frac{8.73}{4} = 2.18$

2 Since the denominator has no exponent, write 2.18×10^{16}.

Problems

a $\frac{1.73 \times 10^{8}}{8.6 \times 10^{6}}$

b $\frac{1.76 \times 10^{-5}}{2.7 \times 10^{-3}}$

c $\frac{4.81 \times 10^{-23}}{6}$

d $\frac{9.2}{6.02 \times 10^{21}}$

Answers

a $2.0 \times 10^{1} \times 20$

b 6.5×10^{-3}

c 8.02×10^{-24}

d 1.5×10^{-21}

Appendix

C

COMPLEX CONVERSIONS

ONE-STEP CONVERSIONS USING UNITS WITH EXPONENTS

In this type of problem units are expressed to powers. The general method is the same as in Chapter 4.

Worked example

Convert 25 cm^3 to cubic inches.

Note The term cm^3 means cubic centimeters and the term in^3 means cubic inches. Both are volume terms.

1 Write 25 cm^3 (original data).

$$25\ \mathrm{cm}^3$$

2 Construct parentheses.

$$25\ \mathrm{cm}^3 \left(\frac{\qquad}{\qquad}\right)$$

3 Place cm^3 in the lower right of the parentheses (i.e., to cancel cm^3 in the original data).

$$25\ \cancel{\mathrm{cm}^3} \left(\frac{\qquad}{\qquad \cancel{\mathrm{cm}^3}}\right)$$

4 Place in^3 (desired units) in the upper right.

$$25\ \cancel{\mathrm{cm}^3} \left(\frac{\qquad \mathrm{in}^3}{\qquad \cancel{\mathrm{cm}^3}}\right)$$

Note If one unit is raised to a power for purposes for cancellation, the other unit used in the conversion factor must be raised to that same power.

5 Place the conversion factors in the proper places.

$$25\ \cancel{\mathrm{cm}^3} \left(\frac{1\ \ \mathrm{in}^3}{2.54\ \cancel{\mathrm{cm}^3}}\right)$$

6 Since the units are cubed, the numbers (both of them) must also be cubed.

$$25\ \cancel{\mathrm{cm}^3} \left(\frac{(1)^3\ \ \mathrm{in}^3}{(2.54)^3\ \cancel{\mathrm{cm}^3}}\right)$$

7 Do the arithmetic:

(a) $25 \times 1 = 25$

(b) $(2.54)^3 = 2.54 \times 2.54 \times 2.54 = 16.4$

(c) $\dfrac{25}{16.4} = 1.52$

Note This calculation can also be done on a calculator. Refer to your calculator manual or your instructor for the method to use on your calculator if you are having difficulty performing the arithmetic.

8 Place the uncanceled units after 1.52.

$$1.52 \text{ in}^3$$

9 Summary:

$$25 \cancel{\text{cm}^3} \left(\frac{(1)^3 \quad \text{in}^3}{(2.54)^3 \quad \cancel{\text{cm}^3}} \right) = 1.52 \text{ in}^3$$

Worked example

How many square centimeters (cm^2) are in a floor whose area is 14,400 in^2?

1 Write 14,400 in^2.

$$14{,}400 \text{ in}^2$$

2 Construct parentheses.

$$14{,}400 \text{ in}^2 \left(\frac{\qquad}{\qquad} \right)$$

3 Place in^2 in the lower right and cm^2 in the upper right of the parentheses.

$$14{,}400 \cancel{\text{in}^2} \left(\frac{\text{cm}^2}{\cancel{\text{in}^2}} \right)$$

4 Place the conversion numbers in the proper places.

$$14{,}400 \cancel{\text{in}^2} \left(\frac{2.54 \quad \text{cm}^2}{1 \quad \cancel{\text{in}^2}} \right)$$

5 Square both conversion numbers, since their units are squared.

$$14{,}400 \cancel{\text{in}^2} \left(\frac{(2.54)^2 \quad \text{cm}^2}{(1)^2 \quad \cancel{\text{in}^2}} \right)$$

6 Do the arithmetic and use the uncanceled unit.

$$14{,}400 \cancel{\text{in}^2} \left(\frac{(2.54)^2 \quad \text{cm}^2}{(1)^2 \quad \cancel{\text{in}^2}} \right) = 92{,}900 \text{ cm}^2$$

Problem

How many square inches (in^2) are in the bottom of a box whose measurements are 12 cm wide and 8 cm long?

Solution: Method 1

1 Calculate the area of the box bottom by multiplying width times length.

$$12 \text{ cm} \times 8 \text{ cm} = 96 \text{ cm}^2$$

Note Since the numbers are multiplied, the units also have to be multiplied (dimensional analysis).

$$\text{cm} \times \text{cm} = \text{cm}^2$$

2 Write 96 cm^2.

$$96 \text{ cm}^2$$

3 Construct parentheses.

$$96 \text{ cm}^2 \left(\frac{\quad}{\quad}\right)$$

4 Place cm^2 on the bottom and in^2 on the top.

$$96 \cancel{\text{cm}^2} \left(\frac{\text{in}^2}{\cancel{\text{cm}^2}}\right)$$

5 Place the conversion numbers in the proper place and *square* them.

$$96 \cancel{\text{cm}^2} \left(\frac{(1)^2 \quad \text{in}^2}{(2.54)^2 \quad \cancel{\text{cm}^2}}\right)$$

Note Since the units are squared for purposes of cancellation, the numbers also have to be squared.

6 Work the arithmetic.

$$96 \cancel{\text{cm}^2} \left(\frac{(1)^2 \quad \text{in}^2}{(2.54)^2 \quad \cancel{\text{cm}^2}}\right) = 14.9 \text{ in}^2$$

Solution: Method 2

1 Convert each dimension from centimeters to inches.

$$12 \cancel{\text{cm}} \left(\frac{1 \quad \text{in}}{2.54 \quad \cancel{\text{cm}}}\right) = 4.72 \text{ in}$$

$$8 \cancel{\text{cm}} \left(\frac{1 \quad \text{in}}{2.54 \quad \cancel{\text{cm}}}\right) = 3.15 \text{ in}$$

2 Calculate the area by multiplying the width (in inches) by the length (in inches).

$$(4.72\ \text{in})(3.15\ \text{in}) = 14.9\ \text{in}^2$$

Note Although method 2 is simpler and is perfectly valid, method 1 should be mastered, since the technique used in method 1 is used in showing how to solve example problems in this book.

Problem

Convert 15yd^3 to cubic feet.

Solution

1 Write 15 yd^3.

$$15\ \text{yd}^3$$

2 Construct parentheses.

$$15\ \text{yd}^3\left(\frac{\quad}{\quad}\right)$$

3 Place yd^3 on the bottom and ft^3 on the top.

$$15\ \cancel{\text{yd}^3}\left(\frac{\text{ft}^3}{\cancel{\text{yd}^3}}\right)$$

4 Place the conversion numbers with the units and cube the conversion numbers.

$$15\ \cancel{\text{yd}^3}\left(\frac{(3)^3\ \text{ft}^3}{(1)^3\ \cancel{\text{yd}^3}}\right)$$

5 Work the arithmetic.

$$15\ \cancel{\text{yd}^3}\left(\frac{(3)^3\ \text{ft}^3}{(1)^3\ \cancel{\text{yd}^3}}\right) = 405\ \text{ft}^3$$

Problems

a Convert 587 in^3 to cubic centimeters.

b Convert 173 cm^3 to cubic inches.

c Convert 81 ft^2 to square yards.

d If a slab of concrete is to be poured which is 15 ft by 18 ft by 2 ft, how many cubic yards of concrete must be ordered from the company?

Answers

a

$$587\ \cancel{in^3}\left(\frac{(2.54)^3\ cm^3}{(1)^3\ \ \cancel{in^3}}\right) = 9619\ cm^3$$

b

$$173\ \cancel{cm^3}\left(\frac{(1)^3\ \ in^2}{(2.54)^3\ \cancel{cm^3}}\right) = 10.5\ in^3$$

c

$$81\ \cancel{ft^2}\left(\frac{(1)^2\ yd^2}{(3)^2\ \cancel{ft^2}}\right) = 9\ yd^2$$

d

$$18\ ft \times 15\ ft \times 2\ ft = 540\ ft^3$$

$$540\ \cancel{ft^3}\left(\frac{(1)^3\ yd^3}{(3)^3\ \cancel{ft^3}}\right) = 20\ yd^3$$

Problem

Convert 10 yd^3 to cubic centimeters.

Solution

1 Write 10 yd^3.

$$10\ yd^3$$

2 Construct parentheses with yd^3 on the bottom and in^3 on the top.

$$10\ \cancel{yd^3}\left(\frac{in^3}{\cancel{yd^3}}\right)$$

Note We could have put ft^3 on the top and then converted to cubic inches by another step.

$$10\ \cancel{yd^3}\left(\frac{\cancel{ft^3}}{\cancel{yd^3}}\right)\left(\frac{in^3}{\cancel{ft^3}}\right)$$

3 Place the conversion numbers with the units and raise these numbers to the third power (cube them), since the units were cubed for purposes of cancellation.

$$10\ \cancel{yd^3}\left(\frac{(36)^3\ \ in^3}{(1)^3\ \ \cancel{yd^3}}\right)$$

Or

$$10\ \cancel{yd^3}\left(\frac{(3)^3\ \ \cancel{ft^3}}{(1)^3\ \ \cancel{yd^3}}\right)\left(\frac{(12)^3\ \ in^3}{(1)^3\ \ \cancel{ft^3}}\right)$$

4 Construct another parentheses and place cm^3 and in^3 in the correct relationship and place the conversion numbers in the correct places.

$$10\ \cancel{yd^3}\left(\frac{(36)^3\ \cancel{in^3}}{(1)^3\ \cancel{yd^3}}\right)\left(\frac{(2.54)^3\ cm^3}{(1)^3\ \cancel{in^3}}\right)$$

5 Summary and answer:

$$10\ \cancel{yd^3}\left(\frac{(36)^3\ \cancel{in^3}}{(1)^3\ \cancel{yd^3}}\right)\left(\frac{(2.54)^3\ cm^3}{(1)^3\ \cancel{in^3}}\right) = 7{,}650{,}000\ cm^3$$

Or

$$7.650 \times 10^6\ cm^3$$

COMPLEX CONVERSIONS

In this type of problem, the original data are given in terms of two (or more) units and one unit or more has to be changed.

Worked example

Convert 15 g/cm^3 to pounds per cubic foot.

Note The approach used here will be to construct conversion factors to change grams to pounds and then to use other factors to change cubic centimeters to cubic feet.

1 Write 15 g/cm^3.

$$15\ g/cm^3$$

2 Construct parentheses.

$$15\ g/cm^3\left(\frac{\quad}{\quad}\right)$$

3 *First,* grams will be changed to pounds. In the parentheses place g in a canceling position (i.e., in the bottom of the fraction) and lb on top. Place the correct conversion numbers in their places.

$$15\ \cancel{g}/cm^3\left(\frac{1\ \ lb}{454\ \ \cancel{g}}\right)$$

4 Second, cubic centimeters will be changed to cubic feet. Construct another parentheses with cm^3 in a canceling position and in^3 in the remaining position below it.

$$15\ \cancel{g}/\cancel{cm^3}\left(\frac{1\ \ lb}{454\ \ \cancel{g}}\right)\left(\frac{\cancel{cm^3}}{in^3}\right)$$

Note Since cm^3 was on the bottom of the unit fraction in the original data, it has to go on top in the conversion factor in order to cancel. Since cm^3 has to go in the numerator, in^3 has to be placed in the denominator.

5 Place the conversion numbers in the proper places and cube the conversion numbers, since the units are cubed.

$$15\ \cancel{g}/\cancel{cm}^3 \left(\frac{1\ \text{lb}}{454\ \cancel{g}}\right)\left(\frac{(2.54)^3\ \cancel{cm^3}}{(1)^3\ \text{in}^3}\right)$$

6 Repeat steps 4 and 5 to convert cubic inches to cubic feet.

$$15\ \cancel{g}/\cancel{cm^3} \left(\frac{1\ \text{lb}}{454\ \cancel{g}}\right)\left(\frac{(2.54)^3\ \cancel{cm^3}}{(1)^3\ \cancel{in^3}}\right)\left(\frac{(12)^3\ \cancel{in^3}}{(1)^3\ \text{ft}^3}\right)$$

Note Again, since in^3 is to be changed and is on the bottom of the previous conversion factor, it has to go on top of the new conversion factor.

7 Obtain the answer by working the arithmetic:

$$\frac{15 \times 2.54^3 \times 12^3}{454} = 936$$

8 Summary:

$$15\ \cancel{g}/\cancel{cm^3} \left(\frac{1\ \text{lb}}{454\ \cancel{g}}\right)\left(\frac{(2.54)^3\ \cancel{cm^3}}{(1)^3\ \cancel{in^3}}\right)\left(\frac{(12)^3\ \cancel{in^3}}{(1)^3\ \text{ft}^3}\right) = 936\ \text{lb/ft}^3$$

Note Since both lb and ft^3 remain uncanceled, they both must appear in the answer in their correct positions (i.e., lb on the top and ft^3 on the bottom). This is read "pounds per cubic foot," which means that 1 ft^3 of this material weighs 936 lb.

Worked example

If an automobile is traveling 90 km/h, how fast is it going in miles per hour?

1 Write 90 km/h.

$$90\ \text{km/h}$$

2 Construct the conversion factor with the correct numbers and units so that kilometers will be changed to miles.

$$90\ \cancel{km}/\text{h}\left(\frac{0.621\ \text{mi}}{1\ \cancel{km}}\right)$$

Or

$$90\ \cancel{km}/\text{h}\left(\frac{1000\ \cancel{m}}{1\ \cancel{km}}\right)\left(\frac{100\ \cancel{cm}}{1\ \cancel{m}}\right)\left(\frac{1\ \cancel{in}}{2.54\ \cancel{cm}}\right)\left(\frac{1\ \cancel{ft}}{12\ \cancel{in}}\right)\left(\frac{1\ \text{mi}}{5280\ \cancel{ft}}\right)$$

Note In this problem, the unit hours should not be changed, since both the original data and the desired answer have hours in the denominator.

3 Obtain the numerical answer by:

$$90 \times (0.621) = 56$$

Or

$$\frac{90 \times 1000 \times 100}{2.54 \times 12 \times 5280} = 56$$

4 Summary:

$$90 \not{km}/h \left(\frac{0.621 \text{ mi}}{1 \not{km}}\right) = 56 \text{ mi/h}$$

Problem

Convert 20 mi/h to meters per second.

Note The abbreviation for miles is mi, and the abbreviation for meters is m.

Solution

1 Write 20 mi/h.

$$20 \text{ mi/h}$$

2 Construct the conversion factors to change miles to meters.

$$20 \not{mi}/h \left(\frac{1 \not{km}}{0.621 \not{mi}}\right)\left(\frac{1000 \text{ m}}{1 \not{km}}\right)$$

3 Construct the conversion factors to change hours to seconds.

$$20 \text{ mi/h} \left(\frac{1 \text{ km}}{0.621 \text{ mi}}\right)\left(\frac{1000 \text{ m}}{1 \text{ km}}\right)\left(\frac{1 \text{ h}}{60 \text{ min}}\right)\left(\frac{1 \text{ min}}{60 \text{ s}}\right)$$

4 Obtain the numerical answer by:

$$\frac{20 \times 1000}{0.621 \times 60 \times 60} \times 8.95$$

5 Answer:

$$8.95 \text{ m/s}$$

Problems

a If there are 100,000 people in each square mile of a city, how many toes (on the average) are in each square yard?

b Convert 200 lb/yd^3 to grams per cubic centimeter.

c Convert 50 lb/gal to grams per milliliter.

d If there are 12 people in a room and each one exhales 2.6 g of CO_2 each minute, how many grams of CO_2 are put into the room in 3 h?

Note This is not just a conversion, but the dimensional analysis method can be used here. Give it a try!

Answers

a

$$100{,}000\ \cancel{\text{people}}/\cancel{\text{mi}^2}\left(\frac{10\ \text{toes}}{1\ \cancel{\text{person}}}\right)\left(\frac{(1)^2\ \cancel{\text{mi}^2}}{(5280)^2\ \cancel{\text{ft}^2}}\right)\left(\frac{(3)^2\ \cancel{\text{ft}^2}}{(1)^2\ \text{yd}^2}\right)$$

$$= 0.32\ \text{toes per square yard}$$

b

$$200\ \cancel{\text{lb}}/\cancel{\text{yd}^3}\left(\frac{454\ \text{g}}{1\ \cancel{\text{lb}}}\right)\left(\frac{(1)^3\ \cancel{\text{yd}^3}}{(36)^3\ \cancel{\text{in}^3}}\right)\left(\frac{(1)^3\ \cancel{\text{in}^3}}{(2.54)^3\ \text{cm}^3}\right)$$

$$= 0.119\ \text{g/cm}^3$$

c

$$50\ \cancel{\text{lb}}/\cancel{\text{gal}}\left(\frac{454\ \text{g}}{1\ \cancel{\text{lb}}}\right)\left(\frac{1\ \cancel{\text{gal}}}{4\ \cancel{\text{qt}}}\right)\left(\frac{1\ \cancel{\text{qt}}}{0.946\ \cancel{\text{L}}}\right)\left(\frac{1\ \cancel{\text{L}}}{1000\ \text{mL}}\right)$$

$$= 6.0\ \text{g/mL}$$

d

$$12\ \cancel{\text{people}}\ (2.6\ \text{g}/\cancel{\text{person}}\cdot\cancel{\text{min}})^*\left(\frac{60\ \cancel{\text{min}}}{1\ \cancel{\text{h}}}\right)(3\ \cancel{\text{h}})$$

$$= 5616\ \text{g or about } 5600\ \text{g}$$

Note This problem can be done in one step if the direct relationships are known or in several steps if necessary.

*This is read "2.6 g per person per minute."

Appendix

D

ELECTRONIC STRUCTURE OF ATOMS: FURTHER STUDIES

QUANTUM MECHANICAL IDEAS

Early in the twentieth century a powerful new theory about the electronic structure of atoms was proposed (see Chapter 6). It was reasoned that if light, which was usually most easily visualized as a wave, could also behave as a particle, then electrons, which were conceived as particles, might also behave like waves. This idea is called the *wave-particle duality theory* of the electron. A mathematical theory called *quantum mechanics* was developed which allowed one to calculate the locations of waves in space. This theory assumes that the electrons are waves traveling within the atom, and by using methods beyond the scope of this text, one can determine the *shapes* and *sizes* of all the possible areas where an electron might be found within the atom. The areas are *not* two-dimensional lines along which the electrons *travel* as the orbits in the Bohr view of the atom are. They are three-dimensional spaces in which there is a high probability of finding an electron. These spaces are called *orbitals*. In this appendix the various shapes and sizes that have been calculated and the standard ways of labeling the orbitals will be discussed, but how the shapes and sizes of the orbitals were actually determined will not be covered. Figure D-1 shows the reasoning that led to the *quantum mechanical* model of the atom, which is now accepted universally. Because the evidence is so overwhelming and the results of these calculations account for various chemical properties, this model is considered to accurately describe the atom.

ATOMIC ORBITALS

In this section the electronic atomic orbitals will be described by means of labels called *quantum numbers*. There are three such labels for each orbital and one label to describe the electron in the orbital. At the end of this discussion, we will have a three-dimensional "road map" of all the orbits where electrons may be found. The quantum numbers are number and letter labels similar to street addresses or highway numbers, not mathematical quantities to be added, subtracted, etc.

The first is called the *principal quantum number*, is symbolized by n, and designates the *size* (diameter) of the orbital. This number can be any positive

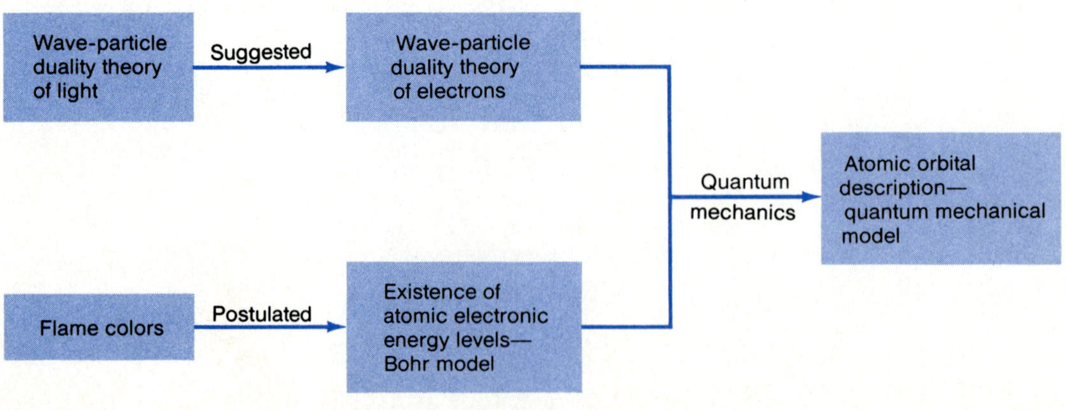

FIGURE D-1

Flow of reasoning leading to the quantum mechanical model.

ℓ (quantum number label)	Rough description	Drawing	Letter label
0	Spherical		*s*
1	Dumbbell		*p*
2	Cloverleaf, etc.		*d*
3	Complex	——	*f*

FIGURE D-2
Shape of atomic orbitals. (The • shows the location of the nucleus.)

whole number; but in real atoms, as they normally exist on earth, n can be any value from 1 to 7 (i.e., there are seven possible sizes for orbitals). When $n = 1$, it means that the orbital being described has the smallest possible size. When $n = 2$, it has the second possible size; etc.

The second quantum number is called the *subsidiary number,* is represented by l, and indicates the *shape* of the orbital. There are four possible shapes, each with a number label and a more commonly used equivalent letter label; see Figure D-2. Thus a 1*s* orbital is a spherical-shaped orbital of the smallest possible size, and a 3*d* orbital is a cloverleaf-shaped orbital of the third possible size.

The third quantum number (m_l) indicates the orientation in space of an orbital. Although it will not be discussed in detail, one should know that there is only one orientation for a spherical (*s*) orbital, and that there are three for a dumbbell (*p*) orbital, five for a *d* orbital, and seven for an *f* orbital. These are shown in Figure D-3. The dumbbell-shaped orbitals, for example, can be either up and down, sideways, or in and out. The number labels for these will not be given, but you should understand the spacial concepts. The following is a complete list of all the orbitals:

First level	1*s*
Second level	2*s*2*p*2*p* 2*p*
Third level	3*s*3*p*3*p*3*p*3*d*3*d*3*d*3*d*3*d*
Fourth level	4*s*4*p*4*p*4*p*4*d*4*d*4*d*4*d*4*d*4*f*4*f*4*f*4*f*4*f*4*f*4*f*
Fifth level	5*s*5*p*5*p*5*p*5*d*5*d*5*d*5*d*5*d*5*f*5*f*5*f*5*f*5*f*5*f*5*f*
Sixth level	6*s*6*p*6*p*6*p*6*d*6*d*6*d*6*d*6*d*6*f*6*f*6*f*6*f*6*f*6*f*6*f*
Seventh level	7*s*7*p*7*p*7*p*7*d*7*d*7*d*7*d*7*d*7*f*7*f*7*f*7*f*7*f*7*f*7*f*

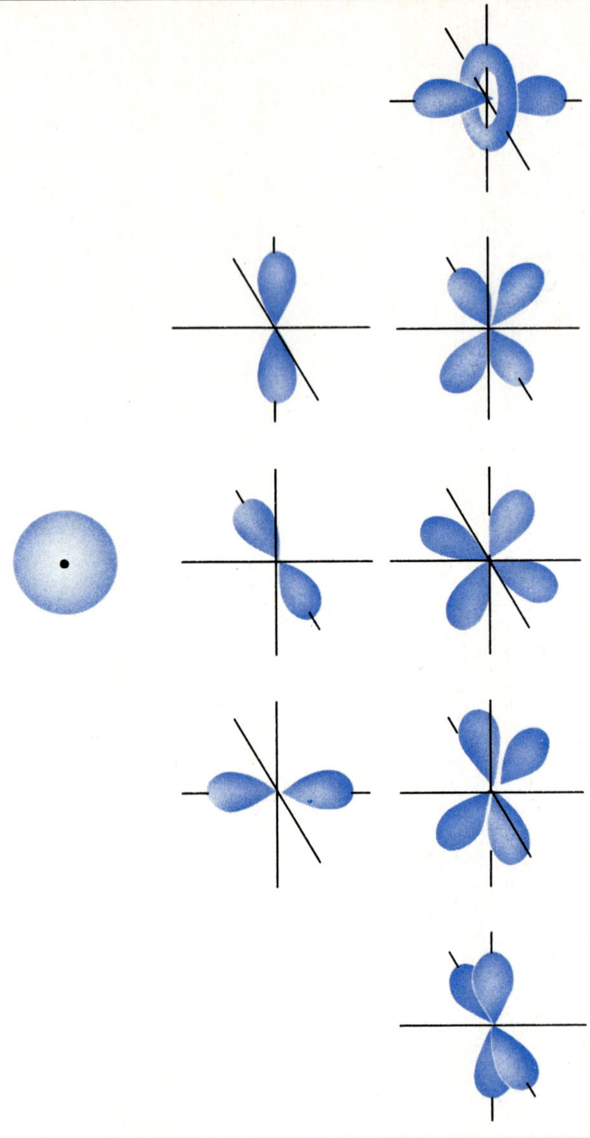

ℓ	Letter designation	Number of orientations	Drawing
0	*s*	one	
1	*p*	three	
2	*d*	five	
3	*f*	seven	—

FIGURE D-3

Orientation of atomic orbitals. (The • shows the location of the nucleus.)

Writing these in sequence, one obtains:

$1s2s2p2p2p3s3p3p3p3d3d3d3d3d4s$
$4p4p4p4d4d4d4d4d4f4f4f4f4f4f4f$, etc.

Notice several features about this list. The first size has only one shape (s), which has only one orientation (see Figure D-3). There is, therefore, only one orbital of the smallest size ($n = 1$). The second size has two shapes, s and p, the former having only one orientation and the latter having three orientations. This means that there is only one spherical-shaped orbital of the second size and there are *three different p*-type orbitals of the second size. Thus, there are a total of four orbitals in the second level (i.e., second size). Likewise, at every size (level) above the first there are three p orbitals, five d orbitals (beginning with $n = 3$), and seven f orbitals (starting at level 4). This list, at least the first part of it (levels 1 to 4), should be memorized.

The fourth quantum number, symbolized by m_s, designates the electron behavior within an orbital. Most simply stated, the electrons can be thought of as having a spin in the clockwise or counterclockwise sense. Actually, they are *not* moving about in orbits, but are found randomly at various places within the orbital. The concept of spinning is useful, however, as will be discussed later.

ELECTRON STRUCTURES OF ATOMS

In this section a method for writing out the electron structures of atoms of each element will be presented. By using the orbitals just described and following a set of rules called the *Aufbau method* (*Aufbau* is the German word for "building up"), it is possible to indicate rather quickly where each electron would be in atoms of any element. The "electron game" should make it easier for students to internalize the ideas of discrete pathways or orbitals and should give students a "feel" for the greater energy needed to occupy certain pathways compared with others. The game can be played by a small group of students, or the instructor may wish to divide the class into groups for illustration or competition. After this game has been presented, the discussion goes on to the Aufbau rules for writing electron structures, and then to ways of expressing these structures.

FIGURE D-4
Court for the "electron game."

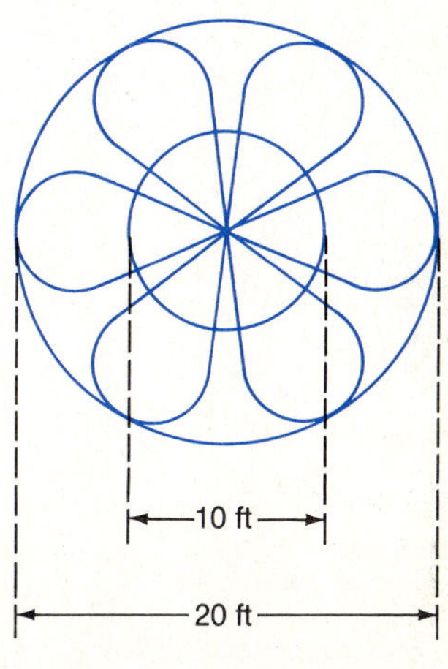

The electron game*

The object of this game is to allow people to run around the paths of a court whose features are shown in Figure D-4. For competition the class is divided into groups of 15 to 20 students each. The groups are then taken to a field on which courts about 20 ft in diameter have been laid off. The court lines approximate the contours of the $1s$, $2s$, and three $2p$ orbitals. The teams compete to see which can most rapidly get 10 players into their respective courts. Because the court paths have some similarity to the electronic orbitals

*F. Wiseman, *Journal of Chemical Education*, **55:**325 (1978).

in atoms, the students are able to develop a feel for the different energy levels available to the electron and the tendency for them to fill the lower-energy orbitals first. For example, it is much easier for a player to traverse the large circle than to traverse the ''figure eight'' paths in the court; thus, the large circle is preferred to the figure eight, just as electrons fill the 2*s* orbital before the 2*p* orbitals. This court can also be laid off on blacktop or cement with chalk or in a classroom (rug or tile) with masking tape or chalk.

The following rules apply.

1 Filling order:
(a) The small circle requires the lowest energy to traverse, the large circle requires more energy, and the three figure-eight paths require the most energy (all three figure-eight paths are equal in energy).
(b) The lower-energy paths must be filled before the higher-energy ones.
(c) When two or more paths have the same energy, each of these paths should be occupied by one player before any path has two players.
(d) A path is full when it is occupied by two players (traveling in opposite directions).

2 Method of play: One player enters the court at a time. An entering person must make two rounds on the path before the next player may enter. After a total of five rounds the player may stop and stand on the path line.

3 If an error is made (i.e., if any rules are broken), all team members must leave the court and start over. (Referees from the opposing team monitor the activity.)

4 The ''atom'' is filled when the tenth runner completes the fifth circuit.

If competition is not desired, students can simply run around a circle and a figure eight to see which is easier (i.e., requires the least energy). Alternatively, atoms can be simulated by placing varying players into the orbitals according to the game rules. For example, if six players enter and one observes the result, the observer would see an approximation of the electron behavior in an atom with six electrons (carbon). Other atoms with atomic numbers of 1 to 10 (hydrogen through Ne) can be simulated.

The Aufbau rules

Now that you have either played the game or read the rules so that you have a general idea of the procedure, let's look at the Aufbau rules for writing the electron structure of atoms. We shall in our own minds strip all the electrons from an atom and feed them back into the ''empty'' orbitals one at a time according to the following rules (similar to the rules of the electron game). When we have fed in the number of electrons which atoms of that element have, we have then indicated the electron structure of those atoms.

The rules are as follows:

1 The electron always goes to the lowest energy level possible. In general the larger and more complex the shape of the orbital, the higher its energy. Remember how you would rather run around a circle than a figure eight. You

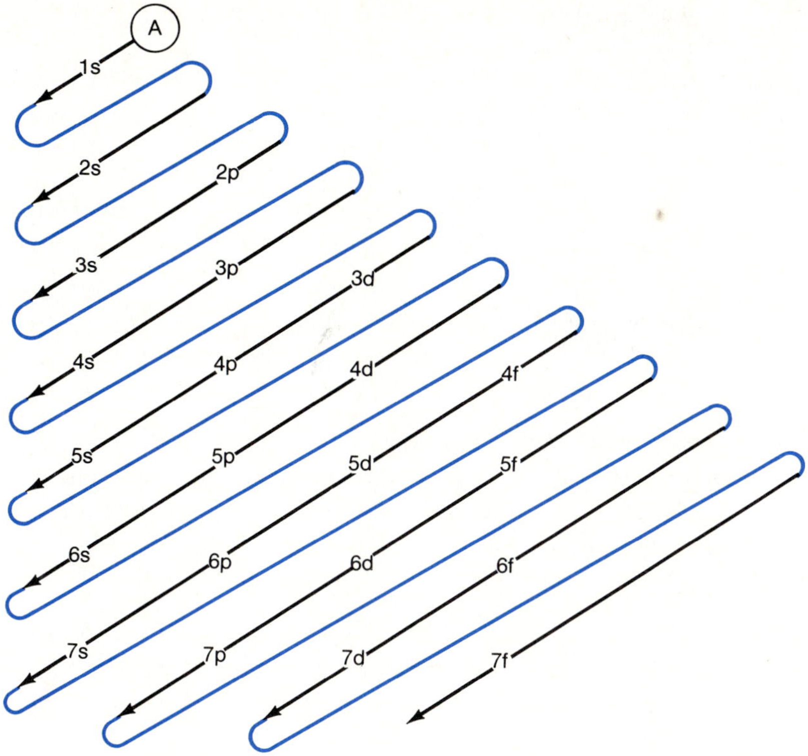

FIGURE D-5
Order of filling for atomic orbitals.

would probably choose the easiest (lowest energy) path if you had a choice. Thus, as we go from $1s$ to the right, energy increases. An exception is that s orbitals are always lower in energy than the previous d orbital. The $4s$ is lower in energy than the $3d$, and thus the $4s$ fills before the $3d$. The mnemonic device shown in Figure D-5 will allow you to easily determine the order of filling. Start at point A and proceed, following the arrows as far as necessary to get in the required number of electrons.

2 Each orbital can hold a maximum of *two* electrons.

3 When there are two or more orbitals of the same shape and size, such as the three $2p$ or the five $3d$ orbitals, one electron is put in each member of the set before doubling the electrons up. These groups are referred to as *degenerate sets of orbitals,* and this rule is called *Hund's rule*. The rationale for this rule is that electrons, being negatively charged, tend to repel each other and would rather be in orbitals by themselves than be doubled up. This preference is small compared with other considerations and affects the electron location only when the energy of the orbitals is identical, as in the case of degenerate orbitals. Remember how much simpler it was for you to traverse a path in the electron game court when you were alone than when someone else was running on the same path!

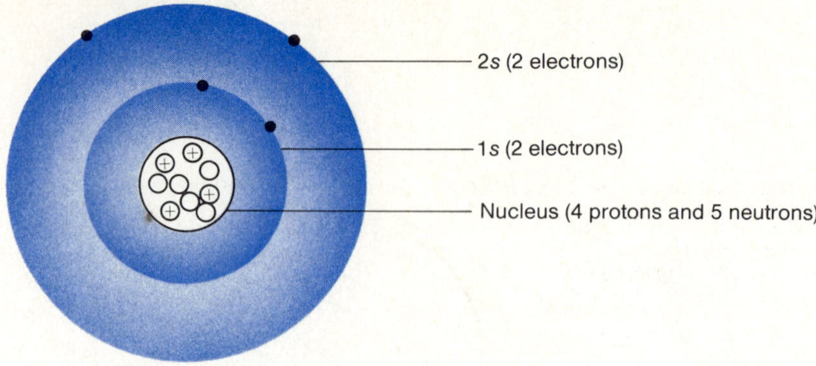

FIGURE D-6
Beryllium (Be) atom.

Expressing electron structures

There are several ways of expressing the electronic structures of atoms once they have been determined. One could, of course, write a prose description, but that is bulky; one could draw a picture (such as that shown in Figure D-6), but that is usually not convenient. The usual ways are as follows.

Energy diagrams To make an energy diagram, one draws lines or boxes on a paper the height of which (on the paper) corresponds to their relative energies. Then one places x's or dots or arrows in these places. For example, the energy diagram of the first two levels is:

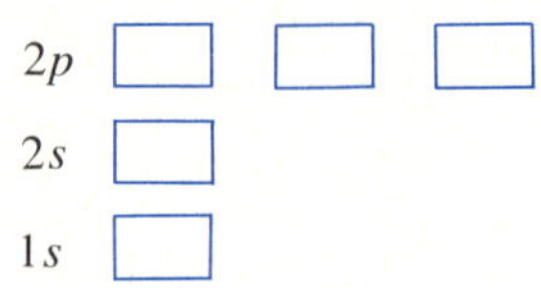

Putting in four electrons (as in Be) gives:

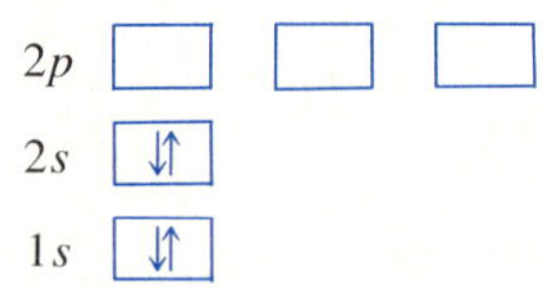

This energy diagram shows two electrons in a 1*s* orbital and two in the higher-energy 2*s* orbital. The physical diagram of this situation is shown in Figure D-6.

Spectroscopic notation Spectroscopic notation is the easiest method. One simply lists the occupied orbitals (the orbitals that have electrons in them) and places arrows or numbers as superscripts to show how many electrons are in each. For Be, this method gives:

$$\overset{\downarrow\uparrow}{1s}\ \overset{\downarrow\uparrow}{2s} \quad \text{or} \quad 1s^2 2s^2$$

This is the method that will be used in the rest of this appendix. The following program will allow you to learn how to write the electron structure of atoms of any element with the atomic number of 1 to 57.

WRITING ELECTRON STRUCTURES OF ATOMS

Worked example

What is the electron structure of an atom of lithium?

1 Find Li on the periodic chart and read its atomic number as 3. This means that there are three electrons in a lithium atom.

2 Write out the first part of the orbital list.

$$1s \quad 2s \quad 2p \quad 2p \quad 2p$$

3 Place one electron in the lowest-energy orbital—the 1*s*. Use the device in Figure D-5 or your memory for this. Use arrows at this stage, and place them above the orbital designations.

$$\begin{array}{ccccc} \downarrow & & & & \\ 1s & 2s & 2p & 2p & 2p \end{array}$$

4 Place a second electron, with the opposite spin, in the 1*s*, since it is not yet full.

$$\begin{array}{ccccc} \downarrow\uparrow & & & & \\ 1s & 2s & 2p & 2p & 2p \end{array}$$

5 Since the 1*s* is now full, place the third electron in the next-lowest-energy orbital—2*s*. All three electrons have now been placed in orbitals.

$$\begin{array}{ccccc} \downarrow\uparrow & \downarrow & & & \\ 1s & 2s & 2p & 2p & 2p \end{array}$$

6 Write the symbols for the *occupied* orbitals only and indicate the number of electrons in each by a superscript.

$$1s^22s^1$$

This is the electronic structure of atoms of lithium expressed in spectroscopic notation.

Worked example

Where are all the electrons in an atom of nitrogen (N)?

1 From the periodic chart, nitrogen atoms have seven electrons.

2 Write out:

$$1s \quad 2s \quad 2p \quad 2p \quad 2p$$

3 Place electrons in the following order:

$$\begin{array}{llllll} & \downarrow & & & & \\ \textbf{(a)} & 1s & 2s & 2p & 2p & 2p \end{array}$$

$$\begin{array}{llllll} & \downarrow\uparrow & & & & \\ \textbf{(b)} & 1s & 2s & 2p & 2p & 2p \end{array}$$

$$\begin{array}{llllll} & \downarrow\uparrow & \downarrow & & & \\ \textbf{(c)} & 1s & 2s & 2p & 2p & 2p \end{array}$$

$$\begin{array}{llllll} & \downarrow\uparrow & \downarrow\uparrow & & & \\ \textbf{(d)} & 1s & 2s & 2p & 2p & 2p \end{array}$$

$$\begin{array}{llllll} & \downarrow\uparrow & \downarrow\uparrow & \downarrow & & \\ \textbf{(e)} & 1s & 2s & 2p & 2p & 2p \end{array}$$

Note The fifth electron could be placed in any one of the $2p$ orbitals, but it is easier to put it in the first one written.

$$\begin{array}{llllll} & \downarrow\uparrow & \downarrow\uparrow & \downarrow & \downarrow & \\ \textbf{(f)} & 1s & 2s & 2p & 2p & 2p \end{array}$$

Note Remember to put one electron in each $2p$ orbital before doubling up electrons (Hund's rule).

$$\begin{array}{llllll} & \downarrow\uparrow & \downarrow\uparrow & \downarrow & \downarrow & \downarrow \\ \textbf{(g)} & 1s & 2s & 2p & 2p & 2p \end{array}$$

4 Since seven electrons have now been placed, we must stop; that is how many electrons nitrogen has. Now write these symbols with superscripts in place of the arrows.

$$1s^2 2s^2 2p^1 2p^1 2p^1$$

This also can be written as $1s^2 2s^2 2p^3$.

Worked example

Write the spectroscopic notation of the electron structure of argon (Ar).

1 Ar has 18 electrons.

2 Write:

$$1s \quad 2s \quad 2p \quad 2p \quad 2p \quad 3s \quad 3p \quad 3p \quad 3p$$

3 Fill in the electron as follows:

1s	2s	2p	2p	2p	3s	3p	3p	3p
↓								
⇵								
⇵	↓							
⇵	⇵							
⇵	⇵	↓						
⇵	⇵	↓	↓					
⇵	⇵	↓	↓	↓				
⇵	⇵	⇵	↓	↓				
⇵	⇵	⇵	⇵	↓				
⇵	⇵	⇵	⇵	⇵				
⇵	⇵	⇵	⇵	⇵	↓			
⇵	⇵	⇵	⇵	⇵	⇵			
⇵	⇵	⇵	⇵	⇵	⇵	↓		
⇵	⇵	⇵	⇵	⇵	⇵	↓	↓	
⇵	⇵	⇵	⇵	⇵	⇵	↓	↓	↓
⇵	⇵	⇵	⇵	⇵	⇵	⇵	↓	↓
⇵	⇵	⇵	⇵	⇵	⇵	⇵	⇵	↓
⇵	⇵	⇵	⇵	⇵	⇵	⇵	⇵	⇵

4 This can be written as $1s^2 2s^2 2p^6 3s^2 3p^6$.

Worked example

What is the electron configuration of calcium (Ca)?

1 Ca has 20 electrons.

2 We already know what the configuration of argon is (see the previous example). Since Ca has two more electrons, we can *start from argon* rather than from the beginning. First list the occupied orbitals of Ar along with their electrons and add several more orbitals.

$$1s^22s^22p^63s^23p^63d3d3d3d3d4s$$

3 Add two more electrons (to get a total of 20). Remember that the $4s$ orbital fills *before* the $3d$. Write:

$$1s^22s^22p^63s^23p^63d3d3d3d3d4s^2$$

Rewrite as:

$$1s^22s^22p^63s^23p^64s^2$$

One can abbreviate the portion that corresponds to argon by the symbol Ar:

Ar → $\boxed{1s^22s^22p^63s^23p^6}$ $4s^2$

This can be written as Ar:$4s^2$.

Note Any *portion* of the electron structure that is the same as the *total* structure of *an inert gas* can be abbreviated using the symbol for that gas.

Problem

What is the electron structure of boron (B)?

Solution

1 Boron has five electrons.

2 Write:

$$1s \quad 2s \quad 2p \quad 2p \quad 2p$$

3 Fill in the five electrons as follows:

$$\begin{array}{ccccc} \downarrow & & & & \\ 1s & 2s & 2p & 2p & 2p \end{array}$$

$$\begin{array}{ccccc} \downarrow\uparrow & & & & \\ 1s & 2s & 2p & 2p & 2p \end{array}$$

$$\begin{array}{ccccc} \downarrow\uparrow & \downarrow & & & \\ 1s & 2s & 2p & 2p & 2p \end{array}$$

$$\begin{array}{ccccc} \downarrow\uparrow & \downarrow\uparrow & & & \\ 1s & 2s & 2p & 2p & 2p \end{array}$$

$$\begin{array}{ccccc} \downarrow\uparrow & \downarrow\uparrow & \downarrow & & \\ 1s & 2s & 2p & 2p & 2p \end{array}$$

4 Write this as $1s^22s^22p^1$.

Problem

Where are the electrons in atoms of silicon?

Solution

1 Si has 14 electrons.

2 Write:

$$1s \quad 2s \quad 2p \quad 2p \quad 2p \quad 3s \quad 3p \quad 3p \quad 3p$$

3 Fill in the electrons to get the following:

$$\begin{array}{ccccccccc} \downarrow\uparrow & \downarrow\uparrow & \downarrow\uparrow & \downarrow\uparrow & \downarrow\uparrow & \downarrow\uparrow & \downarrow & \downarrow & \\ 1s & 2s & 2p & 2p & 2p & 3s & 3p & 3p & 3p \end{array}$$

4 Rewrite this as $1s^22s^22p^63s^23p^2$.

5 Although the above expression is perfectly acceptable, it can also be written in the abbreviated form that was the symbol for an inert gas. The portion in the box has 10 electrons and therefore has the configuration of neon (Ne)

$$\boxed{1s^22s^22p^6}\ 3s^2\ 3p^2$$

This can be written as Ne:$3s^23p^2$.

Problem

Indicate the electron configuration of iron (Fe) atoms.

Solution

1 Iron has 26 electrons.

2 Write a list of orbitals through 4*s* (or further, although they will not be needed).

1*s*	2*s*	2*p*	2*p*	2*p*	3*s*	3*p*	3*p*	3*p*	3*d*	3*d*	3*d*	3*d*	3*d*	4*s*

3 Fill in the first 18 electrons.

⇅	⇅	⇅	⇅	⇅	⇅	⇅	⇅	⇅						
1*s*	2*s*	2*p*	2*p*	2*p*	3*s*	3*p*	3*p*	3*p*	3*d*	3*d*	3*d*	3*d*	3*d*	4*s*

4 Fill in two more in the 4*s* (remember that it fills before the 3*d*).

⇅	⇅	⇅	⇅	⇅	⇅	⇅	⇅	⇅						⇅
1*s*	2*s*	2*p*	2*p*	2*p*	3*s*	3*p*	3*p*	3*p*	3*d*	3*d*	3*d*	3*d*	3*d*	4*s*

5 Finally, place the six remaining electrons in the 3*d* orbitals, placing one in each orbital before doubling up the electrons.

⇅	⇅	⇅	⇅	⇅	⇅	⇅	⇅	⇅	⇅	↓	↓	↓	↓	⇅
1*s*	2*s*	2*p*	2*p*	2*p*	3*s*	3*p*	3*p*	3*p*	3*d*	3*d*	3*d*	3*d*	3*d*	4*s*

6 Rewrite as $1s^22s^22p^63s^23p^63d^64s^2$ or $1s^22s^22p^63s^23p^64s^23d^6$.

Note The first method is preferred because the 4*s* orbital is larger and therefore should be written further to the right.

Problems

What are the electron configurations of the following atoms?

a Rb **b** K **c** Na **d** Cl **e** F

Answers

a Rb $1s^22s^22p^63s^23p^63d^{10}4s^24p^65s^1$

b K $1s^22s^22p^63s^23p^64s^1$

c Na $1s^22s^22p^63s^1$

d Cl $1s^22s^22p^63s^23p^5$

e F $1s^22s^22p^5$

Appendix

E

CALCULATIONS USING CHEMICAL FORMULAS AND EQUATIONS

CALCULATING EMPIRICAL FORMULAS FROM PERCENT COMPOSITION

Methods have been developed by which chemists can determine the percent (by weight) of most elements in a compound. Since these methods are beyond the scope of this book and frequently involve complex instrumentation, we will work only with the *results* of analyses. Oxygen is the only common element that cannot be *quantitatively analyzed* directly when it is found in compounds. This section will explain how to determine the empirical formula of a compound when the percent of each element is known. (See Figure E-1.) Recall from Chapter 5 that the *empirical formula* is the simplest whole-number *ratio* of atoms in a compound.

The following steps are used to calculate the empirical formula from percent composition:

1 Change percent to weight (in grams). For example, 16 percent is changed to 16 g.

2 Divide the weight of each element by its respective *atomic weight* to get moles of each. For example:

$$\frac{16 \text{ g nitrogen}}{14 \text{ g/mol}} = 1.14 \text{ mol}$$

3 Divide the moles of each element by the smallest number of moles. This usually gives whole-number ratios of moles.

4 *If a fraction still exists:*

(a) Change the decimal portion to a fraction.

(b) Change compound fractions to simple fractions.

(c) Multiply all the numbers obtained in step 3 by the denominator of the fraction.

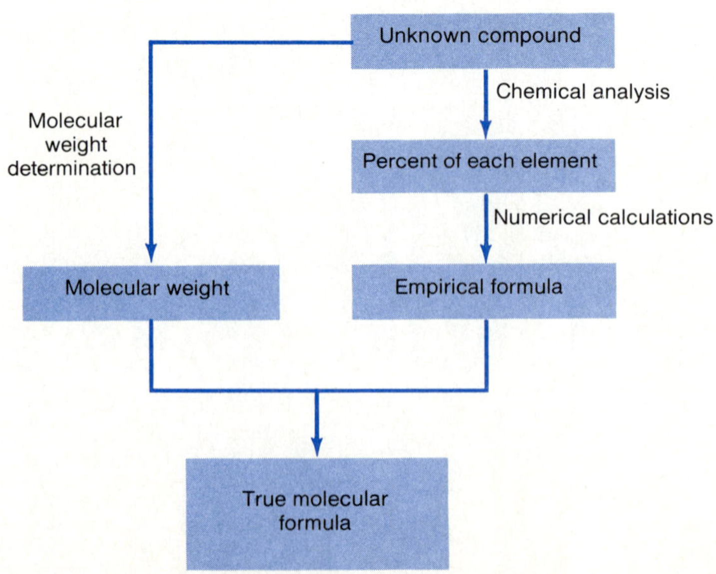

FIGURE E-1 Scheme for determining empirical and molecular formulas.

Note One should check to see if the percents given total to 100 ± 1 percent. If not, the difference between the total and 100 percent is the percent oxygen in the compound. For example, if a compound has 40 percent C and 40 percent N by analysis, since these total to only 80 percent, there must be 20 percent oxygen even though we cannot analyze for it directly. The composition would therefore be:

40 percent C	Given (determined directly)
40 percent N	Given (determined directly)
20 percent O	Calculated to be present

The following program illustrates each of these steps and will help you develop skill in calculating empirical formulas.

CALCULATING EMPIRICAL FORMULAS

Worked example

What is the empirical formula of a compound that is 80 percent carbon (C) and 20 percent hydrogen (H)?

Note Remember that the empirical formula is the *simplest whole-number ratio* of atoms of each type. The molecular formula is the *actual number* of atoms of each type of molecules of covalently bonded compounds.

1 Changing percent to weight (in grams).

$$80\%\ \text{C} \longrightarrow 80\ \text{g C}$$
$$20\%\ \text{H} \longrightarrow 20\ \text{g H}$$

2 Divide the weight of each by its respective atomic weight.

$$\frac{80\ \text{g C}}{12\ \text{g/mol}} = 6.67\ \text{mol C}$$

$$\frac{20\ \text{g H}}{1\ \text{g/mol}} = 20\ \text{mol H}$$

This tells us that there are 6.67 mol of C for every 20 mol of H. But we want simple *whole*-number *ratios*.

3 Divide both numbers by the smallest. Since 6.67 is the smaller of the two, divide both by it.

$$\text{C} \qquad \frac{6.67}{6.67} = 1.0$$

$$\text{H} \qquad \frac{20}{6.67} = 3.0$$

We therefore have the ratio of 1 to 3, and the empirical formula is:

$$CH_3$$

Note If the number is within 0.1 of a whole number, round it off to the nearest number.

Worked example

What is the empirical formula of a compound that is 62.6 percent Cl, 31.9 percent C, and 5.3 percent H?

1 Convert each percent to grams.

62.6 g Cl
31.9 g C
5.3 g H

2 Divide each by its respective atomic weight.

$$\frac{62.6 \text{ g Cl}}{35.5 \text{ g/mol}} = 1.765 \text{ mol Cl}$$

$$\frac{31.9 \text{ g C}}{12 \text{ g/mol}} = 2.66 \text{ mol C}$$

$$\frac{5.3 \text{ g H}}{1 \text{ g/mol}} = 5.3 \text{ mol H}$$

3 Since 1.765 is the smallest, divide each one by it.

$$\text{Cl} \quad \frac{1.765}{1.765} = 1.0$$

$$\text{C} \quad \frac{2.66}{1.765} = 1.5$$

$$\text{H} \quad \frac{5.3}{1.765} = 3.0$$

Since one of these is still not a whole number, use the steps mentioned previously for the fractional number (1.5). Change 1.5 to $\frac{3}{2}$.

$$1.5 = 1\tfrac{1}{2} = \tfrac{3}{2}$$

4 Multiply the moles found in step 3 by the denominator of the fraction (2 in this case).

$$\text{Cl} \quad 1.0 \times 2 = 2$$

$$\text{C} \quad \tfrac{3}{2} \times 2 = 3$$

$$\text{H} \quad 3.0 \times 2 = 6$$

We now have the simplest *whole-number* ratio, and the formula of the compound is:

$$C_3H_6Cl_2$$

Note The order of the elements in the formula is not critical, but usually those compounds containing C and H have those listed *first* and *second,* respectively, in the formula, followed by the other elements. Thus $Cl_2C_3H_6$ is *not* incorrect but is not preferred.

Worked example

What is the empirical formula of a compound that contains 52.17 percent carbon and 13.04 percent hydrogen?

1 The two percents given total to only 65.21 percent; therefore, the remainder (34.79 percent) is oxygen.

2 Change each percent to grams.

$$\begin{aligned} 52.17\%\ \text{C} &\longrightarrow 52.17\ \text{g C} \\ 13.04\%\ \text{H} &\longrightarrow 13.04\ \text{g H} \\ 34.79\%\ \text{O} &\longrightarrow 34.79\ \text{g O} \end{aligned}$$

3 Divide the weight of each element by its own atomic weight.

$$\frac{52.17\ \text{g C}}{12\ \text{g/mol}} = 4.35\ \text{mol C}$$

$$\frac{13.04\ \text{g H}}{1\ \text{g/mol}} = 13.04\ \text{mol H}$$

$$\frac{34.79\ \text{g O}}{16\ \text{g/mol}} = 2.17\ \text{mol O}$$

4 Divide each by 2.17 (which is the smallest).

$$\text{C} \qquad \frac{4.35}{2.17} = 2.00$$

$$\text{H} \qquad \frac{13.04}{2.17} = 6.01 \approx 6.00$$

$$\text{O} \qquad \frac{2.17}{2.17} = 1.00$$

5 Since these are whole numbers, the empirical formula is C_2H_6O.

Problem

Calculate the empirical formula of the compound found to contain 82.4 percent nitrogen and 17.6 percent hydrogen.

Solution

1 The two values equal 100 percent; therefore, *no* oxygen is present.

2 Change percent to grams.

$$82.4\%\ \mathrm{N} \longrightarrow 82.4\ \mathrm{g\ N}$$
$$17.6\%\ \mathrm{H} \longrightarrow 17.6\ \mathrm{g\ H}$$

3 Divide each weight by the appropriate atomic weight.

$$\frac{82.4\ \mathrm{g\ N}}{14\ \mathrm{g/mol}} = 5.88\ \mathrm{mol\ N}$$
$$\frac{17.6\ \mathrm{g\ H}}{1\ \mathrm{g/mol}} = 17.6\ \mathrm{mol\ H}$$

4 Divide each by the smallest (5.88).

$$\mathrm{N} \quad \frac{5.88}{5.88} = 1.0$$
$$\mathrm{H} \quad \frac{17.6}{5.88} = 2.99 \approx 3.0$$

5 The formula is NH_3.

Problem

What is the empirical formula of a compound whose analysis gave the following results: 40.00 percent carbon and 6.67 percent hydrogen?

Solution

1 Since the values add up to 46.67, there must be 53.33 percent oxygen.

2 Change each percent to weight and divide by the atomic weights.

$$\frac{40\ \mathrm{g\ C}}{12\ \mathrm{g/mol}} = 3.33\ \mathrm{mol\ C}$$
$$\frac{6.67\ \mathrm{g\ H}}{1\ \mathrm{g/mol}} = 6.67\ \mathrm{mol\ H}$$
$$\frac{53.33\ \mathrm{g\ O}}{16\ \mathrm{g/mol}} = 3.33\ \mathrm{mol\ O}$$

3 Divide each by 3.33.

$$\mathrm{C} \quad \frac{3.33}{3.33} = 1.0$$
$$\mathrm{H} \quad \frac{6.67}{3.33} = 2.0$$
$$\mathrm{O} \quad \frac{3.33}{3.33} = 1.0$$

4 The empirical formula is CH_2O.

Problem

What is the empirical formula of a compound for which the elemental analysis is 43.7 percent phosphorus (P) and 56.3 percent oxygen?

Solution

1 Since the total percents adds up to 100 percent, no other elements are present.

2 Change percent to weight and divide by the atomic weights.

$$\frac{43.7 \text{ g P}}{31 \text{ g/mol}} = 1.41 \text{ mol P}$$

$$\frac{56.3 \text{ g O}}{16 \text{ g/mol}} = 3.52 \text{ mol O}$$

3 Divide each by the smallest (1.41).

$$\text{P} \quad \frac{1.41}{1.41} = 1.0$$

$$\text{O} \quad \frac{3.52}{1.41} = 2.50$$

4 Since one number is fractional, the ratios are changed to whole numbers as follows:

$$2.5 \longrightarrow 2\tfrac{1}{2} = \tfrac{5}{2}$$

Multiply each of the moles in step 3 by 2 (the denominator of the fraction).

$$\text{P} \quad 1.0 \times 2 = 2.0$$

$$\text{O} \quad 2.5 \times 2 = 5.0$$

5 Therefore the empirical formula is P_2O_5.

Problem

What is the empirical formula of a compound with 57.45 percent carbon and 8.51 percent hydrogen?

Solution

1 Since these values add up to 65.96 percent (57.45% + 8.51% = 65.96%), there must be 34.04 percent oxygen (100% − 65.96% = 34.04%).

2 Convert percent to weight and divide by the atomic weight of each.

$$\frac{57.45 \text{ g C}}{12 \text{ g/mol}} = 4.79 \text{ mol C}$$

$$\frac{8.51 \text{ g H}}{1 \text{ g/mol}} = 8.51 \text{ mol H}$$

$$\frac{34.04 \text{ g O}}{16 \text{ g/mol}} = 2.13 \text{ mol O}$$

3 Divide each by the smallest (2.13).

$$\text{C} \quad \frac{4.79}{2.13} = 2.25$$

$$\text{H} \quad \frac{8.51}{2.13} = 4.00$$

$$\text{O} \quad \frac{2.13}{2.13} = 1.0$$

4 Since one number is not a whole number, the ratios are changed to whole numbers as follows:

$$2.25 = 2\tfrac{1}{4} = \tfrac{9}{4}$$

Multiply each of the moles in step 3 by 4 (the denominator of the fraction).

$$\text{C} \quad \tfrac{9}{4} \times 4 = 9$$
$$\text{H} \quad 4 \times 4 = 16$$
$$\text{O} \quad 1 \times 4 = 4$$

5 Therefore the empirical formula is $C_9H_{16}O_4$.

CALCULATING MOLECULAR FORMULAS FROM EMPIRICAL FORMULAS

While the empirical formula of a compound shows the simplest *ratio* of atoms in a compound, the *molecular formula* designates the actual number of atoms of each element. For example, if a compound has the molecular formula $C_4H_8Br_2$, the molecules of this material have four atoms of carbon, eight atoms of hydrogen, and two atoms of bromine. This is the *actual* number of atoms of each type, and is therefore the molecular formula. The empirical formula in this case is C_2H_4Br. The molecular formula must contain the *same ratio* of atoms as the empirical formula and is always some whole-number multiple of it. The number of each atom in $C_4H_8Br_2$ is *exactly two times* that in C_2H_4Br.

In order to calculate the molecular formula one must know (or calculate) the *empirical formula and* the *molecular weight* of the material. (Refer back to Figure E-1.)

The relationship used in the calculation is:

$$\text{Molecular formula} = (\text{empirical formula})\left(\frac{\text{molecular weight}}{\text{empirical weight}}\right)$$

where the empirical weight is the sum of all the atomic weights in the empirical formula.

CALCULATING MOLECULAR FORMULAS

Worked example

If the empirical formula of a compound is CH_3 and the molecular weight is 45 atomic mass units (amu), what is the molecular formula?

1 The empirical formula is CH_3; therefore, the empirical formula weight is 15 amu (12 + 3).

2 Put the numbers into the above relationship.

Molecular weight

$$\text{Molecular formula} = (CH_3)\left(\frac{45}{15}\right)$$

Empirical formula

Empirical weight

3 Work this out.

$$\text{Molecular formula} = (CH_3)(3) = C_3H_9$$

Note Since the molecular weight is three times the empirical weight, there must be three of the groups of atoms represented by the empirical formula in a molecule.

Note The ratio of molecular weight to empirical weight is always a whole number. Any value that is not exactly a whole number should be rounded off to the nearest whole number.

Problem

If the empirical formula of a compound is NO and the molecular weight is 60 amu, what is the molecular formula?

Solution

1 The empirical weight is 30 amu (14 + 16).

2 Put the numbers into the above relationship.

$$\text{Molecular formula} = (\text{NO})(60/30) = \text{NO}(2) = N_2O_2$$

Problem

If a compound was analyzed and found to contain 92.3 percent carbon and 7.7 percent hydrogen, what is the *molecular* formula if the molecular weight is 78 amu?

Solution

1 Since the percents add up to 100 percent, no oxygen is present.

2 Convert the percents to grams and divide by the respective atomic weights.

$$\frac{92.3 \text{ g C}}{12 \text{ g/mol}} = 7.69 \text{ mol C}$$

$$\frac{7.7 \text{ g H}}{1 \text{ g/mol}} = 7.7 \text{ mol H}$$

3 Since there is the *same* number of moles of both, the ratio is 1 to 1 and the empirical formula is CH.

4 The empirical weight is 13 (12 + 1).

5 Put the numbers into the molecular formula equation.

$$\text{Molecular formula} = (\text{CH})(78/13) = (\text{CH})(6) = C_6H_6$$

CALCULATING PERCENT COMPOSITION FROM FORMULAS

If either the empirical or the molecular formula of a compound is known, the percent (by weight) of each element can be calculated. If a chemist is proposing a particular formula for a new compound that has just been synthesized, the percent of each element can be calculated. The actual analysis can then be done; and if it agrees with the one calculated (within the experimental error, which is usually ±0.3 percent), then there is a high probability that the proposed formula is correct. To do this, one determines the weight (in atomic mass units) contributed by each element. Each weight is then divided by the total weight and multiplied by 100 to give the percent of each element. For example, the percent of element A is:

$$\text{Percent A} = \frac{\text{weight of A (amu)}}{\text{molecular weight (amu)}} \times 100$$

CALCULATING PERCENTS FROM FORMULAS

Worked example

What is the percent of each element in table sugar (sucrose), whose molecular formula is $C_{12}H_{22}O_{11}$?

1 Calculate the weight of each element in sucrose.

C	12 atoms × 12 amu/atom =	144 amu
H	22 atoms × 1 amu/atom =	22 amu
O	11 atoms × 16 amu/atom =	176 amu
	Total	342 amu

Therefore one molecule of sucrose weighs 342 amu or 1 mole of sucrose weighs 342 g. This is, of course, the same procedure used earlier to calculate molecular or formula weights from atomic weights.

2 Use the formula for percent by weight. In the case of carbon: 144 amu of the total 342 amu is contributed by carbon.

$$\text{Percent C} = \frac{\text{weight of C}}{\text{molecular weight}} = 100$$

$$= \frac{144 \text{ amu}}{342 \text{ amu}} \times 100 = 42.11\%$$

Repeat the calculation for H, and then for O.

$$\text{Percent H} = \frac{\text{weight of H}}{\text{molecular weight}} \times 100$$

$$= \frac{22 \text{ amu}}{342 \text{ amu}} \times 100 = 6.43\%$$

$$\text{Percent O} = \frac{\text{weight of O}}{\text{molecular weight}} \times 100$$

$$= \frac{176 \text{ amu}}{342 \text{ amu}} \times 100 = 51.64\%$$

3 As a check, the percents of all elements should total to 100 ± 0.5 percent. In this case,

$$42.11\% + 6.43\% + 51.46\% = 100.00\%$$

Problem

What is the percent of each element in a compound whose formula is $C_8H_{10}N_2O_2$?

Solution

1 Determine the molecular weight.

C	8 atoms × 12 amu/atom =	96 amu
H	10 atoms × 1 amu/atom =	10 amu
N	2 atoms × 14 amu/atom =	28 amu
O	2 atoms × 16 amu/atom =	32 amu
	Total	166 amu

2 Divide the weight of each element by the total weight and multiply by 100.

$$\text{Percent C} = \frac{96 \text{ amu}}{166 \text{ amu}} \times 100 = 57.8\%$$

$$\text{Percent H} = \frac{10 \text{ amu}}{166 \text{ amu}} \times 100 = 6.0\%$$

$$\text{Percent N} = \frac{28 \text{ amu}}{166 \text{ amu}} \times 100 = 16.9\%$$

$$\text{Percent O} = \frac{32 \text{ amu}}{166 \text{ amu}} \times 100 = 19.3\%$$

3 These total to 100 percent.

Problem

A chemist proposed N_2O as the formula of a newly synthesized compound. Chemical analysis showed this material to be 46.7 percent N and 53.3 percent O. Was the proposed formula correct? (Two independent methods of solving this problem are shown below.)

Solution A

Using the method of calculating percent from formulas

1 Determine the molecular weight of N_2O.

N	2 atom × 14 amu/atom =	28 amu
O	1 atom × 16 amu/atom =	16 amu
	Total	44 amu

2 Divide the weight of each element by the total weight and multiply by 100.

$$\text{Percent N} = \frac{28 \text{ amu}}{44 \text{ amu}} \times 100 = 63.6\%$$

$$\text{Percent O} = \frac{16 \text{ amu}}{44 \text{ amu}} \times 100 = 36.4\%$$

3 These total to 100 percent.

4 Since the percent of each element is *different* from those proposed (i.e., there is 36.4 percent oxygen in N_2O, but 53.3 percent oxygen was actually found in the compound), the formula N_2O is *not* correct.

Solution B

Using the method of calculating the empirical formula from the percent composition

1 Change percent to grams and divide by the atomic weights.

$$\text{N} \quad \frac{46.7 \text{ g}}{14 \text{ g/mol}} = 3.34 \text{ mol}$$

$$\text{O} \quad \frac{53.3 \text{ g}}{16 \text{ g/mol}} = 3.33 \text{ mol}$$

2 Since the moles of each element are equal, the empirical formula is NO.

3 This is different from the N_2O proposed; the N_2O was *not* correct, but NO is correct.

STOICHIOMETRY

Chemists frequently need to calculate the amount of material needed to react with a given amount of another material, or they may want to know how much of a product to expect from a certain amount of reactant. These calculations, which are always based on a balanced chemical equation, are called *stoichiometric calculations*.

For example, consider the balanced equation

$$2H_2 + O_2 \longrightarrow 2H_2O$$

A chemist might want to know how much oxygen is needed to react with 10 g of hydrogen or might want to calculate how much water (H_2O) is produced from 251 g of oxygen. This kind of calculation gives the amount expected (theoretical yield), which can then be compared with the amount actually produced to get a measure of the efficiency of the procedure (percent yield). If one expected, by stoichiometry, to obtain 18 g of water but obtained only

16 g, this is only 89 percent ($^{16}/_{18} \times 100 = 89$) of the expected amount. This is said to be an *89 percent yield*. The 18 g expected is called the *theoretical yield*.

The procedure used in stoichiometry is as follows:

1 Write and balance the equation. In this book the equation will be given and only balancing will be required.

2 Calculate the number of moles of one material from the weight given for that material.

$$\text{Number of moles} = \frac{\text{weight}}{\text{molecular weight}} \qquad \text{(see page 171)}$$

3 Calculate the number of moles of the material you want that would be expected from each mole of the material you were given.

$$\text{Number moles of material given}\left(\frac{\text{number of moles of material wanted}}{\text{number of moles of material given}}\right)$$

↑ Mole ratio factor

The mole ratio factor (MRF) is an expression of the number of moles of the material wanted that are contained in 1 mol of a given material or the number of moles of material wanted that can be produced from each mole of the material given or vice versa.

For example, in the compound CO_2, there are 2 mol of oxygen per mole of CO_2. The subscripts represent *ratios* of moles as well as absolute numbers of atoms. There are two atoms of oxygen per molecule of CO_2, but there are *also* twice as many atoms of oxygen as molecules of CO_2 *and* therefore twice as many moles of oxygen as moles of CO_2.

It is in this latter context that the MRF is used, and it is constructed by using the respective subscripts.

$$\text{MRF} = \frac{\text{moles O}}{\text{moles } CO_2} = \frac{2 \text{ mol O}}{1 \text{ mol } CO_2} = 2\,\frac{\text{mol O}}{\text{mol } CO_2}$$

This is read, ''There are only 2 mol O per mole of CO_2.'' (See Table E-1 for examples of the various meanings of the subscripts in the formula.)

Another example is the use of a balanced equation. Consider the following balanced equation:

$$2H_2 + O_2 \longrightarrow 2H_2O$$

There are twice as many moles of H_2O produced as moles of O_2 consumed in the reaction. The MRF in the case where one wants to calculate the moles of H_2O produced and knows the moles of O_2 is:

$$\text{MRF} = \frac{\text{moles of } H_2O}{\text{moles of } O_2} = \frac{2 \text{ mol } H_2O}{1 \text{ mol } O_2}$$

TABLE E-1

Various meanings of a chemical formula

	Na_2	C	O_3
Actual number of atoms	2 ions	1 atom	3 atoms
Actual number of moles	2 mol (of ions)	1 mol (of atoms)	3 mol (of atoms)
Ratios of number of atoms or ions	2 times as many Na *ions* as C atoms	½ as many C *atoms* as Na atoms	3 times as many O *atoms* as C atoms
Ratio of number of moles	2 times as many moles of Na as C ⅔ as many moles of Na as O	½ as many moles of C as Na ⅓ as many moles of C as O	3 times as many moles of O as C 3/2 as many moles of O as Na

TABLE E-2

Various meanings of a balanced equation

$$3H_2 + N_2 \longrightarrow 2NH_3$$

	$3H_2$	N_2	$2NH_3$
Actual number of molecules	3 molecules H_2 (hydrogen)	1 molecule N_2 (nitrogen)	2 molecules NH_3 (ammonia)
Actual number of moles	3 mol (of molecules) of H_2	1 mol (of molecules) of N_2	2 mol (of molecules) of NH_3
Ratio of molecules	3 times as many molecules of H_2 as N_2	⅓ as many molecules of N_2 as H_2	2 times as many molecules of NH_3 as N_2
Ratio of moles	3 times as many moles of H_2 as N_2 3/2 as many moles of H_2 as NH_3	⅓ as many moles of N_2 as H_2 ½ as many moles of N_2 as NH_3	2 times as many moles of NH_3 as N_2 ⅔ as many moles of NH_3 as H_2

One simply uses the balancing coefficients to construct the MRF. (See Table E-2 for various meanings of the balancing coefficients.)

4 The fourth step is to convert the moles of the material wanted to weight using the relationship

$$\text{Weight} = \text{moles (molecular weight)}$$ (see page 171)

One can work each stoichiometric problem in a stepwise fashion, or if desired, one can place all these steps together. In the following example the substance given is A and the substance desired is B.

$$\text{Weight of B} = \left(\frac{\text{weight of A}}{\text{molecular weight of A}}\right)\underbrace{\left(\frac{\text{moles B}}{\text{moles A}}\right)}_{\text{MRF}}\left(\begin{array}{c}\text{molecular}\\ \text{weight of}\\ \text{B}\end{array}\right)$$

Placing all the steps together is mechanically simpler if one is calculating the *weight* of B from the *weight* of A. However, if one is going from the weight of A to moles of B or from the moles of A to the weight of B or some other combination, the stepwise approach is better, since only some of the steps are needed.

STOICHIOMETRIC PROBLEMS

Worked example

If 16 g of O_2 is used, how much H_2O is produced according to the following chemical equation?

$$H_2 + O_2 \longrightarrow H_2O$$

1 Balance the equation.

$$2H_2 + O_2 \longrightarrow 2H_2O$$

2 Convert 16 g of O_2 to moles.

$$\frac{16\text{ g }O_2}{32\text{ g/mol}} = 0.5\text{ mol }O_2$$

3 **(a)** Set up the MRF.

$$\text{moles of }H_2O = \text{moles of }O_2\left(\frac{\text{moles of }H_2O}{\text{moles of }O_2}\right)$$

↑ moles of H_2O: Material we want to calculate
↑ moles of O_2: Material we already know about
↑ ratio: MRF

(b) Put in the data.

$$\text{Moles of } H_2O = 0.5 \text{ mol } O_2 \left(\frac{\textcircled{2} \text{ mol } H_2O}{\textcircled{1} \text{ mol } O_2} \right) = 1 \text{ mol } H_2O$$

Balancing coefficients directly from the equation

Note This means that twice as many moles of H_2O are produced as O_2 used.

4 Convert the moles of H_2O to the weight of H_2O.

$$1 \text{ mol } H_2O \text{ (18 g/mol)} = 18 \text{ g } H_2O$$

Worked example

How many grams of NaCl are obtained if 128 g of Na_2SO_4 is reacted with HCl according to the following equation?

$$Na_2SO_4 + HCl \longrightarrow NaCl + H_2SO_4$$

1 Balance the equation.

$$Na_2SO_4 + 2HCl \longrightarrow 2NaCl + H_2SO_4$$

2 Convert weight to moles.

$$\frac{128 \text{ g } Na_2SO_4}{142 \text{ g/mol}} = 0.90 \text{ moles } Na_2SO_4$$

3 Use the MRF factor.

$$\text{moles of NaCl} = \text{moles of } Na_2SO_4 \left(\frac{2 \text{ mol NaCl}}{1 \text{ mol } Na_2SO_4} \right)$$

$$= 0.90(2/1) = 1.80 \text{ mol NaCl}$$

4 Convert moles to weight.

$$1.80 \text{ mol NaCl (58.5 g/mol)} = 105 \text{ g NaCl}$$

Worked example

Using the following chemical equation, how much NH_3 is produced from 10 g of H_2?

$$N_2 + H_2 \longrightarrow NH_3$$

1 Balance the equation:

$$N_2 + 3H_2 \longrightarrow 2NH_3$$

2 Convert the weight of H_2 to moles.

$$\frac{10 \text{ g } H_2}{2 \text{ g/mol}} = 5 \text{ mol } H_2$$

3 Use the MRF relationship.

$$\text{moles of } NH_3 = \text{moles of } H_2 \left(\frac{\text{moles of } NH_3}{\text{moles of } H_2} \right)$$

$$= 5 \text{ mol } H_2 \left(\frac{\textcircled{2} \text{ mol } NH_3}{\textcircled{3} \text{ mol } H_2} \right)$$

From balanced equation

$$= 3.33 \text{ mol } NH_3$$

4 Convert moles of NH_3 to weight of NH_3.

$$3.33 \text{ mol } (17 \text{ g/mol}) = 56.6 \text{ g } NH_3$$

Problem

Using the equation for the burning of propane (C_3H_8), how much CO_2 is produced from 52 g of C_3H_8?

$$C_3H_8 + O_2 \longrightarrow CO_2 + H_2O$$

Solution

1 Balance the equation.

$$C_3H_8 + 5O_2 \longrightarrow 3CO_2 + 4H_2O$$

2 Convert the weight of C_3H_8 to moles.

$$\frac{52 \text{ g } C_3H_8}{44 \text{ g/mol}} = 1.18 \text{ mol } C_3H_8$$

3 Use the MRF relationship.

$$\text{moles of } CO_2 = \text{moles of } C_3H_8 \left(\frac{\text{moles of } CO_2}{\text{moles of } C_3H_8} \right)$$

$$= 1.18 \text{ mol } C_3H_8 \left(\frac{\textcircled{3} \text{ mol } CO_2}{\textcircled{1} \text{ mol } C_3H_8} \right)$$

From balanced equation

$$= 3.54 \text{ mol } CO_2$$

Note The balanced equation says that three times as many moles of CO_2 are produced as moles of C_3H_8 burned. Thus 3.54 is three times 1.18.

4 Convert moles of CO_2 to weight of CO_2.

$$3.54 \text{ mol } CO_2 \text{ (44 g/mol)} = 156 \text{ g } CO_2$$

Note In this case both materials, CO_2 and C_3H_8, have the same molecular weight (44), but that is only fortuitous.

Problem

How many grams of sodium are in 500 g of Na_2CO_3?

Solution

1 Change the weight of Na_2CO_3 to moles.

$$\frac{500 \text{ g } Na_2CO_3}{106 \text{ g/mol}} = 4.72 \text{ mol } Na_2CO_3$$

2 Set up the MRF relationship.

$$\text{moles of Na} = \text{moles of } Na_2CO_3 \left(\frac{\text{moles of Na}}{\text{moles of } Na_2CO_3}\right)$$

$$= 4.72\left(\frac{\textcircled{2} \text{ mol Na}}{\textcircled{1} \text{ mol } Na_2CO_3}\right) = 9.44 \text{ mol Na}$$

From subscripts

3 Convert moles of Na to weight.

$$9.44 \text{ mol Na (23 g/mol)} = 217 \text{ g Na}$$

Alternatively, one could have used the one-step procedure here, as well as with any of the other problems.

$$\left(\frac{500 \text{ g } Na_2CO_3}{106 \text{ g/mol}}\right)\left(\frac{2 \text{ mol Na}}{1 \text{ mol } Na_2CO_3}\right) \text{(23 g/mol)} = 217 \text{ g Na}$$

This is the *theoretical yield.*

Problem

If a chemist who worked with the Na_2CO_3 in the previous problem obtained only 175 g of sodium, what percent of the theoretical yield was obtained?

Solution

$$\text{Percent Na} = \frac{\text{weight Na obtained}}{\text{theoretical yield}} \times 100$$

$$= \frac{175\ \text{g}}{217\ \text{g}} \times 100 = 80.6\%$$

STUDY GUIDE

Skill practice problems

1 What is the percent elemental composition of epsom salts, $MgSO_4 \cdot 7H_2O$? (This means that there are seven molecules of H_2O with each $MgSO_4$ unit in the crystals.)

2 How many grams of metallic iron can be produced from 1000 g of Fe_3O_4 (iron ore) according to the following reactions?

$$Fe_3O_4 + C \longrightarrow CO_2 + Fe$$

3 A compound being synthesized as a new drug is found to have the following elemental composition. What is its empirical formula?

C	42.9 percent
H	3.6 percent
O	28.6 percent
N	25.0 percent

4 If the compound in problem 3 has a molecular weight of 112 g/mol, what is its molecular formula?

5 Titanium oxide (TiO_2) is used as a white pigment in paint. How much TiO_2 can be manufactured from 100 g of titanium metal?

$$Ti + O_2 \longrightarrow TiO_2$$

6 If the process expressed in problem 5 produced 150 g of TiO_2, what would be the percent yield? What would be the chemical company's percent loss on its investment in the purchase of the titanium?

7 Using the reaction of titanium with oxygen given in problem 5, how many grams of oxygen would be needed to react with 520 g of Ti?

Appendix

F

COMMON CHEMICALS

COMMON NAME	CHEMICAL FORMULA	CHEMICAL NAME	USE	SOURCE
Acetone	C_3H_6O	Dimethyl ketone	Polar solvent	Drugstore, fingernail polish remover
Alum	$KAl(SO_4)_2$	Potassium aluminum sulfate	Pickling and tanning agent	Drugstore, grocery store
Ammonia	NH_4OH	Ammonium hydroxide	Moderately strong base, cleaner	Grocery store
Ammonium chloride	NH_4Cl	Ammonium chloride	Fire retardant	Drugstore
Ammonium phosphate	$(NH_4)_3PO_4$	Ammonium phosphate	Fire retardant	Drugstore
Baking soda	$NaHCO_3$	Sodium bicarbonate	Leavening agent, fire extinguisher, odor absorber	Grocery store
Borax	$Na_2B_4O_7 \cdot 10H_2O$	Sodium tetraborate decahydrate	Mild base, cleaning agent, water softener	Grocery store
Boric acid	H_3BO_3	Boric acid	Very mild acid, eyewash	Drugstore
Calcium chloride	$CaCl_2$	Calcium chloride	Agent for melting ice	Hardware store
Charcoal lighter fluid	C_9H_{20}	Nonane (actually a mixture of hydrocarbons)	Nonpolar solvent	Grocery store
Chalk	$CaCO_3$	Calcium carbonate	Antacid	Drugstore
Chlorine bleach	$NaClO$	Sodium hypochlorite	Bleach	Grocery store
Citric acid	$C_6H_8O_7$		Agent for preventing darkening of fruit	Grocery store
Cream of tartar	$KHC_6H_9O_6$	Potassium acid tartrate	Source of acidity in baking	Grocery store

COMMON NAME	CHEMICAL FORMULA	CHEMICAL NAME	USE	SOURCE
Denatured alcohol (grain alcohol)	C_2H_6O	Ethyl alcohol (ethanol)	Moderately polar solvent, shellac thinner, cleaner, fuel	Hardware store, drugstore
Ethyl acetate	$C_4H_8O_2$	Ethyl acetate	Nonpolar solvent	Drugstore, fingernail polish remover
Ethylene glycol	$C_2H_6O_2$	Ethylene glycol	Permanent antifreeze	Automobile supply store
Epsom salts	$MgSO_4 \cdot 7H_2O$	Magnesium sulfate heptahydrate	Agent for drawing pus from infected wounds	Drugstore
Fruit sugar	$C_6H_{12}O_6$	Fructose	Sweetener	Grocery store
Gasoline	C_8H_{18}	Octane (actually a mixture of hydrocarbons)	Nonpolar solvent, fuel	Filling station
Glycerin	$C_3H_8O_3$	Glycerol	Polar solvent	Drugstore
Lime	CaO	Calcium oxide	Source of basic material, agent for controlling soil pH, plant nutrient	Garden supply store, agricultural supply store
Limestone	$CaCO_3$	Calcium carbonate	Mild base, agent for controlling soil pH	Agricultural supply store
Lye	$NaOH$	Sodium hydroxide	Strong base, oven cleaner, drain cleaner, ingredient for preparing homemade soap	Grocery store
Milk of magnesia	$Mg(OH)_2$	Magnesium hydroxide	Antacid, laxative	Drugstore
Muriatic acid	HCl	Hydrochloric acid	Strong acid, cleaner for masonry and steel	Hardware store

COMMON NAME	CHEMICAL FORMULA	CHEMICAL NAME	USE	SOURCE
Nitrate of ammonia	NH_4NO_3	Ammonium nitrate	Fertilizer	Garden supply store, agricultural supply store
Naphthalene	$C_{10}H_8$	Naphthalene	Moth and small animal repellent	Grocery store, hardware store
Oil of vitriol	H_2SO_4	Sulfuric acid	Motor vehicle battery fluid	Automobile supply store
Oil of wintergreen	$C_8H_8O_3$	Methyl salicylate	Muscle rubbing agent, flavoring	Drugstore, grocery store
para-Dichlorobenzene	$C_6H_4Cl_2$	*p*-Dichlorobenzene	Moth and small animal repellent	Grocery store, hardware store
Rubbing alcohol	C_3H_8O	Isopropyl alcohol	Solvent, muscle rubbing agent, fuel	Drustore, grocery store
Salt	NaCl	Sodium chloride	Flavoring, food preservative, agent for melting ice, abrasive for household cleaning	Grocery store
Sodium triphosphate	$Na_5P_3O_{10}$	Sodium tripolyphosphate	Builder for detergents	
Stearic acid	$C_{18}H_{36}O_2$	Stearic acid	Wax hardener for candles	Craft store
Table sugar	$C_{12}H_{22}O_{11}$	Sucrose	Sweetener	Grocery store
Trisodium phosphate	Na_3PO_4	Sodium phosphate	Cleaner	Grocery store, hardware store
Vaseline (petroleum jelly, petrolatum)	$C_{20}H_{42}$		Nonpolar solvent, lubricant	Drugstore
Vinegar	$HC_2H_3O_2$	Acetic acid	Moderately strong acid, ingredient for cooking, glass cleaner	Grocery store

COMMON NAME	CHEMICAL FORMULA	CHEMICAL NAME	USE	SOURCE
Washing soda	Na_2CO_3	Sodium carbonate	Source of basic material, cleaning agent for laundry	Grocery store, hardware store
Whitewash (slaked lime)	$Ca(OH)_2$	Calcium hydroxide	Mild base, temporary paint, antacid	Hardware store
Wood alcohol	CH_4O	Methyl alcohol (methanol)	Fuel, somewhat polar solvent	Drugstore

Glossary

Absolute temperature scale See *Kelvin scale*.

Acetaminophen A pain-reducing drug (analgesic) which also reduces fever (antipyretic).

Acetylsalicylic acid Aspirin.

Acid A material that ionizes and donates H^+ ions when added to water. Can be recognized by the presence of the letter H as the first letter in the chemical formula.

Acid rain A rain containing acidic materials such as sulfur dioxide, SO_2; sulfur trioxide, SO_3; or nitrogen oxides, $(NO)_x$.

Air inversion An atmospheric condition in which the presence of cooler air on the bottom of a warmer air mass results in stagnation of atmospheric air.

Alchemy The ancient practice of experimenting with matter and bringing about (or attempting to bring about) desired changes.

Alpha (α) The first letter in the Greek alphabet; used in chemistry to designate, among other things, the rays composed of helium nuclei produced by radioactive materials.

Alpha rays (α rays) Particles consisting of helium nuclei that are produced during nuclear reactions.

Amino acids Substances whose molecules contain both an amino group ($—NH_2$) and a carboxylic acid group ($—COOH$). The constituent parts of proteins.

Ammonia NH_3.

Amorphous solid A solid with little molecular orderliness.

Amphetamine A stimulant nonmedicinal drug.

Analysis The process of determining what substances are present in a sample (qualitative analysis) or how much of each substance is present (quantitative analysis).

Analytical chemistry The area of chemistry dealing with analysis (determining what is present or how much).

Anion A negatively charged ion.

Aphrodisiac A sexually stimulating substance.

Applied research Research aimed at producing a new or useful material or process.

Aqueous Pertaining to water.

Aspartame An artificial sweetener.

Atmospheric pressure The pressure exerted by the earth's atmosphere. At sea level the average value is called *one atmosphere* (atm) and equals 760 torr or 14.7 lb/in^2.

Atom The smallest stable individually extant unit of matter.

Atomic line spectrum A display of the component colors of the light generated when atoms are placed in a flame or an electric spark.

Atomic mass The mass of a single atom. Usually expressed on the amu (atomic mass unit) scale.

Atomic number The number of protons in the nucleus of an atom.

Atomic orbitals The volumes within atoms which have characteristic shapes where there is a high probability of finding electrons.

Atomic theory The view that all matter is composed of atoms. Also called *atomism*.

Atomic weight The average mass of atoms of naturally occurring elements. It is the weighted average of all isotopes of that element.

Avogadro's number 6.02×10^{23}.

Background radiation The radiation from cosmic and mineral sources that is constantly present even in the absence of any known radioactive material.

Baking soda Sodium bicarbonate ($NaHCO_3$).

Barbiturate A depressant anticonvulsant drug.

Barometer A device used to measure the pressure of a sample of gas or to determine atmospheric pressure.

Base A material that reacts with an acid or one which ionizes in water and produces hydroxide ions (OH^-).

Benzocain A topical pain reliever (analgesic). The ethyl ester of *para*-aminobenzoic acid. See PABA.

Benzpyrene A polynuclear aromatic compound which is found in cigarette and other smoke. It is highly carcinogenic.

Beta (β) The second letter of the Greek alphabet; used in chemistry to designate, among other things, the rays composed of high-energy electrons produced by radioactive materials.

Beta rays (β rays) High-energy electrons produced in nuclear reactions.

Binary compound A compound containing two elements.

Biochemistry The area of chemistry dealing with the substances and processes of living systems.

Biodegradable Able to be broken down into simpler compounds by microorganisms in the environment.

Biogenic Biologically active.

Bleach A material used to decolorize unwanted stains or other coloration. Usually a mild oxidizing agent that alters carbon-carbon double bonds.

Bond (chemical) The forces holding atoms together in compounds.

Borax $Na_2B_4O_4 \cdot 10H_2O$. A basic material used to neutralize acidic substances such as those found in dirty clothing or in the kitchen.

Boric acid H_3BO_3.

Boyle's law The volume of a gas sample goes down in direct proportion to the increase in pressure on it.

Caffeine A stimulant found in coffee, tea, and chocolate.

Calorie (Capitalized.) The quantity of heat energy needed to raise the temperature of one kilogram (kg) of water one degree Celsius (1°C). Sometimes called a *kilocalorie* (kcal) or *big calorie*. A calorie (lowercase) is the amount of heat needed to raise the temperature of one gram (g) of water one degree Celsius (1°C); it is sometimes called a *small calorie*.

Canal rays Rays composed of relatively large particles. These particles are usually single protons or several protons together.

Canal tube An evacuated glass tube constructed with one solid electrode at one end and one perforated electrode near the center.

Carbohydrates Biomolecules consisting of carbon, hydrogen, and oxygen. Starch, sugars, and cellulose are examples of carbohydrates.

Carcinogen A substance which causes cancer.

Carcinogenic Cancer-causing.

Catalyst A substance whose presence increases the rate of a chemical reaction but which is not itself consumed during the process.

Catalytic converter A device placed in the exhaust system of an automobile which causes the conversion of most of the pollutants into harmless materials.

Catalytic cracking The conversion of high-molecular-weight hydrocarbons (oils) into fuels such as gasoline by the use of heat in the presence of a solid catalyst.

Cathode rays Rays composed of electrons.

Cathode-ray tube An evacuated tube with a solid electrode near both ends.

Cation A positively charged ion.

Cellulose Linear polymeric glucose molecules linked by β-glycosidic bonds.

Celsius scale A temperature scale on which freezing and boiling water have values of 0°C and 100°C, respectively. Also called the *Centigrade scale*.

centi- A prefix meaning one-hundredth.

Centigrade scale See *Celsius scale*.

Chalk Calcium carbonate ($CaCO_3$).

Charge That property of matter that determines electrostatic behavior. There are two types of charge arbitrarily designated positive (+) and negative (−).

Charles's law The volume of a gas increases in direct proportion to its absolute temperature if its pressure remains constant.

Chelation The multiple complexation of a metal ion or atom by an organic ligand. Chelation frequently causes a normally insoluble material to become soluble.

Chemical change A change in matter during which a new substance is produced.

Chemical energy The potential energy stored in chemical bonds.

Chemical properties The characteristics of matter which determine what kinds of chemical changes that material undergoes. They determine how it interacts with other substances.

Chemistry The natural science which involves the study of the materials in the universe, their interaction with each other, and the energy changes associated with these changes.

Chlorine bleach Sodium hypochlorite (NaOCl).

CMC Carboxymethyl cellulose. Used in laundry formulations to help suspend the heavier dirt particles.

Colloidal suspension A suspension of small solid or liquid particles in a liquid or gas. These particles are small enough that molecular motion alone keeps them from settling out.

Color The central nervous system's response to various wavelengths (or frequencies) of electromagnetic radiation.

Combustion Burning; the high-temperature reaction with oxygen that releases large amounts of heat energy.

Compound A pure substance composed of two or more elements.

Condensation The change from a gas to a liquid or a solid.

Contact process A commercial process by which sulfur dioxide (SO_2) is converted to sulfur trioxide (SO_3). The SO_3 is then dissolved in water to produce sulfuric acid (H_2SO_4).

Control rods Rods of cadmium or boron used to absorb neutrons and thus control the rate of fission in a nuclear reactor.

Conversion factor A number that allows the conversion (translation) of a measurement on one scale to another scale.

Copolymer A polymer containing two or more types of monomer units.

Covalent bond A chemical bond formed when two nonmetallic elements share one, two, or three pairs of electrons.

Cream of tartar Potassium acid tartrate (potassium bitartrate, $KHC_6H_9O_6$).

Critical mass The minimum mass of a fissionable material needed to sustain an explosive chain reaction.

deca- A prefix meaning 10.

deci- A prefix meaning one-tenth.

Deduction The thought process by which one goes from general principles to specific facts. In chemistry, it involves the use of a general explanation to predict the results of a specific experiment.

Delta (Δ) The fourth letter of the Greek alphabet; used in science to designate a change in a quantity.

Denatured alcohol Ethyl (grain) alcohol which has been rendered undrinkable by adding small amounts of toxic materials.

Density The mass per unit of volume. The weight of a material for a given size; usually expressed in grams per milliliter or grams per liter.

Detergent A cleaning agent other than soap.

Dextromethorphan An anticough drug.

Diatomic Containing two atoms.

Dissertation A document reporting the results of graduate research. Presented to the faculty of a college or university in partial fulfillment of the requirements for an advanced degree such as an M.A. or a Ph.D.

DOD Departments of Defense.

Drug A chemical substance used to relieve or produce a specific effect in humans.

EDTA Ethylenediaminetetraacetic acid. Used to chelate Ca^{2+} ions in analysis of hard water and to remove toxic metal ions from the body after they have been ingested.

Electrochemical cell A device which produces electricity from a chemical reaction or which uses an electric current to cause a chemical change.

Electrodes Solid structures used to conduct electricity into or out of a system.

Electromagnetic spectrum The continuous list, in order of increasing or decreasing frequency, of all the types of electromagnetic radiation.

Electron A small negative particle which is part of all atoms.

Electronegativity A measure of the attraction of atoms for electrons.

Electronic structure A list of all the occupied orbitals within atoms of a given element along with the number of electrons in each. Sometimes called *electron arrangement* or *configuration*.

Electrostatic precipitator A device placed in smokestacks to remove particulate matter. It consists of charged parallel plates.

Element One of the fundamental materials of which all substances are composed. There are 92 naturally occurring elements and 13 artificially prepared ones.

Empirical formula A chemical formula showing the simplest whole-number ratio of atoms of the elements composing a compound.

Endothermic Energy-consuming.

Energy The ability to do work (i.e., to move mass through distance).

Energy level See *Shell*.

Enzyme A soluble (globular) protein material used as a catalyst in biochemical reactions.

EPA Environmental Protection Agency. An agency of the federal government.

Epsom salt Magnesium sulfate heptahydrate ($MgSO_4 \cdot 7H_2O$).

Equation (chemical) An expression containing chemical symbols representing a chemical change.

Essential amino acid One of the eight amino acids which must be included in human diets, since they cannot be synthesized in the body from other dietary components.

Essential fatty acid The fatty acid (linoleic acid) needed in the human diet, since it cannot be synthesized in the body from other dietary components.

Eutrophication The excessive growth of plants in a body of water that causes some of the plants to die because of a lack of light. The water becomes choked and useless for fish or recreational purposes.

Exothermic Producing or releasing heat energy.

Exponent The number written above and to the right of another number indicating how many times the latter number should be multiplied by itself. Sometimes called a *power*.

Exponential notation The expression of numbers at least partially as some power of another number (usually 10).

Fahrenheit scale A temperature scale on which ice water and boiling water have the arbitrary values of 32°F and 212°F, respectively.

Faith Confidence based on evidence.

Family of elements The groups of elements with similar chemical (and sometimes physical) properties. Members of the group have similar outer-shell electron configurations and are found listed in the vertical columns in the periodic chart.

Fat A class of biomolecules composed of glycerin and three fatty acids. Also called *triglyceride*. Fats are solid at room temperature.

FDA Food and Drug Administration. Regulatory agency of the federal government.

Fellowship The money given to support a student while he or she does graduate research.

Fermentation The conversion of starch or sugar into ethyl alcohol by enzymes in microorganisms.

Fission The splitting of atomic nuclei to produce two smaller atoms. This is accompanied by the release of large amounts of energy as a result of matter being converted into energy.

Food additive Any material added to food to impart desirable properties of taste, texture, or stability or to retard spoilage.

Food chain The progression from plant or bottom-feeding organisms to flesh-eating higher animals.

Forensic chemistry The area of chemistry dealing with legal matters, especially the solving of crimes.

Formula (chemical) An expression containing the symbols of elements along with subscripts indicating the absolute or relative numbers of atoms of each element in the compound, or the number of atoms in molecules of diatomic or polyatomic elements.

Formulation The process of preparing a useful product. Usually it involves mixing several compounds.

Fossil fuel The combustible material formed from once-living matter.

Fractional distillation The process of separating a liquid mixture into its component parts (fractions) on the basis of their differing boiling points.

Free radical An atomic or molecular species with an unpaired electron. A free radical is usually quite reactive.

Freezing The change from the liquid to the solid state.

Freons Halogenated methanes used as refrigeration fluids.

Frequency The number of oscillations per unit time.

Fructose Fruit sugar.

FTC Federal Trade Commission.

Fuel cell A device which produces electricity from a chemical reaction between two continuously supplied fuels.

Functional group An atom or group of atoms in a covalent compound which imparts characteristic chemical and physical properties to the substance.

Fusion The fusing of atomic nuclei to produce larger atoms. Accompanied by the release of large amounts of energy as a result of matter being converted into energy.

Gamma (γ) The third letter of the Greek alphabet. Used in science to designate high-energy, penetrating electromagnetic radiation produced by some radioactive materials.

Gamma rays (γ rays) Electromagnetic radiation produced from nuclear reactions.

Gas A state of matter in which the substance takes the shape of its container and fills it.

Gay-Lussac's law The pressure of a gas in a rigid container increases in direct proportion to its Kelvin temperature.

Geiger-Müller tube A device used for the detection of rays produced by radioactive materials.

Generic name A common name for a drug or other commercially used chemical.

Geothermal energy The energy obtained from hot rocks or from volcanic activity.

Glass A solid material with a low level of molecular order. It is composed of SiO_2 with small amounts of several other materials added to impart particular properties.

Glucose A simple sugar which is the component of most carbohydrates. It is the primary energy source of living cells.

Glycogen The storage carbohydrate of higher animals and humans. Branched polyglucose bonded by α (alpha) links.

Grain alcohol Ethyl alcohol.

Gram atomic weight See *Mole weight*.

Gram formula weight See *Mole weight*.

Gram molecular weight See *Mole weight*.

Greek letters The Greek letters alpha (A α), beta (B β), delta (Δ δ), and gamma (Γ γ) are used in this text. For definitions, see the entries for the individual letters.

Greenhouse effect The retention of the sun's heat energy by the earth as a result of excess CO_2 in the atmosphere. This causes an increase in the earth's atmospheric and surface temperature.

Haber process The commercial process by which N_2 and H_2 form NH_3 in the presence of a catalyst.

Half-life The time required for half of a sample of radioactive material to decay to a daughter isotope.

Hard water Water containing ions of Mg, Ca, or Fe.

HDL High-density lipoprotein. Believed to play a role in preventing cholesterol deposits in blood vessels.

Heat The energy possessed by a quantity of material by virtue of molecular motion.

Heat capacity The amount of heat needed to change the temperature of 1 g of a material by 1°C. Also called *specific heat.*

Heat of combustion The heat energy given off when a given amount (1 g or 1 mol) of material is burned in the presence of O_2.

Heat of fusion The amount of heat transferred during melting or freezing of 1 g of a material.

Heat of vaporization The amount of heat involved in the evaporation or condensation of 1 g of a material.

Heat pump A device for moving heat from one location to another by alternately evaporating and condensing a material such as Freon.

Herbicide A material that is used to control unwanted plant growth.

Heterogeneous Not the same throughout; having a different composition in different parts.

Homogeneous The same throughout.

Hydrated lime Calcium hydroxide, $Ca(OH)_2$, whitewash.

Hyperactivity A state of agitation accompanied by unruliness and inability to concentrate or learn.

Hypothesis An educated guess or reasonable explanation.

Induction The thought process by which one goes from specific facts to generalizations. In chemistry it is used to form general statements from a collection of individual observations.

Inorganic chemistry That area of chemistry dealing with the study of all elements other than carbon, along with their compounds.

Intersystem Between two or more systems.

Intrasystem Within a given system.

Ion A charged atom or molecule. See *Anion* and *Cation*.

Ion exchange resin A polymeric material with the ability to become saturated with ions and subsequently substitute those ions for another type which was formerly in solution. Used to soften hard water.

Ionic bond A chemical bond usually formed by transferring one or more electrons from atoms of a metal to atoms of a nonmetal. The attraction between ions of opposite charge.

Isomers Two or more covalent compounds with the same molecular formula but with different arrangements of the constituent atoms.

Isotopes Atoms of the same element that have different masses. They contain the same number of electrons and protons but different numbers of neutrons.

IUPAC International Union of Pure and Applied Chemistry. Organization responsible for designating correct naming rules for chemical compounds.

Kelvin scale A temperature scale with zero indicating no thermal energy and each degree equal to one-hundredth of the difference in thermal energy between freezing and boiling water.

kilo- A prefix meaning 1000 times.

Kinetic energy The energy possessed by virtue of motion.

Kinetic molecular theory The view of the atomic and molecular nature of gases, liquids, and solids describing the movements, attractive forces, and distance between molecules.

Law A generalized statement of fact.

Law of conservation of energy Energy cannot be created or destroyed; it can only change forms.

Law of conservation of matter The amount (mass) of matter does not change during chemical or physical changes.

Law of definite proportions A given compound is always composed of the same proportion of its various elements.

Law of multiple proportions When two elements form two or more compounds, the ratios of these elements in the several compounds are always whole numbers.

LD_{50} The lethal dose for 50 percent of a population of test organisms.

Lewis dot diagram An array of element symbols along with their valence electrons indicating the arrangement of the atoms and electrons in ions or molecules.

Ligand A species that coordinates (bonds) with a metal ion.

Light That portion of the electromagnetic radiation spectrum detectable by the human eye.

Lime Calcium oxide (CaO).

Limestone Calcium carbonate ($CaCO_3$).

Lipid Any nonpolar biomolecule, especially a fat or oil.

Liquid A state of matter in which the substance takes the shape of its container but does not necessarily fill it.

Liquid crystal A state of matter in which the molecules have an orderly arrangement yet are fluid in nature. The orderliness can be changed by temperature, electric current, etc., and results in color changes. The material has characteristics of both the liquid and the crystalline solid states.

LSD Lysergic acid diethylamide.

Lye Sodium hydroxide (NaOH, caustic soda).

Manometer A device for measuring the pressure of a gaseous sample. It is a U-shaped tube partially filled with a liquid (usually mercury).

Mass A property of matter that causes it to resist a change in motion or to possess inertia.

Mass spectrograph A device used to measure the mass-to-charge ratio of particles by means of the deflection produced by electrostatic or magnetic fields.

MDR The minimum daily requirement of a human nutrient.

Medicine A medicinal drug; a material used to relieve a health problem.

Melanin A dark pigment which produces coloration in human skin when exposed to sunlight.

Melting The changing from the solid state to the liquid state.

Metal An element that conducts electricity and is strong and bendable. Metals usually have a "metallic" luster (shine) and are found in the left part of the periodic chart.

Metallic bonding A continuous cloud of electrons holding atoms of metallic elements together.

Metric system A system of measurement in which each unit is a whole power of 10 times the other units. The basic units are the kilogram (kg) for mass, the meter (m) for length, and the liter (L) for volume.

Micelle A spherical arrangement assumed by molecules which have a polar and nonpolar end. The polar ends all point out, and the nonpolar ends all point in toward the center of the sphere.

Milk of magnesia Magnesium hydroxide, $Mg(OH)_2$.

milli- A prefix meaning one-thousandth.

Mineral An inorganic material required in the human diet.

Mixture A combination of two or more pure substances.

Moderator A material used to slow neutrons in order to produce more effective fission in a nuclear reactor. D_2O or graphite is used, as is ordinary water.

Mole 6.02×10^{23} of anything. In chemistry, usually used to indicate that number of atoms, molecules, ions, or formula units.

Molecular formula The chemical formula indicating the actual number of atoms of each element in a molecule.

Molecular weight The weight of one molecule; the summation of the atomic weights of all atoms composing the molecule.

Molecule A structural unit composed of two or more atoms covalently bonded together.

Mole ratio factor (MRF) The ratio of the moles of material to be calculated to the moles of material known. Used in stoichiometric calculations.

Mole weight The weight of material containing 1 mol of atoms, formula units, or molecules. Called *gram atomic weight, gram molecular weight,* or *gram formula weight* for atomic, covalently bonded, and ionic materials, respectively.

Muriatic acid Hydrochloric acid (HCl) solution.

NASA National Aeronautics and Space Administration.

Natural radioactivity The radiation produced by naturally occurring unstable isotopes

NBS National Bureau of Standards

Neutron A relatively large neutral particle found in the nucleus of all atoms except the most abundant isotope of hydrogen.

Nicotine A stimulant found in tobacco smoke.

NIH National Institutes of Health.

Nonmetals The elements found in the right part of the periodic chart. They are brittle solids or gases or liquids and do not conduct electricity.

NRC Nuclear Regulatory Commission.

NSF National Science Foundation.

Nuclear decay The spontaneous change in unstable isotopes which produces new isotopes along with radiation.

Nuclear reaction Any change, either spontaneous or induced, which produces a new nucleus.

Nuclear reactor A device used to control a fission reaction in order to produce heat energy.

Nucleic acids The genetic material containing chains composed of phosphate and sugar.

Observation A fact obtained by using one or more of the human senses or various instruments, or both.

Octane number An arbitrary scale indicating the antiknock properties of gasoline.

Octet rule Atoms tend to interact with each other so as to have eight electrons in the outer shell of each. Those below element 4 (Be) tend to get two.

Oil A class of biomolecules composed of glycerin and three fatty acids (a triglyceride). Oils are liquid at room temperature.

Optical brightener A fluorescent dye used in laundry detergents to cause cloths to ''glow'' and thus look bright and clean.

Organic chemistry The area of chemistry dealing with covalently bonded carbon compounds.

OSHA Occupational Safety and Health Agency.

OTC Over-the-counter. Usually refers to commonly available drugs.

Oxidation The lost of electrons; an increase in oxidation number.

Oxidation number The charge on an ion or effective charge on a covalently bonded atom; the number of electrons lost (positive charge, +) or gained (negative charge, −) by an atom during reaction.

Oxidation state See *Oxidation number*.

PABA *para*-Aminobenzoic acid. A sunscreen and topical anesthetic. See *Benzocain*.

Particulate matter Pieces of solid matter expelled into the atmosphere that are small enough to remain suspended and be transported by air currents.

Penicillin A complex antibiotic material isolated originally from bread mold.

Percent Parts per hundred.

Period (of elements) The horizontal rows of elements on the periodic chart.

Periodic chart The standard arrangement of the symbols of the 105 elements (along with atomic data) in a way useful to the chemist. It reflects known trends in their properties and serves as an important, easily used reference for these trends. Also called the *periodic table*.

Periodic law Elements exhibit repeating sequences of chemical and physical properties if they are arranged in order of increasing atomic number.

pH scale An exponential scale of acidity: 0 to 7, acidic; 7, neutral; 7 to 14, basic.

Pheromone A chemical causing a powerful and usually species-specific sexual attraction.

Photochemical smog The visible air pollution caused by sunlight shining on stagnant motor vehicle pollutants.

Photon A massless particle of light used to explain the behavior of light in certain experiments.

Photovoltaic cell A device which converts light into electric energy.

Physical change A change in matter involving its physical state or shape but not causing the formation of a new substance.

Physical chemistry The area of chemistry dealing with measuring and correlating large amounts of data and discovering the underlying principles of chemistry.

Physical dependence A state in which the body needs a foreign material to function normally, be pain-free, or feel comfortable.

Physical properties The characteristics of a substance that determine its physical nature. They include melting point, color, and density.

Plaque The deposits, composed mostly of cholesterol, on the insides of blood vessels.

Plasma A state of matter composed of isolated electrons and nuclei rather than discrete whole atoms.

Plastic Any synthetic polymeric material.

Polar covalent bond A chemical bond between two nonmetals in which the electrons are attracted toward one end of the bond. The bond thus has a partially positive end and a partially negative end.

Pollution Harmful, unpleasant, or otherwise undesired material in contact with living systems.

Polyatomic Containing three or more atoms.

Potash of soda Potassium carbonate (K_2CO_3).

Potential energy The energy possessed by virtue of position or chemical or physical state.

ppm Parts per million.

Precision An expression of the care with which a measurement has been made.

Preignition The ignition of fuel in an internal combustion engine due to compression alone before the correctly timed spark-induced ignition.

Properties Characteristic descriptions of a material.

Protein A polyamino acid molecule having 50 or more amino acid units.

Proton A positive particle with relatively large mass which is found in the nucleus of all atoms.

Pure material A substance composed of only one element or compound.

Pure research The research aimed at producing new knowledge with no intent of producing a new or useful product.

Qualitative analysis The process of determining what materials are present in a mixture.

Quantitative analysis The process of determining how much of one or more materials are present in a mixture.

Quantized Able to occupy only certain discrete levels or have certain values. Opposite of *continuous*.

Quat A quaternary ammonium salt used as an antiseptic or cationic surfactant (detergent).

Radioactive Producing ionizing radiation (alpha, beta, or gamma rays).

RDA The recommended daily allowance of a human nutrient.

Reduction A gain of electrons or lowering of oxidation number.

Regulation A rule or law controlling behavior or specifying procedure.

Relative humidity The moisture content of a material compared with the maximum possible moisture content at a given temperature.

Research The process of chemical investigation aimed at producing new methods, new compounds, or new understanding about chemicals or chemical processes.

Runoff The rainwater running off of farmland, which possibly carries dissolved fertilizers and suspended soil particles.

Salt A general term used for any compound containing NH_4^+ or a metal cation along with an anion. Also especially used to indicate sodium chloride (NaCl).

Scale The deposit of $CaCO_3$ formed from hard water during heating.

Scholarship The money given to support a student while he or she studies. Usually awarded on the basis of academic merit.

Science A field of study based on systematic observations. It also includes the organization of those observations and attempts to arrive at explanations of them.

Scientific notation See *Exponential notation*.

Shell A group of electronic atomic orbitals of approximately the same size. Sometimes called an *electronic energy level*.

Significant figures (digits) The number of digits in a measurement or other expression of quantity that has meaning as regards the care with which the measurement was made.

SI system The International System of measurement (*Système International d'Unités*), based on the metric system.

Soap Sodium stearate. Used as a cleaning agent and a surfactant.

Sodium hypochlorite NaOCl. Used in liquid chlorine bleaches.

Sodium stearate The most common type of soap.

Sodium triphosphate $Na_5P_3O_{10}$. Used as a base (to neutralize acid) in laundry formulations.

Softening (of water) The removal of Mg, Ca, or Fe ions from hard water.

Solar collector A device which collects heat from the sun to be used to heat air or water.

Solid The state of matter in which the substance has its own shape.

Solvay process A commercial process used to produce sodium bicarbonate ($NaHCO_3$) from NaCl, CO_2, and NH_3.

Spectroscope A device containing a prism which separates a source of colored light into its component colors.

Stack absorber A device used to remove acidic materials, especially SO_2, from waste gases of industrial facilities.

Starch The storage carbohydrate of plants. Composed of α-linked glucose units.

Structure-function principle The properties of a compound are a direct result of its molecular structure.

Subatomic particle One of the constituent parts of atoms. Subatomic particles include electrons, protons, and neutrons.

Sublimation The change from the solid to the gaseous state without going through the liquid state.

Sucrose Common table sugar (cane sugar).

Sugar A nonpolymeric carbohydrate that is a monosaccharide or disaccharide.

Sulfa drugs A class of antibiotic materials.

Surfactant A material that causes essential homogenization of two otherwise mutually insoluble materials.

Synergistic Referring to an interaction between drugs causing each to exert its effect more powerfully than either one would exert alone.

Synfuel A liquid or gaseous fuel produced from coal.

Synthesis The preparing of new chemical compounds from elements or other compounds.

Temperature A measure of the average kinetic energy of the atoms or molecules of a substance.

Testing The determining of the chemical, physical, and physiological properties of materials with the view in mind of discovering new uses for those materials.

Tg The glass transition temperature. The temperature at which a true crystalline solid changes to a rubbery flexible state.

Theory A tested, generally accepted explanation.

TI Therapeutic index.

Transmutation The converting of one element into another by bombardment with high-energy particles.

Tyndall effect The visualization of a beam of light by a colloidal suspension or micelles.

Unit An arbitrary standard of measurement or a multiple thereof. A whole entity.

Valence electrons The electrons in the outer shell of an atom.

Vaporization The change from the liquid to the gaseous state. Also called *evaporation*.

Vinegar Acetic acid ($HC_2H_3O_2$) solution.

Vitamin An organic material needed in trace amounts in the human diet.

Voltage The tendency or "pressure" for electrons to flow or to be transferred.

Volume Space. The term usually indicates how much space is occupied by a sample of matter.

Washing soda Sodium carbonate (Na_2CO_3).

Wavelength The distance between similar points (i.e., crests) on adjacent waves.

Wave-particle duality The view that light has the characteristics of both particles and waves.

Weight The response of mass to gravity.

Wood alcohol Methyl alcohol.

Xenobiotic See *Biogenic*.

Zeolite A natural ion exchange resin used in water softeners.

Answers to Skill Practice Problems

CHAPTER 1

1 See text, page 16.

2 Organic, physical, analytical, inorganic, biochemical

3 See text, page 3.

4 **(a)** True
(b) False
(c) False
(d) False
(e) False
(f) True
(g) False

5 **(a)** Mixture
(b) Element
(c) Compound
(d) Mixture
(e) Mixture
(f) Mixture
(g) Element
(h) Mixture
(i) Compound

6 **(a)** Chemical
(b) Physical
(c) Chemical
(d) Physical
(e) Chemical
(f) Physical
(g) Physical
(h) Physical

Note There are no skill practice problems for Chapters 2 and 3.

CHAPTER 4

1 **(a)** 5.17×10^{9}
(b) 1.29×10^{-8}
(c) 3×10^{6}
(d) 2.53×10^{-5}
(e) 2.181×10^{2}

2 **(a)** 0.00816
(b) 0.00000306
(c) 671,000
(d) 13,716,000,000
(e) 250
(f) 602,000,000,000,000,000,000,000

3 **(a)** 46 cm
(b) 457 cm
(c) 258 K
(d) 271 K
(e) −40°F
(f) 33 ft
(g) 227 kg
(h) 85 g

4 75 percent

5 5% = 5 parts per 100 mL = 50 parts per 1000 mL
= 100 parts per 2000 mL
2 L = 2000 mL
Therefore in the 2000 mL of vinegar, there are 100 mL of acetic acid or 5% = x/2000 mL × 100, x = 100.

6 14.3 percent of the chemistry students is five students. Six students is 17.1 percent; therefore, one student makes a difference of about 3 percent. 16.7 percent of 30 is five English students. Four students is 13.3 percent, or a difference of about 3 percent.

Since the difference in the two classes is less than the difference one student makes, and on a given day *one* student more or less was absent, the attendance in the two courses was for all practical purposes the *same*.

7 **(a)** 4400 ft/min
(b) 69 blades per square foot
(c) 2.2×10^6 atoms per liter

8 15.6 m/s

9 235 mL milk, 353 mL flour, 118 mL water, 15 mL sugar, 5 mL baking powder, 30 mL cooking oil. Cook at 163°C for 300 s.

10 29.4 cents per liter.

11 8 min

CHAPTER 5

1 **(a)** The number of protons in the nucleus of an atom of that element
(b) An atom of an element having a particular number of neutrons. Isotopes differ by the number of neutrons and therefore the atomic mass.
(c) A mental picture of physical reality
(d) The average weight of naturally occurring atoms of an element
(e) The smallest individually stable unit of matter
(f) The tendency of two oppositely charged entities to move toward each other

2 **(a)** 14 e^-, 14 p^+, 14 n
(b) 9 e^-, 9 p^+, 10 n
(c) 76 e^-, 76 p^+, 114 n

3 **(a)** Dalton
(b) Definite proportions
(c) Isotopes
(d) Deflected toward the negative plate
(e) They would have been deflected even less.
(f) It indicated that almost all the mass was concentrated in a small area in the center of the atom.

4 See text, page 107.

5 (a) Plato did not accept it.
(b) Christians rejected Epicureanism, which included atomism.

6 See text, page 103.

7 There are no known facts which seem to contradict it.

CHAPTER 6

1 **(a)** Sr
(b) Au
(c) Al
(d) Ga

2 F, Si, V, Ca, Cs

3 **(a)** As
(b) Sn
(c) Al
(d) B
(e) Ba

4 Cl, I, Mg, Na, Rb

5 **(a)** Particle, orbit
(b) Red
(c) Violet
(d) Li, Pb
(e) K
(f) VIIA, halogens
(g) Niels Bohr
(h) x-ray
(i) Radio waves
(j) x-ray
(k) P, 5; Si, 4; B, 3

6

	GROUP	PERIOD
(a)	IVA	2
(b)	VIB	4
(c)	IB	5
(d)	VA	4
(e)	VIIIA	3

7 See text, page 131.

8 Mg or Al

CHAPTER 7

1 **(a)** 4
(b) 3
(c) 2
(d) 1
(e) 5

2 **(a)** $CH_4 + 4Cl_2 \longrightarrow 4HCl + CCl_4$
(b) $C_2H_4 + 3O_2 \longrightarrow 2CO_2 + 2H_2O$

(c) $4NH_3 + 3O_2 \longrightarrow 6H_2O + 2N_2$
(d) $6KCN + 2As_2O_3 \longrightarrow 6KOCN + As_4$
(e) $Na_2S_2O_5 + H_2O \longrightarrow 2NaHSO_3$

3 ·S̈e: Since the outer shell has six e⁻, it has to gain two e⁻ to satisfy the octet rule.

·Ċ· It gains four e^- (it is a nonmetal).

·Al· It loses three e^-.

·Ca It loses two e^- to fulfill the octet rule.

4 **(a)** Polar covalent
(b) Pure covalent
(c) Ionic
(d) Pure covalent
(e) Polar covalent
(f) Ionic
(g) Metallic
(h) Ionic
(i) Metallic
(j) Polar covalent

5 It is more soluble in ethylene dichloride (from general knowledge one knows that wax is not removed by water spilling on the floor—the wax must therefore be *nonpolar*).

6 Crayons are composed of nonpolar materials (since they do not dissolve in water). Paint thinner would work; water would not.

7 It is highly explosive.

8 Se^{2-}, anion; C^{4-}, anion; Al^{3+}, cation; Ca^{2+}, cation

9 No. Compounds have different physical, chemical, and physiological properties (including toxicity) from the elements of which they are composed.

10 150 lb = 68 kg
68 kg (4 mg/kg) = 270 mg = 0.27 g

11 270 mg would require 270 L of water.

$$(270\ \text{L})\left(\frac{1\ \text{qt}}{0.95\ \text{L}}\right)\left(\frac{1\ \text{gal}}{4\ \text{qt}}\right) = 71\ \text{gal}$$

CHAPTER 8

1 To be able to "count" atoms and molecules and to be able to think about ratios of atoms or molecules

2 6.02×10^{23} of anything

3 By dividing the weight of that material by its atomic or mole weight

4 $$\frac{168\ \text{g}}{56\ \text{g/mol}} = 3\ \text{mol of Fe}$$

5 $3(6.02 \times 10^{23}) = 18 \times 10^{23} = 1.8 \times 10^{24}$

6 $\frac{10 \text{ g}}{58 \text{ g/mol}} = 0.17$ mol of NaCl

7 $0.17 \text{ mol}(6.02 \times 10^{23}) = 1 \times 10^{23}$ Na^+ ions

8 $MgCl_2 = 95$ g
$ZnCl_2 = 136$ g

9 Chemical synthesis, medicinal dosage determinations, measurements of acidity

10 3.4×10^{11} years

11 1 to 1

12 As long as the chemist knows, for example, that the *ratio* is 1 to 1, it does not really matter what the numbers are.

13 2.86 mol

14 $\frac{3 \text{ oxygen}}{1\ Fe_2O_3} = 3$

15 1.4×10^{22} molecules

16 $\frac{1.4 \times 10^{22} \text{ molecules}}{10{,}080 \text{ min}} = 1.4 \times 10^{18}$ molecules per minute

CHAPTER 9

1 1200 cal

2 40,000 cal or 40 kcal

3 94 kcal

4 Allowing the eggs to sit in hot water for 10 min uses 54 kcal less. Allowing the eggs to sit in hot water would save 57 percent of the energy used in heating the eggs in boiling water for 8 min.

5 659,200 kcal

6 Possible examples are as follows. Gases: Ping-Pong ball in a windstorm. Liquids: Peanuts in a bowl. Solids: ''Lego'' blocks snapped together in a regular pattern.

7 If the pressure inside the can gets too high, the can will explode and can cause serious injury.

8 It means the actual humidity *compared with* what the maximum humidity would be at that temperature.

9 Since the pressure is changed to half the original, the volume will double to 16 L.

10 **(a)** 30°C is 303 K.

(b) 0°C is 273 K.
(c) The temperature (on the K scale) is reduced to 273/303 = 0.9 of its original value. Therefore, the pressure is also 0.9 of its original value.

$$0.9(35\ \mathrm{lb/in^2}) = 31.5\ \mathrm{lb/in^2}$$

11 **(a)** 350°C is 623 K.
(b) The temperature (on the K scale) rises to 623/298 = 2.1 of its original value. Since the temperature (K) has risen by a factor of 2.1, so will the pressure.
(c) 2.1(2 atm) = 4.2 atm
(d) Since this is above the pressure limit of the can, it is likely to explode.

12 Boyle's law; Gay-Lussac's law

13 It slowly contracts. The air molecules bombard it with greater force.

14 It will rise higher, since it is lighter (less dense).

15 It will rise 13.5(760 mm) = 10260 mm = 1026 cm = 10.26 m.

16 37 Calories = 37,000 calories

CHAPTER 10

1 **(a)** Li_2SO_4
(b) H_2SO_3
(c) Na_2O
(d) Cu_3PO_4
(e) CBr_4
(f) $Al(CN)_3$
(g) $AgNO_3$
(h) FeI_3
(i) CS_2
(j) K_2SO_3

2 **(a)** Cobalt nitrate
(b) Iron(II) perchlorate or ferrous perchlorate
(c) Aluminum chloride
(d) Manganese dioxide or manganese oxide
(e) Zinc sulfate
(f) Chromic acid
(g) Uranium fluoride or uranium hexafluoride
(h) Tin(IV) hydroxide or stannic hydroxide
(i) Magnesium cyanide
(j) Silicon tetraiodide or silicon iodide

3 The lime juice is 10 times as acidic.

4 **(a)** 80 ppm = 80 mg/L = 0.080 g of $CaCO_3$ per liter
(b) $\dfrac{0.080\ \mathrm{g}}{100\ \mathrm{g/mol}} = 0.0008\ \mathrm{mol}\ CaCO_3$
(See Appendix E for the method.)
(c) This produces twice as much Na^+, or 0.0016 mol of Na^+.
(d) 0.0016 mol $Na^+ \times$ (23 g/mol) = 0.0368 g = 36.8 mg Na^+

5 **(a)** $50 - 6.7 = 43.3$ mg precipitate per liter.
(b) 1 gal × (4 qt/gal) × (0.946 L/qt) = 3.78 L
(c) 3.18 L × (43.3 mg/L) = 164 mg = 0.164 g $CaCO_3$

6 $CaCO_3 + H_2CO_3 \longrightarrow Ca(HCO_3)_2$

7 pH = 2

8 See text, page 218.

9 Red

10 It is cold liquefied ammonia (NH_3) or ammonia under pressure at ordinary temperatures. No, it should be called *liquefied ammonia*.

11 It produces CO_2, which smothers the fire.

CHAPTER 11

1 **(a)** 3-Chloro-2,4-dimethylpentane
(b) 1-Butene
(c) 3-Bromo-4-chloro-2,2-dimethylhexane
(d) 1,2,3-trichloropropane
(e) Triiodomethane

2 **(a)** 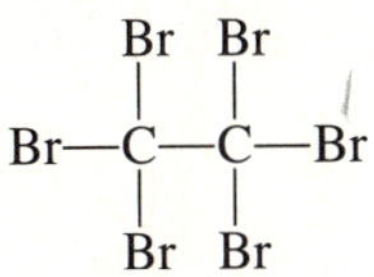

(b) 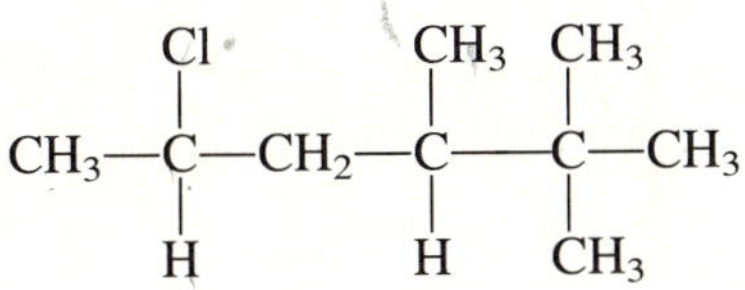

(c) 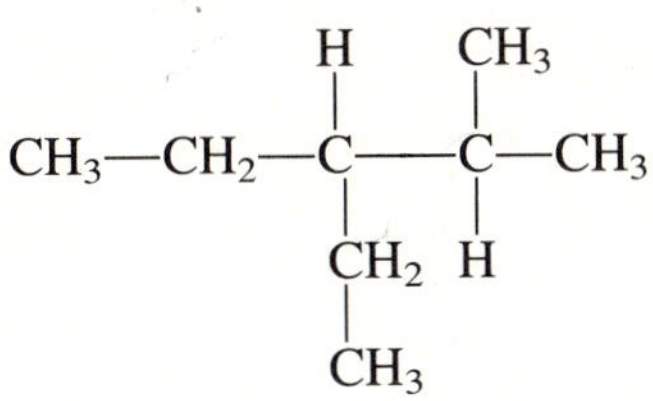

(d) 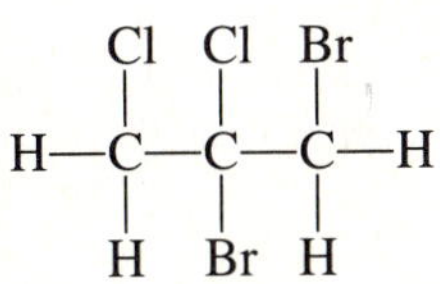

(e) 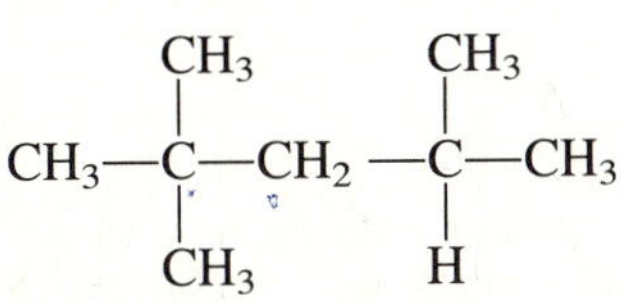

3 **(a)** Ether
(b) Aromatic, ketone
(c) Aldehyde
(d) Aromatic, amide
(e) Aromatic, ether
(f) Aromatic, carboxylic acid
(g) Amine
(h) Aromatic, alcohol
(i) Aldehyde, ketone
(j) Ester

4 **(a)** Welding torch fuel, acetylene
(b) Vinegar, acetic acid
(c) Chewing gum solvent, ethylene chloride
(d) Fuel (gasoline), octane
(e) Solvent (fingernail polish remover), acetone
(f) Oil of wintergreen (flavoring), methyl salicylate

5 **(a)** The process of changing high-molecular-weight alkanes to smaller molecules that can be used for fuel
(b) A compound whose molecules have only carbon and hydrogen and one or more carbon-carbon double bonds
(c) A material that facilitates a chemical reaction but is not consumed
(d) A series of chemical procedures used to prepare a compound from other compounds
(e) Separating components of a mixture by heating in such a way that materials with different boiling points can be collected at different times

6 See text, page 275.

7 To form a polymeric coating

8 b, d

9 **(a)** Not be harmed by blood
(b) Be flexible
(c) Be extremely durable
(d) Be nontoxic

10 No toxic materials are used, and it is very selective. 2,4-D is short-lived in nature.

11 **(a)** Control of tobacco suckers
(b) Control of sprouting in potatoes
(c) Control of growth of grass

CHAPTER 12

1 They are nearer the ground, and the PbO_2 from motor vehicle exhausts slowly settles toward the ground. Thus the children are breathing air that has more PbO_2 suspended in it.

2 **(a)** Not biodegradable—not removed by nature
(b) Harmful to living systems
(c) Most concentrated where most people live

3 **(a)** Acid rain
(b) Corrosion
(c) Bleaching
(d) Lung damage
(e) Crop damage

4 pH = 5

5 **(a)** Basically false. The high standard of living of consumers is responsible. However, individual industries have been irresponsible.
(b) False. Larger ones have more.
(c) True. See Chapter 10.
(d) False. It prevents preignition.
(e) False. It contains mostly SO_3 and other industrial pollutants.
(f) True. See Table 12-6.

6 **(a)** 0.50 ppm = 0.5 mg/kg
(b) 30 μg per day = 0.03 mg per day
(c) 0.03 mg per day ÷ 0.5 mg/kg = 0.06 kg tuna/day = 60 g per day
(d) $60\text{ g}\left(\frac{1\text{ lb}}{454\text{ g}}\right)\left(\frac{16\text{ oz}}{1\text{ lb}}\right) = 2.1\text{ oz per day}$

7 **(a)** Automobile pollution from driving to the golf course
(b) Industrial pollution from producing the steel for the clubs
(c) Gasoline pollution from frequent mowing of the golf course
(d) Manufacturing pollution from the production of golf carts and buildings
(e) Fertilizer runoff

8 The bus holds 25 times as many people but gets one-fifth the gas mileage. This means that it is 5 times as efficient. One would save four-fifths of the energy and pollution, or, to put it another way, pollution and fuel consumption would be reduced by 80 percent.

9 **(a)** One mile would use 1/20 (0.05) gal.
(b) 0.05 gal [4 g $Pb(C_2H_5)_4$/gal] = 0.2 g $Pb(C_2H_5)_4$
(c) $\frac{0.2\text{ g } Pb(C_2H_5)_4}{323\text{ g/mol}} = 6.2 \times 10^{-4}\text{ mol}$
(d) $6.2 \times 10^{-4}\text{ mol } Pb(C_2H_5)_4 \left(\frac{1\text{ mol } PbO_2}{1\text{ mol } Pb(C_2H_5)_4}\right) = 6.2 \times 10^{-4}\text{ mol } PbO_2$
(e) 6.2×10^{-4} mol PbO_2 (239 g/mol) = 0.15 g PbO_2

CHAPTER 13

1 Natural radioactivity, transmutation, fission, fusion

2 Alpha, beta, and gamma rays

3 Directly affect biological function, cause cancer, damage genetic material, and cause deformed offspring

4 To slow down neutrons, so that fission will proceed at a faster rate

5 4000 years

6 5700 years old (one half-life). No; but we can assume that the wood was no more than 100 years old when it was used in the coffin.

7 Manuscripts, paintings and drawings, artifacts from archeological explorations, ancient plants

8 Seek shelter underground for 10 days to 2 weeks, eat washed food and filtered water, and avoid exposure to fallout.

9 It contains less than the critical mass of fissionable material or less than the fissionable concentration of U-235.

10 Advantages: It is relatively safe and produces little air pollution. Disadvantages: It is expensive and produces water heat pollution as well as radioactive wastes that must be stored for a long time.

11 **(a)** $\frac{270 \text{ counts}}{3 \text{ min}} = 90 \text{ cpm}$

(b) 90 cpm − 40 cpm background = 50 cpm

12 $E = (0.01 \text{ kg})(3 \times 10^8 \text{ m/s})^2$
$= 9 \times 10^{14} \text{ kg}\cdot\text{m}^2/\text{s}^2 = 9 \times 10^{14} \text{ joules (J)}$

13 **(a)** Transmutation
(b) Natural radioactivity
(c) Fusion
(d) Transmutation
(e) Fission
(f) Natural radioactivity

14 See text, page 335.

CHAPTER 14

1 It provides greater comfort and higher productivity. No.

2 No. Yes. (See text, page 379.)

3 See Figure 14-6.

4 CCl_2F_2

5 C

6 Increased population and increased per capita energy consumption due to a higher standard of living

7 Possible answers follow. Mechanical: Washing clothing, sawing wood, mowing a lawn. Electrical: Using a pocket calculator, using a flashlight. Heat: Cooking, drying hair. Chemical: Metabolizing food.

8 **(a)** No
(b) Yes—steam turbine
(c) Yes—resistance wire
(d) Yes—photovoltaic cell
(e) No
(f) Yes—internal combustion engine

9 (a) Natural
(b) Requires energy
(c) Natural
(d) Requires energy
(e) Requires energy

10 b

11 Hydroelectric generation

12 Fossil fuel

13 Advantages: Indefinite supply, relatively clean.
Disadvantages: Not evenly distributed.

14 Advantages: It is abundant and cheap.
Disadvantages: It is polluting.

15 Advantages: The supply of fuel is virtually infinite.
Disadvantages: We do not know how to build one yet.

16 False. The production of solar collectors and photovoltaic cells causes some industrial pollution.

17 $2C_4H_{10} + 13O_2 \longrightarrow 8CO_2 + 10H_2O$ + heat energy

18 $2H_2 \longrightarrow 4H^+ + 4e^-$

19 $2MnO_2 + H_2O + 2e^- \longrightarrow Mn_2O_3 + 2OH^-$

20 **(a)** The Fe^{2+} is oxidized to Fe^{3+} at the anode.

21 **(a)** Cl_2 becomes Cl^-; it *gains* electrons.
(b) Pb^{2+} becomes Pb with no charge; it therefore *gains* electrons.
(c) F_2 becomes F^-; it *gains* electrons.

CHAPTER 15

1 False. It is only one factor.

2 **(a)** One lb is 454 g.
(b) 40 mg of vitamin C in a 200-g (from Table 15-6)
(c) $454 \text{ g} \left(\dfrac{40 \text{ mg vitamin C}}{200 \text{ g potato}}\right) = 91 \text{ mg vitamin C}$

3 Proteins

4 Glucose

5 β links cannot be broken in cellulose.

6 d

7 a

8 b

9 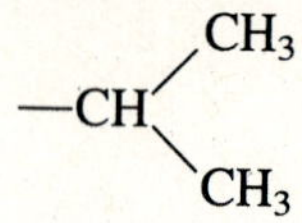

10 Gly-Leu; Leu-Leu; Gly-Gly; Leu-Gly

11 Gly-Gly-Gly; Leu-Leu-Leu; Gly-Gly-Leu; Gly-Leu-Gly; Leu-Gly-Gly; Leu-Leu-Gly; Leu-Gly-Leu; Gly-Leu-Leu

12 See text, page 400.

13 $454\ \text{g/lb}\left(\frac{1\ \text{lb}}{16\ \text{oz}}\right)(1\ \text{oz})(4.5\ \text{Cal/g}) = 128\ \text{Cal (kcal)}$

14 False. Both should be consumed in moderation.

15 5000 IU

16 Recommended daily allowance

17 See text, page 421.

CHAPTER 16

1 Stomach irritation, stomach bleeding, lowered resistance to infection, slow clotting time, hyperactivity

2 150 g = 0.150 kg
1.4 g/kg (0.150 kg) = 0.21 g needed

3 Yes, they are justified because of the expense of developing a drug.

4 Coffee

5 Food and Drug Administration (FDA)

6 Less stomach irritation

7 Use effervescent aspirin or take two tablets with hot water.

8 Rate of ingestion and rate at which the body metabolizes it

9 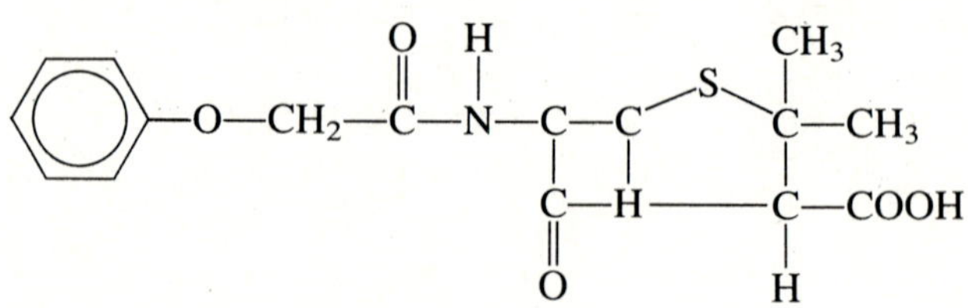

10 It is less irritating to the stomach.

11 c

12 b, d, e

13 Benzocaine

14 A quat such as cetylpyridinium chloride.

15 40 percent

16 Molasses

17 Denatured

18 10 (1 oz)(50%) = 5 oz of alcohol
5 oz (1 oz/h metabolism) = 5 h

19 c and d

20

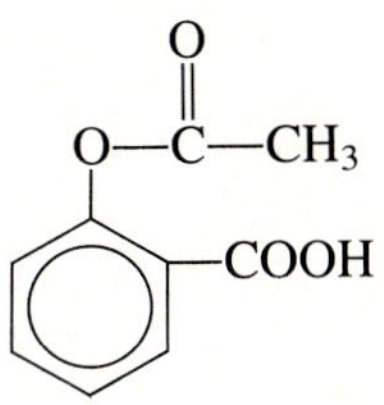

21

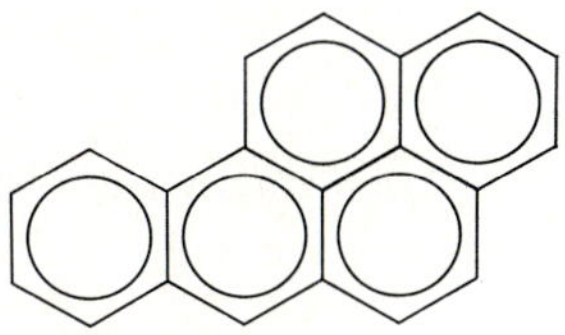

22 Diazepam

23 No. The relative amounts of two enzymes, determined by heredity, are an overriding factor.

24 Yes

25 **(a)** True. It is an aphrodisiac, but it is not a true pheromone.
(b) False. They can stimulate natural opiate formation.
(c) True
(d) True
(e) False

CHAPTER 17

1 See text, page 464.

2 See text, page 464.

3 See text, page 465.

4 Add the H_2O_2 to chlorine bleach; vigorous bubbling of O_2 gas indicates the H_2O_2 is good.

5 Water containing Ca^{2+}, Mg^{2+}, or Fe^{2+} ion

6 It forms "scum."

7 See text, page 469.

8 They have been washed with an optical brightener.

9 By breaking the food into sugar or amino acids which dissolve in the water

10 NaOCl (sodium hypochlorite)

11 2.3 lb per year for one household
230,000 lb per year for the entire city

12 **(a)** Ethylene dichloride or lighter fluid (no flames!)
(b) H_2O_2 (3% hydrogen peroxide)
(c) NaClO (household bleach)
(d) Oxalic acid solution
(e) H_2O_2
(f) Denatured ethyl alcohol (shellac thinner)
(g) NaClO
(h) Rubbing alcohol or shellac thinner
(i) NaClO
(j) Mixture of H_2O_2, NH_4Cl, and alcohol
(k) Ethylene dichloride or, because of possible skin irritation from ethylene dichloride, peanut butter
(l) Lighter fluid
(m) Try paint remover, but if the rug is a synthetic material, paint remover will probably dissolve the rug as well. Buy a new rug or set a plant on that spot.

13 **(a)** 200 g (0.03) = 6.00 g H_2O_2

(b) $\dfrac{6.00 \text{ g } H_2O_2}{34 \text{ g/mol}} = 0.176 \text{ mol } H_2O_2$

(c) $0.176 \text{ mol } H_2O_2 \left(\dfrac{1\ O_2}{1\ H_2O_2}\right) = 0.176 \text{ mol } O_2$

(d) $0.176 \text{ mol } O_2 \text{ (32 g/mol)} = 5.65 \text{ g } O_2$ produced

14 Water running through wood ashes

15 Toxic Cl_2 gas is produced when they are mixed.

16 Sodium ethylenediaminetetraacetic acid (NaEDTA)

17 All but Suave and Johnson & Johnson's

18 b and d. These are the only two pairs in which hydrogen is bonded to O, N, or F in one molecule and the other molecule also contains an O, N, or F.

19 Proteins

20 See text, page 480.

21 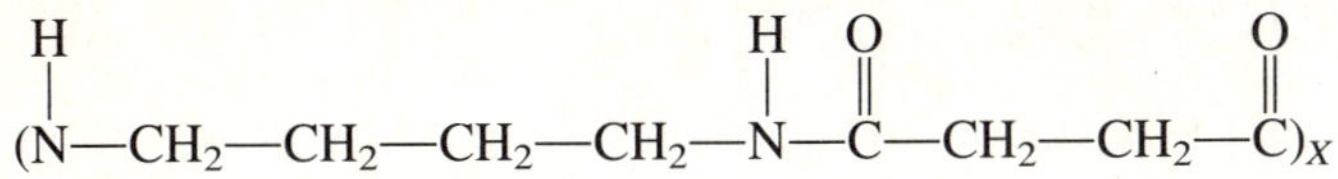

22 See text, page 484.

23 **(a)** Remove the air supply.
(b) Cool the burning material.
(c) Use a blanket of gas or cloth.
(d) Coat the material with something that reacts with free radicals.

24 **(a)** Leavening agent
(b) Fire extinguisher

25 It is acidic and reacts with the basic amines (which cause the odor), rendering them water-soluble and nonvolatile

26 **(a)** It denatures the enzymes that cause darkening.
(b) Other processes include cooking and adding acidic materials like lemon juice or vinegar.

27 It provides the acid with which the $NaHCO_3$ can react to produce CO_2 bubbles.

Note There are no skill practice problems for Appendixes A through D.

APPENDIX E

1

Mg	9.7 percent
S	13.0 percent
O	71.5 percent
H	5.7 percent
	99.9 percent

2 Balance the equations.

$$Fe_3O_4 + 2C \longrightarrow 2CO_2 + 3Fe$$
$$\text{moles } Fe_2O_3 = 4.31$$
$$\text{moles Fe} = 4.31 \times 3 = 12.93$$
$$\text{Weight Fe} = 12.93\ (56) = 724\ \text{g}$$

3 C_2H_2NO

4 $C_4H_4N_2O_2$

5 166 g TiO_2

6 **(a)** 90.4 percent yield of TiO_2
(b) 9.6 percent loss on the investment

7 347 g of oxygen

Note There are no skill practice problems for Appendix F.

Index

THE PERIODIC CHART

I A	II A	III B	IV B	V B	VI B	VII B	VIII B	VIII B	VIII B	I B	II B	III A	IV A	V A	VI A	VII A	VIII
1 H Hydrogen 1.01																	2 He Helium 4.00
3 Li Lithium 6.94	4 Be Beryllium 9.01											5 B Boron 10.8	6 C Carbon 12.0	7 N Nitrogen 14.0	8 O Oxygen 16.0	9 F Fluorine 19.0	10 Ne Neon 20.2
11 Na Sodium 23.0	12 Mg Magnesium 24.3											13 Al Aluminum 27.0	14 Si Silicon 28.1	15 P Phosphorus 31.0	16 S Sulfur 32.1	17 Cl Chlorine 35.5	18 Ar Argon 39.9
19 K Potassium 39.1	20 Ca Calcium 40.1	21 Sc Scandium 45.0	22 Ti Titanium 47.9	23 V Vanadium 51.0	24 Cr Chromium 52.0	25 Mn Manganese 54.9	26 Fe Iron 55.9	27 Co Cobalt 58.9	28 Ni Nickel 58.7	29 Cu Copper 63.5	30 Zn Zinc 65.4	31 Ga Gallium 69.7	32 Ge Germanium 72.6	33 As Arsenic 74.9	34 Se Selenium 79.0	35 Br Bromine 79.9	36 Kr Krypton 83.8
37 Rb Rubidium 85.5	38 Sr Strontium 87.6	39 Y Yttrium 88.9	40 Zr Zirconium 91.2	41 Nb Niobium 92.9	42 Mo Molybdenum 95.9	43 Tc Technetium 98.9	44 Ru Ruthenium 101.1	45 Rh Rhodium 102.9	46 Pd Palladium 106.4	47 Ag Silver 107.9	48 Cd Cadmium 112	49 In Indium 115	50 Sn Tin 119	51 Sb Antimony 122	52 Te Tellurium 128	53 I Iodine 127	54 Xe Xenon 131
55 Cs Cesium 133	56 Ba Barium 137	57 La Lanthanum 139	72 Hf Hafnium 178	73 Ta Tantalum 181	74 W Tungsten 184	75 Re Rhenium 186	76 Os Osmium 190	77 Ir Iridium 192	78 Pt Platinum 195	79 Au Gold 197	80 Hg Mercury 201	81 Tl Thallium 204	82 Pb Lead 207	83 Bi Bismuth 209	84 Po Polonium (209)	85 At Astatine (210)	86 Rn Radon (222)
87 Fr Francium (223)	88 Ra Radium 226	89 Ac Actinium (227)	104 Rf Rutherfordium (261)	105 Ha Hahnium (262)													

58 Ce Cerium 140.1	59 Pr Praseodymium 140.9	60 Nd Neodymium 144.2	61 Pm Promethium (147)	62 Sm Samarium 150.3	63 Eu Europium 152.0	64 Gd Gadolinium 157.2	65 Tb Terbium 158.9	66 Dy Dysprosium 162.5	67 Ho Holium 164.9	68 Er Erbium 167.3	69 Tm Thulium 168.9	70 Yb Ytterbium 173.0	71 Lu Lutecium 175.0
90 Th Thorium 232.0	91 Pa Protactinium 231.0	92 U Uranium 238.0	93 Np Neptunium 237.0	94 Pu Plutonium (242)	95 Am Americium (243)	96 Cm Curium (247)	97 Bk Berkelium (247)	98 Cf Californium (251)	99 Es Einsteinium (254)	100 Fm Fermium (253)	101 Md Mendelevium (256)	102 No Nobelium (254)	103 Lr Lawrencium (257)